GOLD

GOLD

Edited by Agnes Husslein-Arco and Thomas Zaunschirm

belvedere

HIRMER

Art History of Gold

The Cultural History of Gold

The Themes of the Exhibition

Appendix

Preface

Agnes Husslein-Arco

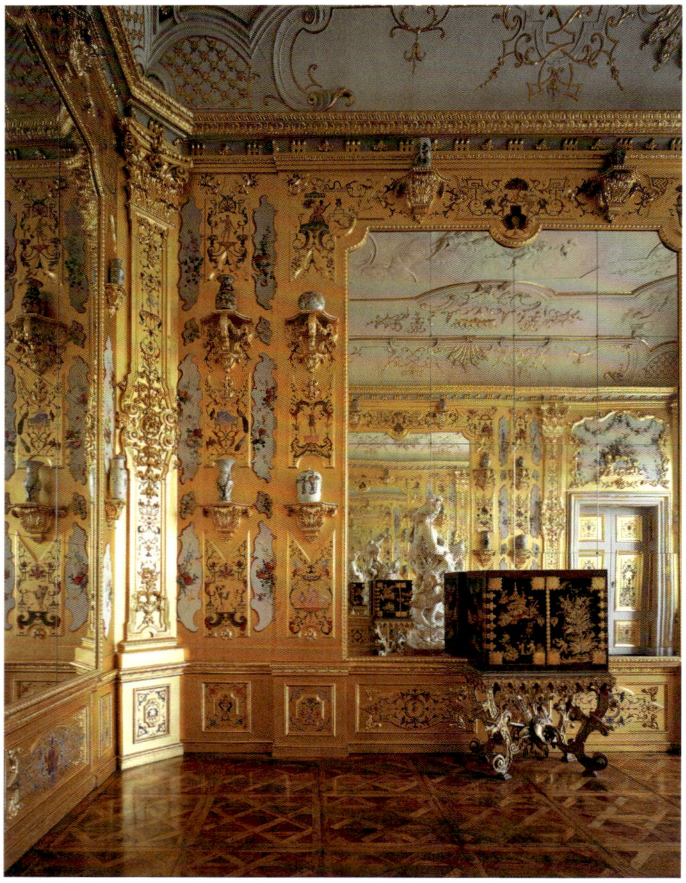

Fig 1 Gold Cabinet, Lower Belvedere

A theme running through a number of the art-historical periods represented in the Belvedere is the use of gold in art. Medieval altarpieces with their gold grounds and gilt sculptures are on display in various parts of the Upper Belvedere and in the Prunkstall. The most famous work in the Belvedere, *The Kiss* by Gustav Klimt, belongs to his 'golden phase'. This art treasure was acquired in the year it was painted, 1908, for what was then known as the Neue Galerie. Gold is, in addition, a dominant element in the decoration of the dual palace; it is programmatically most evident in the two Gold Cabinets, of which only the one originally in the Gold Chamber of Prince Eugen's City Palace is still extant (fig. 1). The Cabinet of Gold and Mirrors from the Upper Belvedere was destroyed by fire in 1950.

For this and many other reasons, the Belvedere is particularly suited to meeting the challenge of staging an exhibition devoted to the inexhaustible topic of GOLD.

The sought-after precious metal is a leading medium in Viennese art. Not just the art of the Secession c. 1900, but also the opulent decoration of the prestige buildings of the nineteenth century on the city's Ringstrasse, and also the Baroque edifices of the seventeenth and eighteenth centuries, have left their mark on how we imagine the capital of the Austro-Hungarian monarchy. Gold, though, is not just confined to past ages. While infrequent in the art of Modernism, it was rediscovered in the mid-twentieth century. In recent years it has become more and more evident that gold is playing an important role for numerous contemporary artists. This is true of Japan just as it is for the USA and Europe.

The Belvedere is the ideal venue for this exhibition also because it confronts traditional manifestations with modern and contemporary art. In this great project, with more than 200 works by some 120 artists from all over the world, the focus is not just on the aesthetic variety of the different stylistic and conceptual approaches; particular emphasis has been placed on the scholarly aspect of this project.

Amazingly, where the question of the meaning of gold in art is concerned, art history is in a state of stagnation, and little development is evident. For decades, the same old prejudices have been trotted out. This is true not just of the assessment of the medieval gold ground, but also of the thesis that gold had disappeared from painting until about 1800. It is therefore all the more pleasing that a series of highly regarded authors have ventured to embark on a new start. The medieval gold ground is subjected to a new examination, as is the cliché of the alleged spirituality of gold in modern art. The catalogue promises to become a compendium of art and culture-historical discourses in gold research. In this sense, the Belvedere is fulfilling to a particular degree its culture-political remit as a scholarly institute in the public sphere.

At a time marked by waves of crises due to global financial speculation, the volatile price of gold reflects only the changing value of the material. Untouched by this quantitative aspect, the amazingly versatile quality of a material that has been of importance for all cultures down to the present day is revealed in its artistic design possibilities. During the preparations for the exhibition, it was clear that the artistic use of gold seems to be subject to a taboo. Many artists are surprised to see themselves subjected to the criticism that the use of gold, particularly gold leaf, is a sign of mere decorative showiness, craft superficiality, or religious kitsch. These clichés are the product of ignorance. A fundamental revision is overdue, in spite of the general interest being shown in gold in other areas of our daily lives these days.

Without the international collaboration on the part of the artists, museums, galleries and collectors, an exhibition occupying the whole of the Lower Belvedere including the Prunkstall and Orangery would be an impossible undertaking. I would like to express my heartfelt thanks to them all.
The curator Thomas Zaunschirm made the selection of artists and works from his extensive archive, and has arranged the exhibits by theme. The individual categories are explained and interpreted in a 'Tour' in the catalogue. I am also very grateful to him and to the other authors who have brought their considerable knowledge to bear on the broad spectrum of issues involved.

Even if recent research seems to show that gold has literally (with meteorites) fallen from heaven, the funding of such an opulent exhibition is no easy task for the museum. Not least, therefore, I should like to express my very great thanks to the sponsors.

Art History of Gold

The Invention of the Gold Ground

Thomas Zaunschirm

The value of gold

The varying religious and economic values attached to gold are a result of changing ideologies and points of view. The fact that sacred objects used to be made of gold once led to the belief that gold itself was a holy metal. In fact gold has no sacred value; rather, it is transfigured by the belief that it owes its properties to higher powers.

The prominent role of gold in most high cultures is due to the indestructible lustre of the precious metal, which predestined it to be a symbol of light like the sun. As such it was directly linked to the gods. In ancient Egypt the working and ownership of gold was the privilege of the Pharaohs, and it was known as the 'flesh of the gods'. The Meso-american term for gold *teocuitlatl* (excrement of the gods) indicates the direct link of the gods with the gold grains found by humans. The Incas considered gold to be the seeds of the gods.

The belief that gold is sacred presumes a tangible connexion to the sun, which has created the world. The portrayal of the sun in ancient Egypt shows rays that end in hands holding the sign of life, Anch ☥, and awaken the world to life. Even when the sanctity of gold took a back seat following the demise of the ancient Egyptian and American cultures, the fascination for the metal continued. Only in 1976 did the International Monetary Fund advise its members to abolish the linking of currencies to gold. The monetary system however did not come crashing down when the quasi-religious foundation of the gold standard was abandoned. Confidence in the value of gold appears undiminished. To this day, it presents a kind of life insurance for citizens and dictators in times of crisis.

The ubiquity of historic examples for the sacredness of gold has overlain art-historical research. The established preconception about the spiritual aspect of this material can hardly be explained. It is not so much the relics and sculptures but the medieval gold ground that is affected by the supposed sacredness of gold. Ironically this has resulted in a taboo on the use of gold in painting. Besides accusations of tawdry religiosity and superficial ornamentation, there was also the fear of the aftertaste of social exclusivity: gold as the material of the gods, rulers and nobility is not well suited as a material for a democratic modern age, which is why the gold of Art Nouveau and the Secession disappeared after the downfall of the Austro-Hungarian monarchy. The gradual return of gold to art during the twentieth century is an interesting chapter in art history. The opulence with which gold has been utilized in art in recent years has surprised art critics but has not led to a revision of iconological prejudices.

New lustre

In public art, gold has already arrived. In the visual arts, gold leaf glittered covertly until recently. Popular sculptures, such as Edmund Hellmer's *Johann Strauß* (1921) (fig. 1) in the Viennese Stadtpark, are an exception. Such statues have a golden sheen, but are mostly made of bronze, or painted. The monument, which is especially popular with visitors to Vienna, is treated as a cultural asset, as a salutation from a bygone age, but it wasn't taken seriously as a modern sculpture even by contemporaries.[1]

But conditions have changed. At the 2009 Venice Biennale the artist duo Michael Elmgreen & Ingar Dragset showcased gilded housemaids in uniform, followed by the gilt figure *Nevermore* (fig. 2). When Richard Wright received the Turner Prize in December of the same year for his ornamental gold-leaf wall paintings (fig. 3), this was tantamount to a blow-off, as many artists felt and still feel their own work with the precious metal to be a breach of taboo. Airily, they were accused of religious kitsch, a regressive desire for decoration, and traditionalism.

The new regard for the countless artists working with gold led to fundamental questions and a historic revision. This is tantamount to the discovery of a new continent. There are hopes

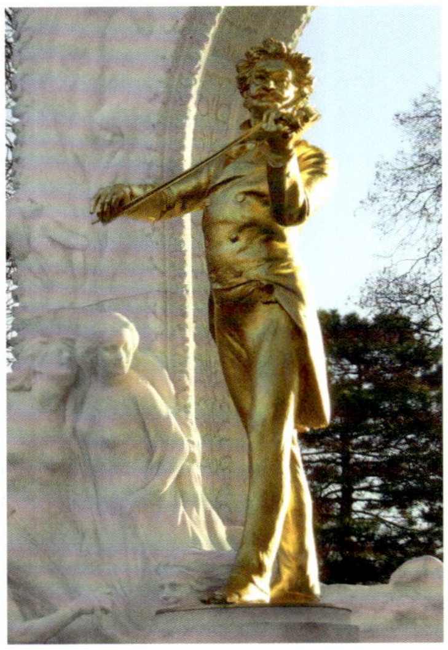

Fig. 1 Edmund Hellmer, *Johann Strauß*, 1921

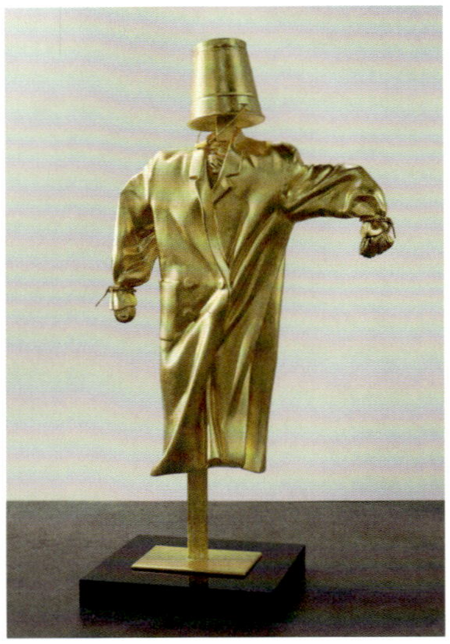

Fig. 2 Michael Elmgreen & Ingar Dragset, *Nevermore*, 2009

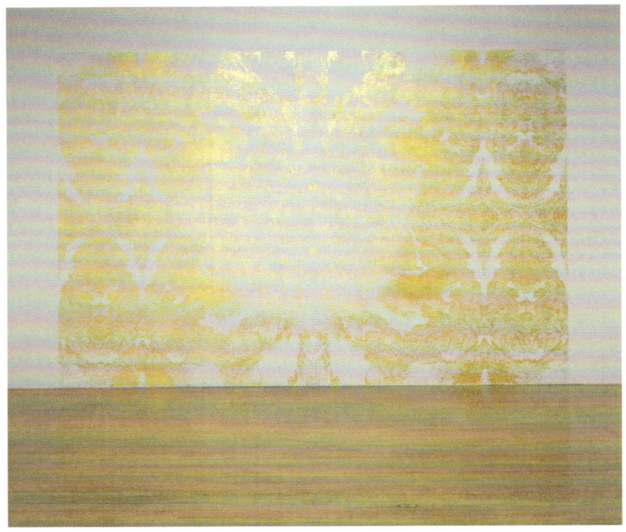

Fig. 3 Richard Wright, *no title*, 2009

that comparison will lead to new bearings. The story of the rediscovery of gold grounds stretches from indifference through rejection and derision to glorification. Every new gold ground immediately evokes the medieval models, but the answer to the question of how far these phenomena can be compared over the ages remains fraught with difficulties. Nobody has risked writing an art history of gold grounds, and there is much uncertainty about possible interpretations. Not least, the debate is marked by misunderstandings.

The autumn of 2010 saw the appearance of the first substantial books on gold in the art of the twentieth century (Anne Schloen, *Die Renaissance des Goldes*)[2] and of the present day (Anne-Marie Charbonneaux (ed.), *L'or dans l'art contemporain*, article by Gérard-Georges Lemaire)[3]. Awareness of this phenomenon was also awakened in several smaller exhibitions. There are plenty of interpretive aids for the understanding of gold in the catalogues of countless travelling exhibitions on non-European and Antique cultures. Now, as a result of the updating of the gold ground by contemporary artists, hopes were raised of a contrastive look at the use of gold in the Middle Ages. The art cannot be separated from a historical view of the period itself. This connexion is generally recognized today. It is clear that the meaning of gold in present-day art does not make sense unless we compare it with gold grounds in the Middle Ages. But are we talking about similarities or differences?

The gold ground—history of the term

The simple question of how long gold grounds have existed can be understood in two ways. On the one hand, since when have gold grounds existed at all? On the other, when did gold grounds as we understand them today first appear? The art-historical term 'gold ground' was coined in the eighteenth century; this does not necessarily mean there was a new interest in the medieval period. Goethe's changing attitude towards gold grounds shows how close approval and disapproval could be. In Weimar, on the 12 October 1801, he penned a letter to the painter Johann August Nahl the Younger regarding the planned decoration of a room (probably for the interior decoration of the refurbished Weimar Palace, which had burnt down in 1774). In this letter he asked for four paintings which were to depict the four virtues of justice, moderation, wisdom, and strength. These were not to be allegories but 'real analogous historic compositions whereby the objects are to be taken, if possible, from the Greek tales, no later than the Trojan War'. Goethe asks for sketches of these pictures and makes the extraordinary request: "As for the manner of painting, we stick to what was agreed: grey figures on gold grounds." It is not known whether this project was ever realized. The instruction 'grey figures on gold grounds' is reminiscent of the silhouettes on gold grounds popular in the last decade of the eighteenth century, for which there even existed textbooks.[4] Another possibility was to cut the silhouette from gold foil (fig. 4) and place it on black paper.[5] There exists a silhouette on a gold ground by Ferdinand Georg Waldmüller, *Self Portrait* (1814), which he probably gave his future wife Katharina Weidner as a present.

In any case Goethe seems soon to have changed his mind. Four years later he describes in his diaries a category in the Weimar competitions "aimed at getting to know artists who, as yet uninfected by the growing misguided enthusiasm for gold grounds, might be inclined to apply their outlook and talent to the good honest life."[6] Nahl was one of the two winners of the year 1801. Goethe's taste had changed rapidly, but what did he mean by 'growing misguided enthusiasm'?[7]

This period includes the failed attempt by Philip Otto Runge to create new effects with the help of a gold ground. Until a few years ago the painting *The Day* (1803) (fig. 5) was seen as the first modern attempt to revive the gold ground. "In 1803 Runge showed his four sheets to Quistorp in Wolgast, who

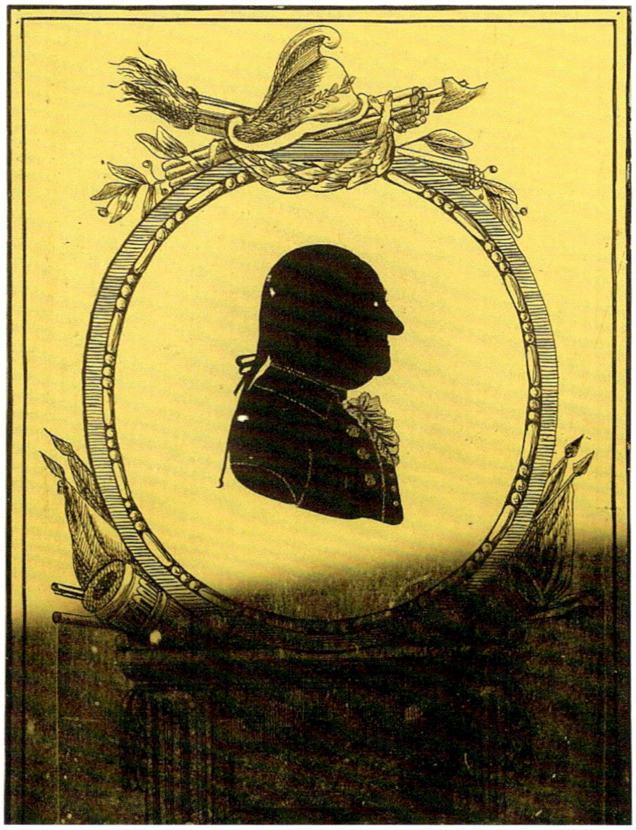

Fig. 4 Spöckberger, *Silhouette of a Rococco man*, um 1790

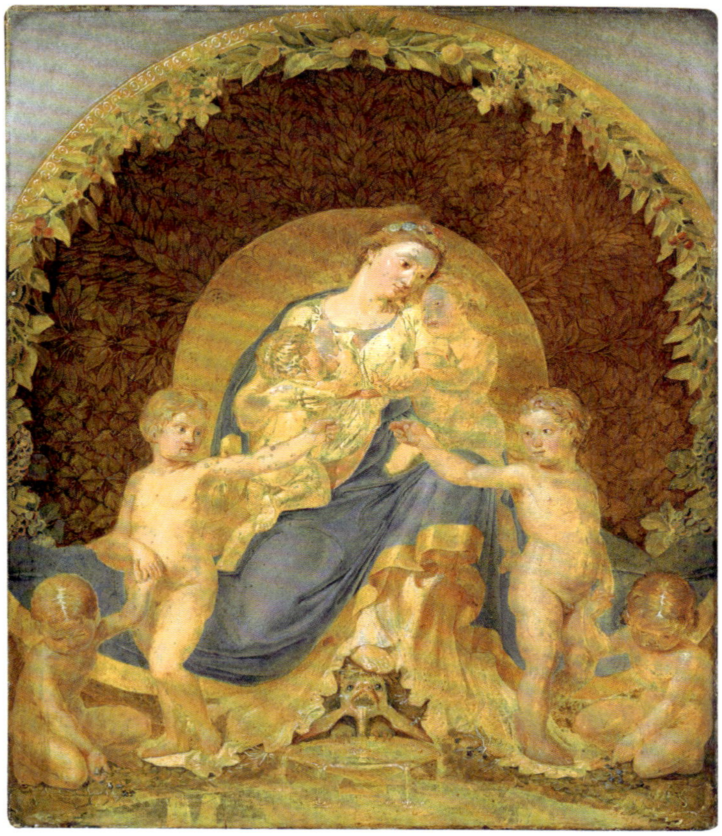

Fig. 5 Philip Otto Runge, *Day*, central group, 1803

recounts that the painter had been burning with impatience to 'execute them in a large format in paint on a gold ground'. Quistorp tried to talk him out of the idea of using 'the barbaric gold ground' because of the 'dirty' colour effects. Runge however was hoping for the exact opposite. [...]. Runge was not so much looking for stylistic echoes of medieval art; rather he thought this technique would make the colours lustrous and luminous."[8]

Johann Gottfried Quistorp was Caspar David Friedrich's teacher in Greifswald. Whilst Runge (independently of his religious interpretation) defined colour quality as a function of the quantity of light captured within it, Quistorp did not see 'the mediating role of earthly colours [...] in their commitment to light'.[9] In the chiaroscuro of modern painting each colour had its place between the poles of light and dark. Colour served to depict light, but was not the bearer of light per se. It is in the nature of things that Runge's system, with a colour globe that goes from a white to a black pole, does not contain gold. This view has been prevalent since Alberti's seminal dissociation of colour from material.[10] Runge now raised an objection. The illusionist depiction of gold by the use of paint had, since Alberti, 'meant' gold. This ambivalence, or dichotomy, played a central role in the interpretation of medieval gold. It appears as a problem for the first time with Runge. For him, gold itself was something so special that the colours were to be elevated through its use. For the traditionalist Quistorp, however, it was only the experience of the illusion that counted. From this point of view gold was damaging to the intrinsic values of the colours.

Within the space of a few years Goethe desired pictures on gold grounds, scoffed at them as a fad, and then, soon after, viewed them again without value-judgements as Gothic works. In his travel journals Goethe mentions pictures on gold grounds many times. In1808 he writes from Schaffhausen about pictures on gold grounds that he had seen in Colmar. In 1814/15 he identifies the link between missing perspective and the gold ground in Cologne Cathedral.[11] Other writers also mention gold grounds. Jean Paul writes, c. 1820, in *Der Komet oder Nikolaus Marggraf*: "In every alley and den there are resplendent pictures of their beauty, painted on gold grounds."[12] There are more than a few such remarks and hints at pictures on gold grounds in the works of travel-writers of the time. Johanna Schopenhauer mentions pictures on gold grounds in her outings to exhibitions (1828 ff.). The sculptor Christian Friedrich Tieck makes the point in a letter to Goethe that gold grounds, when used in a portrait, 'require a very rich coloration' to remain harmonious.[13]

Even though the numerous references to gold grounds in medieval paintings contain notes on (absent) perspective and (garish) coloration, nothing is said about any sacred meaning. This applies not only to writers and artists but also to art historians. Franz Kugler (1808–1858) describes German and Italian pictures painted 'on gold grounds' without attaching any meaning to this.[14] The same can be said about other art historians of the nineteenth century, including Carl Schnaase (1798–1875), Rudolf Eitelberger von Edelberg (1817–1885), Jacob Burckhardt (1818–1897), Wilhelm Lübke (1826–1893). At the same time mosaics and frescoes on gold grounds were created in churches. The Viennese Late Nazarene Eduard Jakob von Steinle (1810–1886) painted angel choirs on gold grounds in the arcade spandrels of Cologne Cathedral in the 1840s.

The entire nineteenth century went back more to the idea rather than the material of gold grounds or mosaics. The painted gold grounds served as a space-sealing foil; where for example Marie Ellenrieder, in her Raphael quotations, did not abandon modern corporeality (fig. 6). Gold grounds link up the individual paintings in the picture cycles of the prestigious Ringstrasse buildings. No distinction is made between the classical mythological scenes (Carl Rahl and August Eisenmenger in the Austrian Parliament building) or artist-apotheoses (Hans Makart in the Kunsthistorisches Museum, fig. 7). The golden ground acts as backcloth to both Albrecht Dürer and Peter Paul Rubens, and can therefore be considered neither as sacred nor symbolic of the Middle Ages. The (false) golden sheen is commensurate with the historical lack of presence.

The interest in gold grounds and mosaics in the churches, palaces and public buildings did not, though, guarantee general praise. Meyer's Encyclopaedia of 1888 still says: "The actual meaning of the same doubtless rests on the aspiration to bestow a greater value on the picture through the use of precious gold, and as such is generally discredited as *barbaric*."[15] This is reminiscent not only of the cited criticism of Quistorp in respect of Runge, but appears to be a typical point of view in the nineteenth century. Hegel too comes to a similar conclusion in his *Lectures on Aesthetics* (1835–1838): "For this reason the oldest painters kept, instead of an elaborate natural environment, a uniform background of gold which the colours of the robes must make head against and parry, as it were; and therefore these colours are more decisive, more crude, than is the case in the times of painting's most beautiful development–not to speak of the fact that *barbarians* in general take their pleasure in simple and vivid colours like red and blue, etc."[16]

There are as few signs of a general 'appreciation' of gold grounds in the last quarter of the nineteenth century as there are with Goethe in 1805. Therefore it is anything but self-evident to think of gold grounds as being no longer assigned to the category 'barbarian' but actually to that of the *a priori* spiritual. Rather there is a confrontation between art and nature. In Conrad Ferdinand Meyer's poem *Auf Goldgrund* ('On a Gold Ground', 1882) gold grounds in art and reality are contrasted:

Ins Museum bin zu später
Stunde heut ich noch gegangen,
Wo die Heilgen, wo die Beter
Auf den goldnen Gründen prangen.

Dann durchs Feld bin ich geschritten
Heißer Abendglut entgegen,
Sah, die heut das Korn geschnitten,
Garben auf die Wagen legen.

Um die Lasten in den Armen,
Um den Schnitter und die Garbe
Floss der Abendglut, der warmen,
Wunderbare Goldesfarbe.

Auch des Tages letzte Bürde,
Auch der Fleiss der Feierstunde
War umflammt von heilger Würde,
Stand auf schimmernd goldnem Grunde.

(I went late today to the museum, where the saints, the supplicants stand out against the gold grounds. Then I strode through the fields, towards the hot glow of the evening, and saw that the corn had been cut today, with sheaves lying on the carts. The wonderful golden colour of the warm glow of evening flowed around the burdens in the arms, around the mower and the sheaves. The last burden of the day, too, the hard work of the final hour, was surrounded by the flames of holy dignity, standing on a shimmering golden ground.)

It is of interest for literary research that Meyer's poem, "as Staiger says, 'is second-degree art, a painting as from a misted mirror which already presents the eye with a dreamy image."[17] It is significant for the art history of the gold ground that Meyer does not even mention 'the museum' in his first version of the poem dating from around 1860, but that instead 'the image is to be seen on the sky's pale background'; the gold only gets mentioned in the second version of 1869 (*Der Erntewagen*, 'The Harvest Wagon'). In the new version of the poem the gold first appears in the sky and only after that in the paintings. In the 'Age of Realism' nature moves into the background in favour of paintings. The connexion of 'holy dignity' and 'gold ground' only becomes established in the final version.

Interest in the gold ground in the historicism of Hans Makart and in the Art Nouveau of his pupil Gustav Klimt ran concurrently with the newly awakening awareness of art history. It is generally assumed that scholarly research into gold grounds only started when an objective detachment from the research subject became the methodical norm with Klimt's contemporary Alois Riegl (1858–1905), so there was no positive or negative prejudice towards the phenomenon (such as in the quote from Meyer's Encyclopaedia). In other words, the awareness of gold grounds and the use of gold in art was awakened late, namely only in a first modern phase at the start of the twentieth century.

It is also possible that the notion of gold grounds as a near-magical medium is not due to the methodical eye of art-history research or the aesthetic use by artists, but instead to a culture-historical worldview averse to differentiated analysis. Oswald Spengler (1880–1936) explains the gold ground in his bestseller

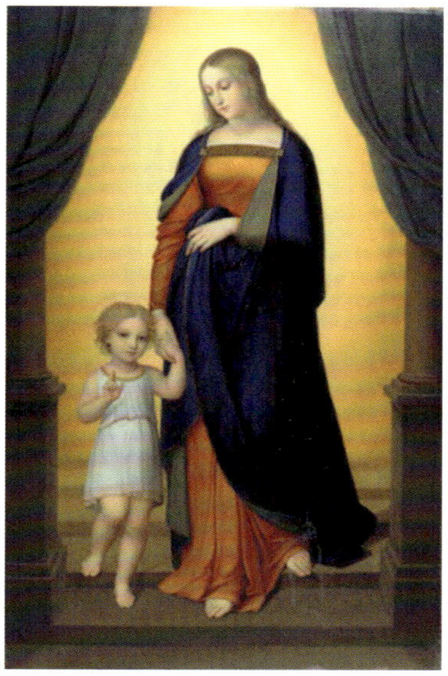

Fig. 6 Marie Ellenrieder, *The Virgin holding the Christ Child by the Hand*, 1824, Staatliche Kunsthalle Karlsruhe

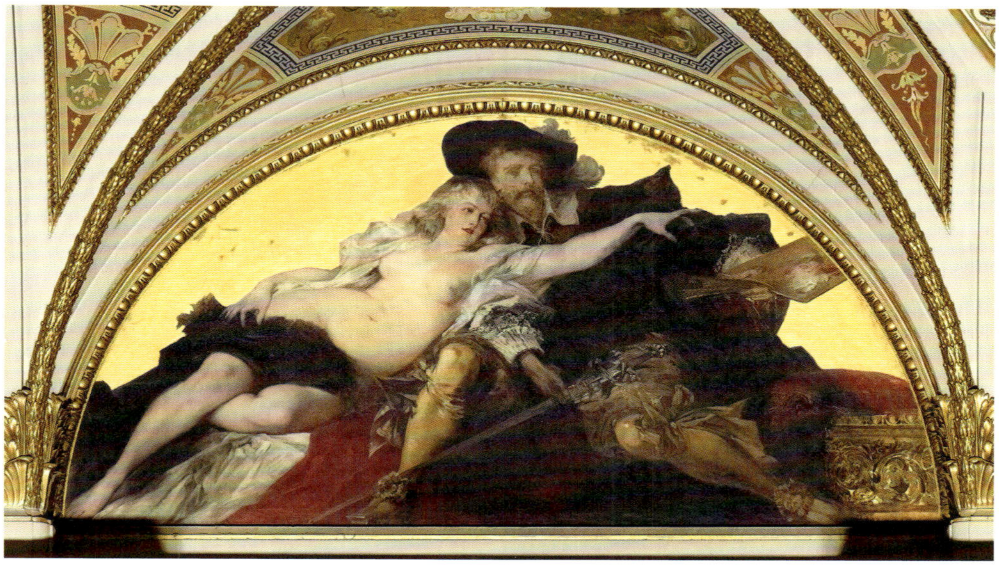

Fig. 7 Hans Makart, *Rubens*, staircase, Kunsthistorisches Museum, after 1881

The Decline of the West, first published in 1918. "It is an express assertion of the existence and activity of the divine spirit. It represents the Arabian form of the Christian world-consciousness, and with such a deep appropriateness that for a thousand years this treatment of the background was held to be the only one metaphysically—and even ethically—possible and seemly in representations of the Christian legend." For Spengler the gold ground is the expression of the 'Magian world-feeling'. "[...] simultaneously with the gold ground of the sacred picture, there vanished from the Councils of the West that Magian, ontological problem of Godhead which had so passionately agitated [...] all the Councils of the East."[18]

Ludwig Curtius said as early as 1919 that 'there was not one thought [in Spengler] whose origins could not be found in the so clever, too clever modern art- historical literature, such as Riegl, Wölfflin, Strzygowski, Pinder, Worringer, Simmel'.[19] That is a polemical exaggeration. When thinking of 're*orient*ation' one might think most readily of Strzygowski's *Orient oder Rom* (1901), but that precisely does not concern the view of the gold ground. Strzygowski was fighting against the prejudice that Rome is the origin of Western art. He hoped that the discovery of the mummy portraits (see below) "would be conducive to totally extirpating the notion of the preponderance of Rome in the field of art."[20] He deals with the gold ground no more than the other art historians mentioned by Curtius.

As early as 1932 Josef Bodonyi made reference to Spengler in his dissertation *Entstehung und Bedeutung des Goldgrundes in der spätantiken Bildkomposition* ('Origin and Meaning of the Gold Ground in Late Antique Pictorial Composition'), submitted in Vienna. He names as an even earlier source the at the time 'completely overlooked' and still largely unknown Danish art writer Julius Lange (1838-1896), in whose *Ausgewählte Schriften* ('Selected Writings'), published in German in 1912, there is talk of the 'symbolic meaning of the gold ground'. In some medieval paintings gold grounds 'serve to give a picture of God's realm, of the fire of heavenly love in which bliss prevails'.[21] Bodonyi omitted to mention in his quote that Lange saw the interpretation of the gold ground as a 'picture of God's realm' merely as one possibility.

Bodonyi read Lange's great essay of 1893 one-sidedly, to his own ends, and apart from Wolfgang Schöne (*Das Licht in der Malerei*, 1954) clearly no gold-ground researcher (Braunfels, Beer, Dittmann etc.) took any note of him. Thus while the symbolic elevation of the gold ground in art history was postulated during the Art Nouveau era, only since Spengler has art history in the modern sense (with the named exceptions) got caught up, to the present day, in the one-sided interpretation.

History of the gold ground

One might think that there is at least a basic consensus concerning the historical framework of with the gold ground. But that is not the case either. A few centuries here, a few centuries there—this seems to be no great matter where gold is concerned. The *terminus a quo* of golden pictorial backgrounds was not Ottonian manuscripts, in spite of what Anne Schloen says in her chapter 'Die Kulturgeschichte des Goldes' ('The Cultural history of Gold'): "The oldest extant manuscript with golden pictorial backgrounds, dates from c. 990 and was an innovation of the Ottonian book illuminators."[22] On the contrary, the phenomenon goes back at least to the *Vienna Dioscurides* (AD 512, the so-called 'Galeonos group', a picture of doctors and allied practitioners, s.p. 33, fig. 7). Ms Schloen seems to be giving credit to the incredibly obstinate prejudices of scholarship. Wolfgang Schöne,[23] Lorenz Dittmann,[24] Christine Sauer[25] and others all base themselves on Hans Jantzen's *Ottonische Kunst*[26] (literature from1947!). This interval of almost half a millennium between the *Vienna Dioscurides* and the Ottonian works may derive from a notion of 'the West' (from AD 900) as defined by

Oswald Spengler. Even so, in his interpretation he spoke of a thousand years of gold grounds. Schöne adds to his Jantzen quote the caveat that what is meant is 'Western painting [in the narrower sense]'.[27] The *Vienna Dioscurides* has been available in facsimile since 1970, and was published with a commentary in 1998.[28]

It is however a puzzle why most historical surveys omit to mention the origin of the gold ground, even though it was already recognized by Bodonyi. The mummy portraits found at the end of the nineteenth century in the Egyptian oasis of Fayum (second to fifth centuries) are the earliest known pictures with a gold ground. Hilde Zaloscer recognized the change from death masks (made of sheet gold) to painted pictures as a paradigm shift. What was decisive, she said, was not the material, but the degree of realism. "The funerary portrait was no longer concerned with the continued existence of the earthly manifestation, of the body,"[29] but with a different idea of the beyond. The obsession of art-historical methodology to derive the later from the earlier, to pursue an evolutionary history, evaporates when it comes to the gold ground. An important feature of the gold ground of the mummy portraits is that it is not present in all of them, but that clearly does not subtract from their function. The question of what distinguishes the gold ground in painting from the gold ground in book illumination is not addressed.

Zaloscer attests a different notion of the beyond in order to see the mummy portraits as precursors of the icon. What speaks against this, however, is not just the fact that only some of the thousand or so extant mummy portraits have any gold, but also that they were painted during the lifetime of the sitter and hung in living quarters; in other words, they are not saints. Gold applications were also used in the portraits for wreaths and garments. The usual question relating to gold grounds generally is clearly inevitable. As with the mummy masks, the non-oxidizing gold should in some magical fashion have prevented decomposition. But in contrast to the usual plaster masks, gilding would have obscured the facial features. "Gilding the background represented an emergency solution, so to speak."[30] But then how does the gold ground protect the face of the sitter against decomposition—or what else does the metal protect? This theory raises more questions than it answers. In the discussion on the use of gold, the mummy portraits play hardly any role these days.

Alongside the delayed introduction of the gold ground to book illumination, Schloen also postulates an interval of several hundred years in post-medieval art in which there is said to have been no use of the gold ground. "Before gold was rediscovered as a material in the visual arts, far more than 400 years were to elapse."[31] Ultimately, she says, it was only the art of the twentieth century that became, through the use of gold or a gold substitute, religious/transcendental/sublime once more.[32] Like many other authors, she refers to Leon Battista Alberti's judgement against gold (1435). If we add 400 years to this date, we should attest a revival of interest in about 1835. But what happened before that? Until recently, art historians knew nothing of Giovanni Battista Tiepolo's fresco cycle of 1757. But Schloen also denies the well-known gold ground of Philip Otto Runge. Runge not only planned a gold ground, but in *The Day* (1803/04) actually executed one (see above). Schoene derives her assertion "But he never carried out the project"[33] from Schöne, but the latter states unambiguously in the same sentence: "(Gold leaf! He also made an experiment)."[34]

Such surveys also fail to take into account the fact that gold continued to be used in sculptures after the Renaissance. If we add the 400 years since Alberti's pronouncement to the aforementioned interval of almost 500 years between the *Vienna Dioscurides* and Ottonian book illumination, there is a truly terrifying chasm in the art-history of gold. How is it possible to blank out the ubiquitous precious metal in this way?

The general consensus is thus deceptive. There is probably agreement that the gold ground is a medieval phenomenon going back to classical models, that it ended with the Renaissance, that then since 1800 or during the nineteenth century there was a renewed interest in the medieval gold ground, that Modernism only made use of gold during the Art Nouveau and Art Deco phases, but that the avantgarde resisted the introduction of the precious material. When in 1915 Malevich presented his Suprematist *Black Square* like an icon in the 'beautiful corner' at the 'last Futurist exhibition 0,10' (fig. 8), this was seen as the deposition of the icon and its gold ground. As the other heroes of abstraction, Wassily Kandinsky and Piet Mondrian, did not speculate with the use of gold, this spirit of abstraction was seen as a period in which gold and the gold ground in particular no longer had any justification. Marcel Duchamp, in one Readymade (*Apolinère Enameled,* 1916/17), even eliminated the gold paint and replaced it with white (figs. 9, 10).[35]

However, maybe we know too little about the circumstances of the lives of great artists. Kandinsky's work is not characterized by a negation or repression of gold; rather, his painting provided no occasion for it. But as a Russian he preserved the memory of the gold of the icons by treating himself in his traditionally furnished apartment to a wall decorated with gold leaf in a niche in his sitting room—as his Bauhaus students reported with amazement. It remained a one-off. In his *Colour Plan for his Dining Room in Dessau* (1926) we can see how precisely he composed the colours of all the other walls. They served as the background to his pictures.[36]

Gold has since then been deployed almost in secret and largely unnoticed. Not until mid-century can it be seen more frequently. It is almost a taboo subject. In the vast literature on Picasso, gold is a blind spot. In Picasso exhibitions, our search for his golden medallions will come to naught. Most of them are round shapes, but Picasso also designed a multi-lobed form and a *faun's* face (fig. 11) frayed at the edges.

There is currently no relevant catalogue raisonné. In the monumental monograph by Werner Spies, while even small

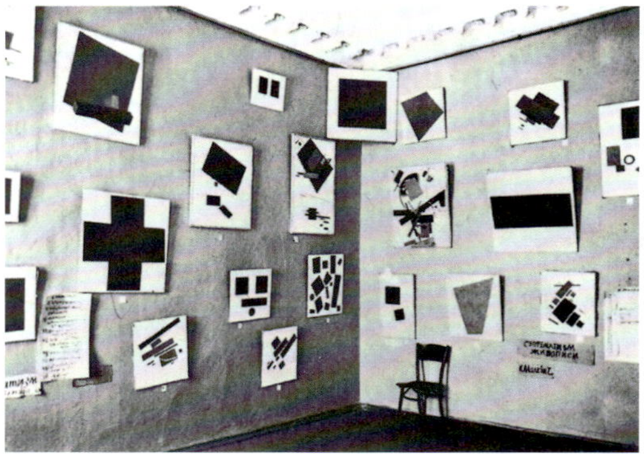

Fig. 8 Exhibition view, *Last Futurist Exhibition 0,10*, 1915

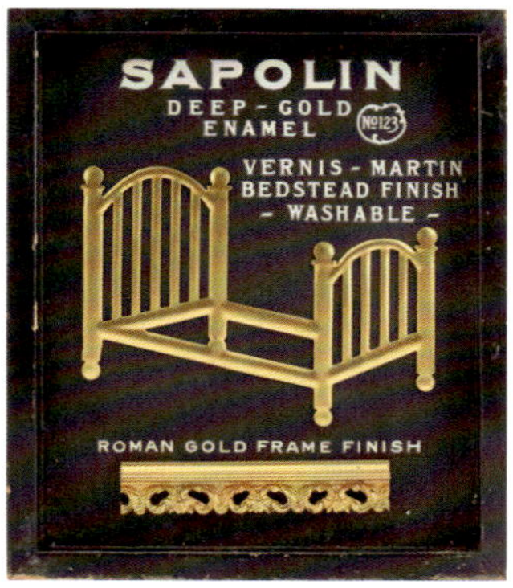

Fig. 9 Advertising sign of the Sapolin Company, Gerstendorfer Bros. New York, c. 1915

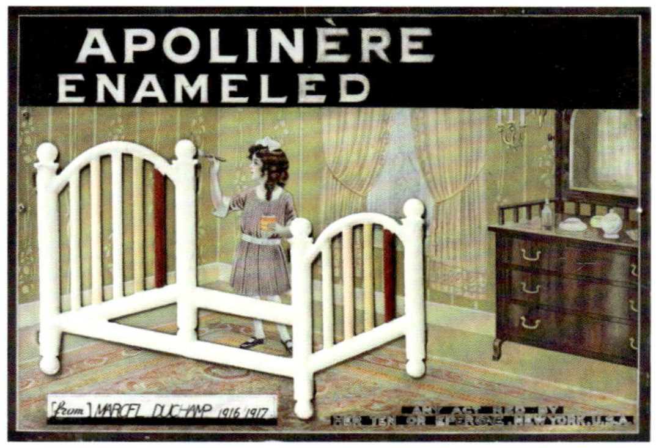

Fig. 10 Marcel Duchamp, *Apolinère Enameled*, 1916/17

scratchings on clay, ceramics, plaster-of-Paris works and drawings on pebbles are listed, there is no trace whatever of the jewellery items.[37] Maybe such craft items were felt to be beneath the dignity of such as Picasso. But Picasso himself knew no such design taboos; he integrated all manner of things into his collages and sculptures, from natural objets trouvés to toys. Maybe this alchemistical aspect, which subjects everything to a metamorphosis, is responsible for the veil of silence drawn across his works in gold. The artist's genius turned everything to gold in any case, or so the reasoning may have gone, so that instinctively nothing was found to honour when he reached for the precious metal. Schloen adduces almost 200 artists who worked with gold in the twentieth century or are still doing so, but Picasso is not among them.[38]

It was only in 2009 that the collector Diana Küppers in Cologne presented some of these forgotten works by Picasso together with jewellery by a number of other artists. The Picasso items comprised a pendant and 16 medallions. Küppers has put her collection on display on her website.[39] In addition there are other works, which for example François-Victor and Pierre Hugo created for Picasso in gold. There was already a debate on whether artists should risk producing jewellery in 1987 on the occasion of the *Schmuck*[40] ('Jewellery') exhibition held in Linz. In his catalogue contribution (entitled 'Is Jewellery Art'), Peter Skubic mentions Picasso as a jewellery artist, albeit without any direct reference or illustration. The obvious disparagement of these miniature sculptures derives from their inappropriate pigeon-holing as decorative, in other words utilitarian, art. Art historians lump such objects together as 'crafts'.

With few exceptions, twentieth-century critics reacted to gold with a lack of interest. This is often a matter of taste, but also depends on the quality of the information. Stimulating as Iris Wenderholm's direct confrontation of Alberti and Yves Klein is, the thematic field dries up straight away in the assumption that Klein's *Monogold* series can 'be seen as the climax and endpoint of artistic concern with the values of gold leaf'.[41] In fact, if anything Klein stands at the beginning of this interest in gold leaf, an interest that continues to this day, undiminished.

One threshold was crossed in 1982 at the documenta 7 exhibition. Even though the subject of gold was not addressed in the two-volume catalogue, and only the illustration of *Tragedia Civile* by Iannis Kounellis gave any hint of his gold-leaf wall, art critics reacted with displeasure. In view of the massive deployment of gold at documenta 7, the critic Wolfgang Max Faust commented: "Since gold no longer offers any security (world currencies are no longer backed by gold), it can hardly have any conviction as a symbol. Not everything hangs on gold."[42] And Georg Jappe spoke scornfully of an 'idolatrous shindig'.[43] Thomas McEvilley took a more neutral view of the use of gold in the art of these years in 1985, presenting the reader with the previous history all the way to Greek Antiquity.[44]

When, a short time later (1991), Tilman Osterwold was preparing his wide-ranging exhibition *Das Goldene Zeitalter* ('The Golden Age') at the Württembergischer Kunstverein Stuttgart (1991), he was inspired by 'developments in international contemporary art'.[45] Now it was less the art of his day than the encyclopaedic aspiration that got the critics excited.[46] *Der Spiegel* spoke of 'deceptive glitz' and thought it was still unproved 'whether a gold vein of contemporary culture had been hit on here'.[47] In view of the 'danger zone [...] of ghost-train effects, swimming in kitsch, of trivially permeated trinkets [...] characteristic of the inadequacy of such an enterprise', the critic writing for the *Kunstforum* closed by asking who if anyone 'in this century in which the *Gesamtkunstwerk* has failed so often still seriously wonders about the universality of gold'. [48]

Iconography

It comes as no surprise, then, that we had to wait another twenty years before anyone dared to come up with an historical overview. What did the two abovementioned publications have to offer? They are two very different approaches, namely a German dissertation on gold in the twentieth century (Anne Schloen)[49] and a French collection of material on contemporary works in gold (Anne-Marie Charbonneaux/Gérard-Georges Lemaire)[50].

After a cursory global history, Schloen presents a series of artists in the form of case histories: Gustav Klimt, Louise

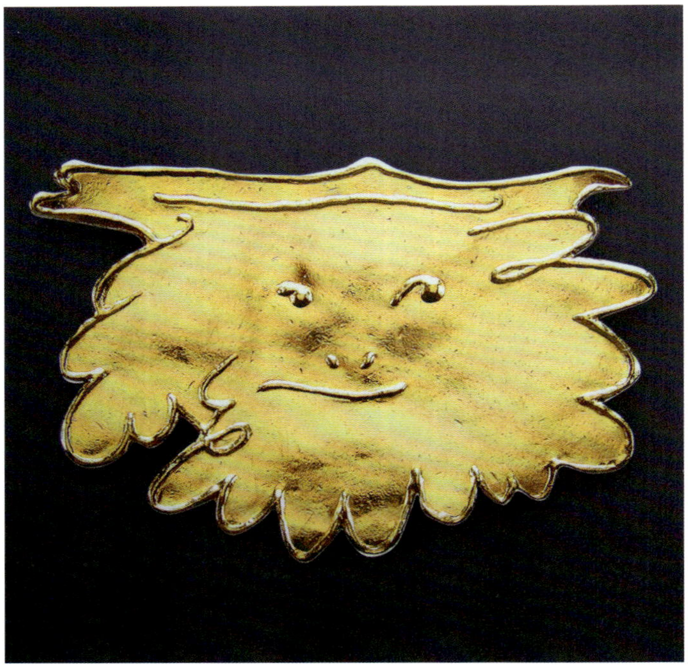

Fig. 11 Pablo Picasso, *Large Faun,* 1972

Nevelson, Andy Warhol, Robert Rauschenberg, Max Bill, Roni Horn, Yves Klein, Marcel Broodthaers, Joseph Beuys, Iannis Kounellis, Constantin Brancusi, James Lee Byars, Marina Abramovic/Ulay, Erich Buchholz, Lucio Fontana, Michael Buthe. Each of these chapters ends inexorably in an outburst of emotion on the part of the author, because she insists that gold in all cultures right down to the present day has had a similar significance: "Its immaterial qualities preponderate, those that made it available as a metaphor if the Golden Age, for example, or as a symbol of spirituality. It stands for transcendental beauty, symbolizes endless space and serves as the symbol of light.[51]

Quite apart from its decorative value, for example, Klimt still associated gold 'with a symbolism of transfiguration and sanctification'. Eroticism was 'almost sanctified'.[52] In the case of Nevelson too, who did not even work with gold, but with gold paint, the everyday is sanctified; she created 'relics for the consumer society'.[53] "The artist exploited the 'magical' charisma of gold to sanctify objets trouvés."[54] As for Warhol's *Gold Marilyn* (1962) "the gold ground is a metaphor of the supernatural light and symbolizes other-worldliness. [...] The gold ground dissolves the this-worldliness of the background, so that the actress is transferred to a different, a supernatural world." With a 'gold ground' (created incidentally of vinyl) 'the deceased has bestowed upon her not only an other-worldly but also a timeless existence'.[55]

All the case studies in this book hang on the iconographical hook of temporal and spatial infinity (Max Bill), the transcendental (Roni Horn), the 'synchronicity of mental and spiritual processes', because 'it is the only material that embodies precisely this duality of materiality and immateriality'[56] (Yves Klein). In the case of Klein, the beholder should plunge into the golden surface and 'submerge himself in the abstract, limitless, timeless golden space and its supernatural light'.[57] The buyer would even experience 'emotional maturation, self-discovery and spiritual rebirth'.[58] No wonder that even in the works of Joseph Beuys 'gold embodies both material and immaterial values', the golden base of his cabinet (for *Palazzo Regale,* 1985) stood not just 'metaphorically for the beyond', but 'removes objects from the sphere of earthly transitoriness and transfers them to a timeless and spaceless sphere'. "Human finitude is lifted by gold into the metaphysical dimension of eternity."[59] As for Iannis Kounellis: "The golden wall is a symbol of redemption."[60]

Confronted with *The Endless Column* (1937/38) of Constantin Brancusi, she says, one experiences 'cosmic consciousness' and the 'feeling of absolute freedom'. In the process, one develops 'a higher degree of awareness', and is thus 'led on to a higher plane of consciousness'.[61] Gradually we are forced to ask ourselves how gold manages to transform the beholder thus, and whether this grace is vouchsafed to all. This spiritual sphere continues in the other case studies. *The Golden Tower* (several versions) by James Lee Byars is said 'to allow people to penetrate into higher, in other words spiritual, spheres'. If this fails, then at least the author attests spiritual salvation for Lee Byars himself: "In the light-metaphysical context, the artist in his golden suit became a heavenly being standing close to the abstract, the logos."[62]

In the case of Marina Abramovic & Ulay, too, we have a 'transcending of space and time: the liberation of spiritual energy'.[63] The metaphysical charisma and spiritual dimension are also present in the works of Erich Buchholz, Lucio Fontana, Michael Buthe. The author asserts, that 'the basic meanings of gold, always similar, and present in all cultures and eras [...] were not questioned in the twentieth century either, but adopted unchanged and sometimes extended'[64].

This overview of the research is designed above all to question the initial assumptions. Methodologically, Schloen says she bases her work on material iconology. In fact, she is pursuing a kind of material ontology. Methodological problems are always cropping up when researching the gold ground. These determine the view. As a result, we are not only faced by a clumsy form of scholarship that lives by copying out what has gone before, but in the nature of things we get interpretations of interpretations—the same statement is understood in different ways, but ultimately, it is always the same clichés that get repeated. People have been content to trot out Alois Riegl's opinion that the gold ground was an 'idealized spatial background' or the thesis (Wolfgang Braunfels, Wolfgang Schöne, Christine Sauer et al.) of a holy light. Only rarely are the manifestations of the various gold grounds examined in more discriminating fashion (Lorenz Dittmann[65], Christian Hecht[66]). Some authors (Josef Bodonyi[67], Hilde Zaloscer[68]) have fallen into oblivion, quite unjustifiably. Already Bodonyi noted unambiguously that there can be no unchanging significance of the gold ground, but that on the contrary, the significance depended on the respective function.[69] The meaning of the gold ground in a mummy portrait was, he pointed out, quite different from that in the mosaics of or in Late Gothic altarpieces. For eight decades, no one has pursued any 'functional analysis' that goes beyond his dissertation.

As for the 'holy light' as the content of the gold ground, it is popular to refer to medieval sources. Just as the end of the gold ground in the *quattrocento* is always associated with Alberti, the attempt is very often made to use Abbot Suger of St Denis to prove the spiritual view. His words, to be read above the entrance to his magnificent abbey church, the first truly Gothic building, also forms the starting point for Anne Schloen's *Die Renaissance des Goldes* (2010). She quotes Abbot Suger extensively. Even the first sentence could have served as a warning: "Whoever you are, if you wish to increase the fame of this portal, do not admire the gold and the expense, but the labour of the work."[70] What other conclusion can be drawn from this admonition other than that it was not the gold that mattered, but the effort that went into designing and building the structure? The inscription continues: "Whoever thou art, if thou seekest to extol the glory of these doors, marvel not at the gold and the expense but at the craftsmanship of the work. bright is the noble work; but, being nobly bright, the work should brighten the minds, so that they may progress, through the true lights, to the True Light where Christ is the True Door [...] The weak spirit rises to the Truth through the material world and having been vouchsafed this sight, rises up to the Light." From this, the author draws the conclusion that what was decisive was the 'metaphorical power' of this '"premonition" of heaven'. In other words, what matters in this view is to transcend the material in favour of the spiritual meaning of its lustre.

However, Hans Belting had already put this view into perspective, because the light symbolism of the 'voluble builder' Suger 'was applied indiscriminately to the architecture, the windows and the huge gold cross from the Meuse, and so had no particular significance for our genre'.[71] Strictly speaking, Suger does not even speak of gold alone, but of everything that contributes to the lustre of the church, in order anagogically to take us to the heavenly light. Between gold the material and 'its meaning' of heavenly light there must be some mediation, or else we are left with its material costliness. During the Middle Ages, this mediation, by analogy with the 'fourfold meaning' employed in scriptural exegesis, was performed on the basis of a 'fourfold meaning' of pictures, which permitted, alongside an historical pictorial statement, also allegorical, moral and anagogic statements.

Bazon Brock pointed to the fourfold pictorial meaning of the Middle Ages as early as 1977, in other words at a time when an interest in gold in Modern Art verged on the risky. In the chapter 'Wieder Gold und Edelsteine?—Farbe und Material in der Bildenden Kunst' ('Gold and Gemstones Again?—Colour and Material in the Visual Arts') in his *Ästhetik der Vermittlung* he writes: "We do not want to make an illegitimate identification between the Middle Ages and the modern view, but only to emphasize the agreement." [72] Regardless of whether the attempt to update the fourfold pictorial meaning was fruitful, the approach is original if only because of its reference to gold. Brock sought to translate the allegorical scriptural meaning into didactics, the anagogic into eschatology, the tropological into the moral, and the material into the autonomous claim of the material itself.

Brock makes two astonishing observations: "I cannot resist the impression that today's painters would think themselves lucky if they could adopt the medieval optimum solutions to the problem: namely gold and gemstones." And: "Gold fulfils [...] in optimum fashion the demand for the fourfold meaning of scriptural exegesis."[73] Brock is a lone voice far and wide in this respect. He is not so naive as to attribute gold a symbolic force. "Colour and material remain simply means with which one works, but in themselves are nothing." Unlike the Middle Ages, he says, we lack 'a view of reality generally regarded as binding [...]'. "The historic process is irreversible. Gold ground and gemstones remain solutions, splendid solutions, of the historical past. Precisely for this reason they are significant for us today."[74] Regardless of how Brock's support of gold as a medium for contemporary art is to be viewed, his observation should have made clear the difference between the medieval and modern gold grounds. In the Middle Ages the anagogic approach allowed a sacred view of material lustre. As our image of the world has changed fundamentally, the significance of the gold ground today lies in its historical manifestations.

Schloen seems to have misunderstood the anagogic principle. "In the anagogic function of gold, the epiphany of the divine became present in a special way."[75] No material has an anagogic function. As a part of the artwork, gold may be the subject of an anagogic interpretation. The question arises as to when the gold was isolated from the picture, and by whom, and separated from it to the extent that the metal took on a significance of its own. In his epoch-making work *Über das Licht in der Malerei* (1954) Schöne was still able to make a clear distinction. To the question of the relationship between the created, pictorial light and the true divine light, his answer was that the medieval pictorial light, as the *Sendelicht* (roughly: 'message light'), contained within it the process of the 'emanation of divine strength', a process that was 'mysterious, not comprehensible by the human mind'.[76] In the 1950s, it was still normal enough to avail oneself of the lofty aspirations of a spiritual view of art. Beer by contrast, in her *Marginalien zum Thema Goldgrund* (1983) spoke out against this 'translation into the sphere of the unreal' (see below).[77] In any case, Schöne does not reduce the 'message light' to gold, but speaks quite generally of Ottonian book illustration, the gold ground, and cathedral windows.

The identification of the earthly metallic lustre with the heavenly light is due to a selective perception on a pre-modern, not to say medieval, basis, in other words in accordance with the then prevalent world image, while at the same time eliminating the anagogic element. For in the Middle Ages it

Fig. 12 Winfried Muthesius, *Golden field*, 2002

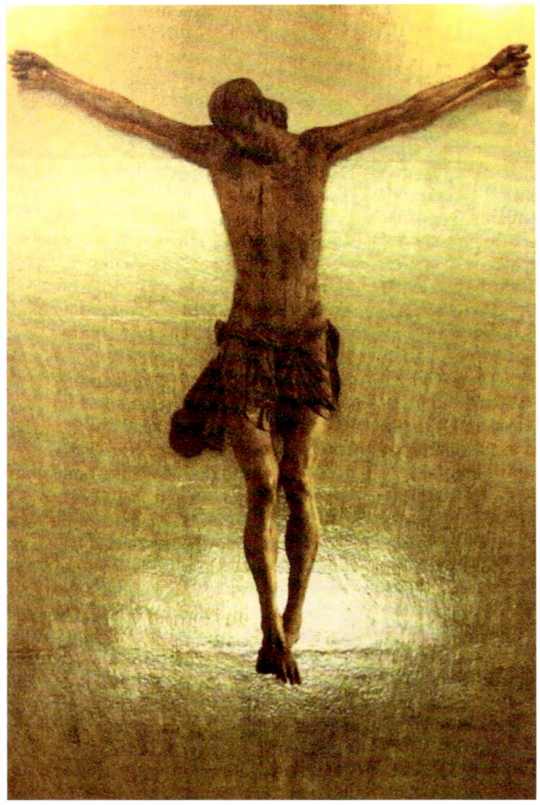

Fig. 13 Veit Königer, *Cemetary Cross*, c. 1720, Bischöfliche Hauskapelle Graz

was known how to bridge the gap between gold (and the other elements in a picture or building) and the heavenly light. Today, in the extreme case, we identify the precious metal with spirituality, without being able to say why. We could also call it quasi-religious kitsch. The frequent characterization of gold as sacred thus represents a misunderstood regression to a closed picture of a geocentric world no more than a few thousand years old.[78]

Both Riegl's 'ideal spatial background' and the popular thesis that the gold ground reflected 'sacred light' have largely left the material basis out of account. That was the reason why Ellen Beer's lecture at the German art-historians' congress in 1982 was seen as a bold new approach. A medieval scholar working in Bern, she attracted the attention of the experts because she asserted that it did not matter what the gold ground 'meant', but what it actually 'was'.[79] This refers to the fact a gold ground in the form of gold leaf is different from a painted gold ground. Beer's complex text directed attention to the circumstance that the gold ground represented less a kind of undefined (supernatural, sacred) space, that the interpretations had not distinguished between the manifestation in terms of colour and material on the one hand and content on the other, that the definitions were fuzzy, that in very few cases was it a 'ground' in any case, in short, that there was not an interpretation of the medieval phenomenon, but an as yet unwritten history of the gold ground.

Ernst Gombrich, in his discussion of Josef Bodonyi's dissertation, already emphasized the unexplained circumstance that gold on the one hand was physical matter, but on the other an imaginary world, 'two facts that in B.'s work are juxtaposed without much connexion [...]'.[80] Otto Pächt had drawn his attention to this. Scholarship sometimes takes strangely circuitous routes, for Ellen Beer insists on this circumstance, evidently in ignorance of Gombrich's review. These scholarly confusions, whose end is not in sight,[81] may only interest the specialist community, but they are already present in the object of investigation—and have a direct association with modern art.

At the same time, we should not expect too much of any material iconology. The way it is applied, it represents no more than the final chapter of the iconology that largely defined the twentieth century. The mere postulate of 'art-worthy materials'[82] is questionable, and what is the 'semantic power of this material'[83] supposed to mean? Time and again, questions are asked about the meaning, the significance. This approach is second nature to every art historian, comparable only to the question of style, which dominated the first decades of the twentieth century. So why should we not ask the great questions of when and where a work was created, and who by, and what it means? And if we did avoid these basic questions, what would we replace them with?

The baleful vicious circle of gold-ground research rests on the assumption that the lustrous surfaces of medieval paintings have to mean anything at all.[84] Peter Cornelius Claussen, emeritus professor in Zurich, has also suggested that questions first be asked about the materiality (of icons) instead of going straight down the interpretation road. It is however noteworthy that following Ellen Beer, a second medievalist has issued this warning, because: "The gold ground remains the projection surface in which the individual medieval pictures are reflected."[85] But the tendency towards ahistorical thinking remains alive and well. Discrimination is prevented by the assertion that ultimately gold always and everywhere has had the same meaning. Thus the new gold grounds are still assigned the same meaning as has been bestowed on the medieval ones with such emphasis.

Different interpretations

The search for the significance of the gold ground has led to a form of automatism. By that I mean the reflex of characterizing the use of gold as transcendental, spiritual, religious, sublime, timeless and eternal. As a result, the boundaries between the categories of view are imperceptibly blurred. If we then consider the gold itself to be eternal and religious, we then get caught up in idolatry. Or it is interpreted as a representation of a different sphere or as a symbol. As a representation of the sacred it depicts this, as a symbol gold represents a different reality, one that is worshipped. Christianity has struggled to distinguish these levels of reality, not just during the Iconoclastic Controversy, but also in the discussion about the relationship between religion and modern art.

If a gold surface (as a picture or a wall) is located in a church, it is easier to comprehend that it elicits not so much an interpretation as a sermon. In a chapel, the priest will not hesitate; he will, upon seeing a non-figural, abstract altar-piece by Muthesius (fig. 12) for example, immediately start preaching: "The gold ground points to God's realm. The witnesses of the Resurrection learn that Jesus did not remain dead, but that He is alive in God's realm and with us. That changes them, they find courage, see their future and their task in a completely new light. Jesus spoke to them personally. Mary Magdalene recognized the living Christ when he called her by her name."[86]

Why, however, a gold surface should indicate God's realm more than, say, a monochromatic black, red or green one, is not a question the devout will ask. A neutral observer will also not see Mary Magdalene and Christ. Since none of what the preacher expresses is displayed in the picture, it is a projection. The gold picture is an empty space par excellence. It is not, however, a space that is emptied in the Suprematist manner, it is a (Christian) sacred space, historically emptied of everything. This reaction to gold is, of course, not an isolated case. Nor does it just affect Christian works. Gold itself knows no sacredness or wisdom. But exactly that is suggested time and again. Thus, Doran H. Ross wrote about 'gold' in a value-free manner, but chose the title *Die Weisheit des Goldes* ("The Wisdom of Gold"). The 'wisdom' alluded to is not actually mentioned in connexion with the precious metal.[87]

Similar observations occur almost daily, especially around Easter and Christmas. At Easter 2009 Bishop Egon Kapellari commented, on the title page of *Furche*, on the gold ground (not just behind the crucified Christ, but in a square beside it) of the episcopal domestic chapel in Graz, refurbished in 2002, in a knowledgeable manner: "The dead body of the Lord is not affixed to the bars of the cross, but to a gilded plate, therefore appearing to be floating in space. A paschal shine thus illuminates Good Friday. Easter, the Resurrection, is the victory of mercy over the gravity of evil and death (fig. 13)."[88] I shall not comment in detail here on how risky the theological content of this interpretation is.[89] If, as the bishop insinuates, the Resurrection already shines into the sacrifice of God's son (through the gold), a sacrifice which vitally determines his incarnation, then a significant element of the Passion disappears, namely the sacrifice of God's son. Crucial for this about-face in the salvation story are the absence of the cross and the spacelessness of the gold ground. The abolition of gravity—as a simile of evil and death—is associated with this spacelessness. But are the expectations placed in the material sur-face of the metal not a little high here? It cannot be the property of this space-negating surface alone that defeats evil.

The train of thought is probably not based on the negation of gravity (which cannot be anchored in the gold). Instead it is more likely based on the tradition of a gold ground, which was standard in the past, before the Renaissance, namely before the spatial construction of perspective was developed. Even where this fact may be self-evident, it cannot be perpetuated at will. In spite of the certainty with which religious dignitaries like to believe in the timeless duration of their doctrines of salvation, this cannot be proved with an ahistorical understanding of how things can be represented. At the time of the medieval gold ground it could have been understood that the 'gravity of evil and death' was defeated by the 'grace of the Resurrection'. In light of the art history of the past five hundred years, this speculation is as up in the air as the Baroque body of Christ without the bars of the cross, mandatory from an iconological perspective, in the compositionally vague disorientation of a gold panel. With regard to the body language of the crucified Christ in the chapel in Graz, the eyes of the interpreter remain closed. Hanging on a cross presupposed gravity in a painful manner, while gravity in itself is nothing 'evil' and it is not 'death'. To perceive the raised arms as a merciful gesture of the Resurrection by dint of the gold ground is questionable. In the Passion story, between the Crucifixion and the Resurrection lie the Deposition from the Cross and the Entombment.

Gravity, given a negative interpretation in the sermon, can also be seen in quite a different light. "In place of monotony came diversity, metaphysics made way for the physical, and truth made way for doubt. Giotto brought spiritual weightlessness to the ground of human life and suffering. The brightness of the gold ground paled against the resplendent lapis lazuli of the earthly sky [...]."[90]

To sum up: the gold ground is supposed to do more for the Baroque crucifix than to imagine an ahistorically preperspective space; it is supposed to suggest the transformation of mortal body into the resurrected body, and thus unite the historical and salvation-story stages. The content of a sermon is, understandably, not to be measured by whether its art-historical interpretation is correct. The crucial question is, what does this gravity-free message, rejoicing in light, orient itself on? Another option would be to quote Wassily Kandinsky: "How many of those who sought God ultimately remained standing before a graven image?"[91] The gold ground may encourage the faithful on their path. Such projections cannot, however, call upon medieval art. It was not the Middle Ages that invented the gold ground as a sacred backdrop, but the historical retrospect. In fact, culture-historically, it was the twentieth century, characterized by destructive ideologies and denounced as materialistic, that thought up this sacred, spiritual interpretation and obstinately stuck to it.

It is not the job of the art historian to criticize the anagogic element of a sermon that uses gold as an argument thus. However, this argumentation leads down a one-way street in which it is not possible to conclude the opposite, namely from the belief in the Resurrection and the heavenly light of Christ to a non-figurative, lustrous gold surface. The danger does not lie in faith but in the art-historical reversion. There is no reflective control in this interpretative pattern. This sort of thing is called a 'dummy system'. Dummy systems do not have any control loops, which means they develop deterministically from a certain starting position to a predetermined end situation. This in turn means that the gold ground automatically leads to belief in the sacred. Communication theologian Norbert Bolz commented laconically: "A scientist can believe, but *as* a scientist, all he can say about religion's claim to meaning is: that's interesting... That means, science has to keep God at arm's length, which it does by making do without the difference between immanence and transcendence. Transcendence transcends precisely that world of distinctions in which science flourishes."[92]

When Richard Wright was awarded the Turner Prize in 2009, too, critics made the knee-jerk link with religious images, in one case with Fra Angelico, when faced with his ink-blot like gold-leaf wall paintings. However, the objection that the seemingly old has been given a new meaning should have made it clear that "You are not suddenly transcending lightly diverting patterns and manifesting timeless authority just because you use a bit of gold leaf."[93]

The unclear relationship between contemporary art and the religious dimension fits our times and also pertains to medieval art. It is not just contemporary art that suffers under the iconodulist ballast of the past, medieval art also runs the risk of not really being perceived in the first place. "Instead, a religious text is poured over the picture like sauce, so that the specifics of the picture are no longer perceptible [...] While at the end of the twentieth century, i.e. between 1980 and 2000, it could still be assumed that religious people were also willing to take in religious images before beating them down with religious abbreviation, this can no longer be assumed today."[94] In summary this means that the same impulse that imagines the entire Passion in abstract gold surfaces also overwhelms compositions when it comes to icons.

There was no way out of this one-way street as long as the preconditions were left unchecked. A crucial reason for this lies in the iconological paradigm of the twentieth century. After the question of style seemed to fizzle out as bland formalism, the question of multiple meanings of pictures became the focus of research. The founder of iconology, Erwin Panofsky, became the definitive art historian of the last century. In the past the amateurish question was: "What style is this and who painted it?", but soon, quite in contrast to the concurrent abstraction in art, the new question became: "What was the artist thinking, what is his message and what do we need to know to understand it?"

The second obstacle is based in the understanding of sanctity. If the sacred is associated with the gold(ground), then the conclusion that art without gold is lacking in sacredness automatically follows. Alberti's judgement against gold is explained by saying: "Art was no longer a product or means of religious orientation."[95]

Hecht showed that it is by no means always possible to determine without ambiguity that the gold ground refers to something heavenly: "Even though the gold ground [...] is used in a completely logical manner, namely in order to depict the golden and luminous heavenly reality, there are plenty of cases in which the visual conditions are portrayed exactly the other way around, i.e. the gold ground is used to portray the earthly scene, while the blue sky is assigned to the heavenly."[96] There is no *a priori* when it comes to the meaning of the gold ground, either in the Middle Ages, or in Modernism. "There is no clear rule for the use of gold ground, because it is always the theme and never the decorative pictorial means or any other elements that make an image a sacred image." [98] The fact that medieval (sacred) works are characterized by a gold ground obviously makes the converse conclusion tempting, namely that the gold ground makes these paintings sacred paintings. The late use of the gold ground in the *quattrocento* had nothing to do with content; it was usually because the clients had certain requests regarding the monetary value of the raw materials. "Questions about the meaning of the gold ground did not play a role."[99] "Since a sacred picture is constituted by its holy theme and not by a specific form of representation, there were no problems with forgoing the gold ground."[100]

Even if the question of expensive materials , i.e. the use of precious metals and gemstones, was important for the late Middle Ages in the sense of "luxury and asceticism", the really crucial question was whether religious images were to be allowed in the first place. The Byzantine Iconoclasm made the question of content, i.e. the theological question, the central one. "It was only when Iconoclasm was overcome that icons obtained their holy character, since from then on a direct connexion was perceived to exist between the depiction and the depicted, with the latter sending 'energies' down on to the icon."[101] It was not the gold that acted as an energy carrier to create the connexion to the timeless, but the belief that the picture was of the same essence (consubstantial) with what it depicted. According to the Iconoclasts, the actual initiator of the heresy of this belief in images was Lucifer, because it was a 'confusion of the divine creation with matter'. "If an icon is merely paint and consequently matter, then those who worship an icon worship an idol."[102] This idolatry causes holiness to be confused with this world's humanness. At the last ecumenical council of the Byzantine church in AD 787 (the Second Council of Nicaea), it was ruled that it was not the icon that was worshipped or venerated, but rather the 'prototype' of the subject depicted. The understanding of the lifelessness of the material was not questioned. That was not what made the gold different from the paint, the wood, the stone. "They are only transformed into something different in the hands of the artist who shapes them with his art."[103] Moreover it is not the artistic freedom that is crucial, but the tradition.

Alternative Materials

Spiritual experiences and religious insights never had anything to do with the precious material (neither for the Iconoclasts, nor the Iconodules), but rather with the possibility of a connexion between the prototype and his/her image. If the very belief in the sacredness of gold is an error, this belief is even more incredible in the face of imitation gold. However, what is the basis of the oft-noted contradiction, already hinted at, between the gold pathos and the simultaneous renunciation of gold in favour of substitute materials?.

It is astounding that investigations based on the method of material iconology do not take gold seriously. Although great attention has finally been paid to gold in art, this interest is contrasted by a simultaneous material blindness. The insights and observations obtained are applied equally and indiscriminately to substitute gold, i.e. to everything that is not gold, but that shines like gold or has the colour of gold. As a result, much that is said about 'gold' is doubtful. How can anyone determine what the use of gold means when other materials are used to symbolize gold? It is easy to get lost in a labyrinth of metaphors and allegorical references.

It is symptomatic that in all retrospectives of pictures that were created using gold, even those without gold are taken into account, as long as the material looks right. Even where no gold is to be seen, just *ersatz* gold, be it in the form of paint or even of vinyl, as in Andy Warhol's *Marilyn* (1962), the aura of the icon is not far away. When Michael Buthe uses gold substitute, viewers are able to have just as much of a spiritual insight because the artist, as a priest, conveys the sacred. The artists' corresponding self-stylization is taken at face value. In Buthe's cultic space of gold (some of it fake), a 'connexion between the human and the divine sphere is created'. Even a little bit of talmi gold then imparts a 'sharing in the divine world'.[104] To say it again, here the metal gold is perceived as having a divine aura and in addition, gold substitute can, it is claimed, achieve the same effect through artistic appropriation.

This was already a characteristic of the first encyclopaedic gold exhibition that also took contemporary artists into account: *Das Goldene Zeitalter—Die Geschichte des Goldes* ('The Golden Age—The History of Gold') at the Württembergischer Kunstverein Stuttgart, 1991. This 'cabinet of curiosities' was not well received among critics,[105] not least because of the dilution of the gold concept. The book's cover depicted a work by Giuseppe Penone with bronze *Potatoes* (1977), the back featured a gold-paper strip based on a design by Daniel Buren. Real gold was the exception among the exhibits. The curator's statement was symptomatic of the exhibition title: "The 'Golden Age' never had anything to do with gold the material, there was no gold there."[106] A fateful misjudgement of gold was betrayed in the characterization of the metal as a material with a 'real transparency', which is why, he said, gold made 'the world phenomenally transparent'.[107]

A lack of gold also characterized the exhibition *Aurum—Gold in der zeitgenössischen Kunst ('*Aurum—Gold in Contemporary Art'*).*[108] Of the 41 artists exhibited, only 16 had worked with gold. The curator Dolores Denaro also jumbled other criteria. In the chapter 'Gold as a Working Material', she suggested the possibility of 'gold being used, like red, blue, yellow, black and white, as a further "pure" colour'.[109] To argue this, she eliminated the need for the metal, if gold were just understood as a colour: "If gold is just used as a colour to expand the palette, then the precious metal itself holds no significance."[110] Instead, as in the case of vinyl, it was about the light-like effect. Gold-coloured acrylic paint also has 'properties that no other colours have'. Whether it is real gold or not becomes secondary, viewers are simply dealing with the artists' 'alchemistical creative method'. Gold-leaf imitation 'is tantamount to a sanctification of the everyday'.[111] Even if we are only dealing with golden glass beads, brass plates or bronze pigments in this substitute swindle, a 'universal significance' can be implied.[112] The curators resemble the first Europeans in Africa, who paid for ivory, gold and slaves (the 'black gold') with worthless glass beads. To summarize, the significance is not necessarily in the metal, it suffices to replace the gold or at least imply that. The *Aurum* text concludes with a look at 'gold from the perspective of contemporary iconography', which is emphasized by a quote from Goethe: "Towards gold throng all / to gold cling all / yes, all! / Alas, we poor!" [113]

Published at the same time as the book *Die Renaissance des Goldes* was the book *L'or dans l'art contemporain*, devoted expressly to gold in contemporary art, as its title suggests, with an article by Gérard-Georges Lemaire.[114] This publication is not free from the fallacy of gold substitutes either. The two mottoes by Friedrich Nietzsche[115] and Ludwig Wittgenstein[116] do not hide the fact that the text nonchalantly brushes over the discussion of the gold ground in the German-speaking world. The text does not quote any German literature.[117] In his article, which describes the history of gold in art from ancient Egypt to the present, and also looks at non-Western cultures, German Gothic, with its gold grounds, is completely absent. Lemaire prefers to look at the gold grounds of the *trecento*. It is obvious to him that the gold ground from Byzantine art found its way into Italian painting via Venice. He also completely omits the tradition of book illumination.

The theory that gold represented light in medieval theology does not prevent the author from pointing out arguments against the use of gold in church art. It was not just the mendicant orders that prohibited golden and silver crucifixes. It always depended on whether the aim was to depict the glory of the Eucharist with precious means or to ban every form of luxury from the church. It is not a question of the metal's properties, but of the relationship between its value and the church's understanding of itself.

Interestingly the chapter 'L'or entre le bien et le mal' ('Gold between good and evil') contains an in-depth reflection on gold's 'evil side' in mythology, religion and art. The author is more interested in gold than in material-iconological speculations. For this reason the work contains many depictions of works without gold by Caravaggio, Lucas Cranach, Frederic Leighton, Nicolas Poussin, Rembrandt, Tiepolo, Titian. This too illustrates a certain lack of interest in gold as a material. He only touches on the nineteenth century, by referring to the use of gold by Gustave Moreau and a few Symbolists. Here too it is a problem that he does not distinguish between gold and gold substitute. Here too *Gold Marilyn* (1962) by Andy Warhol, a screenprint on a gold ground, is cited as an example.[118]

Towards the end of his article, Lemaire poses the rhetorical question whether gold has the power of allowing us to believe in an afterlife. He notes a 'trap' for the understanding here.[119] It is no coincidence that the author chooses to conclude his text with a *Golden Ladder* by Vitantonio Russo, which was exhibited at the Venice Biennale 2009. The interest in economic subjects has, he says, replaced the religious symbolism of Jacob's Ladder, a ladder leading to heaven.[120]

The book is a veritable treasure trove for the use of gold in contemporary art. It is the first to portray the whole scope of the subject, albeit without any (even just alphabetical) order or an index. The more than 100 examples of works open up a fascinating panorama. Unfortunately there are no references in the text to the illustrations, and most of these are not discussed.

Conclusion

What is the basis of the contradiction, which can be observed time and again, between the gold pathos and the simultaneous renunciation of gold in favour of substitute materials? Picking up Ellen Beer's suggestion that it does not depend on what the gold ground 'meant', but on what it actually 'was', then the opposite would hold true for the substitute gold ground generated by paint, plastic or foil. The apparent gold ground is not gold, but it means the same in the sense that it depicts or represents it. It is an allusion to the absent, to the historically exemplary, to the theologically maybe incorrect, but intended anagogic range of meanings that was (also) hinted at with gold. The fake gold is an allegory in the original sense (literally 'I speak other'). As the not-present, it is consistent with the actual task of painting in the sense of the mythical origin of art (according to Pliny the Elder, with Butades of Sicyon), which is to depict that which is no longer present, the absent, that which is past.[121] The converse conclusion is that gold is a material irrelevant to this understanding of painting. Since it does not depict or represent anything, because it is not a colour, the shiny precious metal in an artwork permits an immediate presence of perception. Beyond the false conception, there is not just room for imaginative projections here, but also for understanding and appropriate interpretations.

1 Even when it was first installed, it was criticized by the public for being old-fashioned. The gilding was removed in 1935, re-applied in 1991, and restored in 2011.
2 Anne Schloen, *Die Renaissance des Goldes*, Nuremberg 2010.
3 Anne-Marie Charbonneaux (ed.), *L'or dans l'art contemporain*. Text by Gérard-Georges Lemaire, Paris 2010.
4 *Allgemeines Repertorium der Literatur*, Drittes Quinquennium für die Jahre 1796 bis 1800. Zweiter Band, Zweite Hälfte und zwar die Literatur, XIV. Der schönen Künste, Weimar 1807,159a: *Neues theoretisch-praktisches Zeichenbuch zum Selbstunterrichte für alle Stände.* VI Heft: *Anleitung zum Zeichnen u. Malen der Blumen, jngleichen (sic!) zum Zeichnen der Silhouetten auf Goldgrund*. 1800.
5 I would like to thank Ursula Storch, Wien Museum, for kindly providing this information.
6 "Tag- und Jahreshefte" (1749–1806), in: *Goethe's Werke*, vol. 31, Stuttgart/ Tübingen 1830, p. 218.
7 In the abundant literature on Goethe's aesthetics, no attention appears to have been given to this aspect. Cf.: Sabine Schulze, *Goethe und die Kunst* (exhib. cat. Schirn Kunsthalle Frankfurt/ Kunstsammlungen zu Weimar Stiftung Weimarer Klassik), Ostfildern-Ruit 1994.
8 Jörg Träger, *Philipp Otto Runge und sein Werk. Monographie und kritischer Katalog*, Munich 1975.
9 Ibid. p. 57.
10 See Thomas Zaunschirm, "Alberti's Gold Ground" in the present volume.
11 Johann Wolfgang von Goethe, *Kunstschätze am Rhein, Main und Neckar*, 1814 and 1815, *Goethe's nachgelassene Werke*. vol. 3, Stuttgart and Tübingen 1832/1833, p. 403.
12 Jean Paul, *Der Komet oder Nikolaus Marggraf. Eine komische Geschichte* (written from 1811 on), Berlin 1820, p. 112.
13 Letter from Friedrich Tieck to Goethe from Berlin, no later than 21 Feb. 1824, in Ludwig Geiger (ed.) *Goethe-Jahrbuch*, vol. 17, Frankfurt am Main 1896, quoted from: Elibron Classics series, Adamant Media Corporation, 2006, p. 46.
14 Franz Kugler, *Handbuch der Geschichte der Malerei in Italien seit Constantin dem Grossen*, Berlin 1837, §24, p. 81. (Kleine Schriften und Studien zur Kunstgeschichte, Italienische Studien, Stuttgart 1853, 362–404.)
15 *Meyers Konversations-Lexikon*, 1888. p. 7488. There is nothing about a spiritual significance here. Pictures from the Altlerchenfeld church in Vienna are held up here as a positive example of the gold ground in the art of the time.
16 Georg Wilhelm Friedrich Hegel, *Aesthetics. Lectures on Fine Art*, trans. by T.M.Knox, vol. 2, Oxford 1975, p. 851. Translator's note: Knox's 'background of gold' as a translation of 'gold ground' should probably be replaced by 'gold ground', as 'background' carries inappropriate connotations of depth. The italics are the present author's.
17 Claudia Stockinger, *Das 19. Jahrhundert. Zeitalter des Realismus*, Berlin 2010, p. 107.
18 Oswald Spengler, *The Decline of the West.* Vol. I *Form and Actuality,* trans. C. F. Atkinson, London 1954, pp. 247ff
19 Ludwig Curtius, "Morphologie der antiken Kunst", in: *Logos*, 1919/20, p. 197, quoted from: Massimo Ferrari Zumbini, *Untergänge und Morgenröten. Nietzsche—Spengler—Antisemitismus*, Würzburg 1999, p. 165.
20 Josef Strzygowski, *Orient oder Rom. Beiträge zur Geschichte der spätantiken und frühchristlichen Kunst*, Leipzig 1901, p. 6.
21 Quoted from: Josef Bodonyi, *Entstehung und Bedeutung des gold groundes in der spätantiken Bildkomposition*, diss. University of Vienna 1932, p. 5. This gushing tone on Lange's part is not derived from any pictorial analysis or from any specific picture.
22 Schloen op. cit., p. 36.
23 Wolfgang Schöne, *Über das Licht in der Malerei*, Berlin 1954, p. 22.
24 Lorenz Dittmann, *Farbgestaltung und Farbtheorie in der abendländischen Malerei*, Darmstadt 1987, p. 7.
25 Christine Sauer, "Zum gold ground in der mittelalterlichen Buchmalerei", in: *Das Goldene Zeitalter. Die Geschichte des Goldes vom Mittelalter bis zur Gegenwart* (exhib. cat. Württembergischer Kunstverein Stuttgart), Ostfildern-Ruit 1991, p. 246.
26 Hans Jantzen, *Ottonische Kunst*, Munich 1947, p. 78. *Rowohlts Deutsche Enzyklopädie*, 1959 ff., p. 72.
27 Schöne op. cit., p. 22. Schöne himself has the Middle Ages begin with the fifth century (see p. 20).
28 *Der Wiener Dioskurides, Codex medicus graecus 1 der Österreichischen Nationalbibliothek*, Graz 1998. Commentary by Otto Mazal, p. 21.
29 Hilde Zaloscer, *Vom Mumienbildnis zur Ikone*. Wiesbaden 1969, p. 18.
30 Klaus Parlasca, "Die ägyptischen Mumienporträts", in: *Paula Modersohn-Becker und die ägyptischen Mumienportraits,* (exhib. cat. Kunstsammlungen Böttcherstrasse Paula Modersohn-Becker Museum, Bremen/ Museum Ludwig, Cologne), Munich 2007, p. 56.
31 Schloen op. cit., p. 41.
32 One is reminded of Heribert Illig's 'phantom time' hypothesis. Illig claims the three centuries between AD 614 and 911 never existed. (Heribert Illig, *Das erfundene Mittelalter. Die grösste Zeitfälschung der Geschichte*, 2nd ed. Munich/Düsseldorf 1999.)
33 Schloen op. cit., note 43.
34 Schöne op. cit., p. 217.
35 I am grateful to Walter Holzer, Klosterneuburg, for this information.
36 Clark V. Poling, "Kandinsky in Russland und am Bauhaus 1915–1933", in: *Kandinsky—Russische Zeit und Bauhausjahre 1915–1933* (exhib. cat. Bauhaus Archiv, Museum für Gestaltung), Berlin 1984, pp. 40, 254. This golden wall represents to this day a strange facet of Kandinsky's work. Cf.: Reinhard Krause, "Ein Hauch von Gold", in: *AD. Architectural Digest*, May 2011, p. 72.
37 Werner Spies, *Picasso. Das Plastische Werk*. Stuttgart 1971
38 Schloen op. cit.
39 URL: <http://www.artinjewel.de/picasso.html> [read on 1 Dec. 2011]. Currently 12 gold items are on display.
40 *Schmuck—Zeichen am Körper* (exhib. cat. Oberösterreichisches Landesmuseum Francisco Carolinum, Linz), Vienna 1987.
41 Iris Wenderholm, "Aura, Licht und schöner Schein. Wertungen und Umwertungen des Goldgrunds", in Stefan Weppelmann (ed.) *Geschichten auf Gold. Bilderzählungen in der frühen italienischen Malerei*, (exhib. cat. Gemäldegalerie, Staatliche Museen zu Berlin)Berlin/ Cologne 2005, p. 100.
42 Wolfgang Max Faust, "Documenta: Haare in der Suppe", in: *Kunstforum International*, vol. 53/54, 7–8/82, p. 39.
43 Georg Jappe, "Nullprozentiges Goldwasser. Nachsätze zur d7", ibid., p. 26.
44 Thomas McEvilley, "More Golden than Gold", in: *Artforum*, New York, November 1985.
45 Tilman Osterwold (ed.), *Das Goldene Zeitalter. Die Geschichte des Goldes vom Mittelalter zur Gegenwart* (exhib. cat. Württembergischer Kunstverein Stuttgart), Ostfildern-Ruit 1991, p. 5.
46 "An art exhibition which resembles the sixteenth-century encyclopaedic cabinet of curiosities, a hotchpotch with universal aspirations, with the quarry of a scholarly hunt for gold. Everything to be seen was picked up and taken home."

Elke von Radziewsky,"Ausstellung in Stuttgart: 'Das Goldene Zeitalter'. Vanitas und Pommes", in: *Zeit online*, 51/1991, p. 65. Quoted from: URL: <http://www.zeit.de/1991/51/vanitas-und-pommes> [read on 1 Dec. 2011]

47 "Schlimmer als Eisen", in: *Der Spiegel* 1/1992, quoted from: URL: <http://www.spiegel.de/spiegel/print/d-9272286.html> [read on 1 Dec. 2011]. The author deals somewhat roughly with the curator: " [...] talked as much crap as for example Osterwold in the catalogue.".

48 Martin Blättner, "Das goldene Zeitalter", in: *Kunstforum International*, vol. 117, 1992, 355-357.

49 Schloen op. cit.

50 Charbonneaux op. cit.

51 Schloen op. cit., p. 45.

52 Ibid. p. 55.

53 Ibid. p. 61.

54 Ibid. p. 64.

55 Ibid. p. 67.

56 Ibid. p. 106. Klein had even transformed 'gold into a special state of the soul, in other words into an "inner gold"' (p. 107)

57 Ibid. p. 170.

58 Ibid. p. 108.

59 Ibid. p. 145.

60 Ibid. p. 139.

61 Ibid. p. 164.

62 Ibid. pp. 190, 239.

63 Ibid. p. 194.

64 Ibid. p. 242.

65 Lorenz Dittmann, *Farbgestaltung in der europäischen Malerei*, Cologne/Weimar/Vienna 2010. The book is basically the same as Dittmann's *Farbgestaltung und Farbtheorie in der abendländischen Malerei*, Darmstadt 1987.

66 Christian Hecht, *Die Glorie. Begriff, Thema, Bildelement in der europäischen Sakralkunst vom Mittelalter bis zum Ausgang des Barock*, Regensburg 2003.

67 Bodonyi 1932 (as note 22).

68 Zaloscer 1969 (as note 30).

69 Bodonyi op. cit., p. 117.

70 Schloen op. cit. p. 17.

71 Hans Belting, *Bild und Kult. Eine Geschichte des Bildes vor dem Zeitalter der Kunst*, 6th ed. Munich 2004, p. 339.

72 Bazon Brock, *Ästhetik als Vermittlung. Arbeitsbiographie eines Generalisten*, Cologne 1977, ch. II/5/7.

73 Ibid.

74 Brock op. cit., ch. II/5/7.

75 Schloen op. cit., p. 40.

76 Schöne op. cit., p. 71.

77 Ellen J. Beer, "Marginalien zum Thema Goldgrund", *in*: *Zeitschrift für Kunstgeschichte*, vol. 46., no. 3, 1983, p. 276.

78 Isaac Newton still thought the world was 6,000 years old. And the heliocentric picture of the world was only abandoned in the 1920s. Light, once seen as static, mutated in the twentieth century into a dynamic unit of measurement as a space-time parameter of the mass-energy relationship ($E=mc^2$).

79 Beer 1983 (as note 78), 271-286.

80 Ernst Gombrich, "J. Bodonyi, Entstehung und Bedeutung des Goldgrunds in der spätantiken Bildkomposition" (1932/33), in: *Kritische Berichte zur Kunstgeschichtlichen Literatur*, vol. 5, 1935, 65-75, quoted from: URL: <http://gombricharchive.files.wordpress.com/2011/04/showrev21.pdf>, p. 2 [read on 1 Dec. 2011] (I am grateful to Werner Hofmann for pointing this out).

81 Cf. the attempt by Lorenz Dittmann, *Farbgestaltung und Farbtheorie in der abendländischen Malerei*, Darmstadt 1987, p. 8.

82 Monika Wagner/Dietmar Rübl/Sebastia Hackenschmidt (eds.), *Lexikon des künstlerischen Materials. Werkstoffe der modernen Kunst von Abfall bis Zinn,* Munich 2002, p. 7.

83 Michael Liebelt, "Gold", in Wagner/Rübl/Hackenschmidt op. cit., p. 120.

84 Cf. the article by Viktor Michael Schwarz in the present catalogue.

85 Peter Cornelius Claussen, "Goldgrund", in: *Kritische Berichte*, 35, 2007, 3, pp. 66f.

86 Thus the parish priest Albert Giesener in St Canisius in Charlottenburg in an Easter sermon quoted in the Berlin *Tagesspiegel* (10 July 2007) in light of a picture from the series *Golden Fields* by Winfried Muthesius, which is, by the way, made of composition gold without gold leaf.

87 Doran H. Ross, "Die Weisheit des Goldes", in: *Das Gold der Akan,* Museum Liaunig, Neuhaus/Suha 2008, p. 9. There is an English edition of this work, entitled *Royal arts of the Akan: West African gold in Museum Liaunig*, but the present editor has been unable to ascertain whether the chapter in question actually has the title "The Wisdom of Gold".

88 Egon Kapellari, "Es ist vollbracht", in: *Die Furche*, 9 Apr 2009, p. 1.

89 Cf. Thomas Zaunschirm, "Corpus Christi?", in: *Sphingen, bunte Kühe und eine Ente,* Essen 2001, 199-212.

90 Eva Clausen,: "Blau statt Gold und Zweifel statt Wahrheit. Giotto und das 14. Jahrhundert—eine Ausstellung im Complesso del Vittoriano in Rom", in: *Neue Zürcher Zeitung*, 23 May 2009.

91 Wassily Kandinsky, *Über die Formfrage,* 1912. Here quoted from Kenneth C. Lindsay and Peter Vergo (eds.) *Kandinsky: Complete writings on art*, Cambridge MA, 1994, p. 239.

92 Norbert Bolz, "Kommunikationsprobleme mit Gott", in: *Weltkommunikation,* Munich 2001, p. 174.

93 Matthew Collings, "Why do paintings look nice?" in: *Modern Painters*, 03/2010, p. 27.

94 Andreas Mertin, "Ästhetisierung von Religion? Idolatrie", in: *Tà katoptrizómena. Das Magazin für Kunst, Kultur, Theologie, Ästhetik*, no. 63, quoted from: URL: <http://www.theomag.de/63/am306.htm> [read on 2 Dec. 2011].

95 Schloen op. cit., p. 30.

96 Hecht op. cit., p. 73.

98 Ibid., p. 74.

99 Ibid., p. 75.

100 Ibid., p. 148.

101 Tania Velmans (ed.), *Ikonen. Usprung und Bedeutung*, Stuttgart 2002, p. 7.

102 Gaetano Passarelli,: "Der Bilderstreit—Geschichte und Theologie"*,* in: Tania Velmans (ed.): *Ikonen. Usprung und Bedeutung*, Stuttgart 2002, p. 31.

103 Passarelli 2002 (see note 102), p. 35.

104 Schloen op. cit., p. 230f.

105 Cf. Radziewsky 1991 (as note 47), p. 65.

106 Osterwold 1991 (as note 46), p. 23.

107 Ibid., pp. 44, 46. Here the characterization of gold gets caught up in the conflict between material superficiality, which reflects the light, and symbolic intangible immateriality.

108 Dolores Denaro*, Aurum—Gold in der zeitgenössischen Kunst* (exhib. cat. CentrePasquArt Biel), Nuremberg 2008.

109 Ibid., p. 13.

110 Ibid., p. 14.

111 Ibid., p. 16.

112 Ibid., p. 17.

113 Johann Wolfgang von Goethe, *Faust I,* lines 2802 ff., spoken by Margarete. Here in the translation by G.M.Priest, see http://www.levity.com/alchemy/faust09.html. It should not

be forgotten that the gold mentioned is the jewellery that Mephisto has put down. Margarete's mother gives it to the church however. This gold is not a sacred element, but one from the devil.

114 Charbonneaux 2010 (as note 3).

115 "Tell me, pray: how came gold to the highest value? Because it is uncommon, and unprofiting, and beaming, and soft in lustre; it always bestoweth itself. Only as image of the highest virtue came gold to the highest value. [...]. Gold-lustre maketh peace between moon and sun. Uncommon is the highest virtue, and unprofiting, beaming is it, and soft of lustre [...]". Friedrich Nietzsche, "Thus Spake Zarathustra",trans. by Thomas Common, see http://philosophy.eserver.org/nietzsche-zarathustra.txt. This section is head 'The Bestowing Virtue', but the two passages that talk about bestowing ('schenken', i.e. giving freely) are omitted from the quote (as indicated): "Goldlike, beameth the glance of the bestower" and "a bestowing virtue is the highest virtue"

116 "There is gold paint, but Rembrandt didn't use it to paint a golden helmet." Ludwig Wittgenstein, *Remarks on Colour*, (trans. M. Schättle) Berkeley CA 1978, para. 79 , p. 27. It is difficult to imagine a less suitable quote on the subject, not least because the once famous *Man with the Golden Helmet* has not been attributed to Rembrandt since 1986. Other quotes about gold by Wittgenstein would have been more suitable, such as: "It is easy to see that not all colour concepts are of the same sort, e.g. the difference between the concepts 'colour of gold' or 'colour of silver' and 'yellow' or 'grey'." Ibid. para. 54, p. 7. Both quotes, Nietzsche and Wittgenstein, are used as mottoes without references or comment.

117 German names are also given without any attempt at checking: Johan Heinrich Tuesbein for Johann Heinrich Tischbein, Penhauer for Pembauer.

118 Charbonneaux op. cit., p. 33.

119 "L'or a le pouvoir de faire croire en cet au-delà puisqu'il est inaltérable—mais n'est-ce pas un piège pour l'entendement?" Gérard-Georges Lemaire, "L'or dans tout ses états, de l'Égypte ancienne à nos jours", in: Charbonneaux op. cit. p. 42.

120 "Pionnier dans le domaine de la transposition des thèses des économistes dans le champ spécifiques de l'art, il a récemment introduit l'or dans se compositions, comme eo—cette échelle qui semble bien avoir remplacé l'échelle de Jacob et l'échelle des anges conduisant aux Cieux." Ibid., p. 42.

121 Cf. the story of the daughter of the potter Butades of Sicyon in *Historia naturalis* (AD 77) by Pliny the Elder

The Gold Ground in the Middle Ages
“Don’t ask for the meaning, ask for the use!”

Michael Viktor Schwarz

Truovasi chi adopera molto in sue storie oro, che stima porga maestà. Non lo lodo …
(There are some [contemporary painters] who use much gold in their *istoria* They think it gives majesty. I do not praise it. …)

Leon Battista Alberti, *Della Pittura*, 1435/36

The reservations concerning the use of gold in painting, which the famous Renaissance art-theoretician Leon Battista Alberti formulated in the section of his treatise that follows the above extract, can be summed up as follows. "You can say what you like about majesty (that is not the object of my criticism), but gold destroys the pictorial illusion. Gold is not a colour that clings to an object, but a material, which looks different depending on the lighting and the angle of view, and thus is not part of the picture." Alberti observes, and so can we, that while it might shine out brightly at one moment, we only have to move our heads, or the angle of the incident light only has to change a little, and the brightness turns into a dark spot. The gilded line or area is present as an uncontrollable piece of reality in the artful simulation of the picture.[1] And that was what seemed to Alberti to be wrong.

However, he makes this judgement against the background of an idea of the picture that differed from that of earlier generations and other cultures. His treatise on painting treats the picture as a kind of rectangular cropped reproduction of the visual field[2]—a position as new as it was future-oriented. 150 years later, Johannes Kepler was to define this 'visual field' as an image in the eye. And the photograph, that paradigmatic image of the modern world, is, once chemically fixed, nothing other than a one-eyed visual field: everything in front of the lens also appears in the picture, whether it is important or trivial, wanted or unwanted: the children, the beach, the palms, and the remains of the picnic in the foreground. Indeed, an invention has been attributed to Alberti which belongs to the prehistory of photography. It is a device that allows one to fix on the surface of a veil (velum), not chemically, but with a brush or pencil, the image which one sees through this veil. The apparatus later described by Albrecht Dürer in his book *Underweysung der Messung* goes

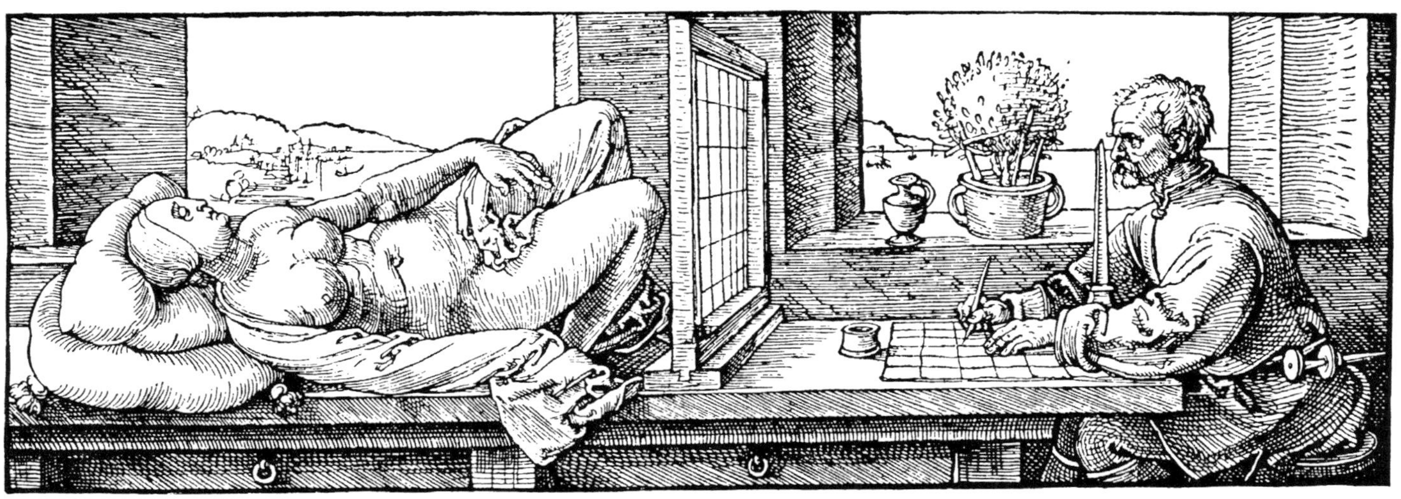

Fig. 1 Albrecht Dürer, *Drawing aid,* Woodcut from: *Underweysung der Messung*, 3rd ed. 1538

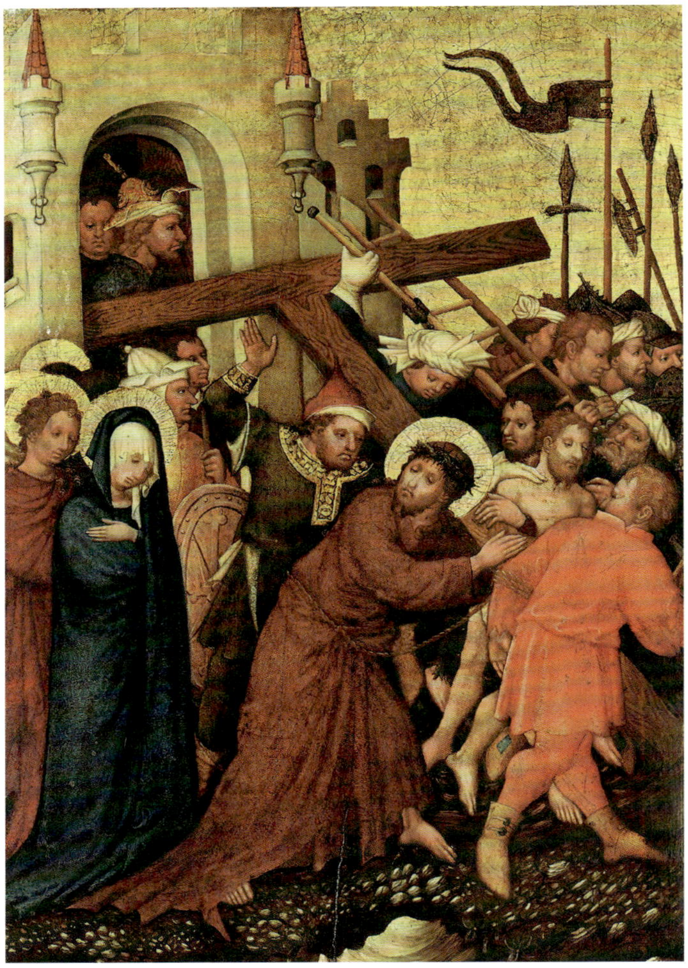

Fig. 2 Master of the St Lambrecht Votive Tablet, *Christ Carrying the Cross*, c. 1430, Belvedere, Vienna

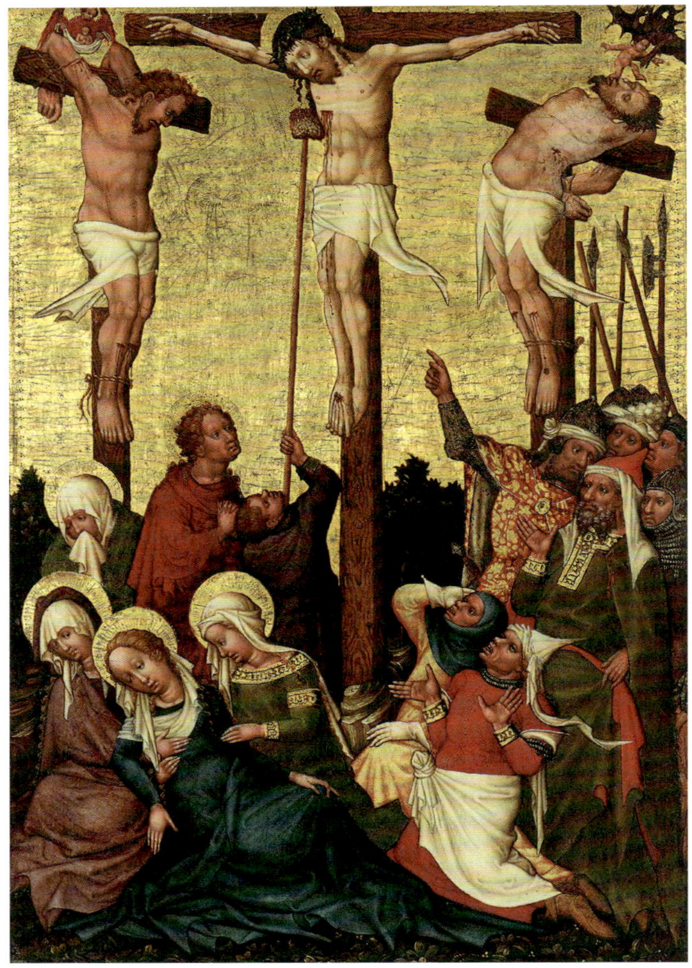

Fig. 3 Master of the St Lambrecht Votive Tablet, *Crucifixion*, c. 1430, Belvedere, Vienna

back to this (fig. 1). Alberti in turn was inspired by the optical devices of his fellow-Florentine Filippo Brunelleschi, devices which served to calculate perspective constructions.[3]

Instead of the reproduced visual field, one can talk about the optical picture. The central criterion is that it is valid as a function primarily of reality. Making pictures is understood as something that the artist does with his eyes and with the instruments and/or rules he uses. It is true that he can omit things or change them and furthermore he can (and this was the normal case) merely simulate the depiction and invent situations; the photographer by contrast can retouch and he can create montages (the possibilities offered by digital image processing these days are boundless). The standard, though, remains the picture that you take, the *mīmiīsis* (imitation) of something that at some time or other has been or still is capable of being seen. It was this standard against which Modern Art rebelled, while for utilitarian images (documentation, advertisement, entertainment) it continues to be valid.

What Alberti's idea of the picture historically deviated from, and what in practice remained alive for a long time after (if it was not *always* alive alongside and beneath the Albertian picture)[4], was the picture that made something appear. It confronts the beholder not with the imitation of a visual field, but with subjects worthy of depiction, be they persons or things. Instead of being associated with visible reality, it is independent of what the painter (or the client) intends, and what he, with the help of the picture, wants to suggest, however and to whomsoever. Depending on the planned use of the picture, much or little may be worthy of depiction, but in any case what appears in the picture is meaningful and wanted: whether this is the half-length figure of Christ in isolation (whose worship the picture is intended to encourage) or a number of figures and implements and topographical details (if one should want to tell stories from his life). A pair of panels dating from c. 1430 show two stations of the Passion[5] (figs. 2, 3). In one case, alongside Christ carrying the Cross, and His entourage, we see a gate, a section of wall and behind it a stepped gable, along with a path strewn with stones—precisely what is needed to make it clear that the procession is leaving the city of Jerusalem by a torturous path (*via dolorosa*). In the other case, alongside the crucified Christ and those accompanying Him there is a meadow in the foreground and hints of shrubs between the figures—elements which locate this event, with just about sufficient clarity, outside of the city. A norm, independent of the media requirement, for completeness of depiction (or lack thereof) does not exist in this kind of picture.

The result of this is that as a rule empty spaces remain around and between the objects and persons. The painter can always fill these spaces with more objects provided he can think of any, and as long as they enrich the picture and enhance its use, rather than distract from it or confuse the scene depicted. Thus one could, in the case of the Crucifixion, add a section of wall above the bushes, and thus locate the scene not just 'outside the city' but, in accordance with St John's Gospel, (John 19,20) 'nigh to the city', as so many painters have done over the centuries (and which was doubtless deliberately not done here in order to avoid repetition of the other panel). But no self-contained (pictorial) world is created in this way. Alberti's concern, namely to protect the world in the picture from foreign bodies such as luminous forms, seems here to be without content. Rather, this kind of picture is concerned to accentuate and shape the empty spaces.

The most popular filler material here was gold leaf, which now brings us to the problem of the gold ground in the pre-Renaissance picture—and only now, for Alberti does not expressly treat the gilded background, although in the Florence of his day it was certainly very widespread and must have been present to his eyes when he wrote his treatise. Apparently beholders were for centuries satisfied with the impression he addresses, that gold ennobles pictures (lends them 'majesty'). This is doubtless more than just a vague feeling; behind it stood the concept of magnificence—a virtue or attitude that has been lost in bourgeois culture, but among the elites of the pre-modern age was ver much alive. It demanded being satisfied only with the best and most expensive: a large theme in, for example, the *Nicomachean Ethics* of Aristotle (IV, ii). As long as this was the ideal, a picture with a gold ground was recognized as one that identified its subject as demanding and the client as magnanimous. Later, at some time or another, the question of meaning in the actual sense arose. This happened c. 1900 when, beyond historicism and positivism, a 'critical' (not to say speculative, but it was in any case concerned with a deeper understanding) art scholarship appeared. On the one hand there was the Danish art historian Julius Lange, who (inspired, as he said, by Dante's *Divine Comedy*) maintained that the gold ground should give, in certain cases, 'a picture of God's kingdom, the fire of heavenly love'.[6] On the other hand the Viennese art historian Alois Riegl, in his groundbreaking work *Die spätrömische Kunstindustrie*, came up with the interpretation that gold was the 'ideal spatial ground'—an artistic means that either replaced depth in the picture (as would have been depictable with painterly means in accordance with the tradition of classical Antiquity) or else represented it vicariously in sublimated form.[7]

Riegl's statement was meant as formal analysis, but could also be understood thematically, or indeed in a Platonic sense, spiritually, and was thus quite compatible with Lange's mystically toned offer. It is on Riegl or Lange or both, directly or indirectly, that the interpretations of the following 100 years are based: the gold is a space-concealing area (Hans Berstl)[8], is light-space (Joseph Bodonyi)[9], is un-space (Hans Jantzen)[10], is potentiated light, open to the phenomena of two and three-dimensional space (Wolfgang Schöne)[11], is timeless space (Thomas F. Mathews)[12], is spaceless world (Robin Cormack)[13] ... Further hyperbole can be found in times hungry for meaning: after the October Revolution the priest and philosopher Pavel Florenskij saw in the gold ground of icons a structural analogy with the divine creation of light (*Die Ikonostase*, 1918–1920)[14], while after the Second World War the art historian Wolfgang Braunfels saw in the gold ground an echo of paradise (*Nimbus und Goldgrund*, 1950).[15]

Apart from the fact that such interpretations hardly work for any picture without bold speculation (just try to apply it to the two Passion pictures), and apart too from the fact that there is not a single known medieval text that deals with the gold ground and could support these interpretations, and finally, apart from the fact that the earliest known appearance of a gold ground was in a non-religious context, namely in two miniatures in a medical manuscript (the *Vienna Dioscurides*, see below): ultimately the attributions are based on a precarious premiss, namely the increasingly dominant certainty (since Alberti) in western Europe that everything in a picture must depict something, just as in a photograph every element represents something, whether we know what it is or not. The grey area in a passport photo may be cardboard, or a section of wall, and 'mean' more or less spatial distance from the portrait head, it represents an extra-pictorial finding. By contrast, in pre-Alberti pictures the gold zone around a figure or around the elements in a story is precisely just what it factually is: the unused surface of a picture support, gilded. Unlike the grey in the photograph, it does not belong to the picture as depiction, but is (like the frame or the *verso*) a part of the picture as object. Correspondingly, for what we call the 'background' (and everything that this term connotes in the way of three-dimensional space), medieval art treatises use neutral terms such as the Latin *campus* ('field', *Mappae Clavicula*, twelfth century) and 'Feldung' (lit. 'fielding', *Göttinger Musterbuch*, fifteenth century).[16] While we also find 'Goltgrunt' or 'Gultgrunt' ('gold ground'), what is meant is not 'gold background', but *fundamentum pro auro*, priming (creating a substrate) in preparation for gilding.[17] For the rest, gold grounds were tacitly interchangeable, for example when a distinction was to be made between first and second-rank types of application, say the inner and outer wings of a hinged-wing altarpiece, or when gilding was too expensive or technically too difficult (which was generally the case with mural painting): in these cases, the *campus* can also be patterned, red, or even blue, by which is not necessarily meant the blue of the sky, but rather lapis lazuli, in other words, that semiprecious stone from which the deepest blue was obtained. Having established this, we must now ask why it had, for centuries, and with such regularity, to be the technically demanding gold ground that ultimately became the paradigm for medieval art.[18] Those for whom the reference to the ethical and materialist motives of magnificence is too unspecific can argue aesthetically, as I shall attempt to do in what follows.

Let us begin with an example from book illumination; Ellen J. Beer has shown a way here by pointing to the decidedly material appearance of many gold grounds. In the thirteenth century, Parisian workshops produced so-called pocket bibles, handy yet highly ambitious editions of the holy scriptures. The painterly illumination consists often of ornamental initials at the beginning of the preface of the individual books of the bible, and of illuminated letters at the beginning of the books themselves. When gold was used, as in the page illustrated (probably from the so-called Aurifaber workshop)[19] the impression is spectacular (fig. 4): 'alluminare' (illuminate) was the standard term in Paris for the illustration and deco-ration of manuscript books, as it still is in English, according to the illuminator Oderisio in Dante's *Divine Comedy* (XI, 81). In light of the initial V, it is not hard to imagine that this has something to do with the use of gold. Not only for the reasons described by Alberti, the gold area on the thumbnail-sized picture that shows the calling of the shepherd-prophet Amos literally stands right out: the substrate (the *fundamentum pro auro*) has been laid on thick, so that together with the varying

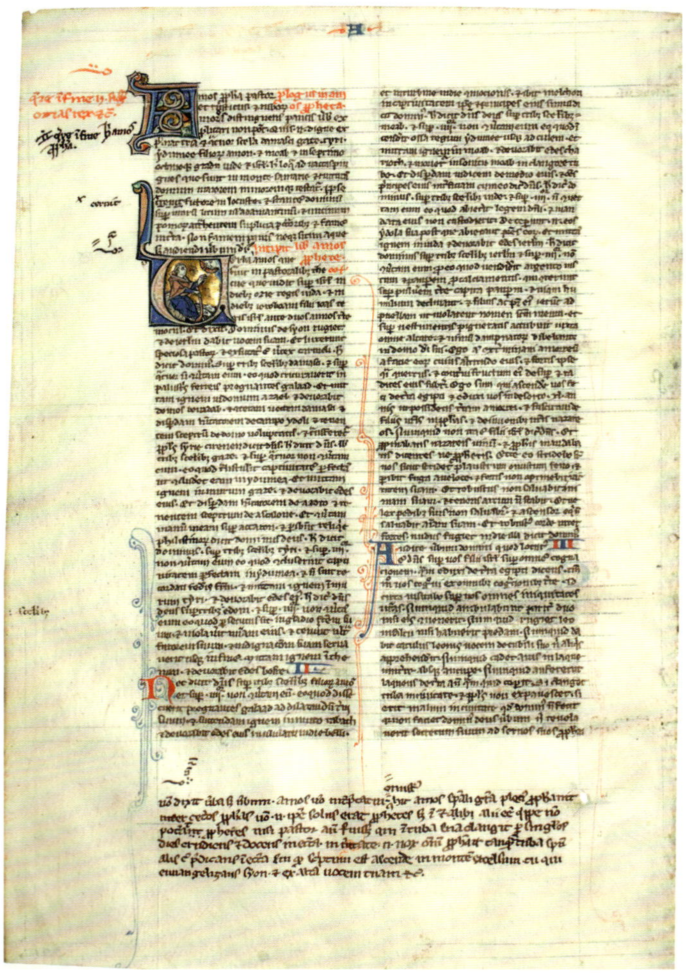

Fig. 4 Beginning of the Book of *Amos* from a Parisian pocket bible, Private ownership, Vienna

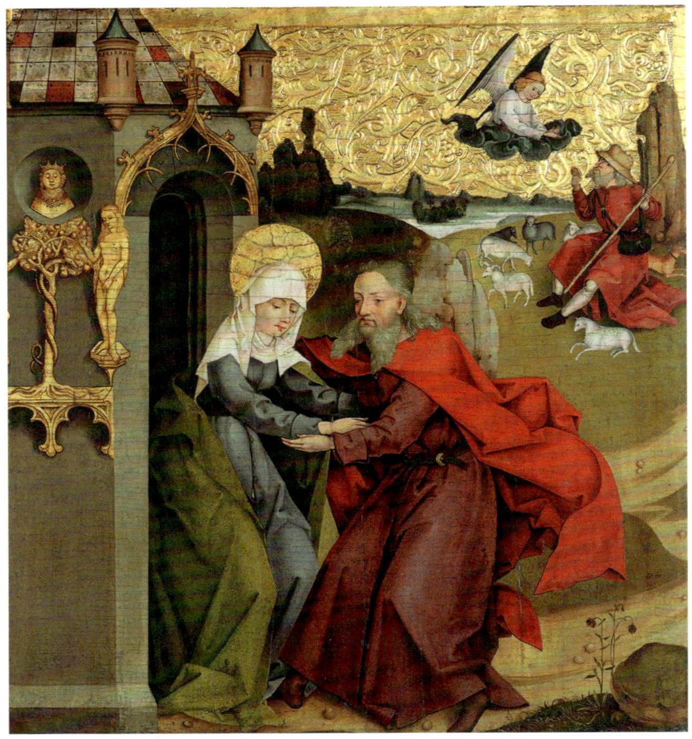

Fig. 5 Master of the Divisio Apostolorum, *Meeting at the Golden Portal*, c. 1490/1495, Belvedere, Vienna

incident light, a striking effect occurs. The ornamental letter lies like a jewel in the book, and rewards the reader who opens it, for the small size and beauty of the script must have made reading more of an effort than we are used to. In any case the gaze is attracted precisely by the optical characteristics of the metal as material. And what is true in the letter field of the pocket bible is similarly true in the dim religious light of a church interior full of pictures, where gold-ground panels could unfold a substantial effect at a distance even when not very large, and legible only close up.

The time when Alberti's idea of the picture began to carry conviction was also the great age of late-medieval religious donations and picture production. Churches filled up with pictures, each of which craved for pious attention. In the process there were curious confrontations between the depiction, which covered ever larger areas of the picture field, and the ground (which depicted nothing), whose gilding was intended to enhance the visibility, the peculiarly pictorial character, of the picture. Two panels from the festal side of a winged altarpiece dating from c. 1490 depict the meeting of Joachim and Anne at the Golden Porch, and the Nativity of their daughter Mary (fig. 5, 6).[20] The first scene presents the narratively necessary city gate and a broad landscape with which the painter recalls the 'story so far', namely how Joachim (like Amos), when he was with his flock, was visited by an angel and dispatched. In the second picture we see Anne surrounded by objects that relate to the profane and spiritual aspects of the natal event, which they place both in late-medieval central Europe and in the context of the salvation story. The scene is backed by a window wall. This

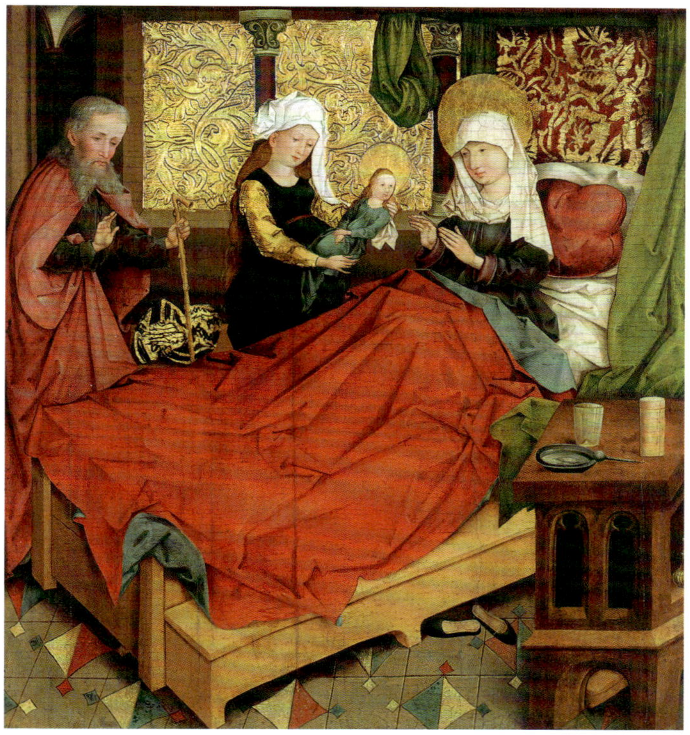

Fig. 6 Master of the Divisio Apostolorum, *Nativity of the Virgin*, c. 1490/1495, Belvedere, Vienna

makes clear the cosiness of the interior and underscores the contrast with the very un-bourgeois circumstances of the nativity of Christ (the depiction of which was certainly not missing from the total cycle). In addition, to also depict the view from the windows seemed to the painter not to take the story any further. No report of the nativity of the Virgin describes the location, so what should the view have shown? Likewise, on the other panel, the painter has placed in the zone above the landscape not sky, but gold: elaborately modelled, there lies beneath the polished gold leaf an agile late-Gothic tendril ornament, whose relief extends in front of the paint layer and whose liveliness stands in provocative contrast to the plasticity simulated by paint in the picture. As long as the gold was in its pristine condition, the fields must have come across as more the products of a goldsmith than a painter. Thus the artist and his client referred in a very concrete fashion to the costliness of the pictures and the magnificence of the altar donation dedicated to Mary or Anne; at the same time they succeeded in making the relief scintillate in almost any light, and thus draw the eyes of the churchgoers to itself.

Another effect of the gold ground can be explained by reference to an example of late-Antique book illumination. In contrast to what is often said, all-over gold does not first appear in manuscripts of the Ottonian period, but already in the *Vienna Dioscurides*, to be precise in two full-page frontispieces bound into the book in 512, among a series of such designed to increase the value of the work, which was intended as a gift to the Byzantine princess Julia Anicia (fig. 7).[21] Where any of the metal surface is still present after one and a half millennia of use, it is, in contrast to the pocket bible, not glossy but matt, even though high-gloss gold does appear a few pages further on in the form of illuminated initials.[22] The pictures each show seven individuals famous in medicine and related fields. To judge by their arrangement, the figures are somehow related, even conversations are hinted at, but ultimately the situation remains open. In fact, they could never actually have met in person. Their (real or assumed) lifetimes are centuries apart. In some cases they are mythical, for example the centaur Cheiron (top centre in the folio illustrated), some are historical, including Xenocrates, who wrote on pharmacology and dietetics in the mid-first century (bottom left). The pictures do not reproduce gatherings, but present (as a prelude to the topic of the manuscript) an ancestral gallery of medicine and pharmacology. The gilt *campus* depicts a nowhere-land: its matt sheen includes on each sheet seven pictures of individuals who are connected with and separate from each other in equal measure. In apse decors, later Christian art often used the indeterminacy of a 'soft', non-glossy, but shimmering gold ground (gold mosaics as a rule come across as a rule as velvety-soft): the first sight shows Christ standing with saints and often with a donor; the second, by contrast, finds clues to a montage of pictorial manifestations, backed by the curved gold-ground of the vault, reflecting the light in a slightly different way when seen from any given standpoint.

A glance at the appendix (written in 1354/55) of a manuscript decretal in Salzburg (Abbey of St Peter) shows how a gold

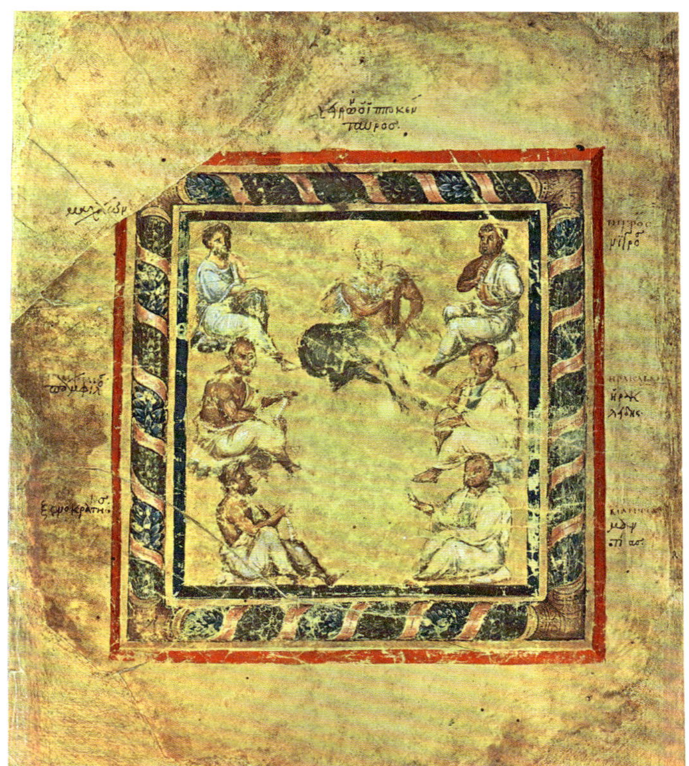

Fig. 7 Picture of Physicians from the *Vienna Dioscurides*, 512 CE, Österreichische Nationalbibliothek, Vienna

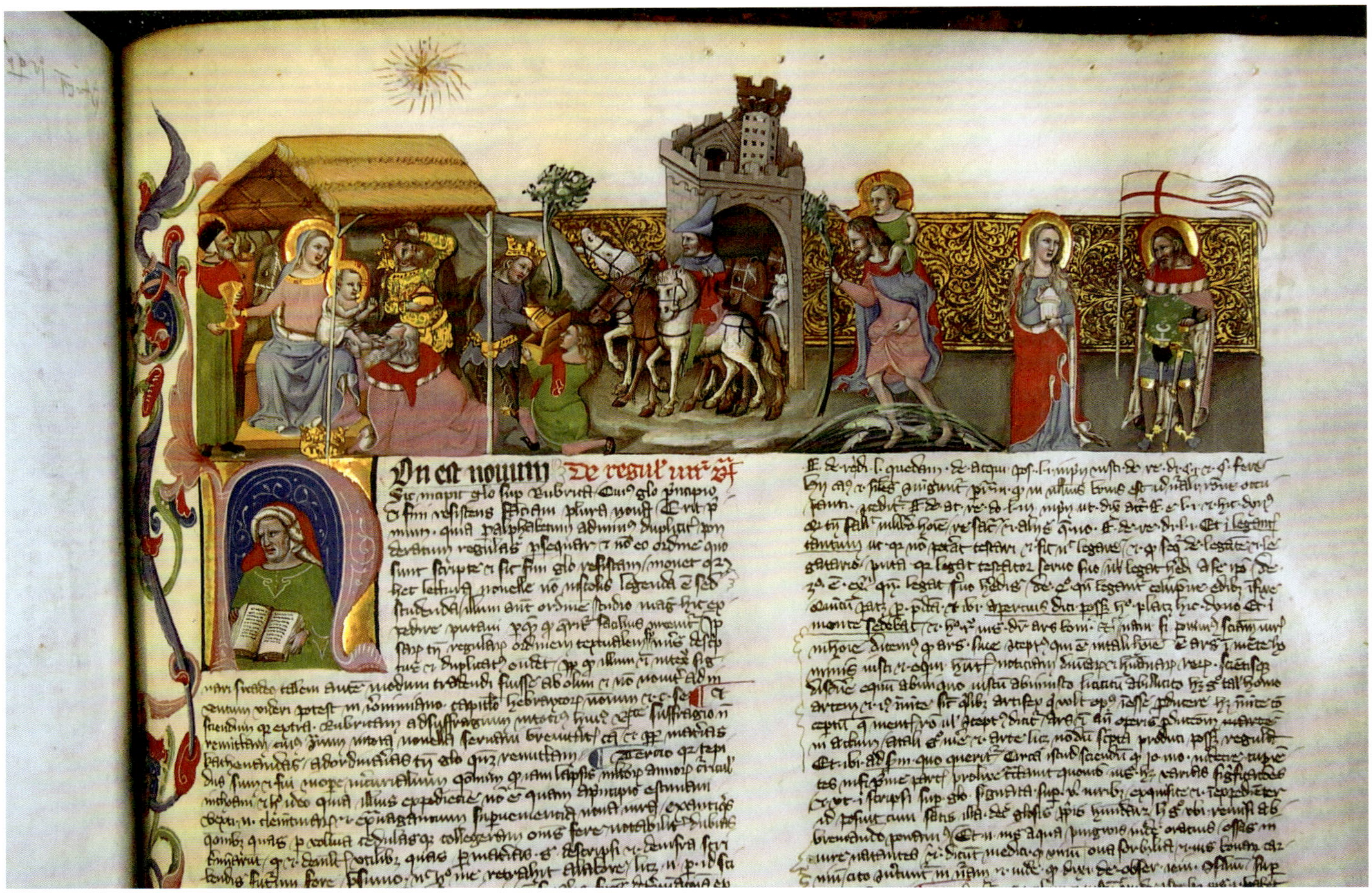

Fig. 8 Niccolò di Giacomo, *Adoration of the Magi with Saints*, c. 1354, Abbey of St. Peter, Salzburg, Cod. A XII 10 (Liber Sextus Decretalium mit Novellen), Fol. 138

ground could be developed for this kind of use:[23] The miniature, created by Niccolò di Giacomo of Bologna, brings together the adoration of the Magi with the appearance of various saints on a horizontal bar that extends across both columns on the page (fig. 8). The whole thing begins on the left with the scene of a thatched shelter in front of a hill that conceals the (invisible) nativity cave. To the right, we see the procession of the Magi; on the one hand, the animals have already reached the hill, while on the other they are only just leaving Jerusalem, which, in its turn, is represented by a gate, a wall, and, behind it, tall buildings. As the route from the palace to the Holy Family's accommodation is of importance to the narrative, the painter has depicted it as rocky ground. Above, in the space between the hill and the city gate, a gold ground of densely intertwined tendrils spreads out, albeit covering the *campus* only up to a particular height, doubtless laid down by the events beneath the thatched roof: star, roof, the crown of a tree, and the crenellations of Jerusalem are all on bare parchment. The gold ground continues at the same height on the other side of the Jerusalem cityscape. There, it surrounds the saints, but is just about adequate for the height of their bodies: here, the head and halo of the Christ child on the shoulders of St Christopher project into the bare parchment, as does St Gereon's banner. Likewise, the painter continues the rocky ground, albeit in semi-abstract form. The transition from rock to a grey surface takes place at St Christopher's feet: the start of the road to Bethlehem serves, so to speak, as the shore of what is pictured by its side as a watery hill. The further continuation is an anonymous grey, to which shading gives a hint of three-dimensionality, and which thus becomes a kind of stage. Like that of the golden strip, the height of the grey one also keeps to what is required by the scene to the left. This is the reason, ultimately, for the narrowness of the gold ground, which appears as more of a stripe than a field.

Niccolò di Giacomo, a decidedly sophisticated and media-conscious book illuminator, transfers to the gold ground firstly the task of creating a hierarchy: that zone of the picture where there are interactions and narrative elements is also the one where he uses a gold ground. Secondly he uses the gold ground to bring together disparate elements, and not just appearances by individuals who lived at different times, as in the two pictures in the *Dioscurides* produced 900 years earlier, but also different media: pictures of saints, who demand veneration, with a scene which needs to be read as a narrative (whether the emphasis is on information or compassion) and functions as a main picture. Less pronounced, by contrast, is the function of the gold ground that we see in the pocket bible: instead of polished gold (as was used in the initial N, in the haloes and in Gereon's armour), behind the pictorial stripe we have intertwined feathers on a brown ground. The brown is best interpreted as 'imperial' purple, a colour which, in combination with gold, does not have a derogatory effect, but rather creates a context: magnificence raised to a higher power. The visual effect is that the gold does not so much come across as the exception as combine with the depicted motifs. Where it appears over the grey 'stage', we have associations of a curtain. Of crucial importance, then, was the task of encompassing, or bracketing, the barely compatible parts of the picture; in the process, the latently figural appearance of the gold stripe, which looks like a strip of cloth, comes across with particular conviction.

A gold ground can also help to concretize the motifs in the picture. It is occasionally asserted of Giotto, who is regarded as the founder of post-medieval painting, that he dismissed the gold ground from his art and thus ushered in the end of medieval artistic practice. It was for just this reason that *Time* magazine did him the honour, somewhat belatedly at the turn of the millennium, of declaring him the 'man of the fourteenth century'. But the opposite is true: Giotto used gold in his panel paintings in just the same way as the painters of earlier generations, only his use of it was perhaps more sophisticated. The great Madonna panel in the Uffizi shows a gathering of saints in an arrangement that is spatially well thought-out; they seem to be almost within reach (fig. 9).[24] The beholder could feel invited to walk right into the picture, were not the step between him and the Virgin all too costly. It is the marbling of this step that introduces an unreal element into the structure of the picture, not the gold that surrounds the figures above. It does not question their presence. And even the gold haloes rob them of none of their reality, but help, rather, to make the spatial arrangement clear. One can even go further: basically, the highly polished golden *campus* liberates the group of figures gathered around the throne, and makes their appearance independent of any Where and How, and thus sovereign in the full sense of the word. If the Virgin, by the standards of older Madonna panels, is of overwhelming presence, there is nonetheless no pinning-down of precisely *where* she is present: in heaven, where a gathering of saints and angels might be expected, or on earth, where she encounters the beholder. The interplay between concreteness and openness contributes to the effect of the picture. The gold ground, incidentally, also has a concretizing effect on the Passion panels in Vienna (whose painter profited from the pictorialness of the Giotto period): the soldiers' lances are much more dangerous against a gold ground than they would be against a painted sky. And in the Crucifixion panel, which has as its theme the sudden transition from execution event to realization of the divinity of the executed man, with the figure of the grieving St John and the arm of the good centurion pointing to Christ, two decisive signals are particularly prominent against the gold ground.

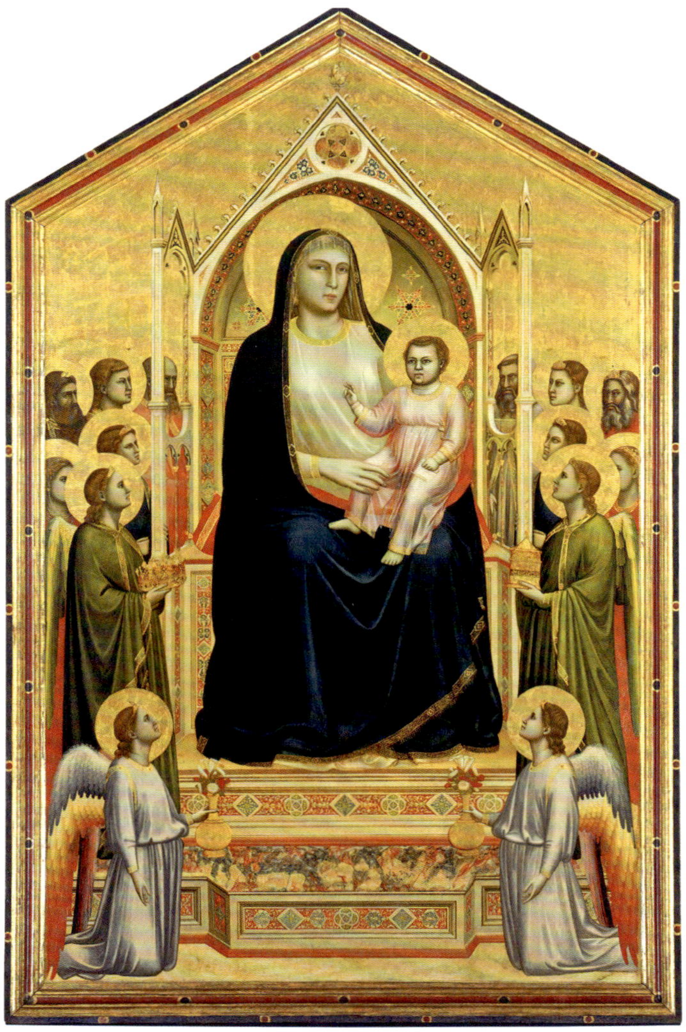

Fig. 9 Giotto di Bondone, *Ognissanti-Madonna*, c. 1310, Florence, Uffizi

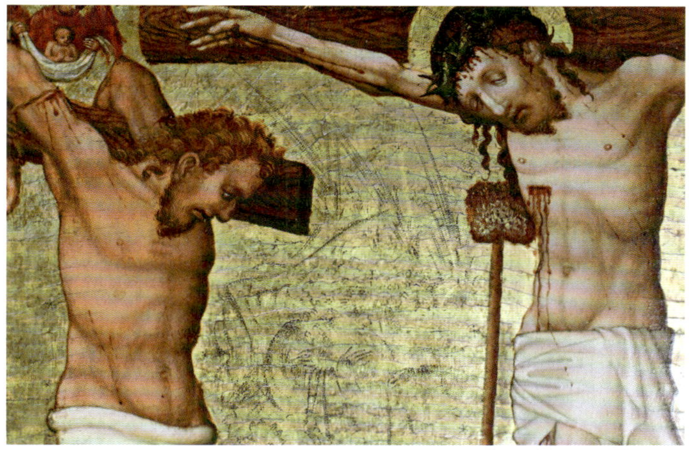

Fig. 10 Master of the St Lambrecht Votive Tablet, *Crucifixion*, c. 1430 (detail from Fig. 3), Belvedere, Vienna

A theme in itself is the internal decor of the gold ground: alongside tendrils in relief, as in the Vienna Life of the Virgin panels, and feathers, as in the Salzburg miniature, there is a whole variety of forms of decoration pressed, painted, engraved and chased into and on to the ground, and all manner of transitional steps between gold grounds, painted grounds, and ornamental grounds. Only in the small format, as in the initials of the pocket bible, and then not always, is an undecorated gold ground the rule. The minimal decoration includes a margin which as the accompaniment to the frame defines the *campus*, as also in Giotto's *Ognissanti Madonna* and in the Vienna Passion panels. The Vienna *Crucifixion* (fig. 3) shows, in addition, a particular form of treatment of the gold ground which was not uncommon in the fifteenth century, and goes beyond mere decoration: on the one hand the area is strewn with embossed stars, not a pattern, but clearly a visualization of the solar eclipse that accompanied Christ's death (Matthew 27, 45; Mark 15, 33–34; Luke 23, 43–44). On the other hand there is, in the area of gold defined on the right by the rod with the sponge and on the left by the good thief, extensive figural embossing. With some difficulty (when the gold surface was still in good condition, maybe with a little less difficulty, but it would scarcely ever have been easy) a grieving angel can be made out, turning to the figure on the Cross with bowed head. To the right of, and behind, this angel there may be one or two others.

The Crucifixion picture is unusual to the extent that it depicts the moment of death explicitly as recounted in St John's Gospel (John 19, 29–30): " Now there was set a vessel full of vinegar: and they filled a sponge with vinegar, and put it upon hyssop, and put it to his mouth. When Jesus therefore had received the vinegar, he said, It is finished: and he bowed his head, and gave up the ghost." The rod with the sponge, the figure of John projecting into the gold ground, protagonist and reliable reporter, and the angel embossed in the gold ground above, to whom it is left to certify the death, therefore belong together. The difference in media is therefore all the more confusing, especially since the painter has depicted the angel above the good thief and the devil, in paint, above the bad thief. However, the grieving angel is identified by the rays emanating from him as a particularly high-ranking being, assigned directly to the throne of God—a seraph or cherub, of a different substance from organisms such as devils, people or ordinary angels. Accordingly, the limited visibility in gold may be indicative of particularly fine matter. But the visual effect must also be considered: perception is delayed, and therefore a temporal element comes into the reading of the picture. Through the broad, seemingly empty zone between the rod and the good thief, and through the latter's gaze, which seems to be aimed at the angel, our attention is drawn to the winged being. Only with these directions does the eye seek and find, and unexpectedly we have one further argument for faith.

"Don't ask for the meaning, ask for the use!" is an apocryphal quotation by Ludwig Wittgenstein from the context of his late philosophy. There, for example on the *Philosophical Investigations,* he is concerned with problems of language and meaning. According to Wittgenstein, language is not understood by replacing, in your head, words and sentences by others you are more familiar with, but by understanding what actions the speaker triggers by this social-acoustic instrument. That the medieval gold ground has the status of a communicative tool of the quality of a word or sentence, and that it is designed to evoke acts of insight, veneration, fear etc., seems to me, for the reasons set out above, doubtful. On the other hand, no more than a picture consists entirely of motifs pregnant with meaning, linguistic communication does not consist of a continuous flow of words and sentences whose meaning is important. There are also non-semantic sounds, soundless moments, and the presence of the speaker. They all structure speech, and modify the intonation, in other words substantially modulate the meaning of what is said and the reaction to it. If we transfer this to the gilded empty spaces of medieval pictures, it seems that, instead of operating with generalizations as has been usual in art history since Riegl (a gold ground 'is' two or three-dimensional space or light or divine or material), we should investigate the effect of the gold grounds in the compositions of pictures and follow up the technical potential of depiction which they embody. When the young Ernst Gombrich in the early 1930s confronted Alois Riegl's intellectual legacy, he realized that figuration and the meaning of the gold ground were not immutable quantities, either in Riegl's sense, or in the sense of his disciples or critics, but would be constantly redefined by the context in which the gold areas appeared.[25] Along with Wittgenstein, I would like to go a step further: a gold ground 'means' nothing and depicts nothing, but has an extraordinary presence which cannot be appropriated to a particular use. Thus it is suitable as a powerful lever which the painter can apply in a whole variety of ways in order to intensify and structure the perception of the picture. The gold ground can, as we have seen, ensure attention, it can separate, connect, emphasize, conceal, create a hierarchy, it can level, accelerate and delay. The same could be said of perspective, which, in addition, can do something else too, namely depict.

1 Albert, Leon Battista, *Della Pittura*. The English translations used here are from http://www.noteaccess.com/Texts/Alberti/ This text is here described as follows: "Alberti, Leon Battista. On Painting. [First appeared 1435-36] Translated with Introduction and Notes by John R. Spencer. New Haven: Yale University Press. 1970 [First printed 1956]".
2 Ibid. The Italian version of the text assumes a simulated visual field, the Latin version an actual one.
3 David C. Lindberg, *Theories of Vision from al-Kindi to Kepler*, Chicago and London, 1976, pp. 148–151.
4 Cf. Norman Bryson, *Vision and Painting: the logic of the gaze*, New Haven 1986
5 Elfriede Baum, exhib. cat. Museum Mittelalterlicher Österreichischer Kunst (Österreichische Galerie, Vienna, cat 1), Vienna 1971, nos. 13, 14.
6 Julius Lange, "Ein Blatt aus der Geschichte des Kolorits (1893)", in: Georg Brandes/Peter Köbke (eds), *Julius Lange's Ausgewählte Schriften*, vol. 2, Strasbourg 1912, p. 138. I am grateful to Thomas Zaunschirm for drawing my attention to Lange.
7 Alois Riegl, *Die spätromantische Kunst-Industrie nach den Funden in Österreich-Ungarn,* vol. 1: *Im Zusammenhange mit der Gesamtentwicklung der bildenden Künste bei den Mittelmeervölkern* (1901), Vienna 1927, p. 14.
8 Hans Berstl, *Das Raumproblem in der altchristlichen Malerei*, Bonn/Leipzig 1920, p. 40.
9 Josef Bodonyi, *Entstehung und Bedeutung des Goldgrundes in der spätantiken Bildkomposition (ein Beitrag zur Sinndeutung der spätantiken Kunstsprache)*, Vienna 1932 (typescript), p. 115.
10 Hans Jantzen, *Ottonische Kunst*, Munich 1947, p. 81.
11 Wolfgang Schöne. *Über das Licht in der Malerei*, Berlin 1954, p. 25.
12 Thomas F. Mathews, *The clash of gods: a reinterpretation of Early Christian art*, Princeton 1993, p. 95.
13 Robin Cormack, *Painting the soul: Icons, death masks, and shrouds*, London 1997, p. 134.
14 Pavel Florenskij, *Die Ikonostase (Urbild und Grenzerlebnis im revolutionären Russland, mit einer Einführung von Ulrich Werner)*, Stuttgart 1988, p. 138. I am grateful to Raphael Rosenberg for drawing my attention to Florenskij.
15 Wolfgang Braunfels, *Nimbus und Goldgrund: Wege zur Kunstgeschichte*, Mittenwald 1979, p. 23. Cf. also Wolfgang Braunfels, "Nimbus und Goldgrund", in: *Das Münster* 3, 1950, 11/12, 321–334.
16 Hellmut Lehmann-Haupt, *The Göttingen Model Book: a facsimile edition and translations of a fifteenth-century illuminators' manual (based in part on the studies of the late Edmund Will)*, Columbia 1972. - Heinz Roosen-Runge, *Farbgebung und Technik frühmittelalterlicher Buchmalerei: Studien zu den Traktaten 'Mappae Clavicula' und 'Heraclius'*, 2 vols., Munich 1967. - Ellen J. Beer, 'Marginalien zum Thema Goldgrund', in: *Zeitschrift für Kunstgeschichte* 46, 1983, p. 283.
17 Thomas Brachert, *Lexikon historischer Maltechniken: Quellen—Handwerk—Technologie—Alchemie* (publication of the Institut für Kunsttechnik und Konservierung im Germanischen Nationalmuseum 5), Munich 2001, 101–103.
18 Peter Cornelius Claussen, "Goldgrund", in: *Mythen der Kunstwissenschaft / Art historical mythologies* Robert Branner, *Manuscript painting in Paris during the reign of Saint Louis: a study of styles*, Berkeley 1977, 109–115.
19 Robert Branner, *Manuscript painting in Paris during the reign of Saint Louis: a study of styles*, Berkeley 1977, 109–115.
20 Baum 1971 (as note 5), no. 69, 70.
21 Ernst Gamillscheg, "Das Geschenk für Juliana Anicia: Überlegungen zu Struktur und Entstehung des Wiener Dioskurides", in: Byzantina *Mediterranea: Festschrift für Johannes Koder zum 65. Geburtstag*, Vienna 2007, 186–195.
22 I am very grateful to Andreas Fingernagel, Ernst Gamillscheg and Christa Hofmann for giving me the opportunity to study the original. What can be seen is a thin, matt bole-gilding. There is no sign that the parchment suffered from a polishing process, as often happens. The polished gold initials of the illuminated title of Fol. 7v have suffered far less from abrasion. As a result, I assume, by analogy with early-medieval gold grounds, a matt surface.
23 Cod. A XII 10, Fol. 138. Hans Tietze, *Die illuminierten Handschriften in Salzburg (Beschreibendes Verzeichnis der illuminierten Handschriften in Österreich 2)*, Leipzig 1905, no. 32.
24 Michael Viktor Schwarz, *Giottos Werke (Giotto Pictor 2)*, Vienna 2008, 458–466.
25 Ernst Gombrich, Review of Joseph Bodonyi *Entstehung und Bedeutung des Goldgrundes in der spätantiken Bildkomposition*, in: *Kritische Berichte,* 5, 1932/33, p. 75.

Mutable Gold

Gold in Late-Medieval Retable Art—Thoughts on the Works in the Belvedere

Veronika Pirker-Aurenhammer

The tour of the exhibition leads through that part of the medieval collection on display in the 'Prunkstall' of the Lower Belvedere. Most of the works of late-Gothic panel painting and sculpture to be seen here are fragmented parts of retables dating from the fifteenth and early sixteenth centuries. There is a great deal of gold to be seen together with colours: panel paintings with a gold ground and golden details alongside multicoloured statues with richly gilded garments. Together with the masterpieces on display in the Upper Belvedere, the total collection reflects a picture, typical of central Europe, of a continuous use of gold leaf from the late Middle Ages until well into the early modern period.

Expensive gold and showy staging

Those retables still preserved in their pristine location confirm this impression. Michael Pacher's *double-winged altarpiece in St. Wolfgang* (1471 - 1481), a climax of retable art in Austria, reveals, when opened to display the feast-day sides, an overwhelming effect of shimmering gold, culminating, in the shrine, in a sculptural mise-en-scène of the Coronation of the Virgin (fig. 1).[1] The financial expense of such large commissions was enormous. Pacher's last known commissioned work, the *high altar* commissioned in 1484 for the parish church (today the Franciscan church) in Salzburg, is, at 3300 Rhenish guilders, the most expensive known retable of its kind.[2] Its extant fragmented panels in the Belvedere[3] suggest a huge altarpiece unfoldable in two ways to reveal different aspects, and originally its total effect must likewise have been richly characterized by the use of gold. There is no other way to explain the price. More modest carved altarpieces from all over Austria are overwhelmingly decorated with gilded elements. The aesthetic alternative of visible wood or partial painting, for which in the late fifteenth century Tilman Riemenschneider in Würzburg and other great woodcarvers north of the Alps decided[4], remained the exception in Austria. They include for example the *Carved Altarpiece* in Mauer near

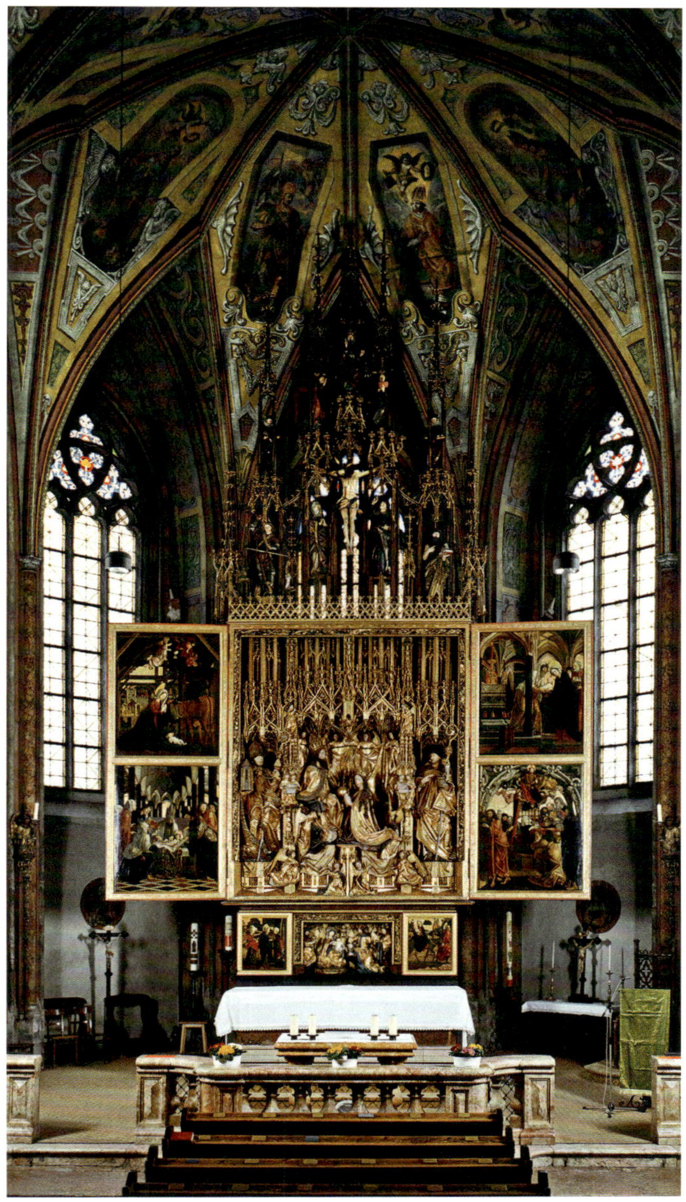

Fig. 1 Michael Pacher, *St Wolfgang Altarpiece*, c. 1471 - 1481

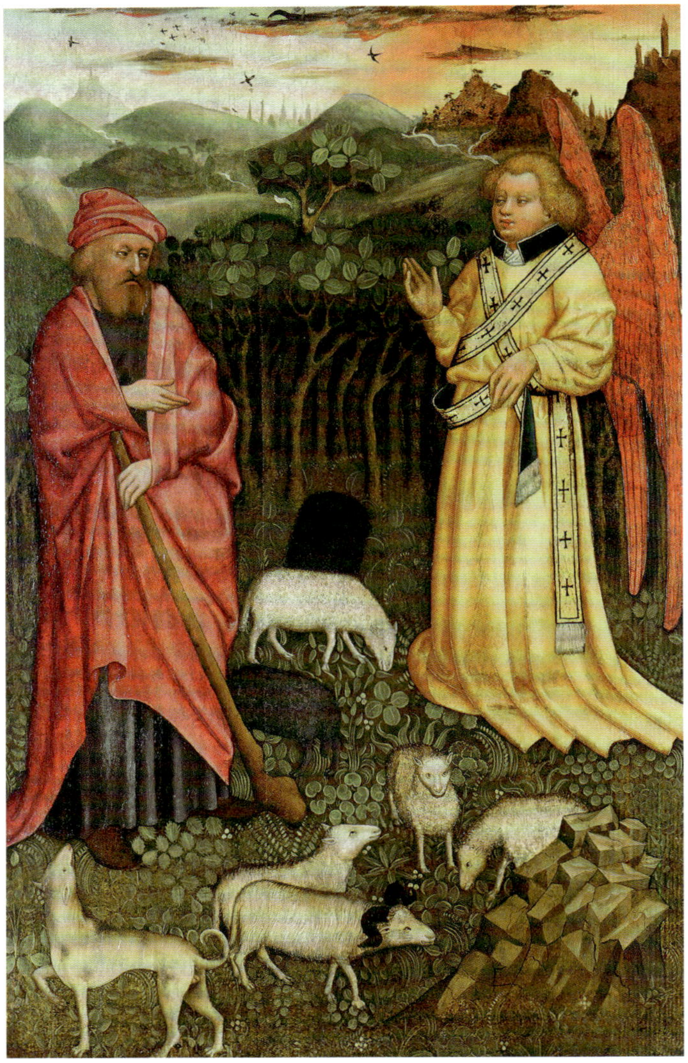

Fig. 2 Master of the Albrecht Altarpiece, *Annunciation to Joachim,* c. 1435/1440, Belvedere, Vienna

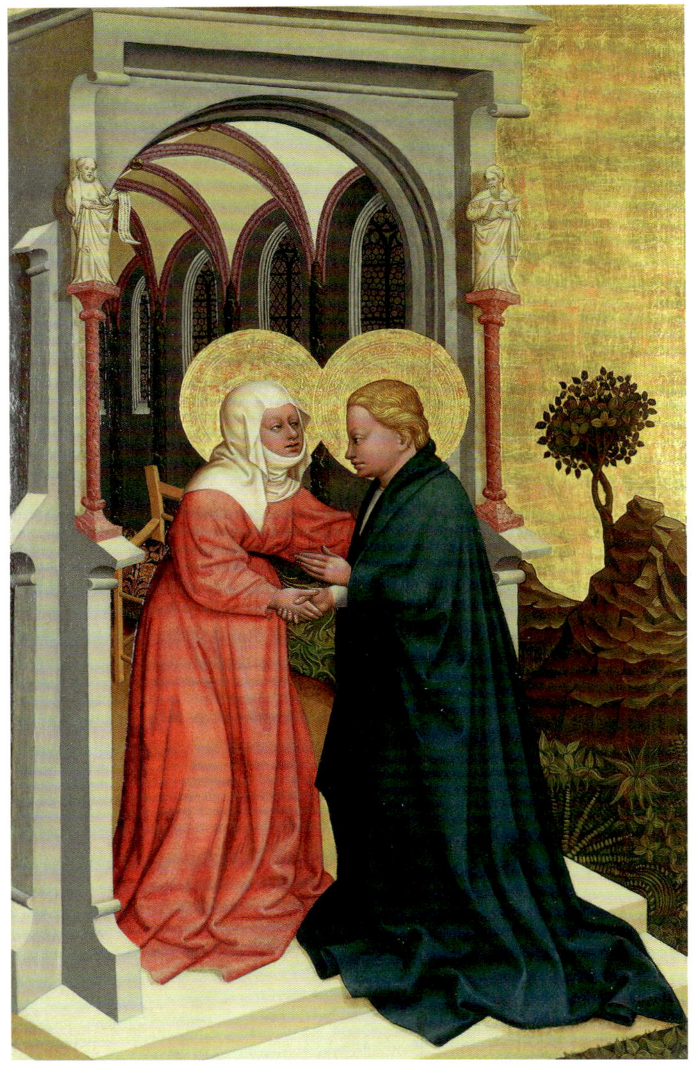

Fig. 3 Master of the Albrecht Altarpiece, *Visitation,* c. 1435/1440, Belvedere, Vienna

Melk (after 1509)[5] and the *Zwettl Altarpiece* in Adamov near Brno (1516–1525)[6].

In the painting medium, by contrast, the abandonment of gold leaf set in as early as the 1430s. This development had its beginnings, as is well-known, in the Low Countries with the new naturalism of Jan van Eyck and Rogier van der Weyden. Phenomena of the visible world, such as the sky, lights, shadows and light-reflecting metals, could now be depicted using painterly means—paint itself could now replace the gold leaf of the grounds and the metallic applications of the motifs. The Viennese 'Master of the Albrechtsaltar in Klosterneuburg' was one of the first generation of artists to consummate this change in central Europe. He already mastered the technique of imitating shiny metal using painted light reflections, and composed atmospheric landscapes in the new naturalistic style,[7] such as the glowing red evening sky over a broad undulating landscape in the 'Annunciation to Joachim' scene (fig. 2). By contrast, the scene now separated from this panel, that of the 'Visitation', still evinces the traditional gold ground (fig. 3). Both pictures were on wings of a small Marian altarpiece dating from the late 1430s: the more costly gold side could be seen when the wings were open, and the other, executed with a painted ground, was visible when the wings were closed.[8] The differentiation between gold ground and paint ground follows the usual system of hierarchical gradation with hinged-wing altarpieces: on the outsides of the wings (ferial side) older altarpieces of the fourteenth and early fifteenth centuries often have neutral colour grounds and, later, landscape views with an atmospheric blue sky. Gold grounds are often reserved to the insides of the wings (festal side), although here too a change took place, with the appearance of illusionist pictorial spaces. While the gold ground in the older panel paintings still had an impenetrable character, from the mid-fifteenth century it was increasingly being combined with architectural set-pieces and land-scapes in front of it, which seems to push the gold back into the distance. But even where gold grounds were dispensed with in painting entirely, the lustrous gold leaf remains present in a carved altarpiece to great effect in the shrine and the framework decoration.

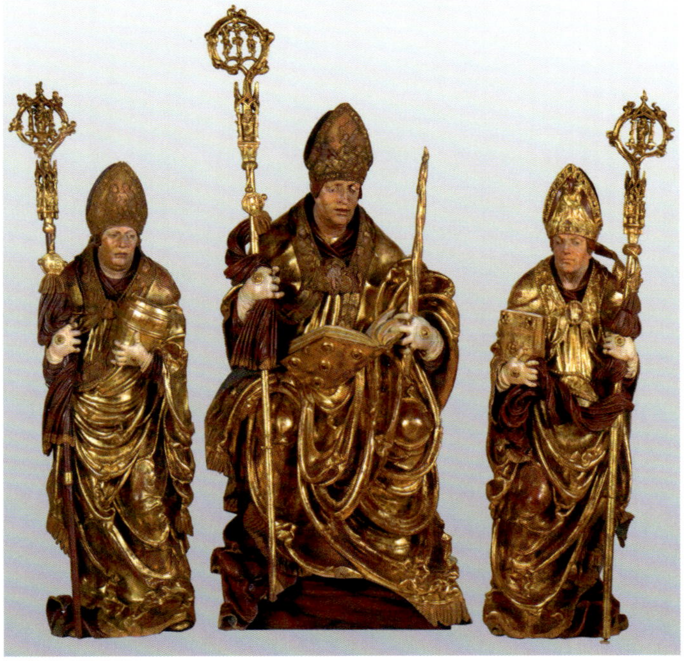

Fig. 4 Andreas Lackner, *St Blaise with SS Rupert and Virgilius,* 1518, Belvedere, Vienna

In short: gold is an integral component of the late-Gothic hinged-wing altarpieces. It served to visually highlight the change from outside to in, and fittingly brings out the function of the retable as a pictorial medium for putting across the message of salvation and as a place of devotion and liturgical acts. The development of a naturalistic form of painting in no way derogates from the practice of gilding, quite the opposite: as Hans Huth noted so many years ago, gilding in the second half of the fifteenth century became "the dominant factor in the total tonal effect. It was only through the gold that the rich sequence of colours their cohesion and only with its help were all those mobile reflecting lights generated that bring out the whole to best effect. Thus the enormous influence on the imagination could be exerted which shifted the event depicted on the altarpiece into a transcendent world, far from all earthly reality."[9] From the point of view of material aesthetics, such an influence doubtless holds true for Pacher's *Wolfgang Altarpiece* and also for the *Cracow Marian Altarpiece* by Veit Stoss (1477–1489), which is ceremoniously opened every day in the presence of masses of curious onlookers.[10].

In this connexion, one should be aware that the light reflections of the gold once came across as much more subtle and lively, as a result of the flickering candles and the stained glass windows of the church interiors, than in the neutral artificial light of a museum. Who these days would want to view art under irregular, fitful illumination—even sacred objects 'before the age of art'?[11] An even illumination, however, vitiates many an effect of subtle metal-working such as the glitter of fine chasing and slightly raised embossed patterns, or the shimmering alternation of polished or matt gold leaf. A fine gradation like this was once evinced, to name one example, by the shrine group of the *Abtenau Altarpiece* by Andreas Lackner, an important late work of Salzburg wood-carving art completed in 1518 (fig. 4).[12] Large areas of the opulently designed insignia of the three bishops are covered with polished gold leaf, while next to them there were discovered, at the last restoration, traces of matt gold and polished and matt (possibly lustred) silver. Such rich metallic decor is also known, for example, from the work of Hans Leinberger, in whose Landshut workshop Lackner may have been trained. The shiny surface enhances the light-and-shade relief of the deeply furrowed folds so that it becomes a 'total picture of moving matter captured in gold'.[13] In this final phase of retable art, we see an artistic virtuosity absolutely typical of the transition from the late Gothic to the Renaissance, to which not least the impressive portrait-like faces of the Abtenau saint-bishops bear witness.

Gold as material and metaphor

Gold has always had a high meaning-potential both in the sacred and secular spheres.[14] By dint of its costliness and its natural properties such as lustre, purity and durability, gold and godliness are not far apart. On the basis of prophetic manifestations of God in the Old Testament and the description of the Heavenly Jerusalem in the Book of Revelation, in Christian interpretation gold has been understood since the earliest days as the light-filled symbol or reflection of God, Heaven and the New Jerusalem.[15] One theme, for example, is the purging of man though fire-resistant gold. On the other hand, gold was often viewed negatively in the scriptures too, in the sense of earthly vanity, the deceptive image of false gods, the human vices of greed and luxury and so on. The chorus of biblical gold-metaphors is polyphonic, and correspondingly polysemous is the metaphorical charge borne by gold in Christian theology and art. In addition there is the tradition of the 'encyclopaedia of nature' that handed down and expanded upon the knowledge of Antiquity.[16] One late example is Konrad von Megenberg's *Buch der Natur*, a German version of a work by Thomas of Cantimpré dating from c. 1350, in which the properties and use of gold are described in terms of its healing power. What it says, in effect, is: gold is heavier than silver, copper or tin, and the noblest of metals, it shines at all times, is pure, durable and immutable, it strengthens, purges and cleanses; it is found deeper in the earth than other noble substances and consequently is nearer to hell, "dar vmb spricht man: Ge zů dem tevfel umb golt"[17]. Such medico-magical qualities open up further dimensions of meaning.

The foregoing only hints at the variety of meanings associated with the use of gold in medieval artefacts. Christel Meier sums up the multiple ambiguity that resides in the related field of gemstone allegory thus: "The expectation of a single meaning (...) would be wrong from the outset, for the concurrent exegetical and poetic interpretation of gemstones and buildings united a number of different approaches and involved a sliding from one to the other."[18] Meier is writing in the context of the jewel-encrusted chapels in Karlštejn Castle and the Wenceslas Chapel in Prague Cathedral: thematically highly complex rooms for Emperor Charles IV, their costly appoint-ments literally giving them the character of treasure chambers with their walls of gemstones set in gold and their dense groups of images.

The broad spectrum of applications for gold in late-medieval art will be sketched out in what follows using the example of three works in the Belvedere in which gold appears in very different semantic contexts. Some basic questions are raised. How is gold deployed by artists to convey particular messages? When does gold start 'to speak', and what about? Where are the boundaries between the decorative and the meaningful, between function and content? At the outset, it must be stressed that one-dimensional interpretation patterns would be out of place, on account of gold's multiple ambivalence. Günther Bandmann, in a first fundamental study of the iconology of the material (1969), spoke out against generalization.[19] The material should also always be seen in the total context of each artwork. Gold may have a significant quality, its light and its 'divine lustre' may single it out for the depiction of heaven. But the use of a material may simply be the result of an unthinking convention on the part of a craftsman, or, in the case of a costly, rare substance, confer a dignity that required no further formulation. "And yet certain tendencies regarding an intention to depict something through the material cannot be overlooked, and in some cases these tendencies have to do with the natural qualities of the substance, [for example] the lustre of gold [...]. The relevance of the material in any particular case can only be judged with any certainty in the context of the individual artwork."[20]

Gold ground and aureole: the Marian panel of Konrad von Friesach

The first panel is the "Throned Madonna" by the Carinthian painter Konrad von Friesach, which makes opulent use of gold and silver leaf (fig. 5)[21]. The panel may once have formed the central section of a winged altarpiece and can be dated to before 1445, as we know from the kneeling donor to the right of the Virgin's feet: he is Johannes Hinderkircher, who from 1445 was the provost of Gurk cathedal, but here he is still shown in the garb of a canon. Mary and the child turn graciously to the donor, the smiling naked Christ-child playfully reaches towards the ribbon with the inscription held by the tiny supplicant figure. Angels surround the throne, which is covered by a cloth of honour: they make music, worship, hold the cloth, while one touches the top of the Virgin's crown. Love of detail is also evidenced by the fish-eye view of a strip of meadow with little birds, scattered strawberries and lilies-of-the-valley, both typical Marian plants of the period. Tellingly, they are absent in the vicinity of the donor, where the grass around the inscription ribbon is curiously shifted. This marks, somewhat clumsily to be sure, the ontological difference between the sphere of the humbly supplicant donor on the one hand, and the Madonna or the cloth of honour on the other. The painter also knows how to manage the subtleties in the metallic areas of the picture, which seem to be layered: at the bottom is the golden relief ground, which shows an already somewhat antiquated brocade pattern with birds in the manner of Italian fabrics of the fourteenth century; above this is the silver-gold cloth of honour, likewise designed as a brocade fabric; and finally the Virgin's aureole, set in a red band of clouds, and filled with finely chased angelic musicians. The aureole, and the cloth of honour, which neutralize the step formed by the throne seat, make the spatial location of the Madonna totally unclear. Seen together with the gold ground, however, they come across as a mise-en-scène of light, which the donor seems to perceive as a vision seen in close-up. The Queen of Heaven, the paradisiac flowery meadow, and the angelic music evoke an other-worldly sphere in which glances, gestures, touches or even the child's playing with the prayer ribbon disseminate a warm-hearted atmosphere. Such mises-en-scène are intended to provide a sensuous stimulation for beholders in their devotions or their *memoria* for the donor. In the case at hand, the Madonna and Child are presented using traditional forms of representation, narrative details and an aesthetically balanced interplay of colour and metallic sheen. Konrad von Friesach is still entirely in the tradition of the *Weicher Stil* ('soft style'). That the Carinthian master was already familiar with a more progressive, illusionist manner of painting is evident from the fringes of Mary's garment and the cloth of honour, which use paint to imitate the shimmering gold fabric, the pearls and the gemstones. The Virgin's dress and the garments of the two angels kneeling on the arms of the throne, by contrast, are still gilded conventionally in gold leaf. Now, the juxtaposition of real gold and gold-imitation painting is a common and highly variable phenomenon in late-Gothic panel painting. In the case of the *Hinderkircher Madonna* the gold leaf is by far the more dominant, and is applied in a subtle alternation of layers of metal and paint, so that it binds the painting strongly to the ground and does not allow any spatial illusion. The aureole comes across at first sight as an area deliberately left unpainted, indeed as a gaping hole in the cloth of honour, but it is revealed on closer inspection as an aura around the Virgin in its own right, connoted as heavenly by the embossed angels (fig. 6, detail). The aureole is connected with the gold ground by means of the Virgin's halo, which overlies both, a gold background for her gold crown.

Viewed historically, the mastery of gilding technique was obligatory for painters in late-medieval workshop practice. Thus for example the Salzburg guild regulations of 1494 demanded as a master-piece (i.e. their 'examination piece') for painters 'ain Mariapild mit ainem kind mit ölfarb gemalt, das vier spann lang sey, auf ain tafl mit ainer vergolten feldung' (a picture of the Virgin with a Child, painted in oils, four spans long, on a panel with a gilt field).[22] Our Madonna panel, together with its original frame, decorated by a gold-and-silver stencil pattern, is an expensive and elaborate example of the highly developed gilding craft of the late Middle Ages.

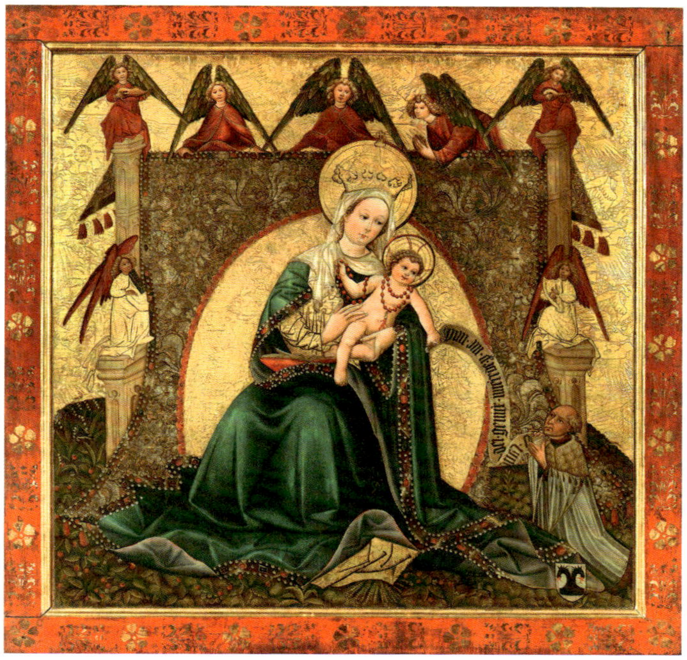

Fig. 5 Konrad von Friesach, *Madonna of Johannes Hinderkircher,* before 1445, Belvedere, Vienna

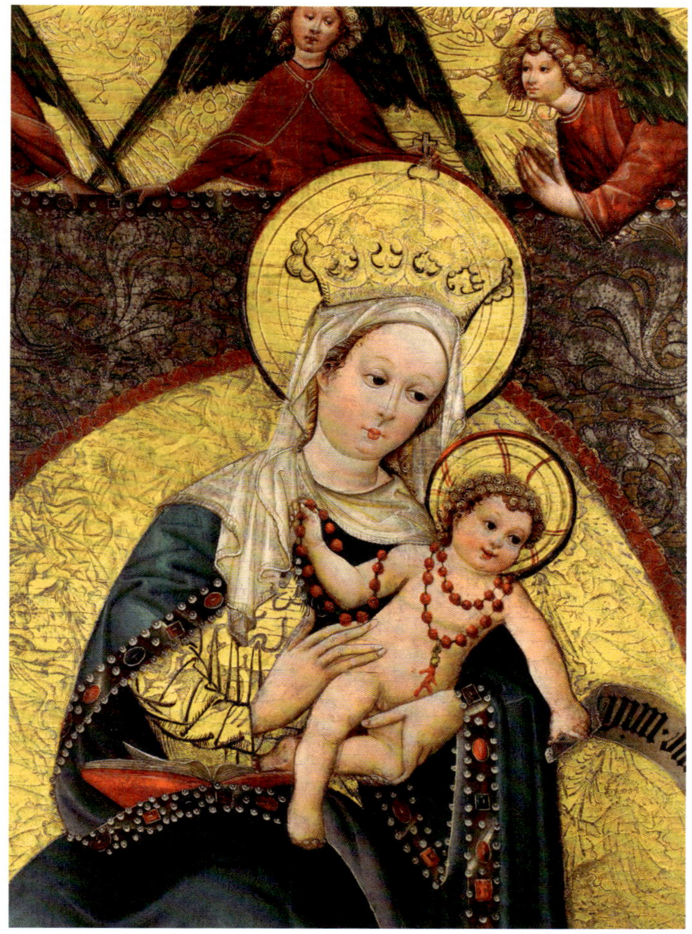

Fig. 6 Konrad von Friesach, *Madonna of Johannes Hinderkircher* (detail), Belvedere, Vienna

From the fourteenth century, starting in Italy, the embossed golden patterned ground was used a great deal. This development may have been encouraged by workshop teams operating on the principle of a division of labour (gilders, painters).

The change from a neutral gold ground to an ornamental ground can be seen as a process of objectification.[23] Various techniques of chasing, tracing, engraving, impressing and pastiglia[24] emphasize the material character of the gold, the brocade patterns are mostly based on real textiles, while other decorative elements of a mostly vegetal nature likewise draw on a generally available stock of craft models. As Ellen Beer has already aptly noted, the embossed, 'hardened' gold ground reinforces the impression of a metallic coating as a 'foreground material', which has an autonomous status vis-à-vis the painting and is in a relationship of tension with the three-dimensionality of the picture.[25] With an analytical gaze, Beer raises the material character of gold quite generally to the supreme principle of Western art throughout the Middle Ages. While she does not entirely dismiss the metaphysical aspect of gold, its connexion with the true divine light in the Neo-platonic sense, she attaches a new importance to the substance as such. Important in our context is the idea that, through the material quality, what is depicted in the picture takes on an 'enhanced presence and higher degree of commitment'.[26] We are concerned here not with the material as bearer of meaning, but with the potential of a 'genuine' material to fetch the content of the picture into the beholder's reality, so to speak. This thought model is, in my opinion, valid for the *Hinderkircher Madonna*. Without a doubt, the gold and silver in the picture constitute a factor which lends a note both of nobility and holiness. At the same time, though, when we look at the brocade pattern, we have a sense of déjà vu of familiar fabrics, whether used for liturgical or non-religious purposes. Such an oscillation between the planes of reality, the ambiguity between the earthly and heavenly spheres, between intimacy and rapture, form the cornerstone of the whole pictorial concept. The patterned ground is part of this pictorial strategy, not a merely decorative accessory. Even the choice of motifs is hardly coincidental: the fluttering birds and the zigzag rays of the patterned ground come across as an echo in the immediate vicinity of the winged angels around the Virgin's throne. To sum up: the *Hinderkircher Madonna* is characterized by a subtle narrative style and aesthetic which manifestly present the transcendent process of the divine act of mercy before the eyes of the beholder. The gold is designed on the one hand as an angelic aureole, and on the other as an ornamental ground. This oscillation of the form and the planes of reality is in my opinion a deliberate ploy on the part of the painter to integrate gold correspondingly ambivalently into the staging of the Madonna with the donor.

The golden St John in the *Znaim Altarpiece*

The second example touches on the question of whether the gilding of statues can contain concrete information, which, in hinged retables, goes beyond a general aesthetic strategy of overpowering the beholder, as addressed above. The

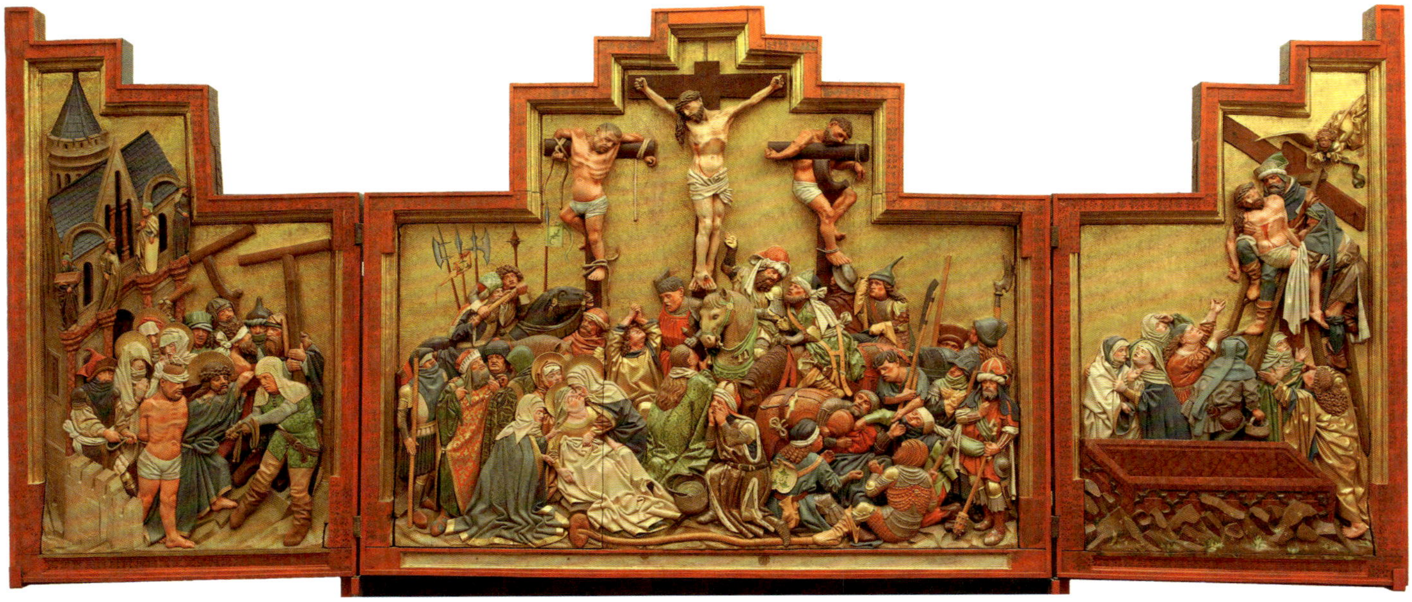

Fig. 7 Viennese woodcarver and painter, *Znaim Altarpiece,* c. 1440/1450, Belvedere, Vienna

question arises in the case of the Passion Retable from Znaim (now Znojmo in the Czech Republic), an unusual and important work of early Realism probably created in Vienna.[27] The open display side depicts Christ Bearing the Cross, the Crucifixion, and the Deposition on three large carved reliefs. The depictions are not in any way squeamish (fig. 7). Densely packed coarse figures form compact masses, which, taken together, create a decidedly pictorial effect. A crucial factor is the largely extant original polychromy. which evinces an extraordinary textural differentiation:[28] fine gradations of tone and lustre bring out the different surface structures of textiles, metal, and leather; the metallic applications consist of gold leaf, zwischgold[29] and silver leaf. Some 40 different cloth and ornamental patterns contribute towards enhancing the realistic effect; in the Crucifixion zone real string was used for the thief on the right. The chased-gold ground is still pristine on the left-hand wing and in the shrine, that on the right-hand wing was renewed later (c. 1500?).

If we now look more closely at the gold leaf on the reliefs, what is striking is that its use is related to the particular objects. Thus haloes, brocades, and smaller sections of garments, such as hems, are gilded, as are parts of suits of armour. It would be quite in order to speak of a 'natural gilding', only used where it literally makes sense. This also applies to the golden angel hovering over the Deposition. Only one figure stands out in this respect as an exception: St John the Evangelist, whose broad golden cloak is conspicuous among the otherwise textile textures. He can be seen in the shrine relief wringing his hands beneath the Cross, and in the Deposition as a figure with his back to us, thrusting himself in agitation on to the scene (fig. 8, detail). Scholars have not hitherto paid any attention to the striking golden figure of St John in the *Znaim Altarpiece*, even though it cannot be explained by any iconographic convention. Maybe it is evidence of a particular devotion to the saint in the place where the retable was originally set up.

Before the *Znaim Altarpiece* was presented in 1825 to Emperor Franz I for the Neo-gothic decoration of Schloss Laxenburg, it was in the Wenceslas Chapel in Znojmo. But it could not have been intended for this location, as the building has only existed since 1520/21. There are no older records of the retable, and the former inscription is also missing (it was in a hollow beneath the Crucifixion). According to Lothar Schultes, the Passion Retable can only have been originally destined for the parish church of St Nicholas in Znojmo, where, he says, it probably served as a nave altar for the Corpus Christi Fraternity. The altar, he believes, was moved to the neigh-bouring Wenceslas Chapel probably in 1688, on the occasion of the transfer of the fraternity, which had its seat in St Nicholas and St John. The small, no longer extant church of St John on the Unterer Platz (now Masarykovo náměstí) in Znojmo, can be excluded as the original location of the retable.[30] For apart from a donation record that does not match the altarpiece, we believe there is a further counter-argument: the little church was probably linked to the Order of St John of Jerusalem, or Knights Hospitallers.[31] If so, it was John the Baptist and not John the Evangelist who was venerated here. It would also be worth examining whether one possible origin of the retable was the Premonstratensian monastery at Bruck (now Louka) near Znojmo: the Wenceslas Chapel in Znojmo was subordinate to this monastery,[32] the church had an altarpiece depicting a Crucifixion, and St John the Evangelist was held in reverential regard by the Premonstratensians (not least, as a strategic counterbalance to the Hospitallers). Louka monastery was rebuilt shortly after its destruction by the Hussites, and a start was made in 1444 on rebuilding the church. Stylistically there is no reason not to date the *Znaim Altarpiece* to the 1440s.[33]

In the *Znaim Altarpiece* a conventional style of painting, rooted in the local Viennese tradition, on the outside of the wings[34] confronts a new, lively and disturbing narrative in the

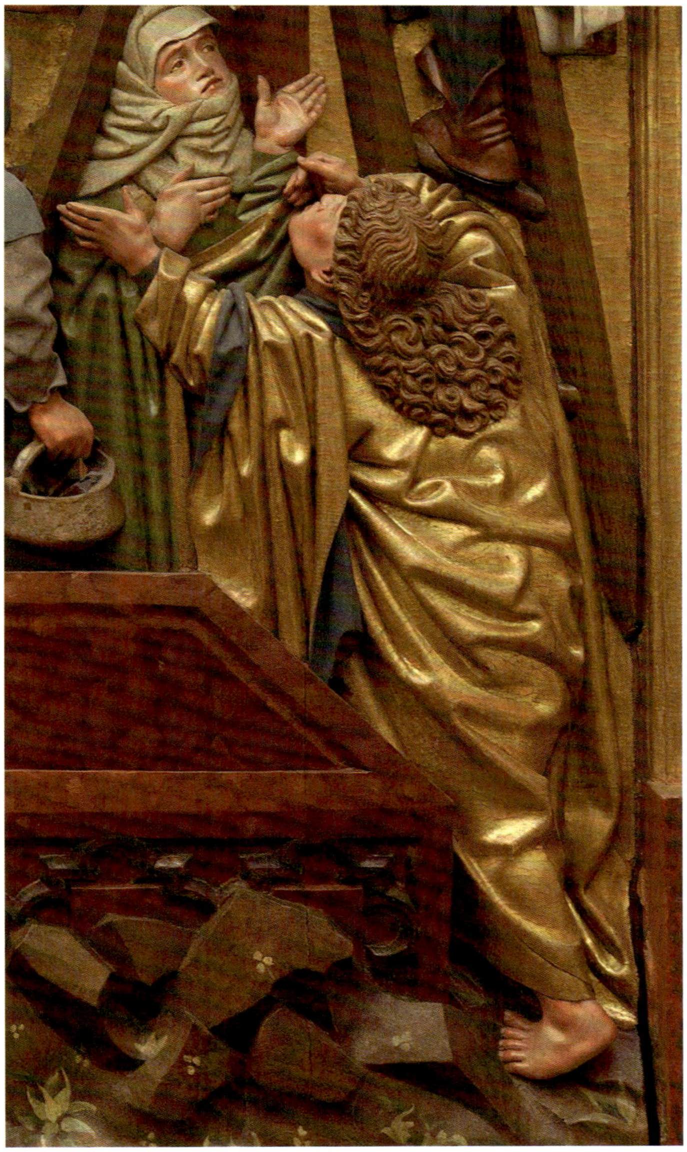

Fig.8 *Znaim Altarpiece* (detail: St. John, the Evangelist), Vienna, 1440s, Belvedere, Vienna

Passion reliefs. The gold leaf, in this highly realistic depiction of brutality and suffering, remains paradoxically present in the picture: as a glow that attaches to the material things, and perhaps also as an appeal to the faithful to find themselves again especially in the sorrow of the 'beloved disciple' of Jesus.

The golden idol in the *Krainburg Altarpiece*

It would be inappropriate our theme, even in a brief sketch of retable art, not to say something about the profane, negative aspects of gold. It takes shape as the stuff of false gods as early as Exodus in the form of the Golden Calf—the archetype of the sin of idolatry, and hence of a breach of the Second Commandment. The idol is nothing but an inanimate deception, and gold a dead material, as the Old Testament declares on a number of occasions, for example: "Their idols [of the heathen] are silver and gold, the work of men's hands." (Ps 115, 4)—"Every man is brutish in his knowledge: every founder is confounded by the graven image: for his molten image is falsehood, and there is no breath in them." (Jer 10,14)

The overthrow of the heathen idol by the Christian faith and contempt for its luxurious, artfully shaped material were widely reflected in Christian allegory and iconography. Both aspects concern questions of the legitimacy of the Christian cult image and the justification of costly materials in church decoration—two great themes, the subject of heated discussion time and again from the early Middle Ages to modern times. Suger of Saint-Denis and Bernhard of Clairvaux took up major opposing positions in the twelfth century in the dispute between luxury and asceticism.

Suger, the powerful abbot and reviver of the royal abbey of Saint-Denis near Paris, argued in the treatise *De administratione*, written in the 1140s, in the context of his vast donations, that gemstones and gold were appropriate as offerings to the Lord. The lustre of the noble material led people to the knowledge of God, he asserted, while contemplation of material beauty had transported him to higher spheres: "Thus, when—out of my delight in the beauty of the house of God—the loveliness of the many-colored gems has called me away from external cares, and worthy meditation has induced me to reflect, transferring that which is material to that which is immaterial, on the diversity of the sacred virtues: then it seems to me that I see myself dwelling, as it were, in some strange region of the universe which neither exists entirely in the slime of the earth nor entirely in the purity of Heaven; and that, by the grace of God, I can be transported from this inferior to that higher world in an anagogical manner."[35]

Bernhard of Clairvaux, at the head of the Cistercian monastic movement, in his famous *Apologia ad Guillelmum Abbatem* (c. 1125) represented the opposite pole to the Cluniac display of magnificence: "Tell me, ye poor men, *(...)* what doeth this gold in your sanctuary? *(...)* To speak plainly, does not the root of all this lie in covetousness, which is idolatry? *(...)* For at the very sight of these costly yet marvelous vanities men are more kindled to offer gifts than to pray. Thus wealth is drawn up by ropes of wealth, thus money bringeth money; for I know not how it is that wheresoever more abundant wealth is seen, there do men offer more freely. Their eyes are feasted with relics cased in gold, and their purse-strings are loosed. They are shown a most comely image of some saint, whom they think all the more saintly that he is gaudily painted. *(...)* The church is resplendent in her walls, beggarly in her poor. She clothes her stones in gold, and leaves hers sons naked. The rich man's eye is fed at the expense of the indigent."[36]

These two prominent utterances show once more that gold stands *pars pro toto* for a 'strategy of overwhelming' in the use of noble materials, both in the positive and in the negative sense.

The pictorial motif of the golden idol, much varied over the centuries, is an easy-to-grasp example of negatively connoted gold.[37] A particularly original version can be found on one of

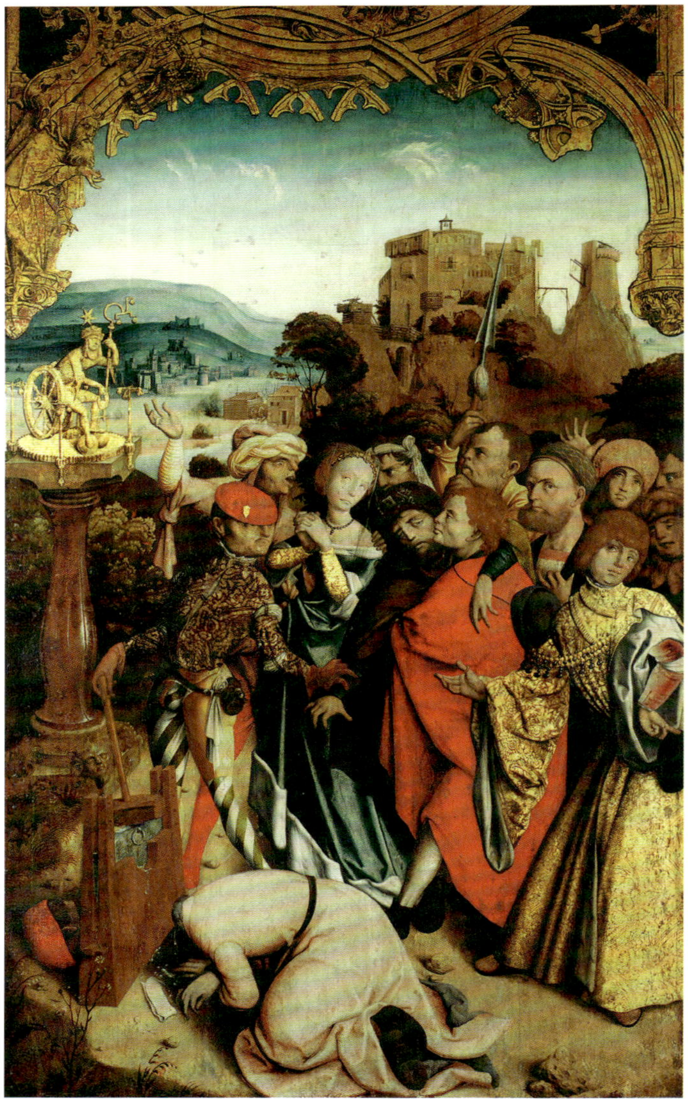

Fig. 9 *Martyrdom of SS. Cantius, Cantianus and Cantianilla*, from the former high altar of St Cantius in Kranj (Slovenia), c. 1510, Belvedere, Vienna

the two hinged panels preserved from the former high altar, dating from c. 1510, of the parish church of SS Cantius, Cantianius, and Cantianella in Kranj (Slovenia).[38] The singular scene of the martyrdom of the three siblings shows, on the left, on a marble column the fully gilded statuette of an idol (fig. 9, 10). According to legend, it is Jupiter, whom the three siblings, having fled from Rome to Aquileia, refused to worship; they were killed as a result. Sitting on a two-wheeled chariot, and with a star on his head, the figure is of a planet-god type of the fourteenth century, such as we find for example by Guariento in the fresco cycle of the Eremitani church in Padua.[39] What is new, though, is the grotesque stylization of Jupiter as an automaton: the floor plate consists of a cog wheel, which, when set in motion, moves rods with little bells; Jupiter's insignia are also fitted with bells. The unusual idea of depicting a classical deity as a tinkling automaton makes manifest the artificial quality of the idol. The painter has skilfully and effectively staged the golden idol against a broad landscape and above the throng of people. Cantianilla refuses desperately to obey the demands of a myrmidon pointing to the false god. In front of her, the garishly clothed executioner is dragging her brother Cantius out of the arms of Cantianus to the axe, where her teacher Protus is already lying, beheaded. To the right the commander Sisinius directs the gaze to the gruesome event. His gold-brocade cloak and the golden idol correspond diagonally across and above the throng and create, so to speak, the visual cornerstones of the action. As a decorative crowning, we have a gilded tracery arch, which simulates, with illusionist modelling in black-stain painting, a carved organ-pipe screen. On the left-hand base of the arch, directly above the idol, is a likewise monochrome-gold prophet-like figure, who seems to be commenting on the scene in gestures. The gold leaf of the idol and the brocade cloak is shaded with black-stain painting, which brings out the material quality of the metal and the fabric respectively. Apart from the golden sleeves of Cantianilla's dress, which serve to give visual emphasis to her gesture of rejection, there is no further gold. The saint is wearing a fashionable, fur-lined cloak, while her two brothers are dressed as pilgrims; there are no haloes.

The Master of the Krainburg Altarpiece was quite obviously interested in achieving contemporary illusionist depiction at all levels: the illusory space with an atmospherically lightened sky, the naturalistic reproduction of materials and everyday things, and also portrait likenesses which already bear witness to a Renaissance image of the individual. Look, for example. at the bearded man behind Sisinius, who is looking directly at the beholder, and is regarded, probably rightly, as a self-portrait of the artist. All this proclaims an epochal change to a modern view of what a picture is, a view which was now, in the last phase before the Reformation, conquering central European hinged retables. The anonymous artist of the *Krainburg Altarpiece* is showing, at this late stage of the development, a logical integration of gold leaf into illusionist painting. It is telling that even in this upheaval phase in the early sixteenth century, there was still no paradigm shift from real gold to a gold imitation by the use of paint. He could have adduced an art-theoretical foundation in Leon Battista Alberti's aesthetic devaluation of real gold in pictures in favour of simulated gold, in which the painter could demonstrate his artistic skills (De pictura, 1435/36).[40] In central Europe, however, this plea for the imitation of gold and gold lustre (written at the same time as the 'ars nova' of the Netherlanders and maybe even as a reaction to this) did not, as is well known, fall on fruitful soil.

Thus the golden idol statuette of the Kranj retable is also a graphic example of the long continuance of real gold in pictures in this country.

A visitation report dating from 1684 describes the old high altar of the parish church in Kranj as having an 'extremely costly' and 'totally gilded' hinged retable with statues of saints in the shrine and a crucifix above.[41] The lost shrine must, therefore, probably have shown the three patron saints of the church depicted on the wings, with carved tracery in the form of the painted decor, and a rich gilding. As usual, an increase in the

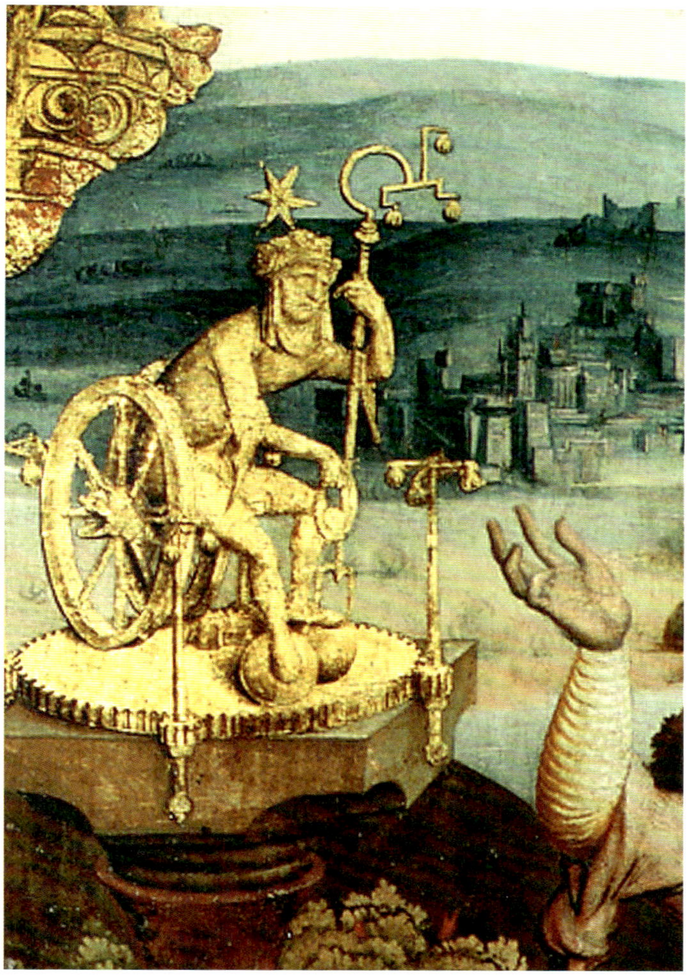

Fig. 10 *Martyrdom of SS. Cantius, Cantianus and Cantianilla* (detail), c. 1510, Belvedere, Vienna

use of gold towards the middle is to be expected. Now, the 'golden rule' of an auratic presentation of gilt figures of saints in the shrine does not coincide thematically with the negative connotation of gold in the martyrdom scene. A positivist art-history may see nothing in this but a meaningless, decoratively effective random approach to the use of gold. Far more plausible, though, is an iconological and material-aesthetic approach, which starts from a polyvalence of gold and its need-oriented aesthetic utilization. For, as is easy to illustrate by reference to the retable of the *Krainburg Altarpiece*: the gold of the idol and the gold of the saint stand next to each other and co-operate as one and the same material with different semantic charges. In this sense, the modern beholder might retain a open mind when interpreting gold, just as the Middle Ages interpreted this noblest of all metals.

The following notes are restricted to a small selection of works and studies relevant to the aspects addressed here. No discussion of the state of research is aimed at or indeed possible.

1 Manfred Koller, *Der Flügelaltar von Michael Pacher in St. Wolfgang*, Vienna /Cologne/Weimar 1998. - Artur Rosenauer (ed.), *Geschichte der bildenden Kunst in Österreich*, vol. 3: *Spätmittelalter und Renaissance*, Munich 2003, cat. 108 (Lothar Schultes) and cat. 216 (Lukas Madersbacher). - Rainer Kahsnitz, *Die grossen Schnitzaltäre. Spätgotik in Süddeutschland, Österreich, Südtirol*, Munich 2005, 76-86, pl. 23-41.

2 To compare: for Pacher's *Wolfgang Altar* 1200 Hungarian guilders was the contractually stipulated maximum. The *Altar of the Virgin in Cracow* by Veit Stoss cost 2808 Hungarian guilders. Riemenschneider's fee of 145 guilders for the carving of the Münnerstadt altar, delivered in 1493, comes across by contrast as positively modest (it was painted and provided with wings by Veit Stoss in 1502-1504 for 220 guilders. - The costs of retables (tabular survey): Michael Baxandall, *Die Kunst der Bildschnitzer. Tilman Riemenschneider, Veit Stoß und ihre Zeitgenossen*, 3rd ed. Munich 1996, pp. 74f. - Still a basic source, the edition of retable contracts in: Hans Huth, *Künstler und Werkstatt der Spätgotik*, enlarged reprint, Darmstadt 1967 (1st ed.. Augsburg 1925). - To date there is no broad cross-media overview; cf. a recent material-rich study of the situation in Burgundy: Renate Prochno, "Materialwert—Geldwert—Kunstwert im späten Mittelalter. Beispiele aus Burgund", in Ulrike Aichhorn (ed.): *Geld- und Kreditwesen im Spiegel der Wissenschaft*, Vienna /New York 2005, 177-216 (on altarpieces: pp. 193-197).

3 *Josef wird in den Brunnen geworfen—Geisselung Christi—Vermählung Mariens*: Elfriede Baum, *Katalog des Museums mittelalterlicher österreichischer Kunst* (Österreichische Galerie Wien, cat. I), Vienna/Munich 1971, 76-78 (cat. 44-46). - Most recently: GBKÖ 3 (as note 1), cat. 217 (Lukas Madersbacher).

4 The oldest-known unpainted carved altarpiece is the retable in Lorch am Rhein dating from 1483 - Kahsnitz 2005 (see note. 1), pp. 36, 122-126, pl. 53-59; ibid., further prominent visible-wood or partly painted carved altarpieces including: Tilman Riemenschneider's *Holy Blood Altar* in Rothenburg ob der Tauber (1499-1505), the *Bamberg Altar* by Veit Stoss (1520-1523) and the *Breisach Altar* by Master H. L. (1523-1526). - Comprehensively on this topic: Georg Habenicht, *Die ungefassten Altarwerke des ausgehenden Mittelalters und der Dürerzeit*, diss. Göttingen 2000, URL:< http://webdod.sub.gwdg.de/diss/2002/habenicht/habennicht.pdf>.

5 Kahsnitz 2005 (see note 1), 324-330, pl. 172-182. - Most recently: Gregor M. Lechner OSB, *Der Schnitzaltar des Meisters von Mauer bei Melk*, Regensburg 2010.

6 GBKÖ 3 (as note 1), cat 132 (Lothar Schultes). - Kahsnitz 2005 (as note 1), 364-370, pl. 196-210. - Bohdana Fabiánová/Zdeněk Vácha (eds.), *Zwettler Altar im Kontext der spätgotischen Kunst Mitteleuropas* (Symposium Schloss Mikulov 2007), Brno/Mikulov 2008. - Manfred Koller, *Die Retabel von Kefermarkt, Mauer und Zwettl: Holzfarbigkeit und Teilfassungen*, ibid. 213-224.

7 Floridus Röhrig (ed.), *Der Albrechtsaltar und sein Meister*, Vienna 1981, esp. Artur Rosenauer, "Zum Stil des Albrechtsmeisters", 97-122. - Most recently GBKÖ 3 (as note 1), 415-418 (Michael and Thomas Rainer).

8 Baum 1971 (as note 3), 39-42 (cat. 17f.) - Röhrig 1981 (as note 7), pl. 29-33. - GBKÖ 3 (as note 1), cat. 190 (Michael and Thomas Rainer).

9 Huth 1967 (as note 2), p. 61.

10 Kahsnitz 2005 (as note 1), 134-142, pl. 60-80.

11 Christoph Trepesch devotes himself to the pre-modern situations of on-site light in *Studien zur Dunkelgestaltung in der deutschen spätgotischen Skulptur. Begriff, Darstellung und Bedeutung des Dunkels*, Frankfurt am Main 1994 (Europäische Hochschulschriften series 28, Kunstgeschichte vol. 210).

12 Baum 1971 (as note 3), p. 135 f. (cat. 92), - Agnes Husslein-Arco/ Veronika Pirker-Aurenhammer (eds.), *Aktuell restauriert. Der Abtenauer Altar von Andreas Lackner*, (exhib. cat.. Österreichische Galerie Belvedere, Vienna), Weitra 2011, esp. the contributions by Veronika Pirker-Aurenhammer, Matthias Weniger and Manfred Koller and the discoveries by the conservators Elisabeth Pfützner and Giovanna Zehetmaier.

13 Anton Legner, "Der Blasiusaltar aus Abtenau, Andreas Lackner und die Salzburger Plastik der Spätgotik", in: *Alte und moderne Kunst*, 27, 1982, no. 180/181, 16-21, here p. 17.

14 An overview: Hans-Jürgen Horn, "Gold", in: *Reallexikon für Antike und Christentum*, vol. 11, 1981, col. 895-930.

15 Wolfgang Schöne, *Über das Licht in der Malerei*, Berlin 1954, pp. 56-69. - Christian Hecht, *Die Glorie. Begriff, Thema, Bildelement in der europäischen Sakralkunst vom Mittelalter bis zum Ausgang des Barock*, Regensburg 2003, esp. pp. 68-77 (on gold and the gold ground). - On Neo-platonic and other positions up to and including the 13th century: Edgar De Bruyne, *Études d'esthétique médiévale*, Bruges 1946, passim, but esp. vol. 2, pp. 69-107. - de Bruyne, *L'esthétique du Moyen Age*, Louvain 1947 (Essais philosophiques 3).

16 E.g. Isidor von Sevilla, *Etymologiae*, XVI, 6, where among other things the word aurum (gold) is derived from aura (effulgence): *"Aurum* ab *aura* dictum, id est a splendore, quod repercusso aëre plus fulgeat. Naturale enim est ut metallorum splendor plus fulgeat luce alia percussus.*"*—Quoted from Edgar De Bruyne 1946 (as note 15), vol. 1, p. 80.

17 Konrad von Megenberg, *Buch der Natur*, VII, 1: Robert Luff/Georg Steer (eds.), *Konrad von Megenberg. Das "Buch der Natur"*, vol. 2, critical edition based on the manuscripts, Tübingen 2003, pp. 511f. The quoted text in archaic German means: "Hence the expression: Go to the devil for gold!"

18 Christel Meier, "Edelsteinallegorese", in Anton Legner (ed.) *Die Parler und der Schöne Stil 1350-1400. Europäische Kunst unter den Luxemburgern*, Cologne 1978, vol. 3, 185-187, here p. 187.

19 Günter Bandmann, "Bemerkungen zu einer Ikonologie des Materials", in: *Städel-Jahrbuch*, N.F. 2, 1969, 75-100.

20 ibid., pp. 94f.

21 Baum 1971 (as note 3), pp. 156f. (cat. 115f.) - On Konrad von Friesach: Janez Höfler, *Die Tafelmalerei der Gotik in Kärnten (1420-1500)*, Klagenfurt 1987, pp. 39f. and 43f. (cat. 7), fig. 16: an altarpiece wing which possibly belongs to the one under discussion.

22 The painters, carvers, sign-makers and glaziers of the city of Salzburg were organized in a guild whose regulations are dated 3 June 1494. - Quote from Johannes Ramharter, *'Weil der Altar altershalben unförmblich und paufellig ...'*. Rechtsfragen zur Ausstattung der Sakralbauten im Salzburger Raum, in: *Fontes Rerum Austriacum*, publ. by Österr. Akademie der Wissenschaften, 3. Abt. Fontes Iuris, vol. 12, Vienna /Cologne/ Weimar 1996, p. 56.

23 A recent survey is Iris Wenderholm, "Aura, Licht und schöner Schein. Wertungen und Umwertungen des Goldgrunds", in Stefan Weppelmann (ed.), *Geschichten auf Gold. Bilderzählungen in der frühen italienischen Malerei* (exhib. cat. Gemäldegalerie Berlin), Berlin 2006, 100-113.

24 Hans Westhoff/ Roland Hahn, "Verzierungstechniken in Malerei und Skulptur", in: *Graviert, gemalt, gepresst. Spätgotische Retabelverzierungen in Schwaben*, (exhib. cat.. Landesmuseum Stuttgart), Ulm 1996, 18-31.

25 Ellen J. Beer, "Marginalien zum Thema Goldgrund", in: *Zeitschrift für Kunstgeschichte*, 46, 1983, 271-286, here p. 279.
26 Ibid., p. 279, a reference to Karl Möseneder, "Lapides vivi, über die Kreuzkapelle der Burg Karlstein", in: *Wiener Jahrbuch für Kunstgeschichte*, 34, 1981, 39-69.
27 Baum 1971 (as note 3), 65-67 (cat. 37). - Robert Suckale, "Das Znaimer Retabel. Zur künstlerischen Herkunft des Bildschnitzers", in: *Österreichische Zeitschrift für Kunst und Denkmalpflege*, 42, 1988, 2-14. - Lothar Schultes, "Zur Herkunft und kunstgeschichtlichen Stellung des Znaimer Altars", in: Suckale (op. cit., as note 27) 14-37. - GBKÖ 3 (as note 1), cat. 86 (Lothar Schultes).
28 Manfred Koller/Giovanna Zehetmaier, "Kunsttechnologische Studien zum Znaimer Altar und anderen Werken des frühen Realismus in Österreich", in: *Österreichische Zeitschrift für Kunst und Denkmalpflege* 1988 (as note 27), 37-44.
29 Translator's note: zwischgold is also known as 'party gold', made by beating gold and silver together.
30 Schultes 1988 (as note 27), 14-17.
31 Insofar as the presumptions in the older literature are correct: Anton Hübner, *Znaims geschichtliche Denkwürdigkeiten, 2. Abth., Znaims Klöster, Kirchen und Bildungsanstalten*, Znaim 1846, 168-171. - See also A. Hübner, *Denkwürdigkeiten der königl. Stadt Znaim*, Znaim 1869, 532-542.
32 Hübner 1869 (as note 31), p. 257.
33 The dating ranges from 1431 (?) in hypothetical connexion with a donation made in that year (Schultes 1988, as note 27) c. 1450 (Suckale 1988, as note 27).
34 The paintings are ascribed to the workshop of the so-called Friedrich Master, see: *Der Wiener Neustädter Altar und der "Friedrichs-Meister". Bedeutende Kunstwerke gefährdet—konserviert—präsentiert*, vol. 10 (exhib. cat.. Österreichische Galerie Belvedere, Vienna) Vienna 1999 and *Der Wiener Neustädter Altar in St. Stephan in Wien. Erforschung und Restaurierung 1985-2004. Bedeutende Kunstwerke gefährdet—konserviert—präsentiert*, vol.19 (exhib. cat.. Österreichische Galerie Belvedere, Vienna), Vienna 2004.
35 http://employees.oneonta.edu/farberas/arth/arth212/liturgical_objects/suger_excerpts.html, where the source is given as Erwin Panofsky, *Abbot Suger on the Abbey Church of St. Denis and its Art Treasures*, Princeton, 1979.
36 Quoted from Robert G. Calkins, *Monuments of Medieval Art*, Ithaca NY, 1979, p. 121
37 See the foundation text: Michael Camille, *The Gothic Idol. Ideology and Image-making in Medieval Art*, Cambridge 1989.
38 Otto Benesch, "Der Meister des Krainburger Altars", in: *Wiener Jahrbuch für Kunstgeschichte*, vol. 7, 1930, 120-200, esp. pp. 120-143, and vol. 8, 1932, 17-67 (repr. in Otto Benesch., *Collected Writings III*, London/New York 1972, 154-237). - Tomislav Vignjević, *Der Meister des Krainburger Altars*, Ljubljana 1997, esp. pp. 25-37.
39 Elisabeth Sears, *The Ages of Man. Mediaeval Interpretation of the Life Cycle*, New Jersey 1986, pp. 109f., fig. 45 (astrological depiction of the moon); see Vignjević (as note 37).
40 Leon Battista Alberti, *De pictura II*, § 49. - A precise analysis of the passage can be found in Hans H. Aurenhammer, *Studien zur Theorie der 'historia' in Leon Battista Albertis 'De pictura'. Themen, Begriff, Funktionen*. Habilitationsschrift Vienna 2004 (unpubl.), 41-58. - Hans H. Aurenhammer, "Malerei im Horizont von Rhetorik und Poesie. Zur Theorie von Leon Battista Albertis 'historia'" in: *Rhetorik*, 24, 2005, 27-42.
41 "Altare majus [...] vetustem quidem, sed pulchrae structurae et maximi pretii, aperibile, ex toto deauratum; in medio habens statuas praefatorum sanctorum; in summitate vero crucifixum", quoted from Vignjević (as note 37), p. 30.

Alberti's Gold Ground

Thomas Zaunschirm

The following concepts are brought up by art historians to mark the threshold between the Middle Ages and the Renaissance: the end of the gold ground, the picture as framed window, central perspective, chiaroscuro, and shadow. We owe the first art-historical reflexions presented by these concepts of modern picture composition to a young polymath, Leone Battista Alberti. He published his book *De Pictura* ('On Painting') in 1435. His influence reaches right down to present-day art history, which not only uses his basic terminology to describe the onset of the Renaissance, but, by analogy, to define the end of the very era that was ushered in by the Renaissance. Then, central perspective was introduced; Modernism abandoned it; then, the picture came to be seen as a window; Modernism abandoned this view, or at least transformed it; then, the gold ground was abandoned, but at the dawn of Modernism it re-appeared. In regard to the crucial paradigms, reference can always be made to Alberti. His importance has certainly not been overlooked; on the contrary. His fame shines out today more brightly than ever. Scholars have worked hard, produced new translations, discussed all the facets, and also pointed to his weaknesses in *De re aedificatoria* ('Ten Books on Architecture', 1452), so that hardly any questions remain to be answered. Alberti is an open book, and anyone can read him.

What continues to be interesting is since when, and for how long, he has been influential in art theory, and to what extent one can speak of a kind of polarity between the beginning and the end of the period between the Middle Ages and Modernism. One might think that this is not a question that needs to be answered. What would be the purpose of invoking Alberti (even as a counterpole) when referring to the use of gold in the 19th century and even more so in Art Nouveau (one need only think of Gustav Klimt)? There was no art writer, who, when discussing Hans Makart, pointed out that this artist had abandoned Alberti's much-quoted condemnation of the gold ground. As theories of perspective and colour had been subject to continual development, every new idea related to its immediate predecessors and not to a work which had allegedly first raised these questions.

In the investigations to hand, reference is made time and again to the corresponding passages in Alberti's treatise. These passages, so often quoted, have mostly been taken out of context, and so what I should like to do is put them back into their context. It is not a question of reinterpreting Alberti, but of concentrating on what sort of cross-references there might be which make one topic emerge from another. The focus is not on pinpointing, for the umpteenth time, the end of the medieval period, or of specifying the start-date of Modernism. Rather, we are concerned with quite simple questions such as: "Why did the gold ground lose its importance?" Or even more naively: "Whatever happened to the gold ground?" To these questions, there are two answers, each of which conditions the other. The first concerns the importance of the gold ground (for example as implying a sacred space), the second concerns those principles of the new style of painting that excluded the gold ground (namely central perspective, shadows, illusionism). In certain kinds of motif, for example landscapes and portraits, this is particularly clear. While here too the undefined space is constructed and 'filled', this is linked to the individual viewpoint, the position of the eye, and the new individual self-awareness.

The examples show that it is not possible to rely on Alberti in all questions of post-medieval self-definition and *weltanschauung* (in the sense of Jacob Burckhardt, who saw the Renaissance as the "discovery of the world and of man"[1]). In one sourcebook on landscape painting, we read: "As to Alberti, we need not say very much in our connexion. Alberti, who not without justification is seen as the real founder of modern art-theory, said a lot about history painting in his 1435 treatise, but absolutely nothing about landscape."[2] In another sourcebook, this time about the portrait, there is no direct quotation of Alberti, but only of a reading of Alberti by Nicholas of Cusa.[3] And this in spite of the fact that we owe to him the first great portrait plaque of the post-medieval era with the first free-standing self-portrait in profile (fig. 1).

Amazingly, Jacob Burckhardt either said nothing about any of Alberti's theses that later became orientation points in academic art history, or else only mentioned them in passing. In his famous 'attempt', as he modestly called his book *The Civilization of the Renaissance in Italy*, he concentrated on legends and anecdotes. He characterized the medieval consciousness not by reference to the gold ground and other features abandoned by the Renaissance, but with a wonderfully poetic periphrasis: "In the Middle Ages, both sides of human consciousness—that which was turned within as that which was turned without—lay dreaming or half awake beneath a common veil."[4] By way of contrast he lays stress on the men of action of the new age. Alberti, the polymath, comes across here, according to his own *Vita*[5], as a *uomo universale*, without parallel as an architect and musician, painter and astrologer, horseman and gymnast, householder, writer and thinker. Burckhardt seems impressed by the episode where, 'in the cathedral, he threw a coin in the air until it was heard to ring against the distant roof' (which was after all 90 metres high) and mentions the 'funeral oration on his dog'.[6] He does not, though, refer to the books on painting, architecture and sculpture, but only to his *Del governo della famiglia* ('concerning the management of the household') published under the pseudonym Agnolo Pandolfini. Here it is related how the author, following his wedding, knelt down with his wife before the domestic altar: "For myself I prayed for wealth, honour and friends, for her blamelessness, honesty and that she might be a good housekeeper."[7] Burckhardt's view has clearly lost none of its power, as a reading of Grafton's major Alberti biography will show.[8]

Gold

Alberti's condemnation of the gold ground, which, in every respect epoch-making, is quoted time and again, can be found at the end of the second book of *De Pictura* (1435) or *Della Pittura* (1436). He writes: "There are some [contemporary painters] who use much gold in their *istoria* They think it gives majesty. I do not praise it. [...] for there is more admiration and praise for the painter who imitates the rays of gold with colours."[9] Pochat commented on this passage thus: "There can be no clearer way of distinguishing the aims of the 'new' painting from the medieval tradition. Since Late Antiquity, gold had indicated the divine and the presence of the divine in the work itself. The sheen of gold served as a guarantee both for its material and metaphysical value. Alberti, however, regarded the artificial imitation of gold as more valuable and more important."[10] Thus there began, according to the conventional wisdom, the illusionism which characterized painting from the Renaissance until the dawn of Modernism.

The argument for a threshold between two eras was so overpowering that it has been repeated like a mantra ever since. Reservations and qualifications have been conveniently brushed aside. Since when has there been this belief that Alberti had spoken out against the gold ground? In 1950, in his investigation *Nimbus und Goldgrund*, Wolfgang Braunfels had, following on from Alberti, asserted: "Paint takes on a significance equivalent to that of the gold ground. The renunciation of the use of gold demanded a growth in art."[11] But within a few years already, Wolfgang Schöne refused to simply accept this interpretation: "This is perfectly true, but must be qualified to the extent that this 'renunciation of gold' no longer related to the original medieval gold ground, which no longer existed, but to the gold ground of the fifteenth century, which in any case tended towards gold paint and in its intrinsic sheen in itself was already heavily restricted in the latter's other-worldly sense."[12]

If we go along with Schöne, we must ask how, conversely, Alberti's remark could have led to the gold ground disappearing from pictures, quite apart from the fact that in those places where it was still used, it would have been a more 'painterly' one. This question relates to the reception of Alberti's treatise among his contemporaries, its influence having been projected back retrospectively. Or, to put it another way: the painters who used the gold ground still saw themselves as craftsmen, while Alberti's friends were trying to liberate themselves from the constraints imposed by the guilds, in order to raise their social status to that of exponents of the *artes liberales*. Leonardo da Vinci, in his *Trattato de la Pittura*, which (in the traditional compilation) begins with a eulogy to the scientific nature of painting, still wrote, in section 71, of a 'certain generation of painters' who felt it necessary to use gold and blue (i.e. lapis lazuli).[13]

The gold ground argument smouldered on in art-historical discourse, even though Schöne played it down, and Sedlmayr, who discusses his book, agreed with him.[14] Alberti is invoked not only in the history of aesthetics (Pochat), but also in the history of the material. Monika Wagner's 'different history of modern art' (the sub-title of her *Das Material der Kunst*) begins with Alberti's view that it is more praiseworthy to imitate the sheen of gold than to use real gold in a picture. "The task [of painting] was, from now on, to create the illusion of all other materials on the picture surface. In other words, the paint had to disappear as a material in order to appear in the picture as gold, as flesh or as any other material. In this materialization of the material was contained a crucial precondition for the success story of the picture."[15] But what does this phrase 'materialization of material' mean? The very statement that "paint had to disappear as a material" is questionable. No material, no paint: what else is it supposed to consist of? Wagner presumably means that coloured materials had to disappear. Should she not then have said '*de*materialization of the material'? The gold is after all not materialized, but on the contrary disappears from the pictures. (It is true that Wagner here is not talking explicitly about the gold ground.)

The commonplace that Alberti had spoken out against the gold ground persists to this day. "Alberti's words encapsulate the central turning-point in the value attached to the gold ground, and were to lead to its gradual disappearance," says the opulent catalogue *Geschichten auf Gold* (2006).[16] This unambiguous finding was doubtless intended to dispel any doubt on the matter. It is all the more surprising, then, that Bätschmann, in the comprehensive commentary in his Alberti edition, which goes into all the terms, should say nothing about gold grounds: "Alberti certainly allows a place for gold and silver for the frame, but he rejects gold areas in the painting, because, depending on the light, they may stand out from the colour context either by shining or by being too dark *(Della Pittura,* 49*).*"[17] Elsewhere he returns to this passage: "In the chapter on paints, he advises against the use of gold..."[18] Now, one might think that by 'gold areas' he means nothing other than the gold ground, and for this reason I shall quote the relevant passage from Alberti here once more.

"There are some who use much gold in their istoria They think it gives majesty. I do not praise it. Even though one should paint Virgil's Dido whose quiver was of gold, her golden hair knotted with gold, and her purple robe girdled with pure gold, the reins of the horse and everything of gold, I should not wish gold to be used, for there is more admiration and praise for the painter who imitates the rays of gold with colours. Again we see *in a plane panel with a gold ground* that some planes shine where they ought to be dark and are dark where they ought to be light. I say, I would not censure the other curved ornaments joined to the painting such as columns, carved bases, capitals and frontispieces even if they were of the most pure and massy gold. Even more, a well perfected istoria deserves ornaments of the most precious gems."[19]

Note the first sentence. In the original it reads "Truovasi chi adoptera molto in sue storie oro, che stima porga maestà." He talks of 'much gold' ('molto ... oro'), but in the translation used by Pochat, this becomes 'gold ground'. As Alberti adduces such examples as Dido's quiver, her knotted hair, her girdled robe, the reins of her horse (and all the rest), it is perfectly clear that he is talking about objects actually made of gold. He finds it more admirable to use paints here to evoke the sheen rather than gold itself. The picture was supposed to be composed with the help of the squared *velum*, with which the perspective was constructed: in other words in space on an empty ground. Alberti was not in the least interested in gold grounds. The '*plane panel with a gold ground*' mentioned in the second half of the above quotation from *Della Pittura* (section 49) is also misleading, as Alberti's Italian text talks of 'una piana tavola' (plane panel) in which we see 'alcune superficie ove sia l'oro' which refers however not to gold grounds, but to individual gold areas. There was no term for 'gold ground' until well into the eighteenth century.[20] In Italian, a gold ground is *sfondo dorato* (literally: gilded background). However this distinction between surface and background is not relevant to Alberti's point, which is that metal surfaces react differently to incident light, and therefore create the wrong effect.

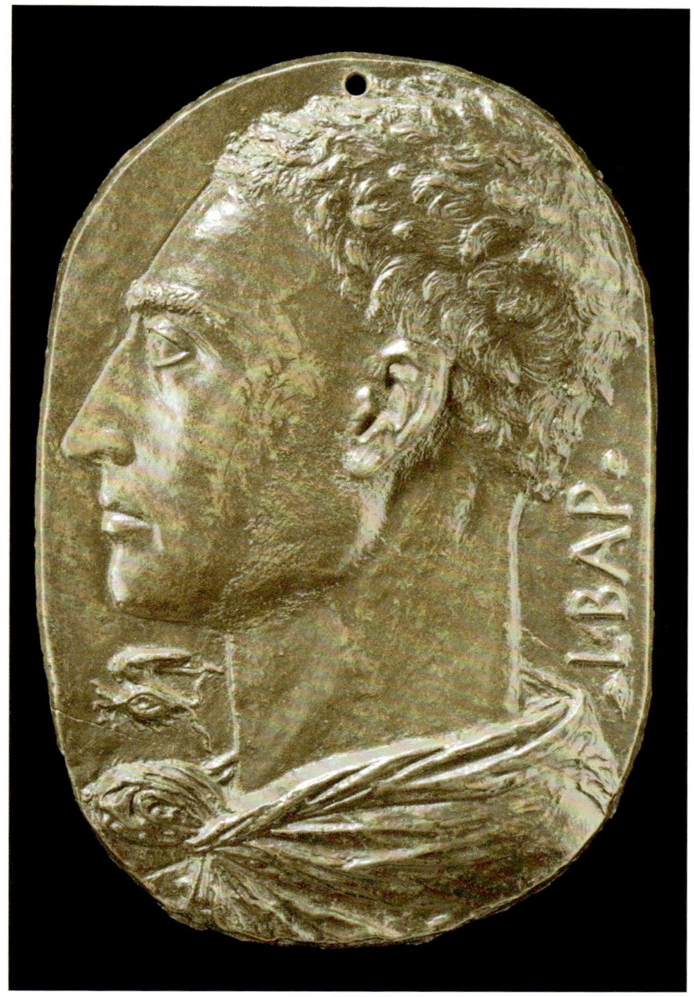

Fig. 1 Leon Battista Alberti, *Selfportrait*, bronce, c. 1435, Samuel H. Kress Collection

In German, at least,[21] the 'gold ground' rendition goes back to the translation by Janitschek. In his 1877 edition, he translated the relevant passage (here rendered more or less literally in English): "Then we see too that on a flat panel with a gold ground, some areas shine when they should be dark, and appear black when they should shine brightly."[22] Generally in the text, the difference in the translations ('gold grounds' or 'gold areas') hardly matters. But it does matter when 'a flat panel with some gold areas' transmogrifies into 'a flat panel with a gold ground', in other words, when a specialist term referring to a characteristic technique of medieval painting is used. Terms develop a dynamic of their own. Alberti's 'gold ground' has been a popular projection of art historians ever since.[23]

Alberti gave the activity of the artist a higher status than it had hitherto enjoyed. In an earlier section (25) of *Della Pittura*, the painterly manifestation is more important to him than the use of costly materials. Here he speaks of gold for the first time: "Ivory, gems and similar expensive things become more precious when worked by the hand of the painter. Gold worked by the art of painting outweighs an equal amount of unworked gold. If figures were made by the hand of Phidias or Praxiteles from lead itself—the lowest of metals—they would be valued more highly than silver."[24] The contrast between noble metals and lead or iron is almost archetypal in one way or another: while Alberti is interested in the artistic value of the relationship, Karl Marx is concerned with the exchange value: "One hundred pounds' worth of lead or iron is of as great a value as one hundred pounds' worth of silver and gold."[25]

Medieval pictures offered no views of the world or insights into it. Only the advent of perspective declared pictures to be windows on the world, and thus (re)constructed reality. With his re-writing of painting, Alberti encompassed a universe, but at the same time removed the basis of a sacredness considered self-evident by dint of the material. Hardly ever have era-thresholds been so directly manifest as in Alberti's demand to dispense with gemstones and gold. Only when artists once more incorporated things into their works, rather than just imitate them, was the next threshold to be crossed, with the start of Modernism and the techniques of the objet trouvé and the readymade. The different use of gold, as a colour and as a material, was to determine the interpretation of medieval gold in art-historical research.

The gold ground proves to be a technical phenomenon which was described and handed down no differently in the Middle Ages from in the twentieth century. What is more, in exhibitions of relevant paintings, the individual stages of gold-ground application are also presented. "Six exemplary panels for the production of a gold-ground painting on the basis of technical descriptions by Cennino Cennini" were prepared for an exhibition at the National Gallery in London in 1989. They have been displayed time and again since, in Berlin or most recently in Vienna.[26]

Black and White

Much has been written about the differences between Cennino Cennini's *Il libro dell' Arte* (c. 1390) and Alberti's *Della Pittura* (1435). As far as the gold ground is concerned, there are technical instructions (ch. 140) in Cennini separate from and independent of the discussion of colours. In chapter 36, he distinguishes seven 'natural' colours in the form of the four 'earths' (black, red, yellow, and green) and the three others that have to be prepared artificially (white, ultramarine or copper blue, and Naples yellow). Further colours are distinguished according to pigment, thus there were seven kinds of red. Unlike Alberti, Cennini included black and white among the natural colours. According to Alberti, however, these are not 'veri colores' (corresponding to the four elements) but 'alteratores', in other words colour-changers. Black and white are, for Cennini, not yet autonomous values that served to change the other colours, as they are for Alberti.

So is there a connexion between colour-system and the attitude to gold? Only when black and white no longer count as colours like the other hues (as they still do for Cennini), but as colourless paints with which the coloured paints could be varied on a light-dark scale (as with Alberti), does chiaroscuro emerge. Black and white harness the colours in the framework of the light-dark polarity, which in turn allows a new relationship of coloration in combination with shading. In this new system, gold is a foreign body. If the painter looks at gold as a material alongside the colours, it is no longer possible to think of gold as a colour. Gold and black-or-white are no longer colours for Alberti, although gold is accorded a function for framing purposes, and black and white the functions of respectively darkening and lightening the actual colours. Gold leaf *isn't* a colour, but it *has* a particular colour. In the age of incipient chiaroscuro, every material can be depicted. Alberti's colour system, with black or white as an instrument for changing the darkness or lightness of other colours, is compatible with the view of the picture as a window. If one goes, both go. No window, no perspective space; no perspective space, no systematic separation of colours.

1 Jacob Burckhardt, "The Discovery of the World and of Man", title of Part IV of *The civilization of the renaissance in Italy*, [1859] trans. L. Geiger, London 1904, p. 283.
2 Werner Busch (ed.), *Landschaftsmalerei. Geschichte der klassischen Bildgattungen in Quellentexten und Kommentaren*, Berlin 1997, vol. 3, p. 65.
3 Gerhard Wolf, "Nicolaus Cusanus 'liest' Leon Battista Alberti: Alter Deus und Narziss (1453)", in: Rudolf Preimesberger/ Hannah Baader/Nicola Suthor (eds.), *Porträt. Geschichte der klassischen Bildgattungen in Quellentexten und Kommentaren*, Berlin 1999, vol. 2, 201-209.
4 Jacob Burckhardt, op. cit. p. 129.
Leon Battista Alberti, *Vita*. See Latin-German parallel text ed. by Christine Tauber, Frankfurt am Main/Basel 2004.
6 Burckhardt op. cit., pp. 136, 137.
7 Ibid., p. 548.
8 Anthony Grafton, *Leon Battista Alberti. Master Builder of the Italian Renaissance,* Cambridge MA, 2002.
9 http://www.noteaccess.com/Texts/Alberti/2a.htm The text is here described as follows: Alberti, Leon Battista. On Painting. [First appeared 1435-36] Translated with Introduction and Notes by John R. Spencer. New Haven: Yale University Press. 1970 [First printed 1956]. See also in this connexion note 21, as the author's argument hinges on a point of translation.
10 Götz Pochat, *Geschichte der Ästhetik und Kunsttheorie. Von der Antike bis zum 19. Jahrhundert*, Cologne 1986, p. 231.
11 Wolfgang Braunfels, "Nimbus und Goldgrund", in: *Nimbus und Goldgrund. Wege zur Kunstgeschichte, 1949-1975*, Mittenwald 1979, p. 24.
12 Wolfgang Schöne, *Über das Licht in der Malerei*, Berlin 1954, p. 100.
13 Leonardo da Vinci, *Trattato della Pittura*, 71: "Vi ha una certa generazione di pittori, i quali per loro poco studio bisogna che vivano sotto la bellezza dell'oro e dell'azzurro."
14 Hans Sedlmayr, "Bemerkungen zu Wolfgang Schöne: Über das Licht in der Malerei", in: *Hefte des Kunsthistorischen Seminars der Universität München*, Munich 1963, no. 5, p. 44.
15 Monika Wager, *Das Material der Kunst. Eine andere Geschichte der Moderne*, Munich 2001, p. 17.
16 Iris Wenderholm, "Aura, Licht und schöner Schein. Wertungen und Umwertungen des Goldgrunds", in: *Geschichten auf Gold. Bilderzählungen in der frühen italienischen Malerei*, ed. by Stefan Weppelmann (exhib. cat. Gemäldegalerie, Staatliche Museen zu Berlin), Berlin/ Cologne 2005, p. 107.
17 Leon Battista Alberti, *Über die Malkunst*, ed. by Oskar Bätschmann/ Sandra Gianfreda. Darmstadt 2002, p. 22.
18 Ibid. p. 30.
19 For the source of this translation, see note 9. But in this connexion, see also note 21. The emphasis is the author's.
20 Translator's note: At this point the author discusses the meaning of the term 'Goldgrund' in German by reference to the Grimms' dictionary; it is clear that in the art-technical sense, it refers to a gold background. See also note 21 in this connexion.
21 Translator's note: The author's intention will be clear to the reader, namely that the 'gold ground' argument is based on a mistranslation of the Italian text into German. It will be clear that the author's point also holds for the (standard) English translation quoted here, although there is no evidence that Spencer consulted the (much earlier) German translation.
22 Hubert Janitschek (ed.), *Leon Battista Alberti: Della Pittura.* ienna 1877, (Quellenschriften für Kunstgeschichte XI), p. 138.
23 Also in popular books, such as *Faszination GOLD. Verheissung des Göttlichen—Fluch der Menschheit*, Munich 2002, this topos is maintained. "In his treatise on painting 'De pictura' ... the famous polymath Leon Battista Alberti rejected the use of metallic gold in painting on principle. In place of the gold ground, he insisted that pictorial backgrounds be painted, and metallic gold be replaced by paints in the picture." (p. 88).
24 http://www.noteaccess.com/Texts/Alberti/2.htm (see also note 9)
25 Karl Marx, *Capital, a Critique of Political Economy* [1867], Vol. 1, part 1. English translation ed. by Friedrich Engels. New York, 2007, p. 44.
26 Alexandra Hanzl, "Die Technik der Goldgrundmalerei", in: *Auf goldenem Grund. Italienische Malerei zwischen gotischer Tradition und dem Aufbruch zur Renaissance* (exhib. cat. Liechtenstein Museum Wien) Vienna 2008, p. 91.

The Use of Gold—
Symptom of a Paradigm Shift in the Material Aesthetics of the Nineteenth Century?

Alexander Klee

Among the prospering middle classes there arose a desire in the later nineteenth century for a form of self-presentation that accorded with their new-found status, a status which, internationally, went hand-in-hand with the economic upturn of the Belle Époque, and in Vienna in particular with the construction of the Ringstrasse.

The painter Hans Makart caught the taste of the times with his scandal-pictures *Moderne Amoretten* (fig. 1). His works found an echo among all social classes, giving rise to pleasant frissons and erotic fascination in equal measure. *Moderne Amoretten* shows, in three pictorial panels, not a collection of comically harmless Baroque putti, but a series of pubescent children and young fauns on the cusp of sexual awareness. The three panels, arranged like an altarpiece, use iconographic symbols, for example a rabbit, to point to sexuality and pure lust. Makart was thus adopting an imagery associated both with medieval religious art and with erotic fancies. The highlighting with a gold, rather than a blue, background also highlights the clash with the erotic subject, for the use of ennobling gold, associated more with the religious sphere, in combination with the triptych form as a religious pathos formula, only reinforces the provocation not only of the erotic scene, but of Makart's combination of religious imagery with a

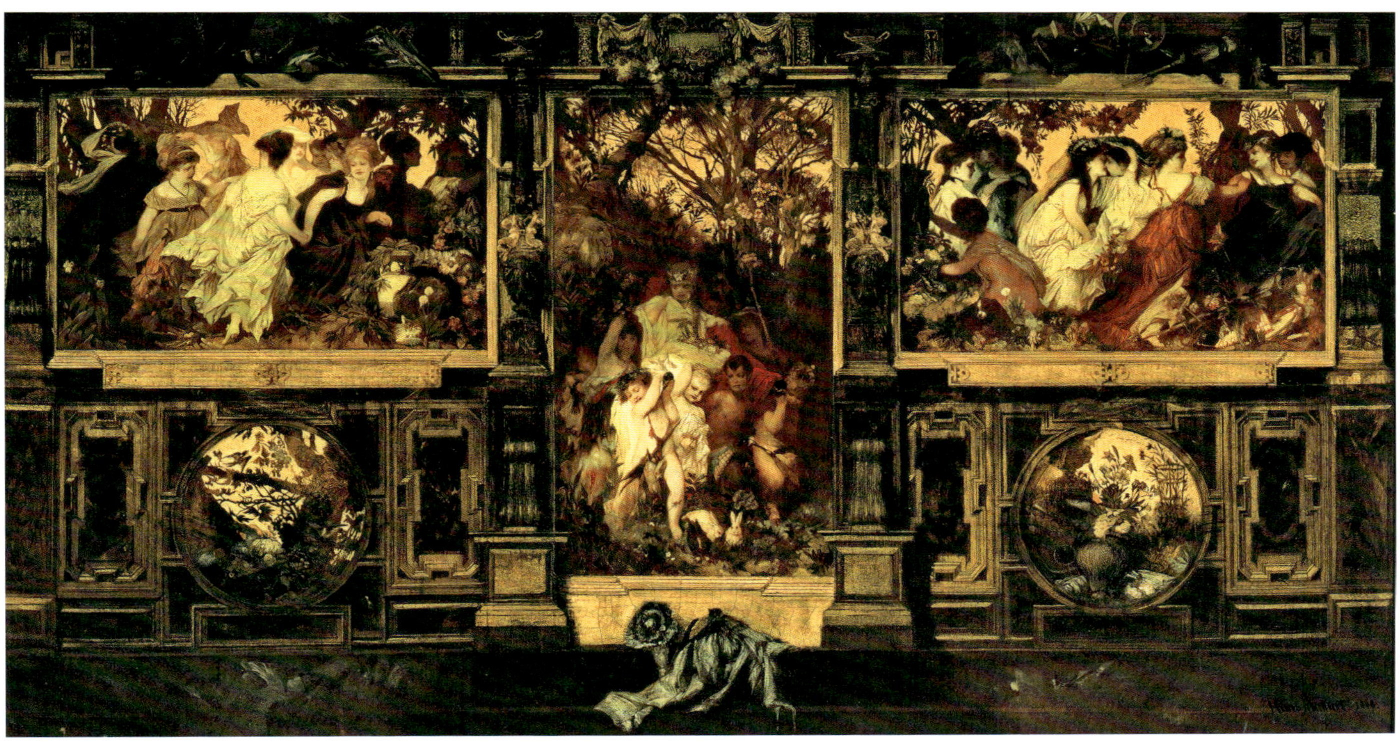

Fig. 1 Hans Makart, Mural Decoration for a Room with *Moderne Amoretten*, 1868, Belvedere, Vienna

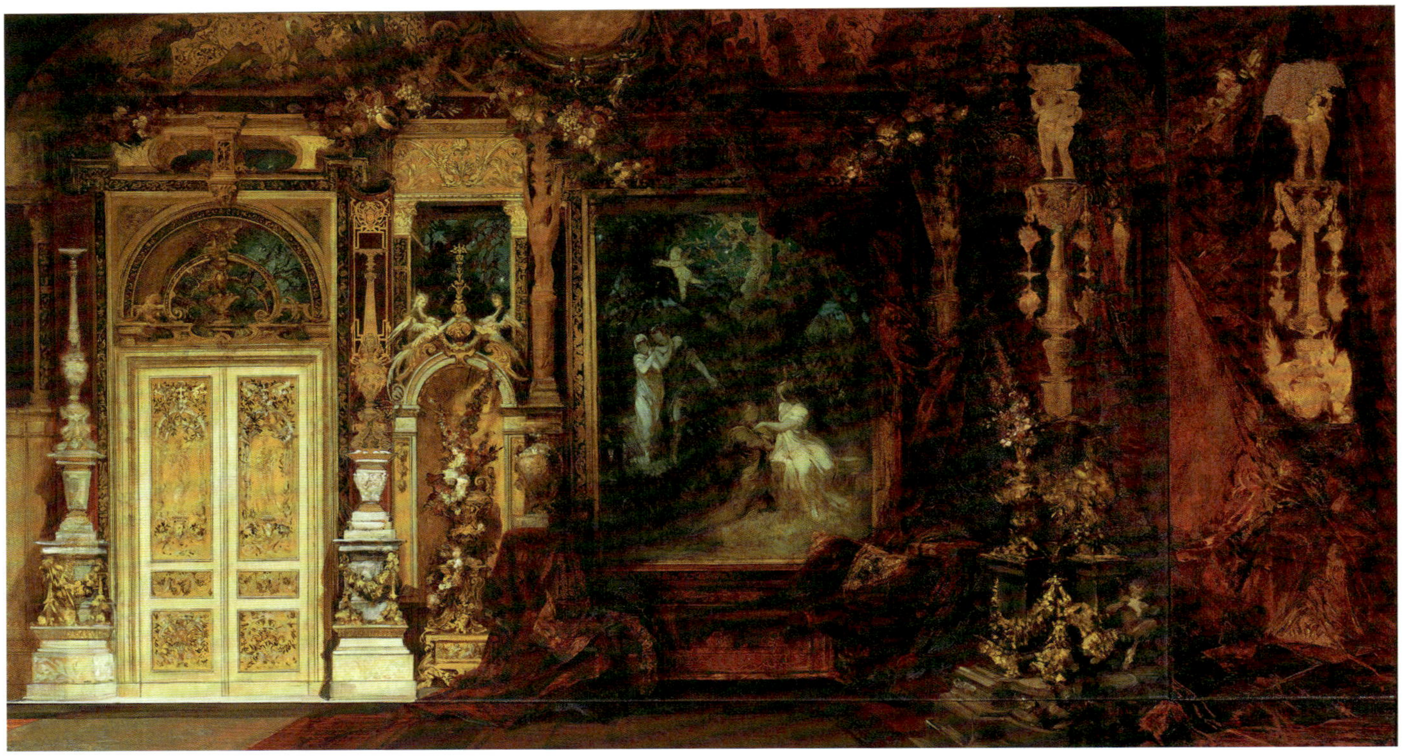

Fig. 2 Hans Makart, *Midsummer Night's Dream*, 1882, Belvedere, Vienna

secular pictorial context. Gold is here being used as an element of a pictorial language, while its material qualities play a marginal role. This is also made clear in the six pictures of the furnishings of Nikolaus Dumba's study, with their nonchalant alternation between backgrounds with and without gold, and between illusionistic interiors and blue-sky landscapes.

Makart used gold paint here in an 'idealistic' sense, in which gold as a "material is only a medium for illustrating an idea, and that for all its great costliness [...]is subordinate to the artistic form. [...] Correspondingly, painted gold is more valuable than real gold, and artificial flowers are more perfect than natural ones..."[1]

Makart used the material aesthetic of gold to realize his pictorial concept. He had no interest whatever in the craft qualities of gold or in the art of gilding. He worked with the materiality of the paint in order to evoke effects and bring spontaneity and movement into his pictures. With the interior designs, whether for Nikolaus Dumba's study or Empress Elisabeth's bedroom in the Hermesvilla (fig. 2), his model was the courtly design which in Vienna was the exemplar and the standard for the up-and-coming bourgeoisie, but at the same time attracted the admiration of young artists such as Gustav Klimt. The latter, after Hans Makart's death in 1884, together with the group known as the 'Maler-Compagnie' ('painters' company') to which his brother Ernst Klimt and Franz Matsch belonged, took over the Hermesvilla commission, for example. The admiration that Makart aroused even long after his death is also reflected in the critiques of Ludwig Hevesi, who described him as a forerunner of the Secession.[2]

While in Klimt's early works Makart's role-model function is still visible, this changed fundamentally in the *Beethoven Frieze* (figs. 3, 4). The *horror vacui* of Makart's interior decor had given way to an understanding of material that allowed for totally untreated wall surfaces. Gold applications were no less possible here than other materials, for example pieces of glass or brass buttons.[3] The 'artistic means' take on a new status in the *Beethoven Frieze*. The materials, even mere buttons, remain recognizable to an observant beholder. Even less conspicuous, but all the more important, is the continued visibility of the wall plaster, something which reveals a new relationship to the material.[4] Doing justice to the material, something which the pupils of Otto Wagner, above all Josef Hoffmann, brought to the Secession, is very evident here. It took the contrast between showiness and austere sobriety (compared with Makart's interiors) to make the upheaval obvious. At the same time, use of the fresco-secco technique reinforced the revival of the medieval or early-Christian arts-and-crafts traditions, without actually copying them. This revival of craft techniques can be seen not only in Klimt's *Beethoven Frieze*, but also in the coffer fields previously beneath the frieze. They had figural wood reliefs by Ferdinand Andri, mosaics of cut and glazed tiles to designs by Baron Felician von Myrbach and Richard Luksch, and embossed copper plates based on motifs by Friedrich König and Maximilian Lenz. Mosaics and cut tiles were used in a design by Kolomann Moser, while Emil Orlik experimented with the use of gold paint. But new techniques were used too, as in the relief by Leopold Stolba, which was enriched by coloured cement with metal, mother-of-pearl and ground-glass inlays. In most of the decorative fields, gold was used

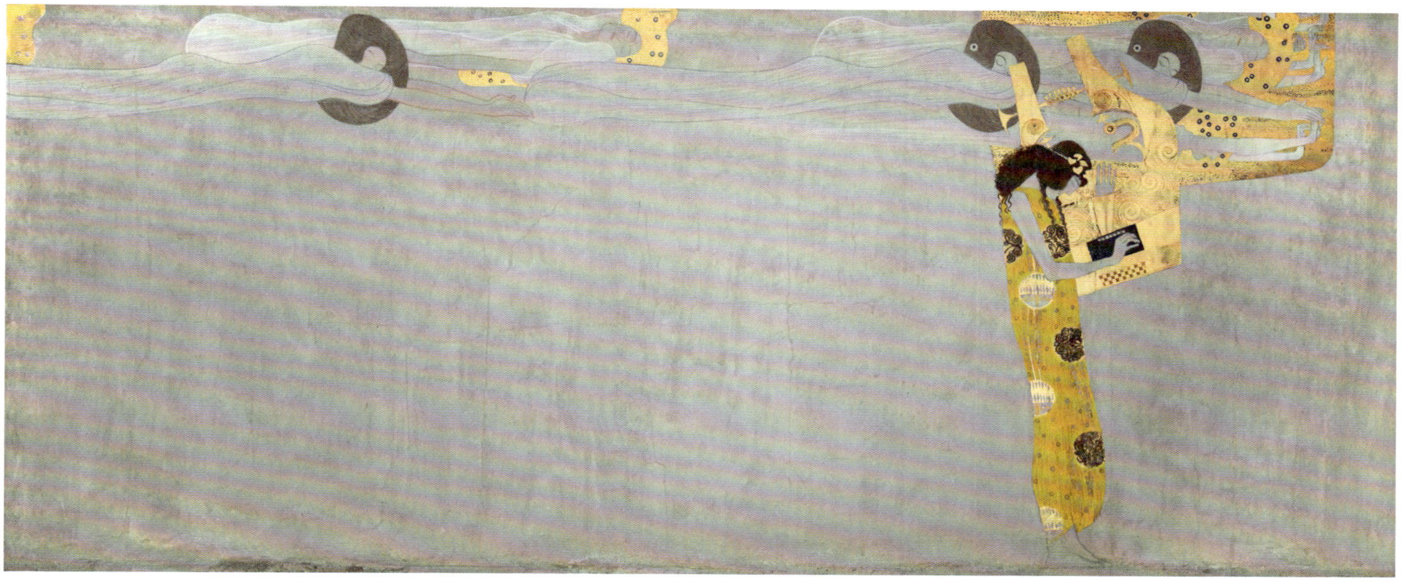

Fig. 3 Gustav Klimt, *Beethoven Frieze*, left-hand long side, 1902, Belvedere, Vienna

as an 'ennobling' material; thus areas of the embossed metalwork were additionally highlighted in gold.[5]

The coffer fields lend the rooms a quasi-sacred dignity, but they were after all conceived as devotional rooms in which to pay homage to Beethoven. The dignity of the artistic means contributed to the 'consecration' of the Beethoven exhibition, which thus became the laboratory for the testing of new arts-and-crafts approaches which the participating artists professed.[6] Materials were mixed, collage-like, and new techniques such as dyed concrete were tried out. In the same way, old techniques such as mosaic were used, demonstrating

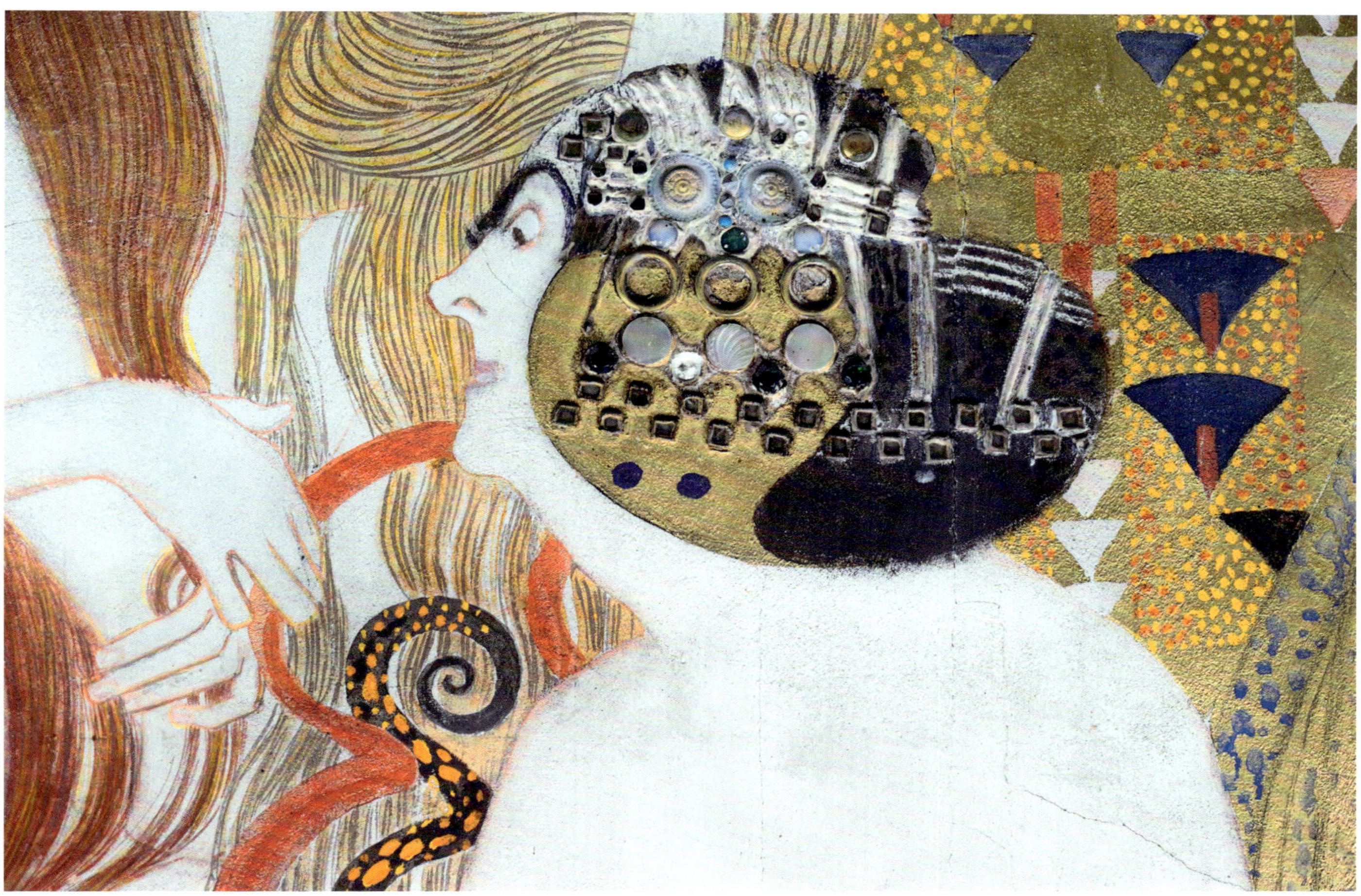

Fig. 4 Gustav Klimt, *Beethoven Frieze: 'Excess'* (detail), 1902, Belvedere, Vienna

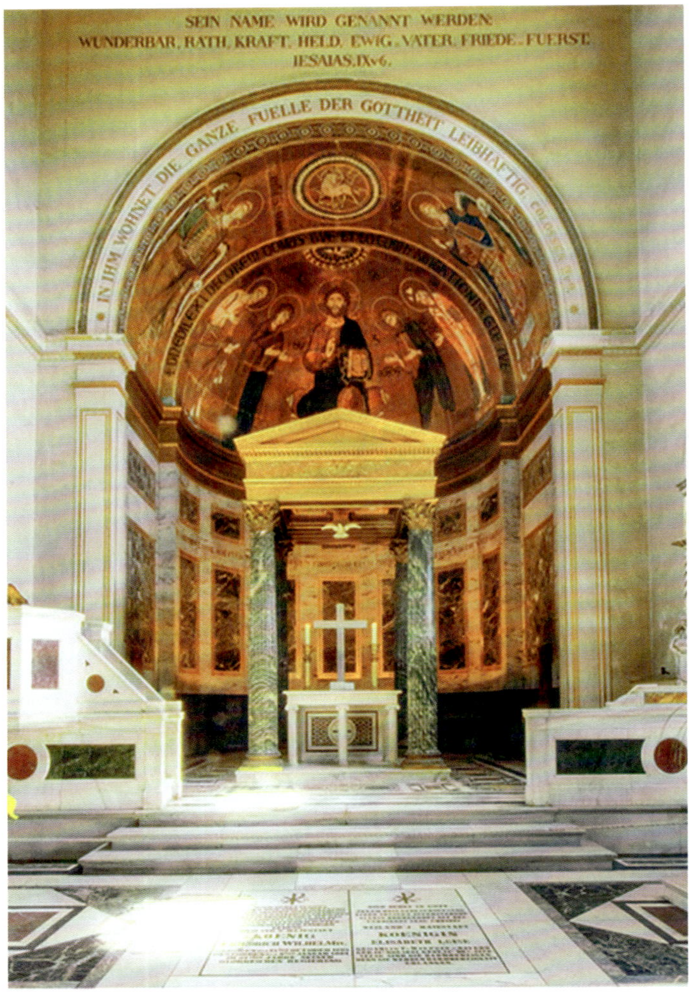

Fig. 5 Church of Peace, Potsdam, with apse mosaic from S. Cipriano on Murano (1st. third of 13th cent.), 1845-48

the new understanding of material. In the construction of the Palais Stoclet, these approaches were brought together into a total concept under the direction of Josef Hoffmann.[7] The new understanding of material was matched by the white sobriety of Josef Hoffmann's famous geometric overdoor in the Beethoven exhibition, which was described in the Secession catalogue as a 'Mörtelschnitt' ('plaster relief') and thus processed and exhibited in a purist or material-appropriate manner.[8]

While Makart still saw himself, as a painter, to be a universal artist at the pinnacle of the visual arts, who, as already in the Renaissance, implemented his aesthetic ideas in architectural fantasies, in the Secession the impetus towards a new understanding of material and a new conception of space came, conversely, from the architects. Otto Wagner, the teacher of the Secessionists Josef Maria Olbrich and Josef Hoffmann, put his stamp on the emerging new attitude to material and form when he said: "Our feeling must tell us today already that the load-bearing line, the slab-like plane, the greatest simplicity of conception and an energetic accentuation of construction and material will dominate the new art form in the future; modern technology and the means available to us will demand this."[9] These changes thus contributed (in the end crucially for the new, revolutionary concept of 'artistic means') to the use of gold with its specific material quality precisely as an artistic means within art.

At the same time, the quasi-sacred character aimed at by the 14th exhibition of the Viennese Secession in general and of the *Beethoven Frieze* in particular points towards Romanticism with its dark, threatening aspect, which comes across in the stories of E. T. A. Hoffmann or in the pictures of Henry Fuseli. Thus in the *Beethoven Frieze* the sinister and unconscious side of the human essence is represented through the menacing-looking giant with his daughters, the Gorgons, and by figures which relate to sickness, madness, death, lust, excess, unchastity and nagging worry. Hovering above humanity are humanity's personified yearnings and desires. In the frieze, the fears and threats are ultimately resolved in people's love for one another. The Secessionists shared with the Romantics the desire to link the visual arts with music; thus Philipp Otto Runge sought to compose his main work, the series of *Times of Day*, as a symphony.[10] Part of this series is a draft for *Day*[11], in which he wanted to use gold leaf to highlight the radiance of the colours, which corresponds to a process used above all in the production of icons. Although in his sketch for *Day* Runge primarily used the gold ground for its particular material characteristics, [12] in other words simply to reinforce the luminosity of the colours which it underlay, what comes out most strongly is the mystical character of the material. The motive doubtless lay in Runge's interest in the work of the mystic and philosopher Jakob Böhme entitled *Aurora oder Morgenröte im Aufgang*, dating from 1612, which also inspired the aforementioned series *Times of Day*. Runge's reference to Böhme's writings suggests such an interpretation, which corresponds to comparable experiments with gold leaf by William Blake, who similarly strove for a mystical vitalization of art.

The Romantic understanding, such as we see in the work of Runge, of a life-cycle characterized by mystical-Christian ideas is not reflected in the Viennese Secession. The Secession used elements with mythological meanings, like the Gorgons or Typhoeus in the *Beethoven Frieze*, as symbols of general human fears and desires. The Secessionists used gold freely as a design instrument, without any iconographic preconditions, by combining it on equal terms with paint and materials.

The material applications in the *Beethoven Frieze*, like the mosaics in the decorative fields below it, suggest a reference to Byzantine art. Interest in mosaics, though, had begun far earlier, with the Byzantinism of the first decade of the nineteenth century, and can be traced without a break right up to the time of the Secession. The enthusiasm shown by King Ludwig II of Bavaria for the Byzantine style and in particular for the Byzantine mosaics in Ravenna and in St Mark's basilica in Venice furthered this interest, which soon turned into an international fashion.[13] This went so far that King Frederick William IV of Prussia had a thirteenth-century mosaic from the Benedictine church of San Cipriano on Murano,

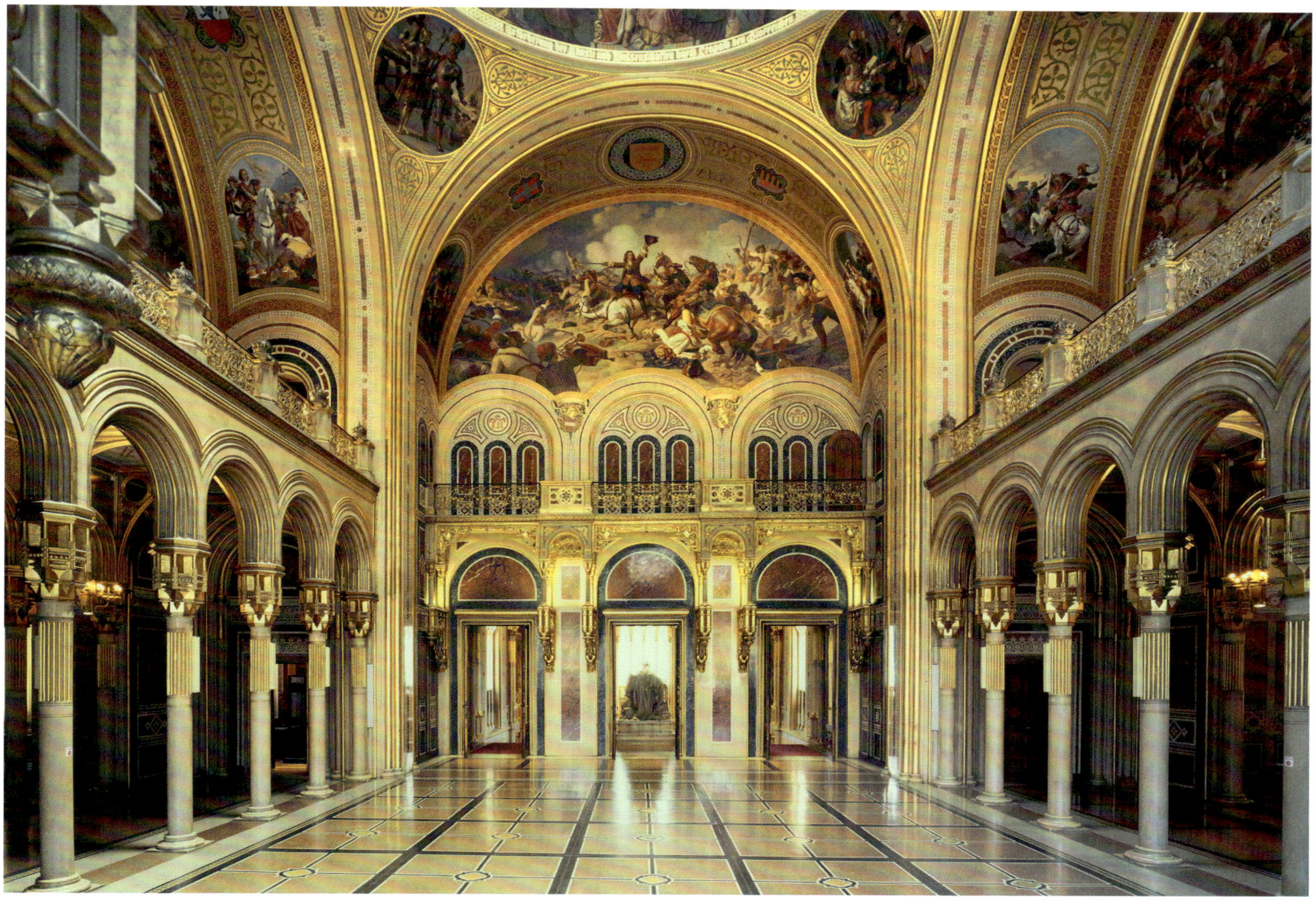

Fig. 6 Ludwig Foerster and Theophil Hansen, *Hall of Fame in the Heeresgeschichtliches Museum*, 1850-58

which he had acquired in 1834 while he was crown prince, incorporated into the Church of Peace in Potsdam, built between 1843 and 1848 (fig. 5).[14] Mosaic art received a further boost to its popularity as a result of the invention of new techniques by the Venetian glassmakers Antonio Salviati and Lorenzo Radi in the late 1850s. [15] These allowed quicker and cheaper production and ensured the transportability of the mosaics.[16] In Vienna, Byzantinism was introduced by the architects Christian Ludwig von Förster, Theophil von Hansen, the architect of the Byzantine-style Arsenal (fig. 6) and by Hansen's protégé Eduard van der Nüll, interior architect of the Altlerchenfeld Church (fig. 7, who also worked on the Arsenal), the teacher of Otto Wagner,[17] who in his turn was the teacher and role model for the guiding light of the 14th Secession exhibition, Josef Hoffmann.

The quasi-sacred lustre of gold, the apparent permanence of the mosaic, and presumably also the claim by the imperial regime to represent the heritage of Byzantium as the historic bastion of Christianity, all eased the reception of Byzantine art in the Habsburg dominions.

The Secession attached special importance in this connexion to craft skills, from the revival of which they expected new artistic approaches. For this reason, the technique patented in 1840 by Richard Prosser for the more rapid production of mosaic tesserae, which required that the latter be of uniform size, was rejected.[18] With the combination of painting and material art in the *Beethoven Frieze*, the Secession struck out along a new path. The harsh contrast between painted areas and materials, especially between the gold and the empty white spaces, reflects a new understanding of material, which can also be seen in the context of the confrontation with Japanese art,[19] as is also reflected in the gold lacquer works of Emil Orlik. While the centrepiece of the exhibition, Klinger's Beethoven monument, evinces a combination of a whole variety of materials, this very combination still follows an iconological concept.[20] The intensive use of gold supports Otto Wagner's demand for a 'slab-like structure of the plane', that is to say, the renunciation of illusionist three-dimensionality in the composition of the wall. The light seems to gather in the bright sheen of the gilt surfaces, which stand out in front of the matte, light-absorbing white wall covering. This reduction, and the rigorous use of the artistic means, come across as a sober analysis of their characteristics and potential.

Such an analysis was pursued from a scientific perspective by Ernst Wilhelm von Brücke in his work *Die Physiologie der Farben für die Zwecke der Kunstgewerbe* ('The Physiology of Colours for the Purposes of Practitioners of Art'). This is

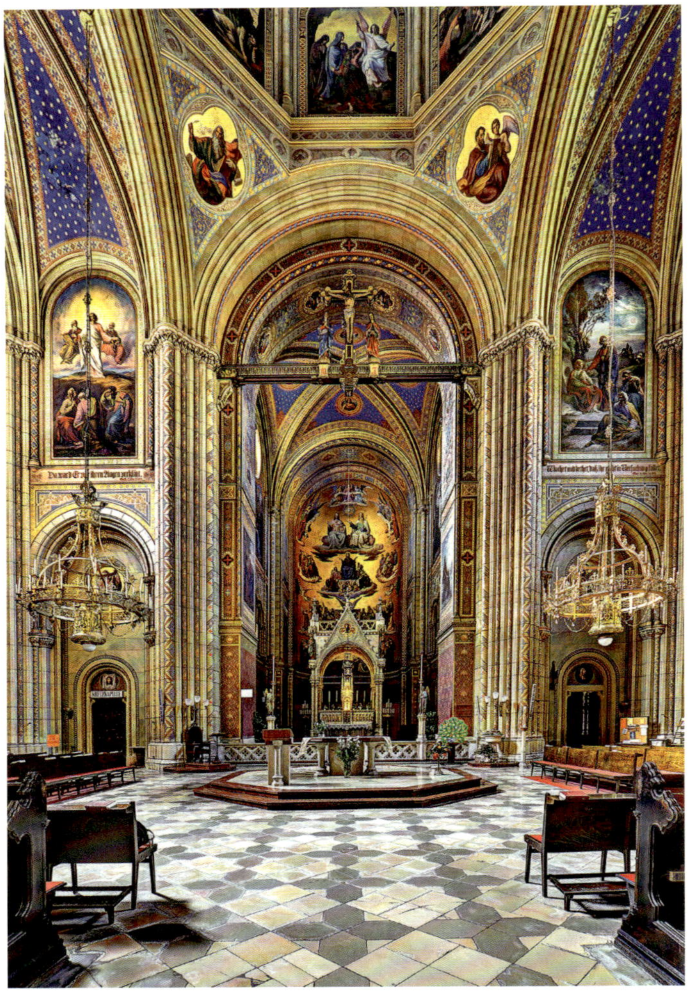

Fig. 7 Johann Georg Müller, Eduard van der Nüll et al., *Altlerchenfeld Church*, 1848-61

interesting in our context to the extent that Brücke was a member of the board of trustees when the Royal and Imperial Austrian Museum of Art and Industry was founded in Vienna in 1864, and was writing his handbook (which appeared in 1866) during precisely this period, on the instigation of the museum's directors Rudolf von Eitelberger and Jacob von Falke. He devotes not only one whole chapter to the particular properties of the colours of metals, but also concerns himself in some detail in another chapter to the colour effects of gold.[21] Long before art-historians confronted it, he analysed the effect of gold, distinguishing between gold leaf and gold paint, which cannot create the same effect. The reason lies in the lustre of gold leaf, which unlike gold paint, needs no varnish coating for depth and luminosity. The effect of pigments is vitiated by surface reflections. Gold, though, especially gold leaf, develops its lustre only when light reflects from it, and thereby increases its colour effect and brightness. The lustre of gold likewise allows the free use of colours, "it reflects the colours of its surroundings, modified by its own." Alongside further favourable characteristics of gold, Brücke also describes the movement that causes the confusion produced by the reflection in the eye of the beholder, and which gives gold its shimmering effect. Brücke meticulously analyses the materiality of gold, the cause of its particular suitability for picture frames, and the appropriate or inappropriate effect of matt or shiny gold in sculpture. As a scientist, Brücke was not interested in the significance of gold as a colour reserved to the religious or auratic sphere, or with its associations of splendour and wealth. For him, gold was a colour with particular physical properties.

The change in the perception and significance of the material, in other words of the 'artistic means', reveals a paradigm shift, which Günter Bandmann describes as a change away from the idealist aesthetic which sought to distance itself from nature by denying the material, [22] and towards an aesthetic of the material, whether natural, such as wood, stone or gold, or man-made, such as steel, concrete or plastic.

1 Günter Bandmann, "Der Wandel der Materialbewertung in der Kunsttheorie des 19. Jahrhunderts", in: *Beiträge zur Theorie der Künste im 19. Jahrhundert*, ed. by Helmut Koopmann and Josef Adolf Schmoll gen. Eisenwerth, Frankfurt am Main 1971, vol. 1, p. 132; Alfred Lichtwark, *Makartbouquet und Blumenstrauß*, Hamburg 1894, pp. 33f.

2 Ludwig Hevesi, "Hans Makart und die Sezession", in: *Acht Jahre Sezession (März 1897-Juni 1905), Kritik—Polemik—Chronik*, Vienna 1906, 265-269; Gerbert Frodl, "Begegnung im Theater, Hans Makart und Gustav Klimt", in: *Mitteilungen der österreichischen Galerie*, 22/23, 1978/79, H. 66/67, 9-36; Agnes Husslein-Arco/Alfred Weidinger, *Gustav Klimt und die Künstler-Compagnie* (exhib. cat. Belvedere, Vienna), Weitra 2007.

3 Alfred Weidinger (ed.), *Gustav Klimt, Werkverzeichnis*, Munich 2007, p. 271, no. 153.

4 Different qualities of plaster are put to different uses in the frieze, depending on the intention of the painting: coarser or finer, smooth or open-pored. Cf. Ivo Hammer, "110 Jahre Beethovenfries von Gustav Klimt", in: Agnes Husslein-Arco/ Alfred Weidinger, *Gustav Klimt—Josef Hoffmann—Pioniere der Moderne* (exhib. cat. Belvedere, Vienna), Munich 2011, pp. 141f.

5 Illustrations and the exact technical descriptions can be found in the Gustav Klimt / Josef Hoffmann catalogue, Husslein-Arco/ Weidinger 2011 (as note 4), pp. 74 ff.

6 "A uniform room should first be created, and painting and sculpture then ornament it in the service of the idea of the room. What is to be done here, in the given proportions, within strict limits, is to subordinate the parts to the effect of the whole. The inexorable logic forces immersion in the character of the room and adherence to a guiding idea. All this is required in tasks of monumental art, and the highest and best that people have been able to offer in all ages developed in temple art. [...] We wanted to experience the blessings of a work which had a purpose and an aim. We wanted to learn. This was now to be realized in an exhibition within the narrow confines of a building in which nothing permanent can be created, as an exhibition must necessarily swallow up the work of the previous one." Ernst Stöhr, "Unsere XIV. Ausstellung", in: *Katalog der 14. Ausstellung Secession*, Vienna 1902, 9-12 (here in translation).

7 "The limited nature of the available means and the self-evident duty to use genuine material, to energetically avoid lies and mere appearance, likewise demanded adherence to the greatest simplicity in material and formal vocabulary." Josef Hoffman, in: *Katalog der 14. Ausstellung Secession*, Vienna 1902 (here translated).

8 Hammer 2011 (as note 4), p. 141.

9 Otto Wagner, *Die Baukunst unserer Zeit*, ed. by Eva Winkler, Vienna 2008, p. 185.

10 Klaus Lankheit, "Die Frühromantik und die Grundlagen der "gegenstandslosen" Malerei", in: *Neue Heidelberger Jahrbücher*, Neue Folge 1951, p. 68.

11 Jörg Traeger, *Philipp Otto Runge oder Die Geburt einer neuen Kunst*, Munich 1977, p. 83.

12 Ibid. p. 82

13 Dorothea Müller, *Bunte Würfel der Macht—Ein Überblick über die Geschichte und Bedeutung des Mosaiks in Deutschland zur Zeit des Historismus,* Frankfurt am Main 1995, pp. 28ff.

14 Ibid., p. 63ff.

15 Peter Springer, "Modernisierung einer alten Kunst", in: *Wände aus farbigem Glas, Das Archiv der Firma Puhl & Wagner—Gottfried Heinersdorff—Mosaik und Glasmalerei*, ed. by Helmut Geistert/ Elisabeth Moortgat (exhib. cat. Martin-Gropius-Bau, Berlin) Berlin 1989, p. 96.

16 J. Barrie Bullen, *Byzantium Rediscovered*, London 2003, p. 142f.

17 Ibid., pp. 46 ff.

18 Springer 1989 (see note 15), p. 99.

19 The Secession exhibited Japanese art shortly afterwards, in its 16th exhibition.

20 Bandmann 1971 (see note 1), vol. 1, p. 141.

21 Ernst Wilhelm von Brücke, *Die Physiologie der Farben für die Zwecke der Kunstgewerbe*, Leipzig 1866, pp. 105ff §12 ("Die Farben der regulinischen Metalle") and p. 222 § 24 ("Wesshalb man das Gold freier anwenden kann, als die gelben Pigmente").

22 Bandmann 1971 (as note 1), vol. 1, p. 157.

Gold Leaf and Golden Leaves: on an Unimplemented Idea by Frank Lloyd Wright for *Fallingwater*

Joachim Driller

"Gold and silver, lead and copper, tawny iron ore;
all [...] flow obedient to the hands of the architect;
all become pawns to human will in the plan of the human mind. And jewels, happy discoveries. [...] All this I see as the architect's garden, his palette. [...] Materials!
What a resource."[1]

Frank Lloyd Wright, 1928

Fallingwater, Frank Lloyd Wright's famous residence over a waterfall in Pennsylvania (figs. 1, 2), is considered to be the architect's most magnificent example of his ability to integrate a building into the natural environment.[2] However, the building, completed in 1937, would probably have sat even more subtly in its surroundings and their changing seasonal moods if Wright's original idea for the façade's design had been implemented. It was rejected by the clients, and not just for financial reasons.

The son of the couple who commissioned the building, Edgar Kaufmann jr., recalled in his memoirs on the creation of *Fallingwater* that Wright suggested the building's interior and exterior concrete surfaces to be covered in gold leaf, especially the balustrades of the boldly projecting balconies and the 'floating' roofs. The client felt this idea was unsuitable for a country house, which, in their minds, should have an

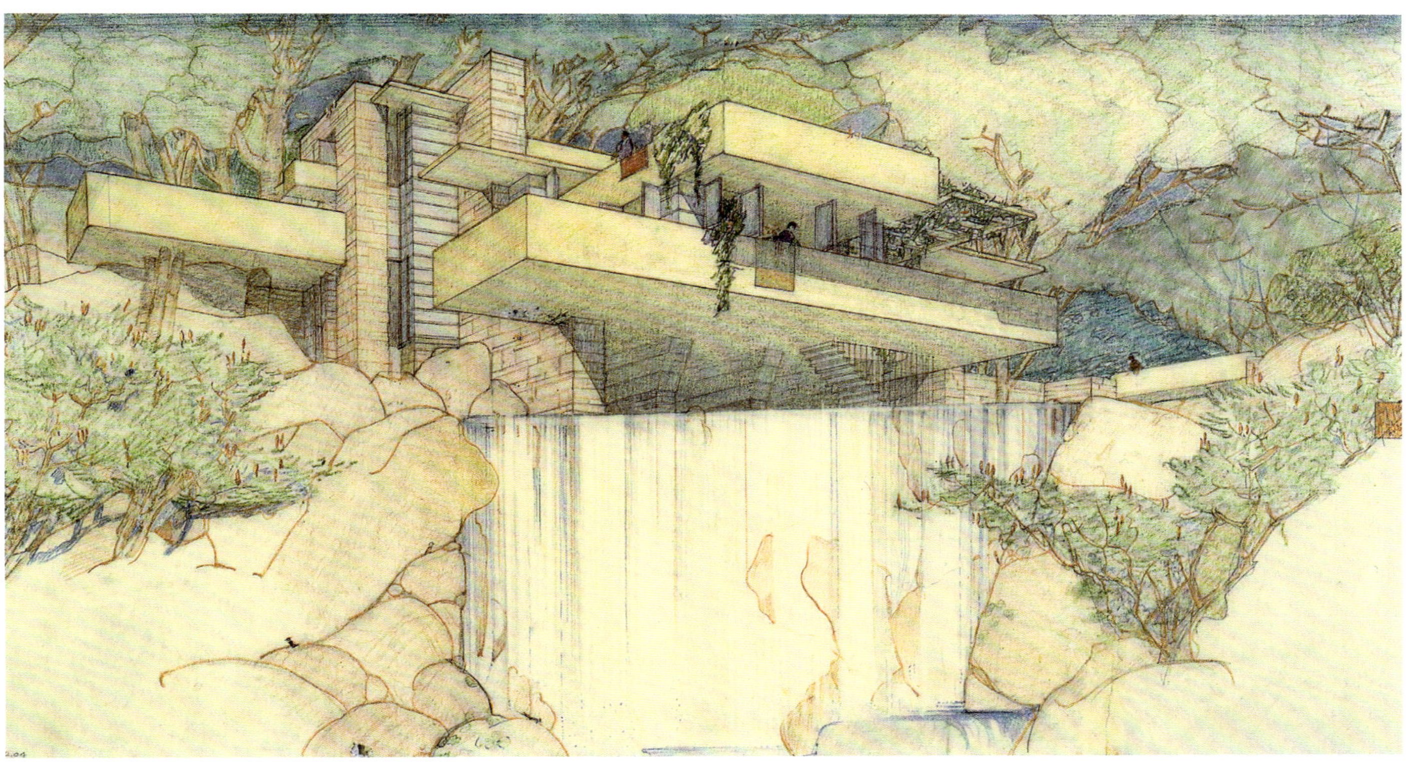

Fig. 1 *Colour sketch of Fallingwater House*, The Frank Lloyd Wright Foundation, Scottsdale (AZ)

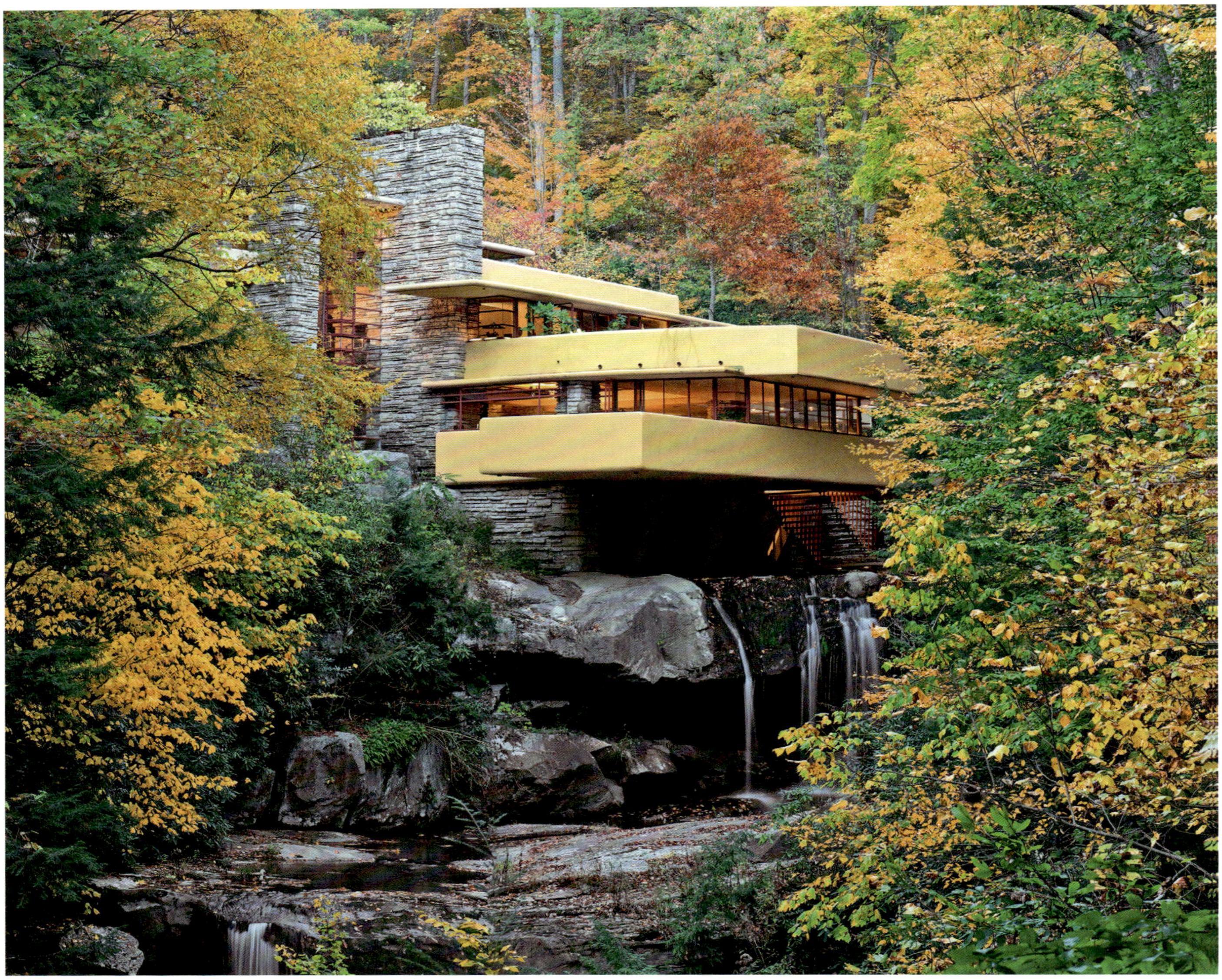

Fig. 2 Chris Little, *Fallingwater House*

unobtrusive design and provide a setting for an 'easy-going country life'.[3] According to Kaufmann the idea of using gold leaf had its origin in the architect's 'Japanese experience', a claim confirmed by statements by Wright and one of his assistants. This idea will be picked up again later; however, first of all a few other possible sources of inspiration which Wright could have drawn from will be briefly introduced.

John Ruskin and Louis Sullivan

In 1849 John Ruskin advocated gilding in architecture in his chapter 'Lamp of Truth', no less, in his book *The Seven Lamps of Architecture*. Probably with Venice's medieval buildings in mind, which he revered so much and some of which featured lavish gold decoration, the English architectural critic delivered a surprising reason why gilding did not constitute a break with the constructive honesty and eschewal of imitation materials in architecture which he usually demanded with great vehemence. There were, he claimed, exceptions which had become legitimate, through frequent use for example. These exceptions included the gilding of surfaces, as this could not lead to any deception in architecture, since gilding could never be misunderstood as (solid) gold. The situation was thus completely reversed for jewellery, where gilding had to be considered deceptive for this very reason.[4]

Ruskin's writings achieved widespread attention in the United States in the late nineteenth century. Louis Sullivan, in whose office, run together with Dankmar Adler, Wright had been employed for several years at the start of his career, also seems to have confronted them intensively and critically.[5] Maybe this was one of the reasons why Sullivan ennobled his typically rich ornamental apparatus with gold paint and gilding in some of his most famous buildings, in at least one case even the exterior façades. The temporary *Transportation Building* (fig. 3) designed by Sullivan for the World Fair in Chicago in 1893, one of the office's last projects that Wright was also involved in, featured as an entrance a huge round arch adorned with gold leaf.[6]

Thus Ruskin and Sullivan alone provide two possible inspirational sources for Wright's extravagant idea of gilding *Fallingwater*. Sullivan, whom Wright referred to as 'Lieber Meister' (dear master) during his lifetime, may be even more significant in this regard. Sullivan had, after all, expressly pointed out in his famous 1892 essay, *Ornament in Architecture*, that an 'organic' system of decoration, in which the structure and the applied ornament were perceived as an inseparable unit, would naturally have to comply with the individual 'voice' of a building and its 'character'.[7] Consequently large structures with a powerful presence required a different decorative treatment than more delicate structures.

Wright attempted to implement this demand in a very idiosyncratic and original manner for *Fallingwater*: the idea of using gold leaf demonstrated the conclusion he had come to long before, namely that the 'organic' decoration of large structures and surfaces, as demanded by Sullivan, could not be obtained by piecemeal ornamentation, but rather by using contrasting material, form and colour effects—in the case at hand by the interplay of horizontal, seemingly floating, projecting 'colour bands' and the powerful vertical slab walls of quarry stone. This interplay and his dialogue with the surrounding 'untamed' nature, which already exceeds any ornament invention through its diversity of forms, generate the 'emotional expression' whose significance for architecture Sullivan had particularly highlighted in his 1892 essay.

According to Sullivan, such an expression was inherent in every building that could be considered a true work of art—an 'emotional expression' that, in the case of *Fallingwater*, was not just based on Sullivan's theories, but also on Wright's 'Japanese experience' as mentioned by Edgar Kaufmann jr.

Wright had been shaped significantly by the impressions of his long sojourns in Japan between 1913 and 1922, during which he had worked on designing and building the *Imperial Hotel* in Tokyo. It is thus to be suspected that the architect was not inspired to incorporate gilding into his design for *Fallingwater* from the 'quiet gold of Japanese folding screens' alone, to which he himself referred in this context.[8] Further references to Japanese art and architecture possibly present in the idea of covering surfaces in gold leaf are certainly conceivable.

Japan

The gilding planned for Fallingwater can also be interpreted as a reference to the famous fourteenth-century Golden Pavilion (Kinkaku-ji) (fig. 4) in Kyoto.[9] Keeping Wright's quasi-religious wor-ship of nature into account, it seems quite likely that the architect intended to 'consecrate' his country house, situated as it was in a striking, isolated location in the forest, by creating the relevant association with a gold-shimmering (Japanese) temple.

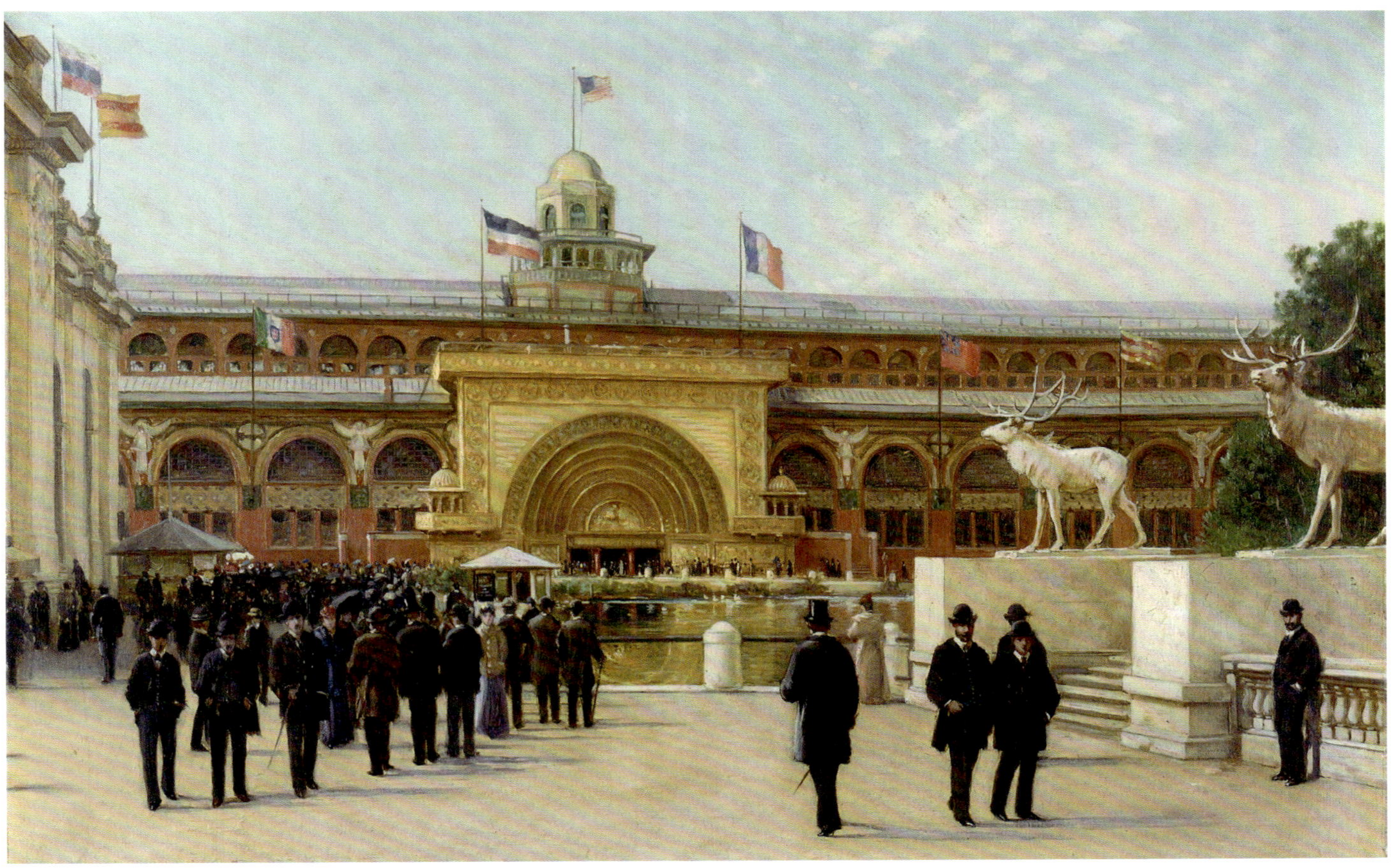

Fig. 3 Frank Russel Green, *The Golden Doorway of the Transportation Building, World's Columbian Exposition of 1893*, Chicago History Museum

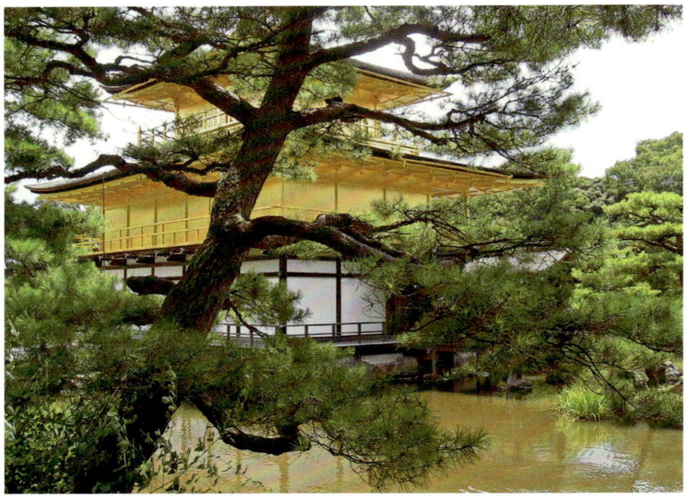

Fig. 4 Kinkaku-ji (Golden Temple), Kyoto

In addition various interpreters have already highlighted Wright's great passion for Japanese coloured woodblock prints, of which he had compiled an extensive and valuable personal collection. The prints by Hokusai (fig. 5) were cited in particular and especially so a view of the Ono Falls, which probably dated back to the 1830s and was in Wright's possession.[10]

It depicts the majestic waterfall, which seems to be frozen into a blue-white column of ice, a powerful vertical pictorial element, around which, emerging diffusely and vaguely from clouds and mist, plant-covered rock formations are grouped. Near the waterfall stands a house on a rocky outcrop, in front of it a bridge spans a stream. The similarity between the motif and the topographical and architectural situation of *Fallingwater* is striking. The possibility that Wright got the idea for his well-known presentation drawing for the house from this or other prints by Hokusai cannot be excluded. In this drawing he depicted the rugged cascades of Bear Run as a closed 'wall of water', completely in line with the Japanese woodblock prints.

A comparison between Hokusai's depiction of the Ono Falls with Wright's presentation drawing also already exists in the literature.[11] By contrast, a possible connexion between Wright's interest in Japanese woodblock prints and the planned gilding of *Fallingwater* does not seem to have been recognized hitherto, although this ought to be fairly obvious. By covering the balcony balustrades and roofs with gold leaf, the technoid character of the overhangs, which is created by the pale paint on the concrete surfaces on the house as actually built, could have been avoided; this technoid character is not sufficiently concealed by the cream-coloured paint and it led to *Fallingwater* being called the 'great machine in the forest'.[12] The gilding would have softened the acute profile of the overhangs relative to the surrounding environment; in addition the depth-perception of the building as a whole would doubtless have been made 'fuzzy', not least as a result of the light reflections produced by large areas covered in gold leaf. Depending on the angle and direction of the sunlight, the protruding balconies would have seemed like de-materialized, floating light-and-mirror surfaces. This 'fuzziness' would have compositionally 'blended' the stereometrically complex architecture with the environment to a much greater extent than the actual building does, quite in line with the Japanese prints admired by Wright and their geometrically subtly balanced, atmospheric depictions of nature, whose (landscape) details seem to float in the picture while at the same time lacking spatial depth.

These golden days …

Such an optical 'blending' of *Fallingwater* with it surroundings would undoubtedly have been at its greatest during the annual, spectacular natural event that is the 'Indian summer'. Wright is said to have compared the planned gilding, which he believed would acquire a patina over time as a result of the damp nature of the location, with the colours of the (autumn) forest.[13] Quite in line with the 'organic' unity of architecture and nature that he constantly evoked, the building's gilded overhangs would thus have become one with the 'golden' foliage of the trees into which they protrude. The mysticism of nature and the romantic force of the removal

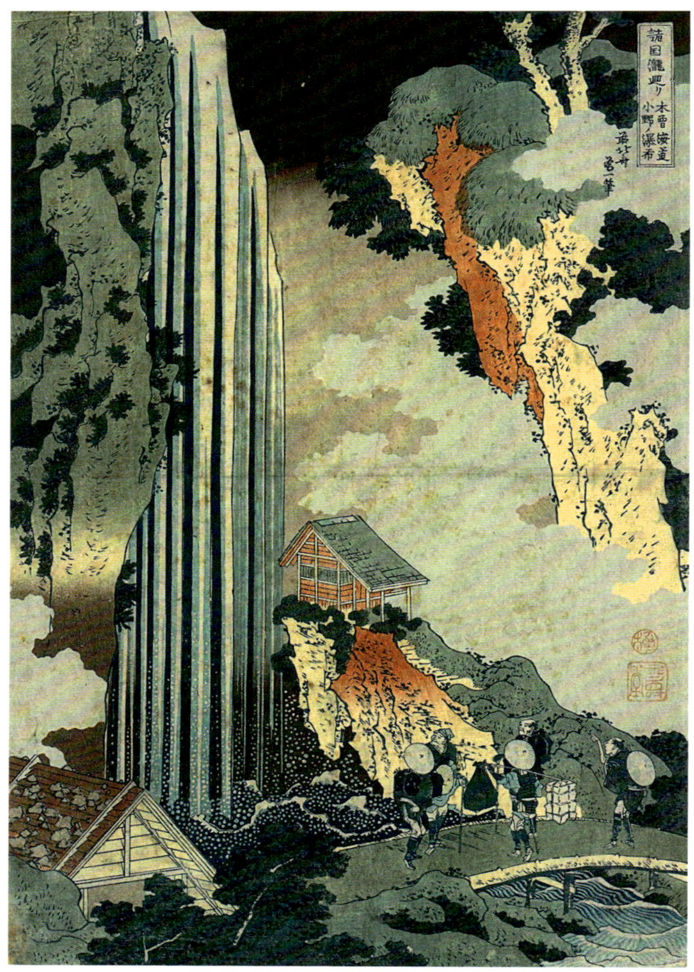

Fig. 5 Hokusai, *Ono Waterfall on the Kiso Road,*
MAK—Museum für angewandte Kunst/Gegenwartskunst, Vienna

of clear boundaries thus manifested recall the authors of American transcendentalism, with whose ideas Wright too was familiar. More than 70 years before *Fallingwater* was built, Henry David Thoreau wrote:

> "Far in the woods, these golden days,
> Some leaf obeys its Maker's call;
> And through their hollow aisles it plays
> With delicate touch the prelude of the Fall."[14]

1 Frank Lloyd Wright, 'The Nature of Materials', in: *The Architectural Record*, October 1928. Quoted from: *Frank Lloyd Wright: Writings and Buildings*. Selected by Edgar Kaufmann and Ben Raeburn, New York 1974, 222–224, here p. 223f.

2 The literature on Wright in general and *Fallingwater* in particular has reached vast proportions. The most recent publication was Lynda Waggoner (ed.), *Fallingwater*, New York 2011. The present brief article mainly refers to the following three publications: Donald Hoffmann, *Frank Lloyd Wright's Fallingwater. The House and Its History*. New York 1978—Edgar Kaufmann, jr., *Fallingwater. A Frank Lloyd Wright Country House*, London 1986—Franklin Toker, *Fallingwater Rising. Frank Lloyd Wright, E. J. Kaufmann and America's Most Extraordinary House*, New York 2003. Unavailable for this article was, among other publications, Lynda S. Waggoner, *Fallingwater. Frank Lloyd Wright's romance with nature*, New York 1996. The only interpretation of the gilding of *Fallingwater* known to the present author can be found in Neil Levine, *The Architecture of Frank Lloyd Wright*, Princeton 1996, p. 237f and 250f. Levine refers, among other things, to the analogy between the gilding and the colour of 'dying leaves' and (in turn) portrays *Fallingwater* as a metaphor for the eternal creation and decay in nature.

3 Cf. Kaufmann 1986 (as note 2), p. 104.

4 Cf. John Ruskin, *The Seven Lamps of Architecture*, London 1849. The comments in this article are based on the reprint of the 2nd edition of 1880, New York 1989, p. 35.

5 Cf. Lauren S. Weingarden, *Louis H. Sullivan and a 19th-Century Poetics of Naturalized Architecture*, Farnham 2009.

6 Cf. Hugh Morrison, *Louis Sullivan. Prophet of Modern Architecture*, New York/London n. d., (reprint of the original edition of 1935), p. 136.

7 Cf. Louis Sullivan, 'Ornament in Architecture', 1892. Reprinted in: Robert Twombly (ed.), *Louis Sullivan. The Public Papers*, Chicago/London 1988, 79–85. Also available online under http://www.ahameri.com/cv/Courses/CU/Arch%20in%20Theory/Sullivan-Ornament.pdf [last read 12 Dec. 2011]

8 Cf. Hoffmann 1978 (as note 2), p. 52.

9 Cf. the reference to a corresponding theory in the literature in Toker 2003 (as note 2), p. 442.

10 Cf. Mark Girouard, 'The House and the Natural Landscape: A Prelude to Fallingwater'. Preface to the book by Kaufmann 1986 (as note 2), pp. 14–23, here p. 22f. On Wright's confrontation with the composition and aesthetics of Japanese woodblock prints, cf. also Edgar Tafel, *Frank Lloyd Wright persönlich,* Zürich/Munich 1981, p. 104.

11 Cf. Toker 2003 (as note 2), illustrations.

12 Ibid., p. 156f.

13 Cf. Hoffmann 1978 (as note 2), p. 53. Further commentaries in the 2nd, revised edition of the book, New York 1993, p. 61.

14 Henry David Thoreau, verse from the poem *The Fall of the Leaf*, n. d. Quoted from: Carl Bode (ed.), *Collected Poems of Henry Thoreau*, extended edition Baltimore 1969, 236–238, here p. 237.

The Lustre of Gold
The Self-staging of a Metal in Art since 1945

Ulrike Gehring

At the 2009 Biennale a golden light permeated the otherwise dark exhibition room in the Arsenale from the floor to the ceiling (fig. 1).[1] When one changed one's position, ever new strands of gold appeared from out of the darkness, only to disappear again shortly afterwards. In diagonal, interpenetrating light beams, the reflected light condensed into a golden mist, such as can also be seen in Gothic cathedrals when the light shines brightly through the diaphanous walls and relieves the architecture of its optical density.[2] Lygia Pape's installation *Ttéia I* (2002) did not permit any analogous comparison between earthly material lustre and heavenly light, however, as the technology of the lighting was staged no less prominently than the light effect itself.[3] Garishly dazzling ceiling spotlights projected their concentrated light on to gold threads, anchored in square plates and running parallel with the floor (fig. 2). Only where the cone of light fell directly on to the metal threads did the gold reflect sufficiently strongly to outshine the physical support and evoke the impression of pure light. Outside of this area, the light was dissipated and the material lost its metallic sheen. Here alone the gold came across as what it was: a metal, which in spite of all the theological and teleological interpretations, related to nothing but its own physical essence. It was not the communicative qualities of gold that were in the foreground here, but the lack of any alternative way of seeing it. Although visitors were aware that they were confronted with gold, the material, they could only partially perceive that this was so.

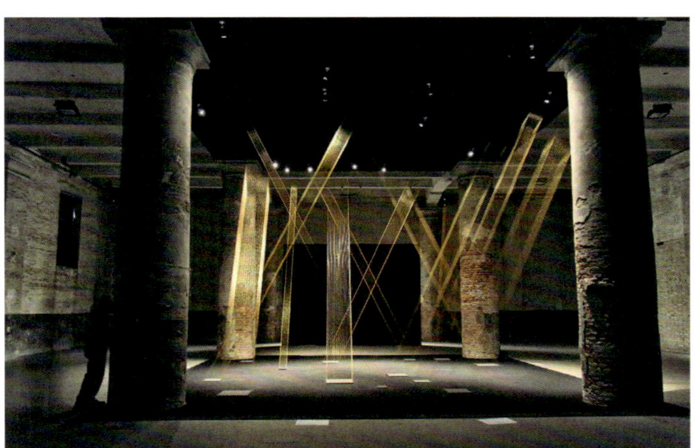

Fig. 1 Lygia Pape, *Ttéia I,* 2002 , 53rd Venice Biennale, 2009, Arsenale

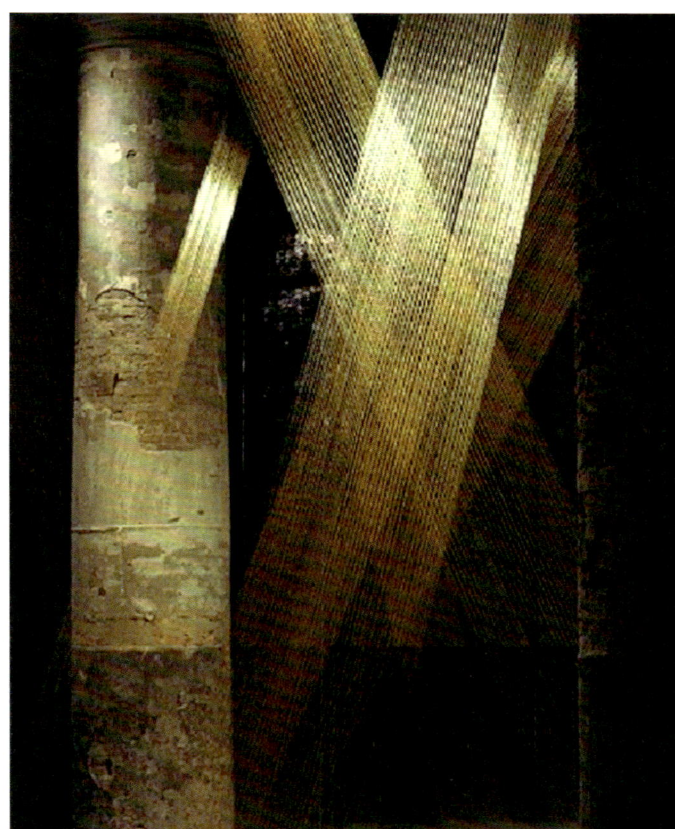

Fig. 1 Lygia Pape, *Ttéia I,* 2002 , 53rd Venice Biennale, 2009, Arsenale, Detail

Fig. 3 Yves Klein, *MG 8 (Monogold),* 1962, Private collection

Lygia Pape thus addressed a core problem of the material. It is all too easy for people to be tempted by its dichotomous appearance into attributing some transcendental power to its coloured sheen. That it is precisely the materiality of gold, however, that inspires artists such as Lucio Fontana, Robert Rauschenberg, Yves Klein and Roni Horn, is what it is my purpose here to demonstrate.[4] I shall focus on works in which the metal is not the bearer of meaning, but itself creates meaning.

Let us begin with Yves Klein. To this day, the scholarly analysis of his works has been crucially determined by the self-staging of the artist. This is true in particular of the *Monogolds*, dating from between 1959 and 1961, which explore the boundary between material and ephemeral texture.[5] In their interpretations, Sidra Stich[6] and Andre Charlet[7] go further when they pursue Klein's eclecticism and see in a Buddhist, Catholic and Rosicrucian weave of meaning the basis for an 'alchemistical transformation from the material to the spiritual'.[8] In view of such considerations, one should ask whether the panels coated in gold leaf do not, in their excess, critically analyse the communicative function of gold (fig. 3).

Klein had learnt the technique of gilding from the frame-maker Robert Savage in London in 1949, and deepened his knowledge with a second visit in 1957.[9] He knew how to manage the material and also recognized the 'textile' effect that resulted from pressing the square gold leaves on to a wooden support over which gauze had been stretched. The support material influences Klein's monochromes in three ways: firstly, via the textile substrate, whose structure comes through in the gold; secondly, via the grid structure resulting from the juxtaposition of the square gold leaves; and thirdly via the narrow ridges that result from the overlapping of the gold even when it is only fastened on one side and otherwise hangs loose from the picture support.

The emphatically material character is opposed by the communicative qualities which Klein sought to explain by the enhanced light effects of the reflecting metal. Gold, he wrote in 1959, was a 'celebration of light'.[10] And indeed, the quantity of light that gold reflects is far greater than that reflected by a comparably bright yellow pigment. What appeared to the artist in his blue-gold-red triptych (1960) as the 'apotheosis' of light, however, was the physical focus of a colour trilogy which, while it could have religious connotations, did not demand them (fig. 4). Even though the golden central field evinced the highest luminosity, it did not create the illusion of transcendent illumination. Rather, it was the source of the light emanating from it.

With this anti-illusionism, Klein was putting himself in the tradition of Alexander Rodchenko and Robert Rauschenberg. Rodchenko also categorically denied any semantic charge in the colours in his *Triptych. Pure red colour, pure yellow colour, pure blue colour* (1921) (fig. 5). In a sort of manifesto, he stated: "I brought painting to its logical end, and exhibited three panels: red, blue, and yellow. I declared: This is the end of painting. These are the primary colours. Every pictorial surface is only itself, and there will be no more depiction."[11] Rodchenko thus denied that the material had any content, and also that it had any sensuous or emotional effect. Thus painting was 'emptied out', which had a lasting effect on modern art. Painting lost its power of reference the moment the texture of the material moved into the foreground. In opposition to the maxim, in force since the fifteenth century, that the skill of a painter depended above all on his ability to imitate the subject as faithfully as possible, and to 'imitate the sheen of gold through paint', Rodchenko substituted the gold of the icons with a profane yellow that was owed to the Constructivist utopia.[12]

In the middle of the century, and in the full knowledge of the potential of this gesture, Rauschenberg took up once more the idea of the 'end of art' and in 1953 confronted the notion of the pure materiality of paint with that of the pure materiality of gold. With his recourse to gold leaf, however, he did not affirm the traditional material hierarchy, but rather questioned once again the aura of gold: thus in the *Gold Paintings*, in which he applied the paper-thin material in several layers on to a glue-coated fabric (fig. 6). Where they stuck, the texture of the support showed through; where the gold leaf and the cloth parted company, the intrinsic metallic character of the former was emphasized. Quite obviously Rauschenberg was not concerned with composing a golden picture, but with the gold in the picture. Thus the *Gold Paintings* represented a logical development of the *White Paintings*. While the white

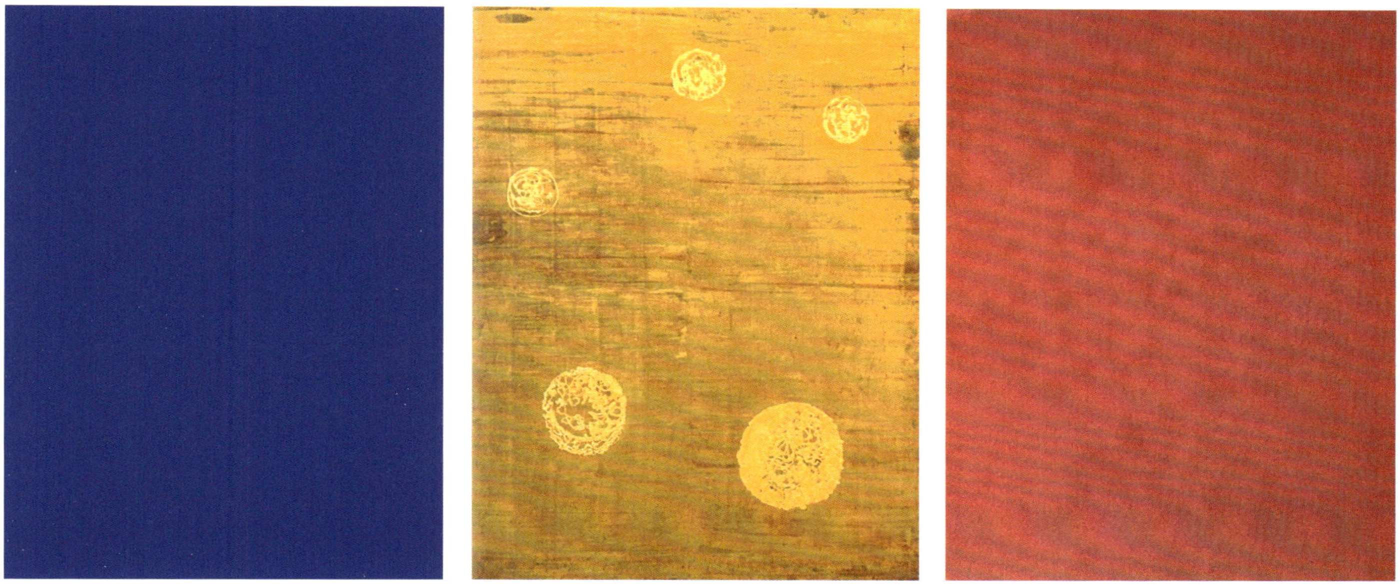

Fig. 4 Yves Klein, *Monochrome Blue, Monogold and Monopink*, 1960, Louisiana Museum of Modern Art, Humlebaek (Denmark)

monochromes, dating from around 1950, still served as a projection surface for the light, the *Gold Paintings* required no further supplementation. They developed their light-and-shade effect predominantly from themselves, from the velvety soft sheen of the gold.[13]

The effect of this sheen, or lustre, was investigated by the perception psychologist David Katz, whose results, published in *World of Colour* (1935)[14] and *Psychological Atlas*[15], enjoyed a favourable reception, not least in America. He noted in respect of metals: "Even in their non-lustrous or dully lustrous parts, metals do not appear as simple surface colours. We perceive the surface of the metal, it is true, but its colour seems to lie behind this surface. [...] we feel that we are boring through the surface into the colour of the metal."[16]

The question of how strongly the colour is bound up with the effect of light, or vice versa, of how inseparably the effect of light goes hand in hand with the surface colour, is also demonstrated by Roni Horn's *Gold Mats. Paired—For Ross and Felix* (1994/95) (fig. 7). The oversize luminous yellow, paper-thin 'mats' are of virtually pure (99.999%) gold leaf. Each consists of a kilogram of the metal, which was rolled mechanically into eighteen strips and then compressed into a flat rectangle.[17] In the next stage of the process, the gold was hammered for as long as it took to give each 'mat' a thickness of 0.02 mm and an area of 124.5 x 152.5 cm. The production process is still visible on the surface, thus pointing to a further important property of gold: its extreme softness and resulting malleability. Numerous creases, furrows, and ridges break up the light, to which the metal is opaque, almost completely.

Fig. 5 Alexander Rodchenko, *Pure Red Colour, Pure Yellow Colour, Pure Blue Colour,* 1921, whereabouts unknown, Rodchenko Archive Moscow

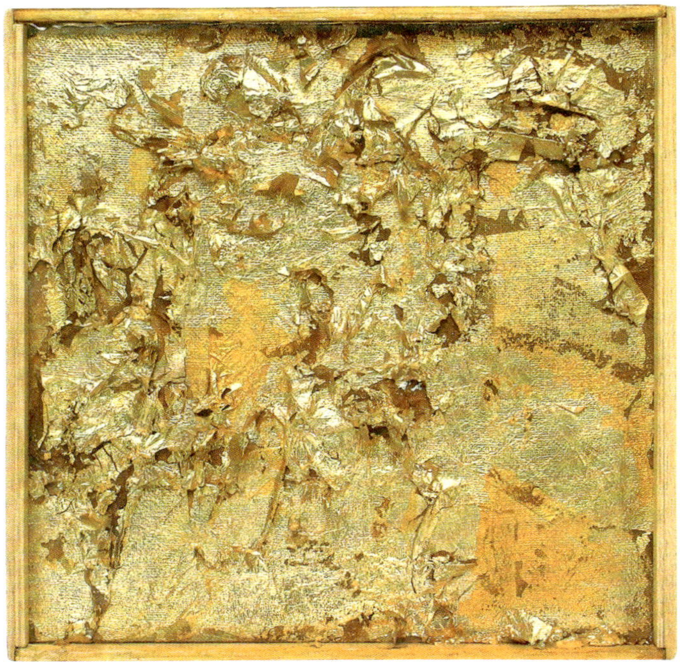

Fig. 6 Robert Rauschenberg, *Untitled (Gold Painting)*, c. 1953, Solomon R. Guggenheim Museum New York

As the light reflection exceeds the yellow colour of the gold in brightness and is perceived as an autonomous light source, we have, particularly in the hollows, the impression of a coloured light without a source. It is that yellow light that can be reflected in this way only by gold, and in its hue, it points back to the material of the source. Roni Horn seems to have fulfilled his aspiration to the extent that the material points to nothing but itself. The installation represents pure gold and nothing else.[18]

Regardless of the conceptual and aesthetic differences, the works presented are linked by the common aspiration for a worldly interpretation, which stands opposite the spiritual, timeless and sacred reading on equal terms. The 'alchemistical transformation of the material into the spiritual' thus confronts the perception-psychological process of materialization and dematerialization. Both explanatory models have their origin in light, that specific light which in the works of Robert Rauschenberg, Yves Klein, Roni Horn and Lygia Pape does not overcome the materiality of gold, but confirms the physical essence of the underlying picture support.

At the same time it cannot be denied that artworks made of gold also always pose the question of their own material value. However, this implicit question has an explicit answer when Andy Warhol or Sylvie Fleury choose gold, by dint of its costliness and exclusivity, to be the ironic representative of a consumer world characterized by excess (fig. 8). By coating product advertisements with imitation gold leaf (Warhol) or presenting gilded shopping trolleys in the museum as fetishes (Fleury), they compare and contrast the value of the gold as commodity with that of the artwork.[19] However, between the materialist and the material-aesthetic approaches there opens up a chasm which is no less profound than that between the beholder who values gold for its material properties, and the one who reveres the gold lustre as the image of the heavenly light. While the former reaction still falls into the category 'viewing', the latter is idolatry.

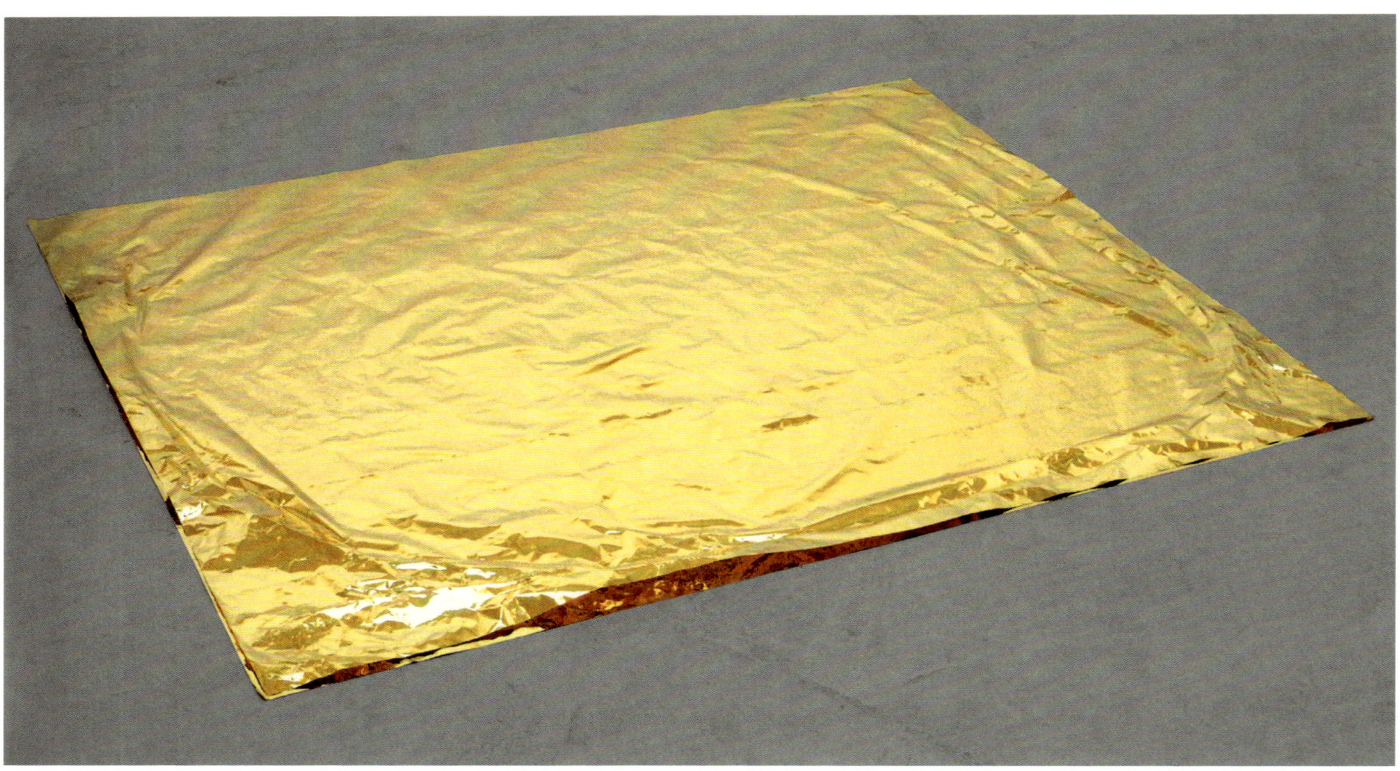

Fig. 7 Roni Horn, *Gold Mats. Paired—For Ross and Felix,* 1994/95, Gallery Hauser & Wirth

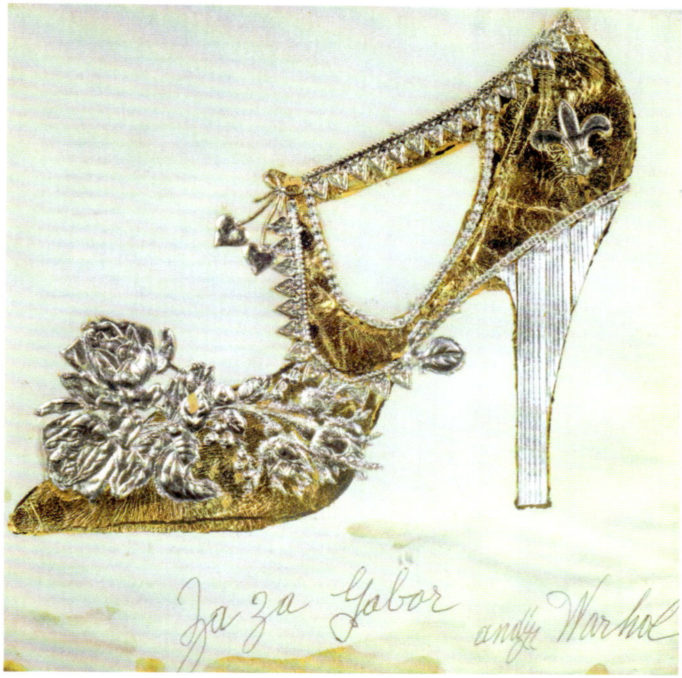

Fig. 8 Andy Warhol, *Shoe, Za Za Gabor,* 1956, Privat collection, New York

1 For more information, see: Daniel Birnbaum (ed.), *Making Worlds* (exhib. cat. 53rd International Art Exhibition, La Biennale di Venezia), Venice 2009.
2 The word 'Diaphanie' [in German; it corresponds to 'diaphany'] was used by Hans Jantzen in 1927 to describe those Gothic wall constructions which aim at an enhanced light effect. See Hans Jantzen, *Über den gotischen Kirchenraum und andere Aufsätze*, Berlin 1951, pp. 7–20.
3 In this point, the present article does not support the opinion of Daniel Birnbaum, who believes an almost 'biblical' element is realized in the mise-en-scène. See: Daniel Birnbaum's interview with Heinz-Herbert Jocks, "Auch die Behauptung des Unpolitischen ist politisch", in: *Kunstforum International*, 53. Biennale Venedig: Making Worlds, vol. 198, 2009, p. 124.
4 Numerous art-historical surveys, catalogues and illustrated volumes are devoted to this aspect of gold. By way of example one might name: Hans-Gert Bachmann, *Mythos Gold. 6000 Jahre Kulturgeschichte*, Munich 2006. A good survey of the use of gold in present-day art can be found in: Anne-Marie Charbonneaux, *L'or dans l'art contemporain*, Paris 2010.
5 Valeria Varas/Raul Rispa, *Le Monochrome de Malévitch à aujourd'hui*, Paris 2004, pp. 104 ff.
6 Sidra Stich, *Yves Klein (*exhib. cat. Museum Ludwig, Cologne/ Kunstsammlung Nordrhein-Westfalen, Düsseldorf), Ostfildern 1994.
7 Nicolas Charlet, *Yves Klein*, Munich/New York 2000, pp. 190–199.
8 Stich 1994 (see note. 6), p. 193.
9 Thomas Mc Evilley, "Yves Klein: Conquistador of the Void", in: *Yves Klein (1928–1962), A Retrospective (*exhib. cat. Institute for the Arts, Rice University Houston), Houston/New York 1982, 19–88.
10 Yves Klein, *Le Dépassement de la problématique de l'art*, La Louvière 1959, p. 11; Ernst Beyeler (ed.), *Farbe zu Licht* (exhib. cat. Fondation Beyeler, Basel/Riehen), Ostfildern 2000, pp. 110 ff.
11 Alexander Rodchenko, quoted from: Beate Epperlein, *Monochrome Malerei*, Nuremberg 1997, p. 76; see also: *Artstudio*, no. 24, Printemps 1992, pp. 82 ff.
12 In his 1435 treatise *De Pictura* ["On Painting"] Leon Battista Alberti states that "there is more admiration and praise for the painter who imitates the rays of gold with colours" than there is for the use of actual gold. Ever since, no independent value has been attached to the colour as material. The task of the artist now consisted primarily in creating the illusion of materials on the surface of the picture. Cf.: Monika Wagner, *Das Material in der Kunst. Eine andere Geschichte der Moderne*, Munich 2001, p. 17 and the same author's "Farbe als Material. Authentizitätsvorstellungen in der zeitgenössischen Kunst", in: Karl Schawelka/Anne Hoormann, *Who's Afraid of (...). Zum Stand der Farbforschung,* Weimar 1998, 159–193.
13 Walter Hopps/Susan Davidson, *Robert Rauschenberg, Retrospektive* (exhib. cat. Museum Ludwig, Cologne), Ostfildern 1998, pp. 44 ff.
14 David Katz, *The World of Colour*, London 1935.
15 David Katz, *Psychological Atlas*, New York 1948.
16 David Katz, as note 14, p. 26f. Cf. also: Wolfgang Schönem, *Über das Licht in der Malerei* (1954), Berlin 1989, pp. 253ff., and Karl Schawelka, *Farbe. Warum wir sie sehen, wie wir sie sehen*, Weimar 2007, pp. 25–39.
17 Cf. Anne Schloen, *Die Renaissance des Goldes,* Nuremberg 2010, p. 96.
18 Ibid..
19 Cf. Tim Osterwald (ed.), *Das Goldene Zeitalter. Die Geschichte des Goldes vom Mittelalter zur Gegenwart* (exhib. cat. Württembergischer Kunstverein Stuttgart), Ostfildern 1991, pp. 53 ff. Also: Heiner Bastian, *Andy Warhol. Retrospektive* (exhib. cat. Nationalgalerie, Staatliche Museen zu Berlin), Cologne 2001, pp. 84–85.

The Cultural History of Gold

The Sultan's Vow:
"We shall meet again by the Golden Apple"

Jürgen Borchhardt

From Mehmed Fatih (1444–1481) onwards, the girding of the new Ottoman ruler with the sword (*kılıç alayı*), a ritual comparable to the coronation of Christian monarchs, took place on the Golden Horn in Eyüp, at the mausoleum of Abû Ayyûb, who is said to have died outside Constantinople as the standard bearer of the prophet Mohammed in 672 CE.[1] After the ceremony, the sultan would pass the Janissaries' barracks on the way back to the palace. A senior officer offered him a drink from a bowl. The sultan accepted, drank, and returned the bowl after it had been filled with gold coins by his arms-bearer. The sultan took his leave with the formula "Kızıl Elmada görüşürüz" ("We shall meet again by the Golden Apple").[2] This vow gave the Muslim warrior the chance to die in the Holy War as a *Şehid*, a 'blood witness', and thus to go direct to paradise, or else to return home as a *gazi*, a victorious hero. Sociologically, this vow confirmed the interdependency between the high command and the ordinary soldier in the achievement of victory, and the sharing of renown, honour and booty on the battlefields of the Christian West.

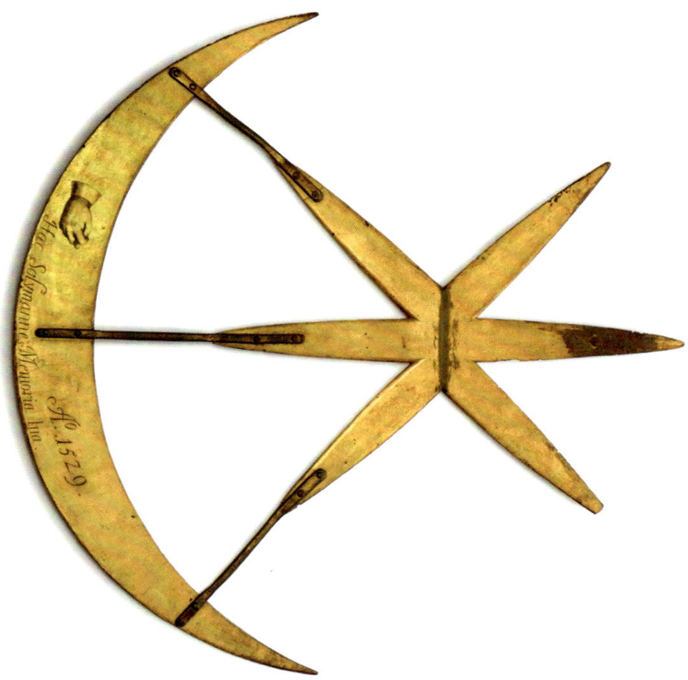

Fig. 1 *Old Steeple Pinnacle of St Stephen's Cathedral in Vienna*, 1519, Wien Museum, Vienna

Numerous sources testify that both in the sultan's family and in the army there was a conviction for many years that Vienna would open its gates to Islam without any destruction of its riches in the process. According to Evliyâ Çelebi, Schloss Neugebäude in Simmering was interpreted as the 'faithful image' of Sultan Süleyman's encampment during the siege of Vienna in 1529. In glowing colours, he describes the beauties of the architecture, the parks, the botanical gardens, the menagerie, and the pheasantry, and finally breaks out into prayer: "But if Allah the Sublime wills it, all of this was ultimately built for us: may Allah grant that it falls to Islam at last."[3]

Evliyâ Çelebi invokes Christian historians when he asserts that Şem'ûn-ı Safâ (Simon Peter), Jesus's disciple and apostle, had stood by the Mençâriyan (Hungarians) as founders of Vienna and prophesied to them: "And when, in the year which according to the calendar of the future prophet corresponds to the word Zılle, the great Süleyman comes forth from Constantinople, fear not: you must only make the fortress strong. But when in the year of the word Gânim the noble Sultan Yûsuf Mehmed comes, then take care to make peace with the Mohammedans!"[4] Evliyâ Çelebi, a member of Kara Mehmed Pasha's Ottoman delegation to Vienna in 1665, continues: "With these significant oracular words, Şem'ûn-ı Safâ issued many instructions for the construction of Fortress Vienna. Even today, these words, full of onomantic power, proclaim loudly and clearly in St Stephen's church in Fortress Vienna, and in all the Hungarian and Latin and German and Greek works of history, that one day the Ottomans will seize the Golden Apple of Vienna and also the Golden Apple of the

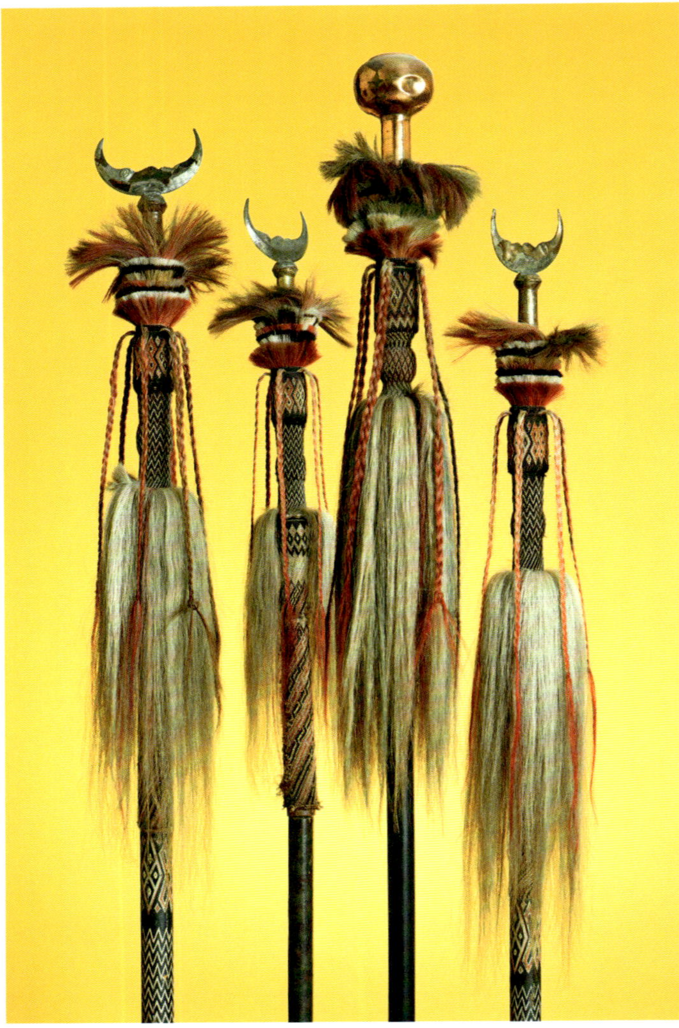

Fig. 2 *Four Ottoman Horses' Tails with a Recumbent Crescent Moon*, 2nd half of the 17th century, Wien Museum, Vienna

Pope in Rome." Evliyâ Çelebi closes his praise of the hospitals, doctors, and soup kitchens in Beç (Vienna) with the prayer: "But may the eternal Lord God protect this city from the evils of war! For if the enemy surges into the city fighting, then it will be destroyed. May Allah grant therefore that it fall to Islam without a struggle!"[5] Evliyâ Çelebi claims to have seen on the Dominican Bastion a huge kıltır tree, that is to say an elder, 'a true wonder of God', the metamorphosis of that staff which Şem'ûn-ı Safâ (Simon Peter), to reinforce his prophecy, had planted into the soil when the city of Vienna was founded.[6] Just as Süleyman in 1529 did not want to destroy St Stephen's cathedral by bombarding it with cannon fire, wishing to transform it into a mosque on the model of Hagia Sophia following his hoped-for occupation of the city, it is said that Kara Mustafa Pasha, too, had hoped in 1683 to seize the Golden Apple of Vienna by surrender. According to the Ottoman laws of war, as commander-in-chief he would have done very well out of the distribution of the city's riches. By contrast, if he had had to storm the city, he would have exposed it to three days of looting, in which much valuable booty would inevitably have been destroyed.

We might reasonably doubt whether Evliyâ Çelebi (1611–1683), the son of a leading court Goldsmith and a lady-in-waiting from the Caucasus, drew exclusively, in the tradition of oriental story-tellers, on the rich store of folk and soldiers' tales which he had heard during his training and on his travels in the Balkans, Crimea, Mesopotamia, Syria, the Aegean and Egypt. The prophecies of the non-violent occupation of Vienna, the Golden Apple, doubtless penetrated far into Turkish folk tradition, but without the sophisticated diplomacy of the Sublime Porte (a synonym for the Ottoman government) they would be barely imaginable. In the Arabic script, in which letters are also used as a substitute for figures, the characters of the word *zılle* (cloud) have a numerical value of 900+30+5; this corresponds to the Muslim year 935 AH, which in turn corresponds to the Common Era year from 15 September 1528 to 4 September 1529. According to the prophecy of Simon Peter, Sultan Süleyman was not destined to, and could not, capture Vienna; by contrast, Sultan Mehmed IV, whose given name was Yusuf, was certainly destined for success in the year of the word *gânim* (booty-laden) with the numerical value 1091 AH, in other words 1680/81 CE or 1100 AH = 1688/89 CE.

A further weighty justification for the Ottoman claim to Beç (Vienna) was the asserted existence of a *meşhed*, a site bearing witness to the death of a *Şehîd*, a holy warrior. Such a grave had *baraka*, that is to say, magic powers of blessing. Süleyman's orderly retreat following the unsuccessful siege of Vienna in 1529 could only be secured by the sacrificial death of Kasim Dede and his 40,000 warriors. In Vienna, this fictitious monastery is said to have been erected in his honour. A *türbe*, his tombstone, is said to have stood beneath the main dome of the *tekke*, that is to say, of the Bektashi dervish monastery.[7] The monastery itself is said to have had 1700 domes and 14,103 porphyry columns. Over the tombstone, a five-ton Golden Apple is said to have been suspended: "And beneath this dome, in the place called *Kızıl Elma*, prayers of blessing for Islam are recited night and day."[8] According to the Ottoman version, King Ferdinand (1503–1564) is also said to have honoured *Dayı Çerkes*, an heroic warrior in Süleyman's army, in the city. The acceptance of these fantasies among Muslim listeners resulted from the population policy of the sultans following the taking of Constantinople in 1453. Alongside the Muslim majority, Greeks, Armenians, Italians and Jews were tolerated with their cults.

To gain some understanding of Islamic claims to world domination, it is instructive in addition to cast an eye on the circular layout of Baghdad at the time of the Abbasids, with its four gates oriented to the four points of the compass, with the palace complex in the middle, including the mosque and the Golden Portal. This shape was known as *Madinat as-Salam* (city of peace) or *Dar as-Salam* (house of peace), and embraced the whole of the Islamic world. By contrast, the non-Islamic world, which still had to be conquered, was known as *Dar al-Harb* (house of war).[9] For the Ottomans, there was a duty to secure affluence and peace in the

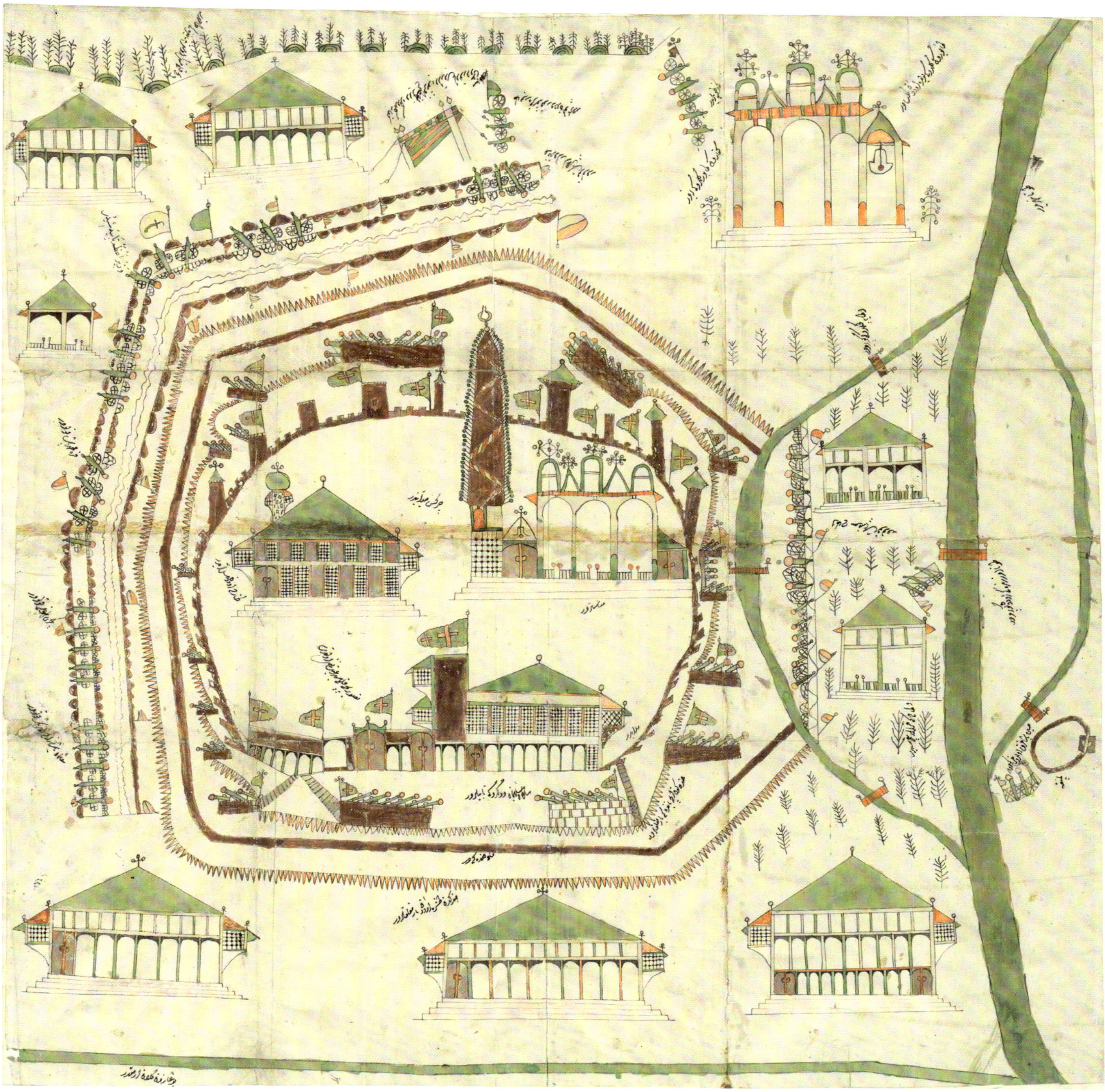

Fig. 3 *Ottoman Plan for the Siege of Vienna*, 1683, Wien Museum, Vienna

Dârülislam (territory of Islam) and, in the *cihâd*, the Holy War, to conquer the rest of the world, the *Dârülharb* (house of war), for Islam. One has the impression that the Sublime Porte banked not only on cannon, but also on the long-term effect of prospective mythology.

The Ottoman prophecies and descriptions of honours bestowed by Christians involve a religiously motivated claim to possession, and served at the same time as instruments of psychological warfare: the intention was to give the soldiers the courage to accept the dangers of war. They might, it was true, die as martyrs, but the chances of their returning home as victorious war heroes with rich booty were far higher, for a number of reasons. Firstly, the Kasim Dede monastery was, after all, in Fortress Vienna (the suggestion was that it was a complex in the Leopoldstadt district) and proved by its very existence the power of the sultan and the fear felt by the infidel. A martyr's death would match the death of Eyyûb and would guarantee the future conquest of Vienna. Secondly, it was said that not far from Vienna forty survivors of Kasim Dede's troops had been pardoned. They went on, it was said, to marry local girls and to be given the right to live in a newly founded suburb, where they built seventy mosques and 1,000 chapels. But, the story went on, the Muslim community had

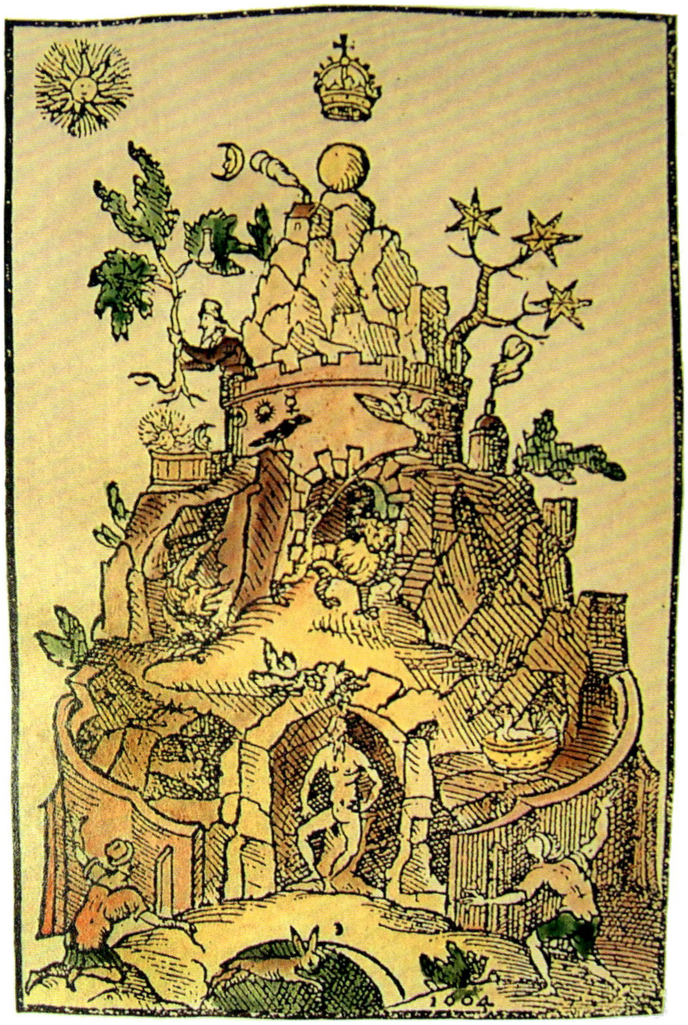

Fig. 4 *Mons Philosophorum*, Altona 1785–1788

been exploited as taxpayers, and had, for this reason, to be rescued.[10] This Ottoman fairy-tale doubtless had something to do with the actual custom of converting Ottoman prisoners-of-war to Catholicism through the so-called 'Turks' baptism'.

The War of the Moon and the Stars: from the Imperial Orb [11] to the *kızıl/altin alma* (Red/Golden Apple)

In his *seyâhatnâme* (travel diary) Evliyâ Çelebi has the *dramatis personae* of the 1529 siege of Vienna enter a fantastical competition for the pinnacle of St Stephen's. Sultan Süleyman had, he said, not allowed the bombardment of the lofty bell-tower from the deep conviction that "one day this tower would be a minaret for the Mohammedan call to prayer, part of a Muslim place of worship. Therefore it is to bear my symbol."[12] Before the gates of the city, in his encampment, he had, said the diarist, had two gold globes made, each weighing a hundredweight, and had them sent to King Ferdinand in the beleaguered city with a specific command: "The infidel king had the golden globes attached to the highest point on the church spire that very night, and since then, Fortress Vienna has been known, by dint of these golden globes, as 'the Golden Apple of Germany and Hungary'."[13]

In the Ottoman propaganda, Archduke Ferdinand, who had been crowned king of Hungary in Székesfehérvár, had been demoted to the status of vassal, who, while not daring to remove the sultan's Golden Apple, had altered it: "Later, however, when Sultan Süleyman had abandoned the siege of the Fortress and departed, King Ferdinand had a golden moon and a silver sun placed on the golden globe, Sultan Süleyman's Golden Apple. When Sultan Süleyman learned that this presumptuous *giaour* had placed other objects on his globes, he threatened: Be prepared, I am going to war against thee!"[14] According to Evliyâ Çelebi, following the taking of Baghdad in 1638/39 CE, "the German emperor [...], this baleful king, had broken the peace treaty concluded with Sultan Süleyman, by planting a wrong-headed golden cross on his globe. When he was reproached with having placed a cross-shaped thing on top of Sultan Süleyman's Golden Apple, he replied: 'All we have done is to place a cross-shaped weather-vane up there, to stop the birds perching on the golden globe.' And verily, no bird can perch on this wheel with the cross, which spins like a top when the wind blows. But certainly one day another warlike and powerful ruler like Sultan Süleyman will appear to fetch down this cross-wheel from the globe."[15]

If Çelebi with his own eyes only saw a cross on the golden globe in 1665, then the addition must have been removed in the meantime, in other words after 1529. The reality, though, is more complicated. On 28 July 1519 the spire of St Stephen's received a new pinnacle[16]: an eight-pointed star, half set in an upright crescent moon, beneath a pommel, clad in eight gilt copper plates, symbolizing the sun (fig. 1). The potentially misleading hanging in the Wien Museum and the illustration in the literature with the star above a recumbent crescent moon, get in the way of a proper understanding of Evliyâ Çelebi's description. In the Ottoman description, the 'moonlight' mutates as a result of rapid rotation into an illusory addition of a wheel with eight spokes, with a scarecrow function. The upright crescent moon became the rim, and the star the spokes of a wheel.

The Viennese designers would not have known that since Sultan Selim (1512–1520 CE) the recumbent crescent had been used to surmount standards (fig. 2). Only during the siege of Vienna in 1529 did attackers and defenders alike notice the ambiguity of the symbol of sovereignty, which in turn led on both sides to the formation of the myth as regaled by Evliyâ Çelebi. The petition addressed by Vienna City Council on 15 February 1530 to Emperor Ferdinand I to be allowed to take down the crescent moon and replace it with a figure of St George as flag-bearer was not acted upon. The petition stated "that several years ago [1518 or 1519] on Saint Stephen's spire here a star was thoughtlessly attached and a crescent moon, gilded. But as we subsequently found on his tents, when the Turk attacked Rhodes [1522], also the crown of Hungary [1526] and now the city of Vienna, he used the

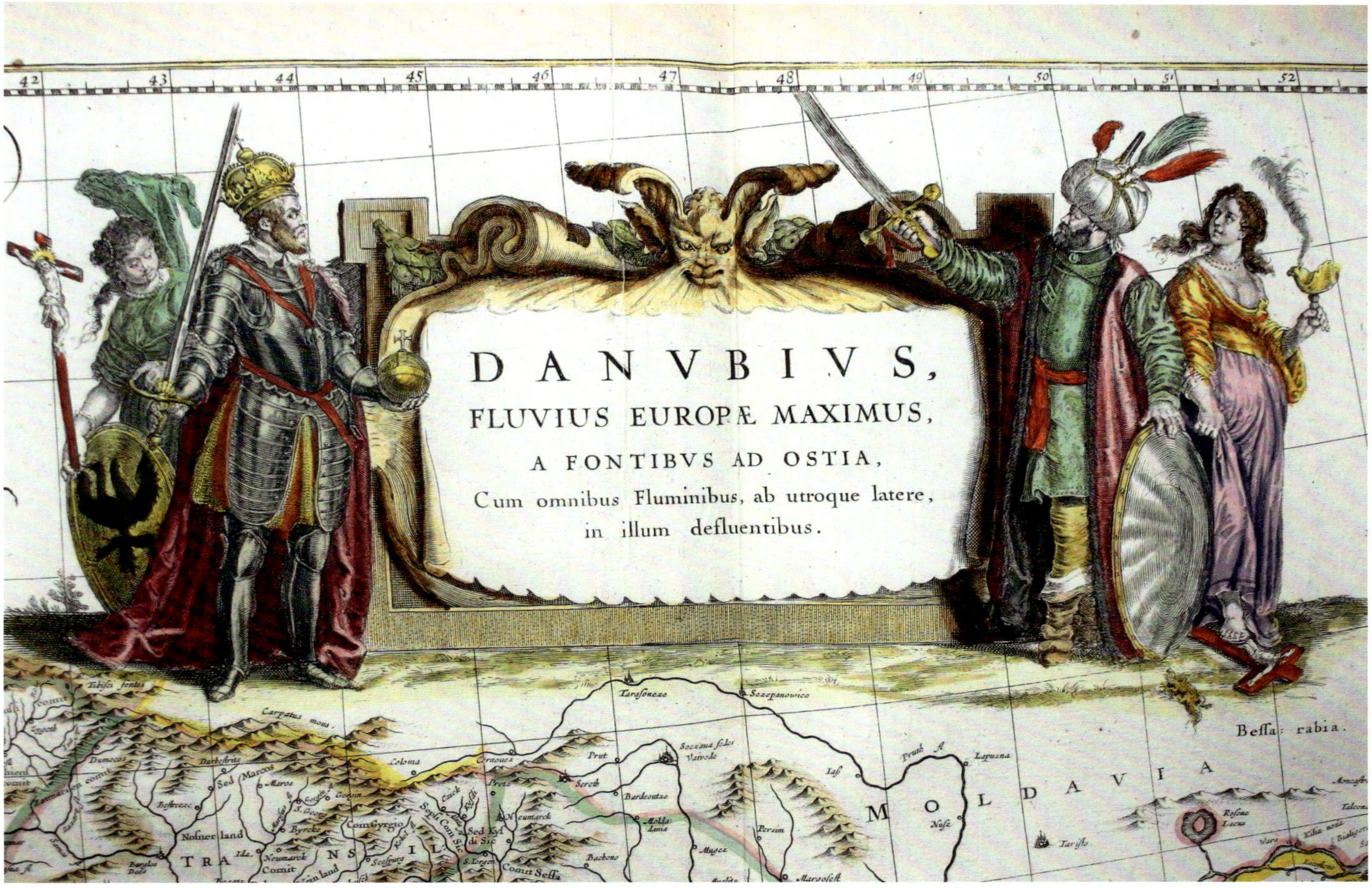

Fig. 5, Jan Jansson, *Map of the Danube*, 1635, from: W. Blaeu, *Theatrum Orbis Terrarum*, Amsterdam 1735

said star and crescent as his insignia. [Therefore] we thought to take down the said star and crescent [...] and instead to place there the Saint George flag [i.e. a banner with the city's arms, as St George had on his lance, namely a silver cross on a red field]."[17] The petition goes on to admit that an ambivalent sign had been placed on the city's landmark: "Should we leave the aforementioned heathen sign or take in down and replace it with a different, Christian sign?"[18] On the 1683 Ottoman siege map of Vienna, the flag with the cross is blowing over the defensive installations. Atop the spire of St Stephen's, the crescent moon has been turned through 90° in anticipation: the star appears above the recumbent crescent moon (fig. 3). The wish of the citizenry to have the symbol of authority, now perceived as Ottoman and Islamic, exchanged for a Christian cross was not fulfilled until 1686.

When a Christian king at his coronation takes the orb in his hand, this symbolizes his claim to earthly power. If the orb is crowned by a cross, it legitimates the claim to spiritual leadership. Turcologists presume that *kızıl* refers both to blood and to gold. Already in the prophecy of Euliyâ Çelebi it could be seen that for the Ottomans, not only Vienna was the 'city of the Golden Apple', but also Rome. Constantinople had already been designated thus. Seen from Bursa, the city on the far side of the Sea of Marmara at the entrance to the Bosporus must have seemed like half of a golden apple on the Golden Horn, destined one day to fall from the heaven of history into the hands of faithful Mohammedans. After all, had not the Byzantine author already praised the beauty of Hagia Sophia and particularly the hemispherical dome of Justinian, dating from 537 CE: "[the dome] does not appear to rest upon a solid foundation, but to cover the place beneath as though it were suspended from heaven by the fabled golden chain".[19] The decoration of Christian churches with gold mosaics is not understood as expensive display with a valuable material, but as a metaphysical striving to create a non-place, a utopia, of transcendence. Mehmet Fatih respected this miracle edifice, and after the capture of the city in 1453 transformed it into the imperial mosque. Beneath the dome he had a golden globe suspended, with a capacity of 50 bushels of wheat.

If Evliyâ Çelebi had the infidel Ferdinand place 'a golden moon and a silver sun' on St Stephen's cathedral, he not only misunderstood the meaning of the symbols, in other words the moon, sun, and their materials, but also totally distorts the fundamental connotations of sun (= sulphur, red, hot energy, male principle) and moon (= white, female, virginity, cold energy, mercury). According to the ideas of the alchemists, the union of base metals could result in gold, the highest stage of spiritual enlightenment.[20] The laborious path up to the light, to the sun, moon and stars, is represented as the ascent of the *mons philosophorum* (fig. 4). The spherical

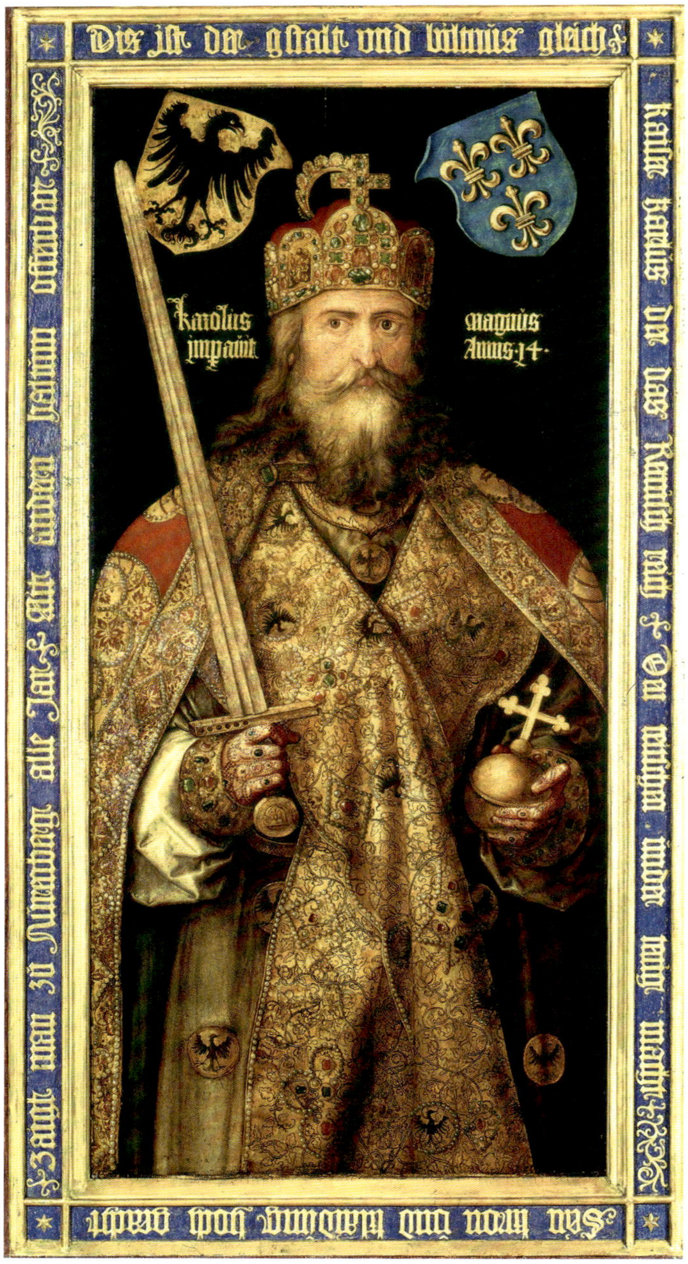

Fig. 6 Albrecht Dürer, *Fantasy portrait of Charlemagne*, 1511/12
Germanisches Nationalmuseum, Nuremberg (on loan from the City of Nuremberg)

element at the pinnacle is supposed to symbolize the pearl, the symbol of the colours of the rainbow, which emerges from the struggle between the sun and the moon, and their mutual annihilation in the nigredo, the black putrefaction, the blackness of blacknesses.[21] The tower is also adorned by the crescent moon, sun, and sign of mercury.

On a Dutch map, the conflict of interests on the Danube between the Christian West and the Islamic East is personified by Charles V on the left and Selim II on the right (fig. 5).[22] Only superficially is equal status aimed for. For while the allegory behind the emperor, holding shield and cross, identifies the bearer of crown, sword and orb as the protector and enlarger of the Christian West, the allegory of heathendom is trampling

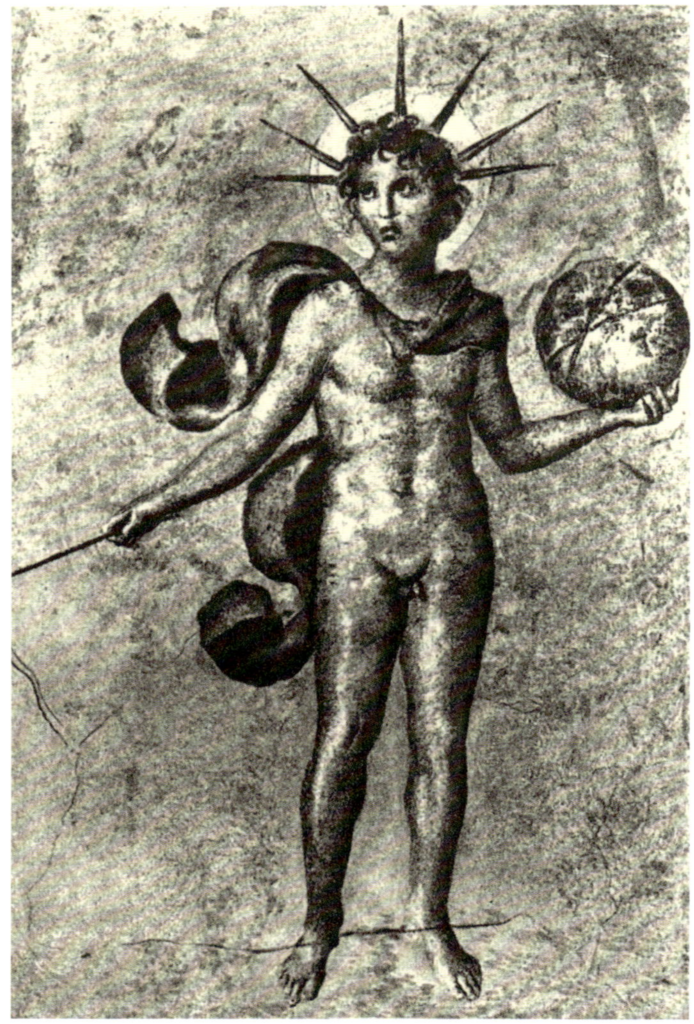

Fig. 7 *Apollo as Cosmocrator*, 1st cent. CE.
Museo archeologico nazionale, Naples

the cross underfoot. The emblem on the sultan's round shield may go some way to explaining Evliyâ Çelebi's enigmatic statement. A silvery crescent moon covers a golden sun with five red beams. The orb in Charles V's hand matches the twelfth-century orb now in the Treasure Chamber in Vienna, which was first mentioned as being used at the coronation of Emperor Henry VI (1165–1197); it was made of gold and gold filigree, and surmounted by the *crux gemmata*, the jewelled cross. In 1512 Albrecht Dürer was commissioned by the city of Nuremberg, which had charge of the imperial insignia, to create a fantasy portrait of Charlemagne with the imperial crown, sword and the golden orb in his left hand (fig. 6)[23].

The orb as part of the insignia is said to have been introduced by Constantine the Great. In unmistakable fashion, Roman art portrayed Helios/Apollo as 'cosmocrator' (fig. 7)[24]. He floats through space, the whip in his right hand identifying him as the driver of the heavenly chariot. The beardless youthful face is illuminated by a halo. In his left hand he holds the *sphaira* with the clearly marked lines of the celestial equator and the ecliptic or zodiac, the assumed

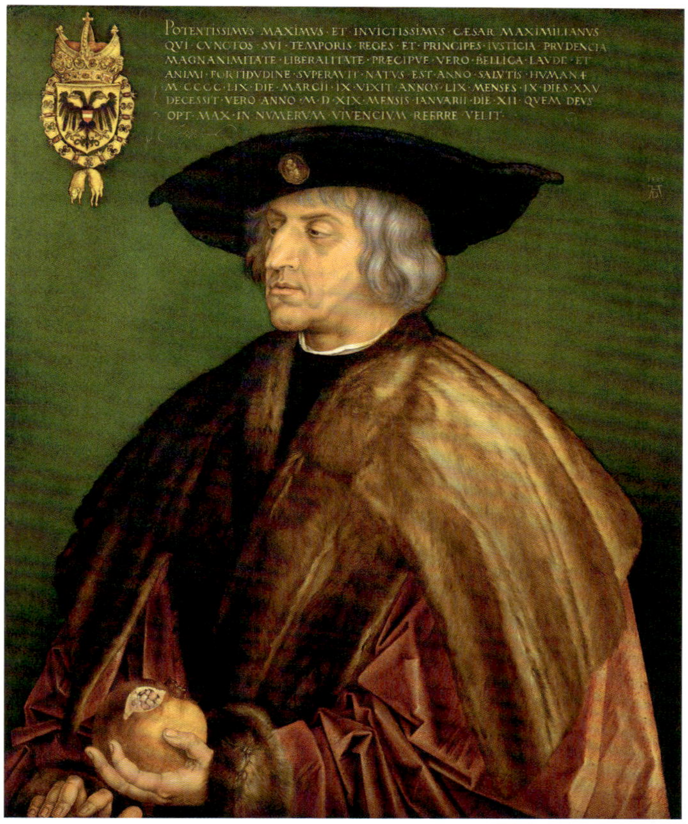

Fig. 8 Albrecht Dürer, *Emperor Maximilian I,* 1519
Kunsthistorisches Museum, Vienna

path of the sun around the earth, the two lines intersecting at the equinoxes. Byzantine art adopted the globe as the symbol of worldly power.

Percy Ernst Schramm convincingly argued that medieval symbols of rule were necessarily always being re-interpreted: "[…] although perhaps no one any longer understands the original meaning, e.g. of the imperial orb. For the fact that it is an image of the cosmos, represented through the idealized form of a star-studded globe, which had been placed in the hand of the ruler of the universe as a sign of his power, was during the Middle Ages secondary to quite different ideas now valid."[25] The globe (Latin: *sphaera*) thus represents the basic idea. Who rules the globe, rules the world. Already Plato had said of the creator of worlds: "wherefore He wrought it into a round, in the shape of a sphere [σφαιροειθές], equidistant in all directions from the center to the extremities, which of all shapes is the most perfect and the most self-similar."[26]

We should draw attention to certain aporia in this context. Why is the Golden Apple in the sultan's vow not called *altın elma*? Presumably the red apple turned into a golden apple in the sixteenth century, and the two expressions were used synonymously. The dated source in the State Library in Ankara reports the sacrificial death of Kasim Dede and his 40,000 warriors in Kızılalma and mentions the *tekke* erected in his honour. His tomb is said to have been beneath the main dome: "Furthermore the *türbe* of Kasim Dede is beneath the main dome. The Danube flows beneath that *türbe*; its roar can be heard 48 hours away. What is called the Danube […] is in the form of a golden apple weighing eighty-four kantar [= almost five tons]. This is why [the city] is called Kızılalma. And [beneath] this main dome, in the place called Kızılalma, prayers of blessing for Islam are spoken night and day."[27]

The Ottoman letter-writer's comparison of Vienna to an apple may superficially recall the panoramas of the city at the time of the first siege in 1529, which evokes analogies, but also the idea of a round fruit on a bough in the form of a river, conceived as masculine, and opens up associations with the semantic field of family-tree and son-and-heir, and indirectly the emphasis on the *oikos*, the tent, the house, and the dynasty. In Turkish, gold plays a major role. A gold coin, for example, is an *altın sikke*, a gold medal an *altın madalya*. The word *kızıl* is a paraphrase for a blood-red material, for example *Kızılay* (Red Crescent) or *Kızıldeniz* (Red Sea). The changing colour of the fruit of the pomegranate tree (*Punica granatum* L.) from golden yellow to red may have been an inspiration.

When he painted *Emperor Maximilian I* with a golden pomegranate in 1519 (fig. 8), with its skin open to allow a view of the flesh, did Albrecht Dürer know that in the second millennium BCE there had already been, in the eastern Mediterranean, a golden cult object that very realistically imitates the form of a pomegranate with its characteristic 'crown' (fig. 9)?

In the question of the origin of the orb as a symbol of rule in the Christian West and the Islamic East, a picture painted in Iran in the first quarter of the fourteenth century could be

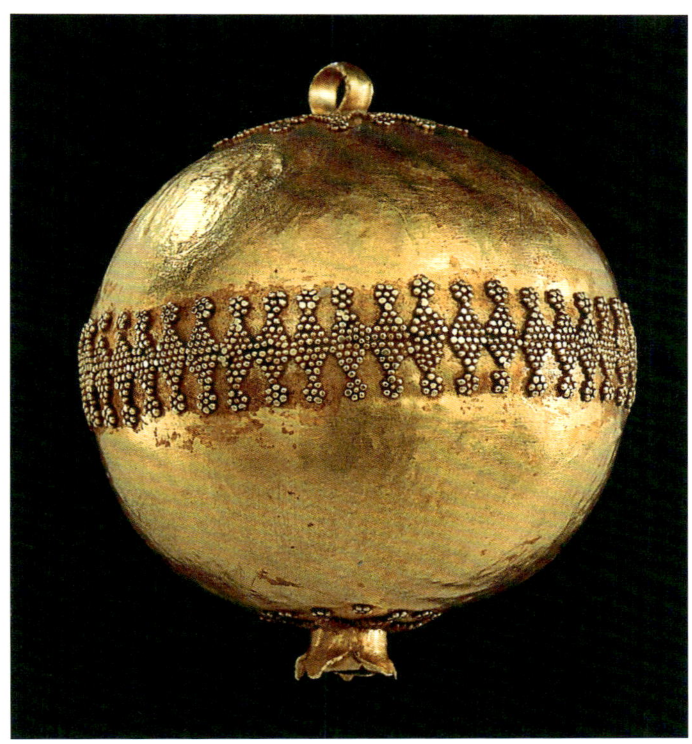

Fig. 9 *Golden pomegranate*, 2nd millennium BCE
Bible Lands Museum, Jerusalem

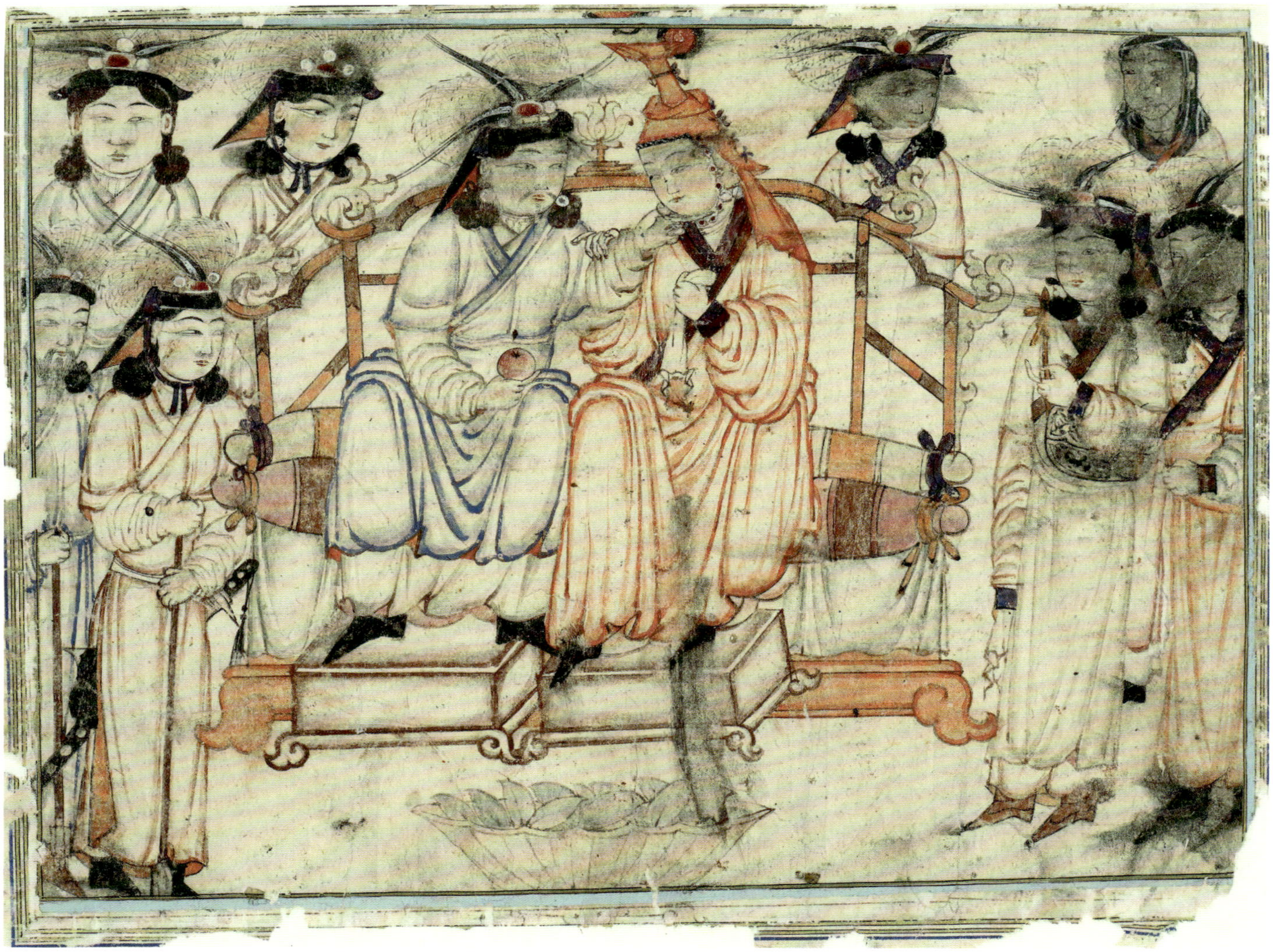

Fig. 10 *Khan and Khatun in conversation*, 14th cent., Staatsbibliothek zu Berlin—Preussischer Kulturbesitz, Oriental Dept.

revealing. The khan and the khatun sit on thrones on a dais surrounded by standing courtiers engaged in private conversation[28] (fig. 10). In his right hand, he is holding a reddish-yellow apple. Has an ancient Iranian custom been reactivated here, or were the Mongol elite passing on Turkish ideas? Only experts can judge.

The result of our inquiries is as follows. There is no simple, clear, convincing answer to the question of what the Ottomans understood by an object or city 'of the golden or red apple'. The Byzantines already used the word μήλον (apple) synonymously with σφαῖρα (globe). Not surprising, the theory that the Ottoman concept derived from the Byzantine 'globe' or from an equestrian statue of Justinian I outside Hagia Sophia, holding a golden orb in his hand, also has significant adherents among turcologists. Ebû-Eyyûb el-Ausari, who was killed at the siege of Constantinople in 668/69 CE, is said to have acted as a mediator between the Byzantine and Ottoman empires, for, it is reported, he held in his right hand, wrapped in a saffron shroud, a bronze *mühre*, i.e. a spherical object, the Golden Apple.[29] If we accept the opinion of the best expert on the sources, then the Eyyûb legend on the Golden Horn can be linked to the Kızıl Elma myth of the Golden Apple: Eyyûb as Holy Warrior made possible the conquest of Constantinople, and, by preserving Justinian's 'Golden Apple', the sultan's aim of conquering the Christian West for Islam was legitimated.

The description Kızıl Elma is seriously ambiguous. It can be reduced to the material sense of the crowning pinnacle of a Christian cathedral, but one can also understand by it a spire, like that of St Stephen's cathedral, looking from a distance like a sceptre topped by a sphere and crescent moon, or else as the capital city of a great crowned Christian ruler. To seize it would in any case mean the enormous extension of the *Dârülharb* (house of war) for Islam and at the same time included rule by the Ottomans within the succession of empires as widely understood in East and West alike. We still have no answer to the question posed by turcologists, "Should we understand the process of myth-formation more as a Turkish feature on a Western substrate, or did a Western symbol amalgamate with a Turkish substrate?" If this question is to be answered, we need the co-operation not only of turcologists, but also of archaeologists and art-historians.

1 Klaus Kreiser, *Geschichte Istanbuls—Von der Antike bis zur Gegenwart*, Munich 2010, p. 48, 51.
2 Erich Prokosch/Karl Teply (eds.), Richard F. Kreutel, *Im Reiche des Goldenen Apfels—Des türkischenWeltenbummlers Evliyâ Çelebi denkwürdige Reise in das Giaurenland und in die Stadt und Festung Wien anno 1665*, Vienna 1987, pp. 33f.; Kerstin Tomenendal, *Das türkische Gesicht Wiens—Auf den Spuren der Türken in Wien*, Vienna/Cologne/Weimar 2000, p. 16.
3 Ibid., p. 112
4 Ibid., p. 126
5 Ibid., p. 181 f.
6 Tomenendal 2000 (as note. 2), p. 79
7 On the importance of the order, cf. Suraiya Faroqhi, *Der Bektaschi-Orden in Anatolien vom späten fünfzehnten Jahrhundert bis 1826*, Vienna 1981.
8 Çelenklioğlu (ed.), "Hikâyet-i Kızılalma" [= The History of the Kızıl Elma], Cönk 44 der Ankara Millî Kütüphanesi, Türk Kültürü 185, Mart 1978, 309-310, p. 309.
9 Günter Kettermann, *Atlas zur Geschichte des Islam*, Darmstadt 2001, p. 35.
10 Çelenklioğlu 1978, p. 310 (see note 8); Karl Teply, *Türkische Sagen und Legenden um die Kaiserstadt Wien*, Vienna 1980, pp. 135 ff.
11 Translator's note: the German word for *orb* is *Reichsapfel*, literally 'imperial apple'.
12 Prokosch-Teply 1987 (as note 2), p. 167
13 ibid.
14 ibid.
15 op. cit., p. 168.
16 Georg Kugler in Wilfried Seipel (ed.), *Kaiser Ferdinand I. 1503-1564. Das Werden der Habsburgermonarchie*, Vienna 2003, pp. 413 f., cat. V. 21.
17 Walter Hummelberger, "Wiens Belagerung durch die Türken 1529", in: *Militärhistorische Schriftenreihe*, vol. 33, 1981, p. 56 f.
18 ibid., p. 57.
19 Procopius, *De Aedificiis* [c. 550 CE] http://www.fordham.edu/halsall/source/procop-deaed1.asp
20 David Fontana, *The Secret Language of Symbols: a Visual Key to Symbols and their Meanings*, London and San Francisco 1997, p. 146.
21 Erich Hornung, *Das esoterische Ägypten: Das geheime Wissen der Ägypter und sein Einfluss auf das Abendland*, Munich 1999, p. 44.
22 Tomenendal 2000 (as note 2), p. 11, fig. 1.
23 Paola Rapelli, *Symbols of Power in Art*, Los Angeles 2011, p. 111.
24 Percy Ernst Schramm, *Sphaira, Globus, Reichsapfel, Wanderung und Wandlung eines Herrschaftszeichens von Caesar bis zu Elisabeth II*, Stuttgart 1958, p. 10.
25 Percy Ernst Schramm, *Herrschaftszeichen und Staatssymbolik—Beiträge zu ihrer Geschichte vom dritten bis zum sechzehnten Jahrhundert III*, Stuttgart 1956, p. 8
26 Plato, *Timaeus*, 33 b. trans. W.R.M. Lamb, Cambridge MA, London, 1925
27 Teply 1980 (see note 10), pp. 135 ff.
28 Jutta Frings/ Claudius Müller, *Dschingis Khan und seine Erben: Das Weltreich der Mongolen* (exhib. cat. Kunst- und Ausstellungshalle der Bundesrepublik Deutschland GmbH, Bonn), Munich 2005, p. 262.
29 Teply 1980 (see note 10), p. 64.

Apollo's Apartments in Versailles
Decoration in Gold between Material Display and Allegorical Function

Markus A. Castor

"Le prince, en tant que prince, n'est pas regardé comme un homme particulier, c'est un personnage public, tout l'Etat est en lui, la volonté de tout le peuple est renfermée dans la sienne [...]. Sous un prince sage tout abonde, les hommes, les biens de la terre, l'or, l'argent ; le bon ordre amène tous les biens."[1]

In his *Politique tirée des propres paroles de l'Ecriture Sainte*, Jacques Bénigne Bossuet, from 1670 the private tutor of the Dauphin of France, the son of Louis XIV, is here describing—in allusion to King David, who in the Versailles *Chapelle Royale* is depicted as a vicarious representation of the French king in the gilded bas-relief of the altar—the true prince, thus laying the foundation of the *autorité royale sacrée*. Just as every perfection and virtue was present in God, so the strength of every individual was united in the person of the prince. It was also the prince that made the gold gleam.

If we were to seek visible proof for this absolutist claim, in which gold played a corresponding role, then it would be in the place which gave the prince space for the visual expression of his authority. The *appartement princier* is *the* early modern location which represents in the highest concentration the power, the self-image, and the claims of the sovereign.[2] The use of gold is expected, and is a self-evident part of the obligations of princes to implement the display required of them in the weave of courtly competition. How could the *Roi soleil* be imagined, if not in gold?

In Gérard Corbiau's film *Le roi danse* (2000), when the young king Louis XIV, as the god of light, Apollo, directs the orbits of the planets in the cosmos of his court, then body, symbol and metaphysically enhanced materiality become one. The king is golden from head to toe, and from him alone emanates the glory that moves and illuminates the world between peace and war. Even though, in contrast to the cinematographic projection, the king did not appear half-naked, he appears coloured in the painted portraits and mostly white in the sculptural images, each depiction of the king taking on an

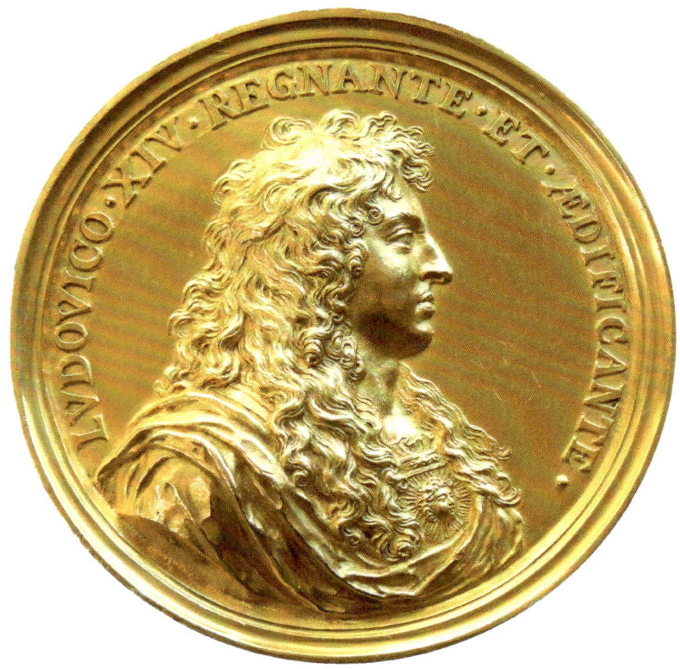

Fig. 1 Jean Varin, *Histoire Métallique du règne, Portrait of Louis XIV*, 1665, Bibliothèque nationale de France, Paris

important vicarious function. In Versailles the material nature of gold is coupled to a symbolic charge which, in its labyrinthine complexity, makes it difficult to properly classify the function and significance of the noble metal. A close inspection reveals the opposite of what, from the point of view of historical detachment, may draw 'King—Sun—Gold' into a fuzzy congruence of meaning.

Less ephemerally than in Lully's *Ballet royal*, the great deeds of the monarch throughout his reign are communicated in the *Histoire du Roi* with its 334 medallions (fig. 1).[3] Both the celebration of the ruler in gold, and also the 'hard' currency of the Louis d'or (from 1640) were imitated, for example in Bavaria, Prussia and Saxony (Max, Friedrich and August d'or

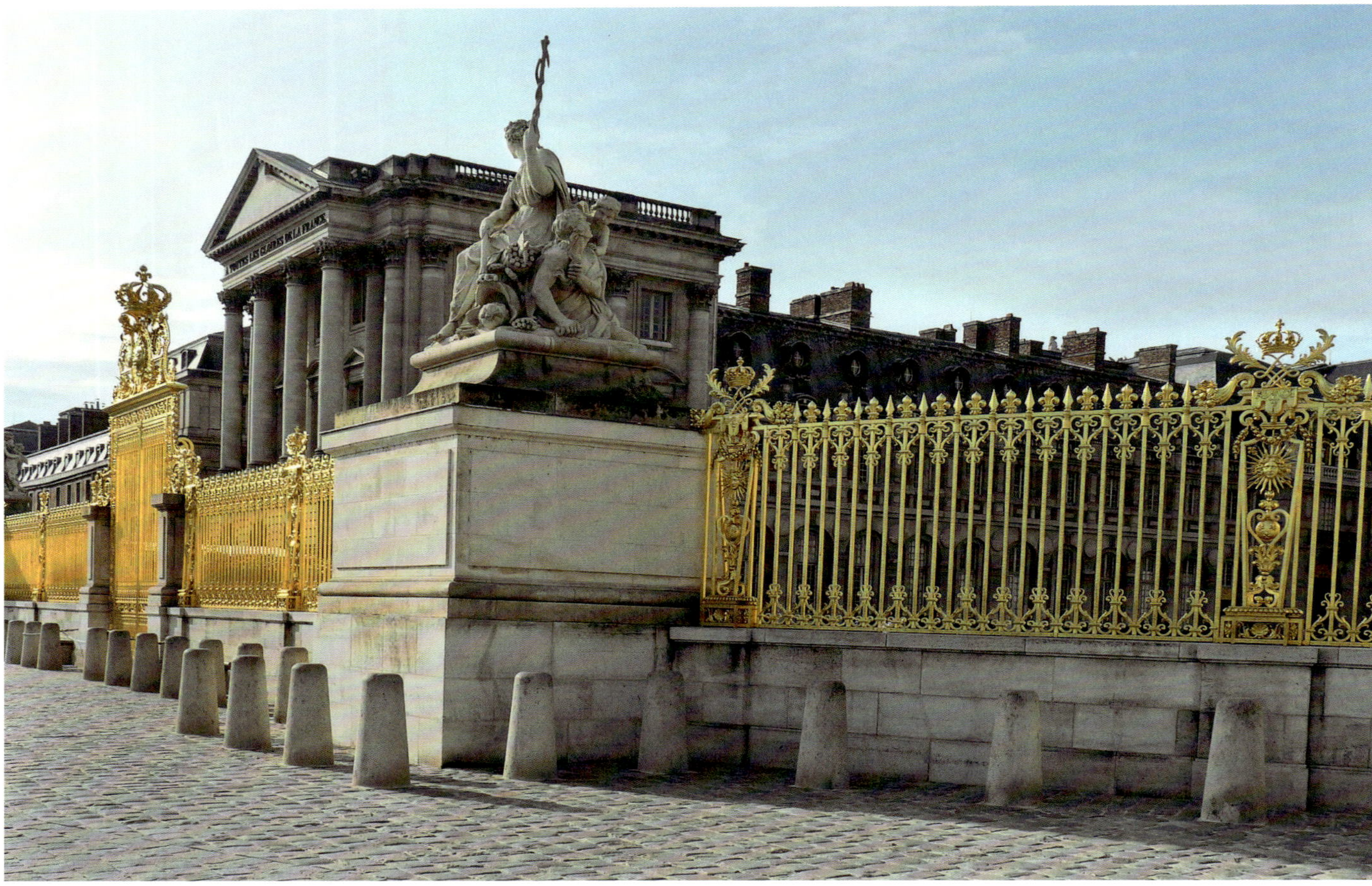

Fig. 2 Palace of Versailles, *Grille royale*

respectively) and in Vienna, with Leopold I and the rival medallion-history of the Habsburg emperor. So nothing here is as exclusive as it looks. Already Emperor Maximilian I (*reg.* 1508–1519) and his successors claimed the sun as their own.[4]

But French absolutism saw a long-term re-casting of patterns of thought and pictorial memory insofar as they concerned the arts, their history, and their perception. What could be more obvious, then, than to illuminate this immediate environment of the *Roi soleil*? There was much to choose from, for example a comparison with the Baroque Chapelle Royale in Versailles with the medieval royal chapel on the Ile de la Cité and their respective gilded heavenly Jerusalems. Or with the church of Les Invalides, dedicated to St Louis, and often enough instrumentalized by Louis XIV for the legitimation of his descent and his divine right, whose crowning gold dome is a kind of *architecture parlante*, the embodiment of a king, embracing Paris in his arms, with Versailles and his army at his back. Even for the *Bâtiments du Roi*, there are preconditions. And for all the excess of Versailles, all the *grandes machines*, the major buildings in and around Paris, had been influenced since time immemorial by the import of Italian exemplars, for example in the Louvre, with the execution of the ceiling paintings in gilt stucco frames.

However that may be, Versailles established the European model of a ruler's residence. The Olympus of the Apollonian figure of light,[5] the architectural introit of the era of absolutism, distributed on all the channels of royal propaganda, left its mark for centuries as a quasi-sacred temple of power on the idea of the *non plus ultra* of glittering gold magnificence. Versailles, La Fontaine's *Palais d'Apollon*, is a place where expectations come true, as they did in 1660 with the *Entrée solennelle*, the ceremonious entry into the city of Paris, in which Louis played the peace-bringing Apollo.[6]

The talk of the Golden Age, which for Dutch painting, with its absence of gold, was doubtless aimed more at economic success, and as the *siglo de oro* referred not least to the heyday of Spain's economic power and foreign-policy influence, nourished by imports of gold from South America, makes the 'golden century' label look suspect as a projection. The time known in France as the *siècle classique*, a bourgeois neologism,[7] is glorified in the panegyrics of the seventeenth century, for example by Racine, as the *Siècle Louis le Grand* and as the return of the golden age.[8] The rule of the miracle-working king, and the public historiography itself, nourished the sun myth to which the Versailles decor subscribes. Where, if not here, did the heavenly power have to be illustrated?[9] Thus the court historian André Félibien explained in 1674:

"Comme le soleil est la devise du Roi, l'on a pris sept planètes pour servir de sujet aux tableaux des sept pièces de cet appartement, de sorte que dans chacune on y doit

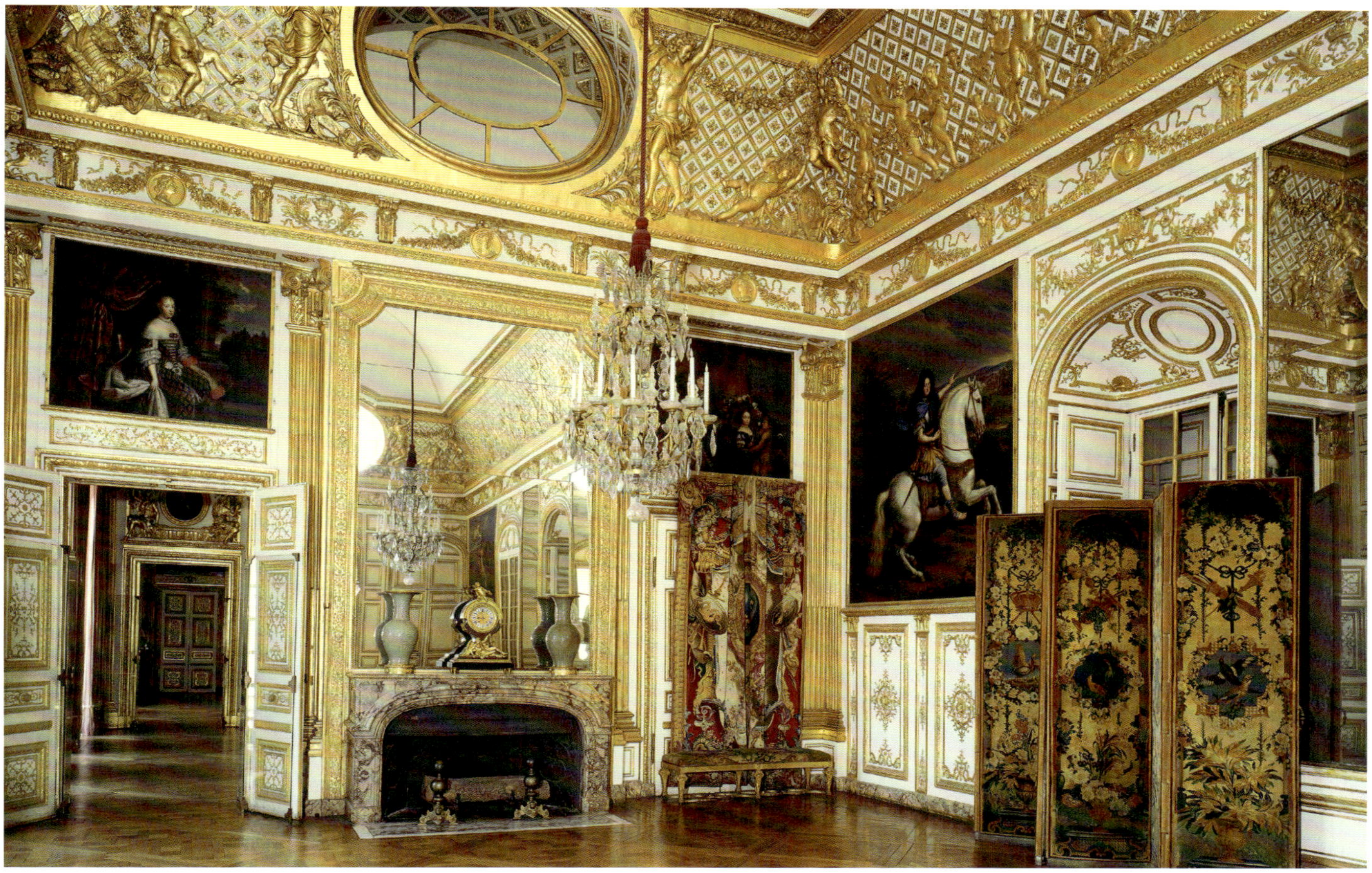

Fig. 3 Palace of Versailles, *Salon de L'Œil de Bœuf*

représenter les actions des Héros de l'Antique qui auront rapport à chacune des planètes et aux actions de Sa Majesté. On en voit les ornements symboliques dans les ornements de sculpture qu'on a faits aux corniches et dans les plafonds."[10]

For all the expectation and the unquestioning attitude, it is surprising that gold is not mentioned in this text. Gold has its place in this correct order of things. But precisely which concretely visible places did it now occupy?

The modern gold-plating of the century—the prefigured view

The extent to which the wish and the greed for real or fake gold, self-gilding and the noble external coating have remained the textbook preference was demonstrated in 2008 by the inauguration of the so-called *Grille royale* in the Palace of Versailles (fig. 2). The ahistorical *Grille dorée*, a purely imaginary 'reconstruction' thought up and executed *ex nihilo*, and gilded almost all over at a cost of five million euros, promises the innocent visitor passage into the radiant gold centre of past power and grandeur, a peep, so to speak, that reflects long-held expectations.[11]

The storm disaster of 1999 triggered a restoration campaign in Versailles, which once again showed how complex the *grande machine* was. What could then be more appropriate, in view of the restoration measures and their focus on re-gilding, than to enquire about the significance of 'gold' for psychohistory, for the history of awareness; about what, with ambassadors' reports and descriptions, became the symbol of *majesté* throughout Europe?[12] Especially in the state apartment, all the arts enter into a complex system of supplements and rivalries. Visual and architectural arts, decor and its use in the context of ceremonial, utilized the highest level of artistic display.

The Appartement du Roi—significance-enhancement in gold

Opulence in decor and furnishings expressed the dignity of the *Roi de France*. Grandeur required, without fail, the *splendeur des matériaux*. In line with ceremonial and etiquette, the splendour of the apartment grew as the state bed was approached. In the words of Félibien: as one passes from one room to the next, one sees more and more riches, be it in the use of marble or in the sculptures and paintings that decorate the ceilings. This is true also of the dimensions and variety of the gilding that leads to the *Chambre du Roi* and rhythmically heightens expectations.[13]

The very moderate and reticent decor of the *Salle des Gardes*, the first room of the apartment, originally had a wall hanging

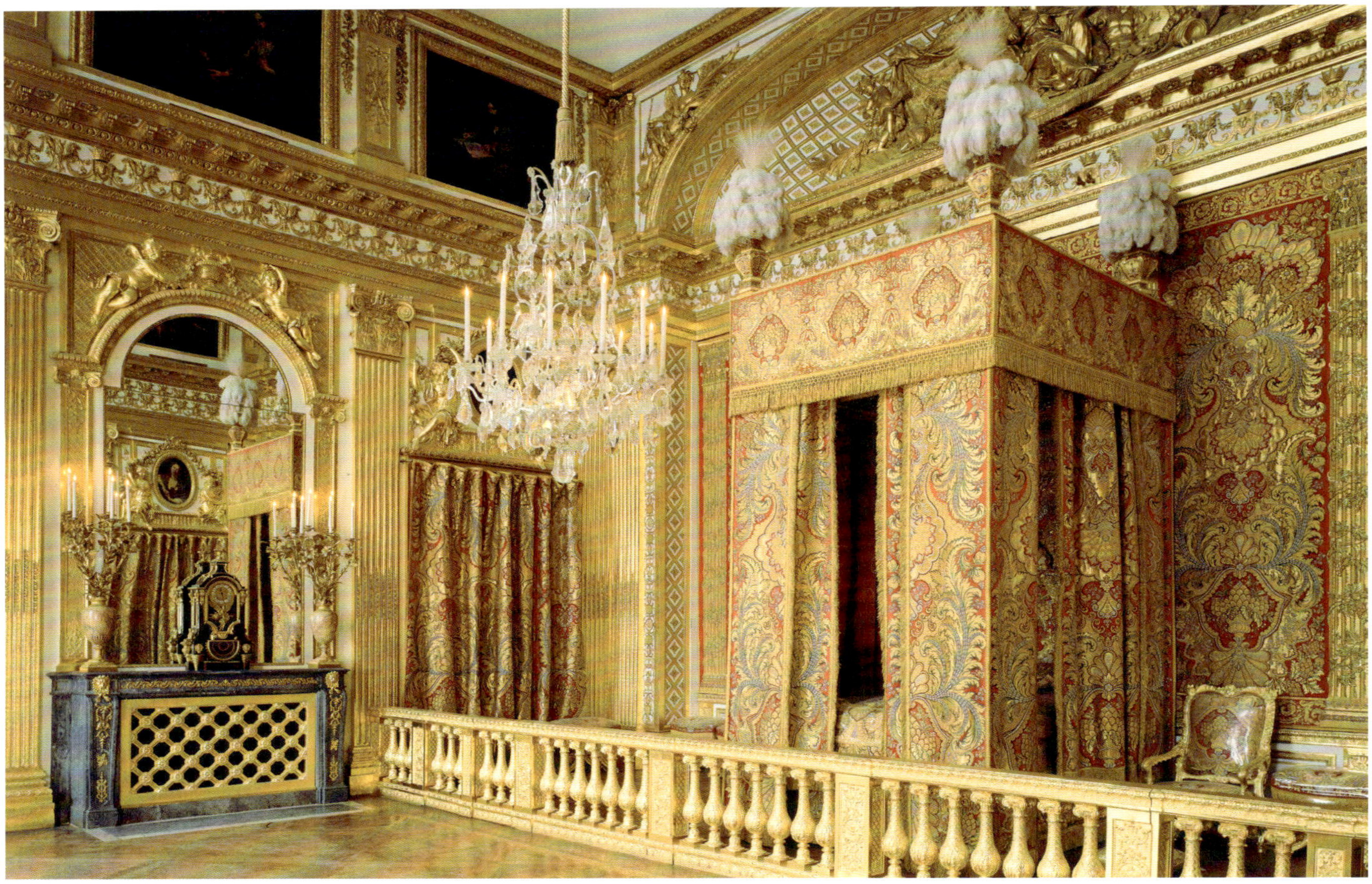

Fig. 4 Palace of Versailles, *Chambre du Roi*

of leather with gold embossing. The cornices of this guard-room are decorated with sculpted helmets and trophies. The furniture, not mentioned in the 1708 inventory, must have been sparse. The room was lit by two four-armed bronze chandeliers.

The decor of the First Antechamber, with its twelve ancient battles painted by Joseph Parrocell, has panelled walls, and the cornice is now gilded. The only sculpted element is the ornamentation with insignia of kingship, the crown and the sceptre between the *fleur de lys*. The eight-armed crystal chandelier and the fireplace with its varieties of marble both evince reticence. The furniture here is limited to four benches. The Antechamber, accessible to almost everyone, was where the king took his public meals when the queen was absent.[14]

The enfilade was rebuilt after the death of the queen, with the result that the *Salon des Bassans* was united with the original *Chambre du Roi*, which itself was transferred to the next room and thus into the focus of the complex. Admission to the resulting enlarged Antechamber, known as the *Salon de L'Œil de Bœuf,* was reserved for the Haute Noblesse, who prepared themselves here, in the immediate approach to the *Chambre du Roi*, for the Levée. Its more opulent decoration includes sculpted, and now completely gilded, panelling and a gilt stucco frieze, while even the grooved pilasters here are gilded (fig. 3). The room was given a set of mirrors, a so-called *cheminée à la royale* with a large mirror on the mantel, and paintings by Veronese, while the furniture consisted of twenty-four tabourets, whose red brocade was embroidered with gold and silver flowers. The increased need for light, resulting also from the function and size of the room, was met with three crystal chandeliers.

We reach the culmination of the enfilade in the *Chambre du Roi*, which was moved to the exact centre of the palace in 1701. The room, which faces on to the *Cour de Marbre*, forms the nucleus, so to speak the cella, of the Temple of Apollo. What catches the eye in the king's bedchamber, ahead of any iconography, is the sheer opulence of the material. The gilding covers the sculpted panelling and the frames of the mirrors, and lends emphasis to the paintings. Ambassadors, for example the envoy from Brandenburg, Ezechiel Spanheim, stress time and again in their reports the financial power which found its expression here.[15] Apollo and Jupiter as allegories of Louis are seen here at the windows. They and the royal cipher are reinforced by the allegory of France in the alcove occupied by the state bed. All of this speaks of a concept of monarchy that is first and foremost kingship.[16]

The state bed at the heart of Versailles bestows *majesté, justice royale* and *présence du Roi*. Crowned by Nicolas Coustou's gold relief of France triumphant, it dominates the

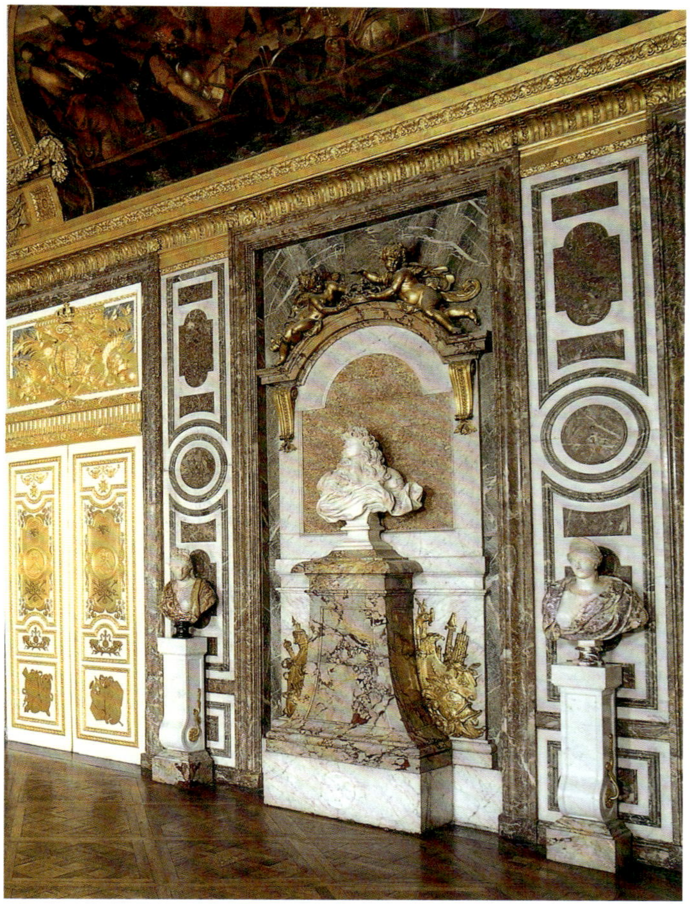

Fig. 5 Palace of Versailles, *Salon de Diane*

room and creates a vicarious corporeality of its own, which, with the quasi-sacred balustrade, suggests the untouchability of the king (fig. 4). In a configuration of rood screen and holy of holies, the chance of a glimpse of the king becomes a distinction, and access becomes an exclusive elevation. But there is no religious allegory in the decor of the *Chambre du Roi*. The mythological reference system is nominally desacralized. Only the painting links the king to the salvation story, when he is reflected in Domenichino's *King David*, supplemented by Caravaggio's *John the Baptist* and Guido Reni's *Mary Magdalene*; Louis's predilection for the Italian masters is evident.

The rich gilt furniture and fabrics, which continued to be used by Louis XV, but in 1785 on the orders of the Directeur du Garde-Meuble de la Couronne were burnt in order to recover the value of the metal, complete the impression of a house of power pervaded by gold. But in the gold gestalt of the room, one can hardly see more than a basic effect achieved solely by perception, not via interpretation. The absence and then the sparing use of gold in the sequence of the first rooms of the apartment are a perceptible design element without any preconditions, and shape expectations right up to the *Chambre du Roi*. The effect of the gold can be experienced directly. It needs no scholarly interpretation.

The Sequence of Rooms, Gold Decor and Light

The *Grand Appartement* is one of the few sequences of rooms whose fixed wall furnishings are still largely preserved in their pristine condition. The suite, which was first entitled *Appartement des Planètes*, was accessed from the *Escaliers des Ambassadeurs* and served for receptions and parties, only the most important diplomatic receptions being held here.[17] The *Appartement du Roi* in the opposite wing was the regular venue for court ceremonial.

When Jean-François Félibien comes to talk about the *Appartement en haut*, the *Grand Appartement*, and the sequence of rooms, what he finds most important is the marble decor in each of the salons:

"La première, est un salon qui a cinq toises & demie de long sur cinq toises de large. Les bandeaux des portes & des fenêtres sont de marbre jaspé de blanc & rouge. Les embrasures des portes [...] sont de marbre blanc remply." This is followed by a description of the parquet flooring and the ceiling paintings: "Les plafonds doivent être enrichis de peintures par les meilleurs Peintres de l'Académie Royale. Et comme le Soleil est la devise du Roy, l'on a pris les sept Planettes." And even when he is talking about the decoration, he does not mention gold, but rather the programme: "On en voit les figures symboliques dans les ornemens de sculpture qu'on a faits aux corniches & dans les plafonds."[18]

In Pierre Rainssant's report, too, his *Explication des tableaux de la galerie de Versailles, et de ses deux salons* dating from 1687, the gold is treated in a strangely offhand way in the description of the *Grande Galerie*. Certainly it is given no symbolic interpretation:

"Toute cette architecture est de marbre de differentes couleurs, à l'exception des Bases & des Chapiteaux, qui font de bronze doré, aussi-bien que les Trophées, les Peux de lion, les Festons de lauriers & de fleurs, les Soleils Rhodiens, & les Roses, qui ornent les arcades, & les entre-deux des pilastres." And as to the ceilings: "Toute la Galerie est voûté d'un berceau en plein cintre, enrichi d'une composition d'architecture en perspective de divers marbres, avec des compartiments d'or; ... Les bordures de tous ces Tableaux sont de stuc doré, avec des ornemens, qui ont rapport aux sujets."[19]

How do we see this today? The ascent to the *Grand Appartement* via the *Escalier des Ambassadeurs* (created from 1672, destroyed in 1752) formed a mise-en-scène independent of the apartment itself, a theatre of its own with the king at the centre.[20] The field of view is dominated by the red, green, white and grey of the marble, contrasting with the gilt bases and capitals, the gilt balustrades of the staircase,[21] and the gilt beam zone. With the gilt stucco decor of the dome opening, the elements frame the pure white marble bust of the king by Jean Warin, commissioned in 1665, in the central axis, backed by a large gilt niche. The Ionic orders alternate with battle paintings and trompe l'œils, which make the peoples of all

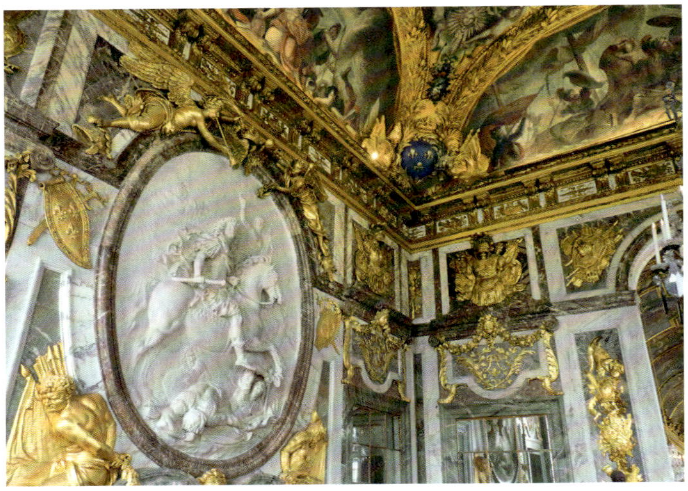

Fig. 6 Palace of Versailles, *Salon de la Guerre*

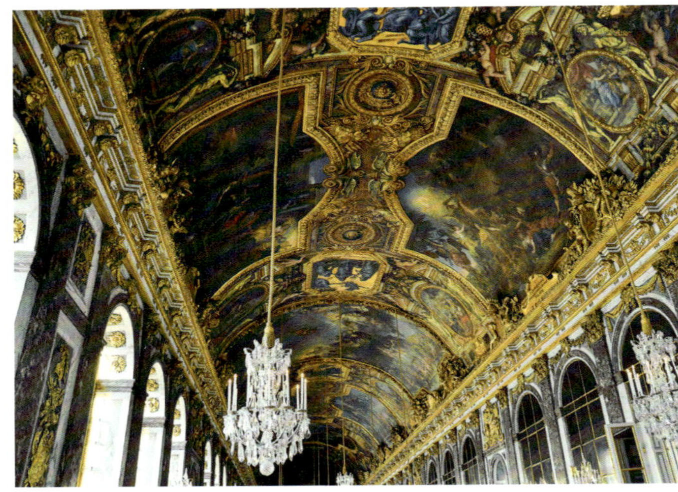

Fig. 7 Palace of Versailles, *Salon de la Guerre*

the continents spectators of the spectacle. The balustrade and beams take on, as architectural elements, a visual guide function, by accompanying the ascent or descent as a framed movement past the king's portrait. They are further emphasized by their gold finish.

After the *Salons de Venus et de Diane* with their function of a vestibule between the staircase and the apartment, visitors would enter the *Salon de Mars*, a guardroom, and then an antechamber (*Salon de Mercure*), the *Grande Chambre du Roi* (*Salon d'Apollon*), the *Cabinet de Jupiter* (today *Salon de la Guerre*), a *Petite chambre du Roi* (or *Chambre de Saturne*) and finally the *Cabinet du Roi*. After the death of the queen in 1684 and the establishment of the *Appartement du Roi* in the old part of the palace, the *Grand Appartement* became the venue, on which no expenditure was spared, for *grandes réceptions* and what were known as *Appartements*.[22]

Under the direction of Charles Le Brun and the supervision of Jean-Baptiste Colbert,[23] who initiated the so-called *Petite Académie* especially for the purpose, there was created a 'planet programme' at the royal residence as the result of collective reflexion, and in its ornamentation nothing was left to chance. Without exception, all the details, including the decoration of the doors, supported the theme announced by the central ceiling painting.

Let us take the *Salon de Diane* (fig. 5). In this antechamber the presence of the marble of the wall surfaces comes up against its limit (also in the architectural sense) in the cornice zone. The gilt architrave marks the beginning of the ceiling zone. As in the function of a tectonic structural skeleton, consisting of beams and frames, the gilt elements form a kind of co-ordinate system defining the empty spaces which create room for the paintings and sculptures. The doors are designed the same way. Set in a heavy marble frame, the door comes across as a colourless blank space, announcing the transition, and promising passage, to the next salon. What the gilding of the door decor refers to, thematically, are the symbol bearers standing out among the carvings, which, as emblems of power, together with the golden elements of the frame, represent tangible, highly crafted materiality (the Order of the Holy Ghost, which rivalled the Habsburg Order of the Golden Fleece, sceptre, *fleur de lys*, Louis's crowned insignia, the sun borne aloft in swags, and the signature LVDOVICVS XIV—ANNO 1682). Here it is the white ground that opens up the abstract space. The gold remains in the object. Only in this combination of white and gold is the hinge of the room-blocking and room-opening door made to speak in a promising manner.

The element of temporality created by the design of the doors—in the horizontal plane with the change in rooms—is repeated when one looks upwards, with the transition from the here-and-now, the marble load-bearing structure, to the sculpted, painted and gilded vault of heaven. The function of the gold seems to lie in pointing to this transition as ontological detachment. But this is achieved neither by transparency attributable to the material (the opposite is the case), nor with any characteristic suggesting space and infinity. Gold is always found where highly concrete, suggestive contents are prominent. The 'alchemy' of gold, which after all nowhere reinterprets the object which it clothes, seems to lie in the function of the room, which to start with points to nothing transcendental.[24]

Heaven and Earth

It is tempting to want to divide the spatial structure of the salon, in particular the literally weighty beam zone, which acts as a visual divider, into heavenly and earthly spheres, decorated with stone and gold respectively, but that certainly does not mean there is no gold in the earthly sphere. The permeability of the two zones lies in the formal and thematic reference made here by the colour gold. By dint of its sensory characteristics, in the contrast of material and light, its connective property becomes more evident the further one progresses through the sequence of rooms. Embedded in a

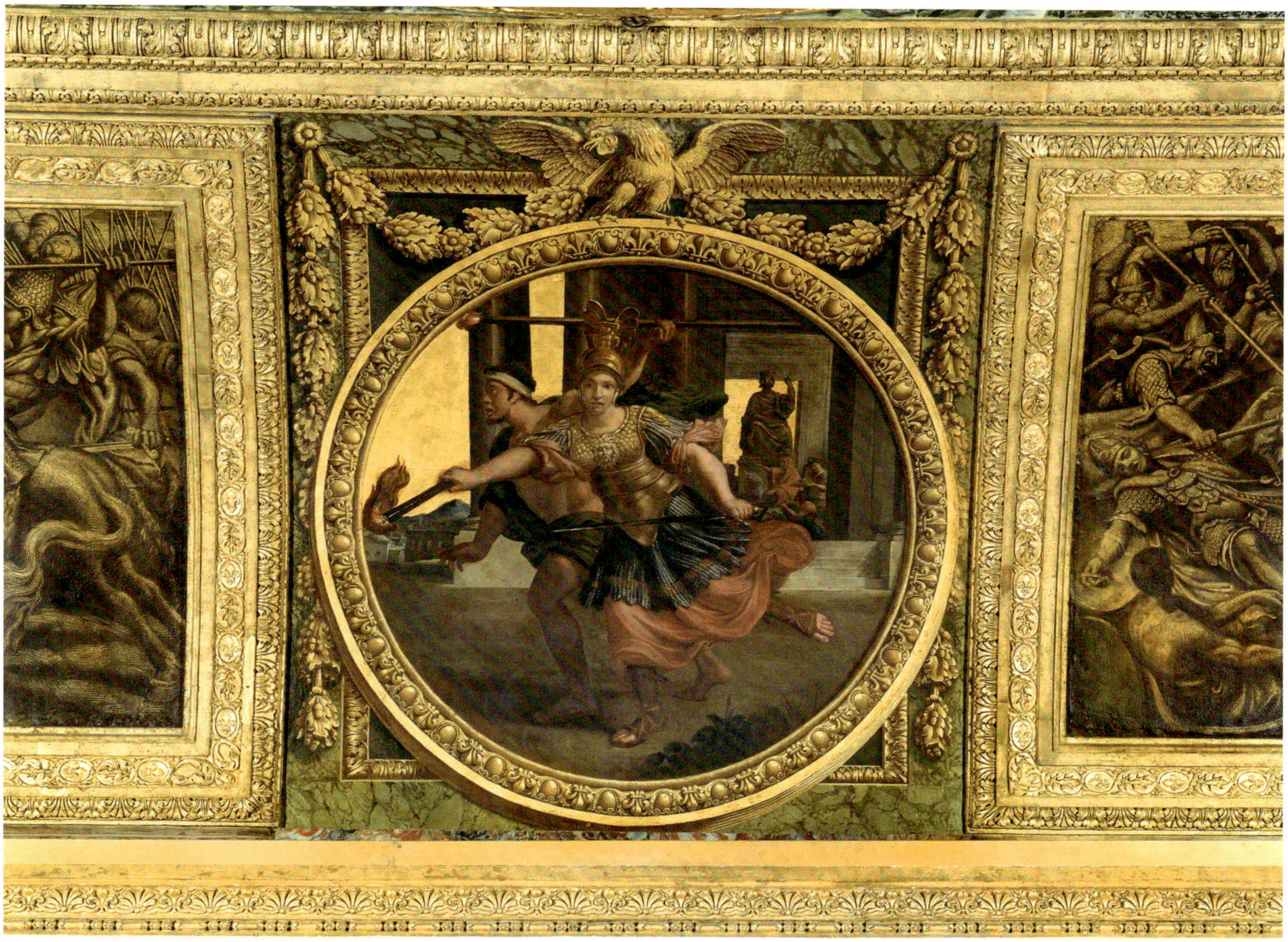

Fig. 8 Antoine Paillet, *La Fureur et la Guerre*, 1672

complex system of correspondences, it thus shares in the symbolic value of the iconography, which, with a view of the sun, does not need to dispense with the physically formal qualities of fire-gilding, the natural link with light. In other words, the charging of the gold with symbolic meaning in the context of the decorative programme is automatically joined by these scientific aspects. What we have here after all is an astrological programme of moved and moving planets with the sun at the centre, at a time when the influence of the stars on earthly destinies was not in doubt. Incorporated into the logic of the system were the astronomical discoveries of Giovanni Domenico Cassini, summoned by Louis to the new Académie des Sciences in 1669.[25] The image of the ruler on earth, by the grace of God, beneath the decor of the vault, favoured however a reproduction in pure white marble, in the form of the bust on the staircase, and the one by Bernini in the *Salon de Diane*, decorated with gilt bronze by Balthasar Keller, and the standing statue by Jean Warin in the *Salon de Venus*.

The *Salon de la Guerre*, a transitory corner room, which with a change of direction leads on to the *Grande Galerie*, had the task of preparing the climax. The marble wall-surfaces with applied gilt-bronze trophies by Pierre Ladoireau,[26] are rivalled by mirrored surfaces whose costliness and effect is not surpassed by the other materials.[27] The dominant, radiant-white medallion, the bas-relief of the triumphant king by Antoine Coysevox, with a red marble frame, is interpeted by gilt effigies of Victory and fettered slaves. Across the gallery and the entire wing of the palace, this room has its counterpart in the *Salon de la Paix*.

The *Grande Galerie*, begun in 1678 at the end of the Dutch War, is, with its 'art totale' (Milovanovic), the culmination of the sequence of rooms, and with the adjoining rooms (*Salon de la Guerre, Salon de la Paix*) occupies the whole garden front.[28] The history of its creation has always been interpreted as the symptom of a fundamental philosophical change from mythology to history.[29] While the first draft provided for a metaphorical Apollonian iconography, of the kind which reached its climax c. 1670 in the picture of the *Roi soleil*, eventually this was changed to a programme featuring Hercules, a Gallic hero invoked copiously by Henry IV in particular.[30] The designs already made by Le Brun however were discarded shortly before they were due to be executed, and on the orders

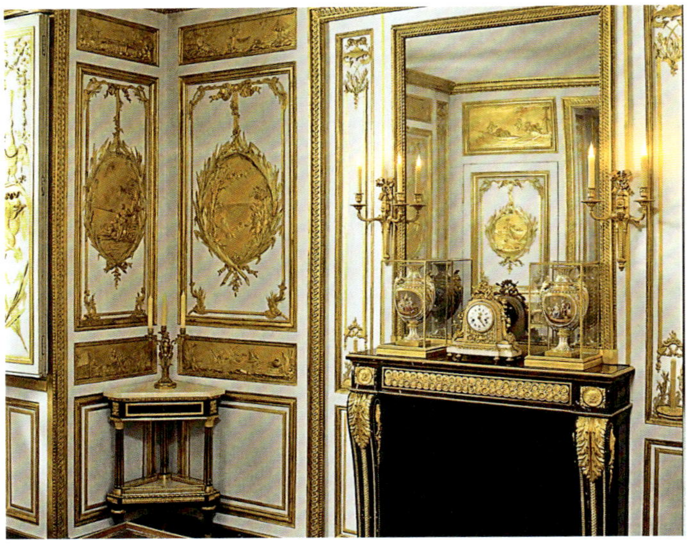

Fig. 9 Palace of Versailles, *Salle de Bains*

of the king replaced by a programme emphasizing contemporary deeds. But the change from a mythological to an historical focus should not be thought of in too categorical terms. The programme, which concentrates on the martial feats and benign rule of the king, is backed up by a decor which mixes allegory and history in the layering of numerous pictorial planes and references (fig. 6). A part in this delirium is played not least by the gold decor. The interweave of thematic meanings is matched by the increased variety of materials (coloured marble, 357 mirrors, gilt-bronze bas-reliefs, the French order of columns,[31] gilded stucco figures in the corners of the room, and Le Brun's ceiling paintings). Just as the themes are woven into a system of *images savantes*, consisting of emblems, mottoes and medallions, the formal design as a multitude of materials and modes is an aid to interpretation and a formal guide to seeing. This is done, to start with, in a highly strident manner, with a series of legends in gold letters which now, since the restoration, are even more striking.[32] Originally composed in Latin, but then replaced by a French version by François Charpentier, the legends are a prerequisite for a proper understanding of Le Brun's pictures, the 'emblèmes héroiques des actions du roi', as Nicolas Boileau called them. In the vaulting, which is only indirectly illuminated, they had to stand out.

So what role is played by gold here? With Le Brun's ceiling paintings there unfolds the story of the king's deeds (the crossing of the Rhine, the occupation of Maastricht, the Peace of Aix-la-Chapelle, the reconquest of the Franche-Comté etc.). The radiance of the king, who in the paintings as the unmoved mover not only metaphorically determines the course of the planets but also, and in fact, the course of events in the world, is emphasized by a frame and a sculpted, but nonetheless still gilded, arsenal. The revelation in the painting is prepared, both formally and thematically, by the golden decor; or, conversely, the compartments and planes of reality reflect the charisma of the king. In the ceremonial act, in the rhythm of striding forward, stopping and explaining, the sum of the king's feats becomes comprehensible. What we have here, however, is by no means a linear picture story to be understood in one direction. The decor shifts the mise-en-scène in a gilded amalgam on to overlapping planes of significance. In each case with the painted king in silver-gilt armour, the repetition and constant variation increases the forcefulness and credibility in a Cartesian sense. What is initially incomprehensible is rehearsed in the similar, the comparable, and the re-conjugated. It is here that the mystery is created, but then it requires scholarly interpretation. In his paintings, Le Brun balances the similar postures (the king, in Augustan-Apollonian calm, directs the actions of those in the picture) with the variation of the individual pictorial solution in accordance with the different, yet related, themes. By contrast, and in spite of the layering of the planes and techniques, the almost serial concept of the ceiling ornamentation represents a total system. How could a better balance between the parts and the whole be achieved? The three different frame formats, robust and sculptural, bring out the pictures to such an extent that with Le Brun's emphatic primary colours, bright again after the most recent cleaning, the painting remains unrivalled as *inventio*. The ceiling decor repeats in each (fictitious) bay the same decorative forms in gold: the cartouches with their legends, the shields held aloft by putti, along with their trophies, and at the ridge of the ceiling the sequence of trompe l'œils of gold-coloured medallions, which show Apollo's mask beneath the royal crown against the royal-blue background of the globe. In an entirely practical fashion, the gilding here has the function of illumination, with the constantly changing, constantly regenerated fragmentation and multiplication of the light.

The seeing eye is challenged by the interlaced ceiling levels and planes of significance, differentiated only by the different use of gold, leading to endless amazement. Gilded sculpture, gilded stucco, gold-coloured grisailles, the trompe l'œil of the gold-coloured coffered ceiling of the triumphal-arch architecture (they repeat the embrasure of the window arcades on the garden side) and finally, at the summit of the ceiling, and alternating with Louis's cipher in the wreath of Apollonian masks, the military feats in grey-black grisaille. These proclaim, almost like silhouettes against the glowing light of the gold ground, the repetition of the earthly deeds in the eternal heavens. Here it is the gold ground that is abstracted from the site of the action, but in spite of this transcendence it remains totally material. The further extension of the room, unclear in the gold ground in view of the lack of indications, is however due to Le Brun's paintings, which open up the view of heaven, and it is the painting that lifts the king out of the earthly sphere, in compositions which transcend the earthly actors with mythical pictorial ones, just as Rubens had done in the gallery of Marie de' Medici.

Gold Ground between Two and Three Dimensions

In the combination of the various modes of representation of gold in the *Grand Appartement*, two aspects are striking: the ceiling fields with their gold ground, and the figures executed in 'grisaille' or as *camaïeu d'or*. Let us take, from the medallions in the ceiling of the *Salon de Mars*, the painting *La Fureur et la Guerre* by Antoine Paillet (fig. 7). It is mentioned in the 1681 description by Sieur de Combes: 'un tableau rond, de coloris à fond d'or',[33] with an iconographic explanation that relates to the customary form of declaration of war employed by the Romans, according to Livy's History. The medallions in the room's short axis set off the series of horizontal grisailles portraying ancient battles along with the completely gilded trophies in the corners of the room.[34] That the paintings above the cornice are executed in gold hardly points in itself to greater material significance, one which would give the gold a specifically thematic and enhancing function. In Noël-Nicolas Coypel's paintings the technique of highlighting in gold, and the linear execution, lead to a dematerialization of the depicted subject, which in the way the light is reflected leads in turn to an alternation between the corporeal and the immaterial. But here, in the intermeshing of a whole variety of techniques, we see, rather, a demonstrative mastery of the medium in its variations and astonishing effects. It may be that the respective technique in gold supports the power and timeless validity of the classical example right into the present age, but this does not seem to be a necessary conclusion. Rather, we prefer to perceive a spectacular interplay of craft skill with the laws of physics and optical phenomena. Here, the gold colour assigned as a matter of course to the ruler helps, over and beyond its physico-technical, optical relationship to the light, to shape the light *metaphor*. It needs, however, the references which are in principle independent of it, defined by the symbolic meaning of the object itself.

The king himself preferred to be portrayed in marble. While gold, here, has a function in the context of the consecration of the royal person, it also symbolizes the power of the ruler from whom the light emanates. There is much in the way of symbolism here, doubtless also of physics according to the ideas of the time, but nothing magic. The reflection of the gold is part of a functional apparatus which is directed at what really matters, namely the pictorial representation of the king as making the propaganda come true. As an applied coating, the gilding has the function of ennobling the art which it supplements, surrounds or reflects. It cannot manage without an object; it is always the gilding of *something*, and it never appears without a form. In this way it does not simply point, in the context of its interpretability, to the transcendental, but lends glory to that which is celebrated in the here and now, a glory which it derives, at a distance, from the enlightened one. It is part of a constant *célébration de la personne royale*, reflecting permanently but ever anew the effulgence of the king. This sanctifying celebration is the result of the concert of the multifariousness of execution and genres. But depiction is depiction, only it is gilded to boot. Without the preceding, at times strenuous, attempt to interpret, it signifies no more than our historically conditioned, mostly unconsciously levelled projection.

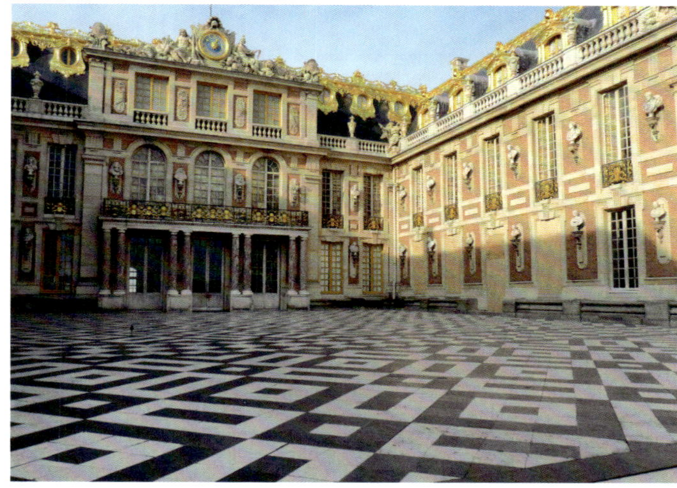

Fig. 10 Palace of Versailles, *Cour de Marbre*

Art versus Material and the War Coffers

The furnishing and decoration of the rooms had a crucial role in their total effect and the way this was enhanced as one passed through the sequence of *chambres*. The importance of the gilding becomes even clearer in the report of the monthly *Mercure galant* of December 1682, which lists the goldsmithery and gilding. The text gives a detailed survey of the objects, the swags 'portées par des guéridons dorés', the vases and the items of furniture, the cartouches and borders, the chandeliers and mirrors. It is easy to match the items with the entries in the inventories of the *Mobilier de la Couronne*.[35]

Both the inventories and the *Mercure galant* stress the quantitative aspect, above all by stating the size and weight of the items. There is no suggestion of any symbolic function of the silver and the gilding. And *Descriptions* à la Félibien are interested in the unfolding of the theme undertaken by the leading medium of painting.

In 1689 the Marquis de Sourches, the overseer of His Majesty's household, reports the equanimity with which the king had consented to the destruction of a large part of the decoration, a measure by which he had hoped to raise six million livres to top up the war chest.[36] Almost the total solid-silver content of the decoration was melted down and replaced by objects with cheaper gilding, for example the balustrade of the *Chambre du Roi*. Silver and gold seem interchangeable. The exchange here simply followed fashion and economic requirements; after all, they were only ornaments.

In the sources there is no sign of any particular semantic emphasis which might ascribe to the gilding any layer of meaning over and beyond wealth and splendour. What is emphasized is the technically feasible, the craft skills, and the development of numerous gilding techniques. We get the same impression from reading the *Conférences* of the *Académie Royale de Peinture et de Sculpture*, the institution from which the artists for the decoration of the *Bâtiments du Roi* were recruited. In the lectures, which deal with basic artistic questions, there is no mention of gold. In the biographies of the painters and sculptors, by contrast, the painting works in gold are mentioned not as particularly *significant* techniques of gilding, but as examples of craft skill, and of the rich ornamentation of the residences and *Hôtels particuliers*.[37]

The impression that we get here is not deceptive. The wealth of the nations of early modern Europe can also be measured by the quantity of precious metal. In the view of the seventeenth century and far removed from any abstract concepts of growth, the stock of gold and silver was almost constant. The amount of gold available rose and fell only through exchange and the balance of trade.[38]

Dis-illusion

In the conception of the royal apartments, the pictorial expression of the figure of light by the grace of God, gold does not represent the sacred element of rule. It is the formal compositional factor of this structure of significance. How far this deconsecration of gold goes—whatever previous sacred function we might presuppose—can be seen in the decoration of the *Salle de Bains* of Louis XV, restored a few years ago in exemplary fashion thanks to the sponsorship of Oréal (fig. 8).[39] In accordance with the most fashionable designs of the age, everything here, down to the bolts on the doors, is gilded. Nothing has been spared its coating of gold. The indiscriminate use of the colour gold leaves scope only for the white ground to create a boundary-less spatiality.

Gold, in our context, needs a contrast with the surrounding space from which it derives its significance in thematic charge. One might speak of mutual induction, with an attribution of significance that ultimately always derives from interpreters. What is absent is the golden key, which might answer the desire for unambiguous meaning. The attempt, which ends in disillusion, to see anything more in gold is ultimately left with nothing but amazement. One might in the end see in gold a merely general transcending function, a luminous colour leading heavenwards, which expressly excludes the body of the king, but which, after the restoration, once again includes with the *Cour de Marbre* the exterior form of the *temple* (fig. 10). Like the seam with the sky, the gilding rests heavily on the roof, with only sparse, appropriate resonance in the apartment storey, and surmounted by Apollo and the course of the sundial (only allegedly immutable). The sunset announced for 1700 took place in 1715. The year in which the king died marked the end of the court of Versailles. When the regent Philippe d'Orléans moved to Paris, the function of gold would have to be reconsidered.

1 Jacques Bénigne Bossuet, *Politique tirées des propres paroles de l'Ecriture Sainte*, Paris 1725, p. 238. In this text Bossuet repeatedly says that the authority and person of the king are sacred.
2 The apartment as shaped in post-medieval France with the ceremonial sequence of *Salle, Antichambre, Chambre du Roi* and *Cabinet*, had its first culmination with the Apartment of Anne d'Autriche and the new apartment for Louis XIV in the Louvre. Cf. Alexandre Cojannot/Alexandre Gady, "Les appartements du Louvre au lendemain de la Fronde—1652-1654", in: *Revue de l'Art* 142, 2003, 13-29.
3 *Médailles sur les principaux événements du règne de Louis le Grand, avec des explications historiques*, Paris 1701, new ed. 1 723. Cf. Claude-François Ménestriers *Histoire du roi Louis le Grand par les médailles, emblèmes, devises, jetons, inscriptions, armoiries et autres monuments public*, Paris 1689, new ed. 1691.
4 Cf. most recently: Hendrik Ziegler, *Der Sonnenkönig und seine Feinde. Die Bildpropaganda Ludwigs XIV. in der Kritik*, Petersberg 2010.
5 On the Apollonian metaphor see Jean-Pierre Néraudau, *L'Olympe du Roi-Soleil. Mythologie et idéologie royale au Grand Siècle*, Paris 1986.
6 On the entry of the newly wed royal couple into the city cf. Karl Moeseneder, *Zeremoniell und monumentale Poesie. Die 'Entree solennelle' Ludwig XIV. 1660 in Paris*, Berlin 1983.
7 Hartmut Stenzel, *Die französische "Klassik"—Literarische Modernisierung und absolutistischer Staat*, Darmstadt 1996, pp. 6ff.
8 With reference to Virgil's Fourth Eclogue ("Tu modo nascenti puero, quo ferrea primum / desinet ac toto surget gens aurea mundo / casta fave Lucina: tuus iam regnat Apollo").
9 On the mythico-historically represented figure of Louis XIV in the context of Versailles cf. Gérard Sabatier, *Versailles ou la figure du roi*, Paris 1999.
10 André Félibien, *Description sommaire du château de Versailles*, Paris 1674, p. 14.
11 More than half of the cost was borne by a Franco-American company that otherwise makes construction machinery. The 'reconstruction' of the *grille* by Jules-Hardouin Mansart is based largely on drawings by Jean-Baptiste-Alexandre Le Blond, but actually serves to control the streams of visitors. The sources, e.g. Jean-François Félibien in his *Description sommaire de Versailles* of 1703 (p. 32), speak of a 'grille de fer'.
12 The important contemporary descriptions relating to the apartment are: the issues of the *Mercure galant* of December 1682 (pp. 6-13), December 1684 (pp. 1-84), April 1687 (pp. 14-56); François Charpentier, *Explication des tableaux de la galerie de Versailles*, Paris 1684 (Ms Bibliothèque municipale de Versailles, ms G 138, fol. 121-126); Pierre Rainssant, *Explication des tableaux de la galerie de Versailles et de ses deux salons*, Versailles 1687; Claude III Nivelon, *Vie de Charles Le Brun et description détaillée de ses ouvrages*, c. 1698, Ms Biblioteca Laurenziana, Florence, Ms. Ashburnham 1723, and Jean-François Félibien des Avaux, *Description sommaire de Versailles ancienne et nouvelle avec des figures par Monsieur Félibien des Vaux, historiographe des Bâtiment du Roi*, Paris 1703.
13 Between 1665 and 1669 the *Appartement du roi* was furnished and appointed in Versailles from the north wing to the middle of the central wing, and it was extended between 1670 and 1680.
14 Cf. most recently Nicolas Milovanovic, *L'Antichambre du Grand Couvert—Fastes de la table et du décor à Versailles*, Montreuil 2010.
15 Ezechiel Spanheim, *Relation de la Cour de France*, Paris 1973.
16 This would be the opposite of the anecdotal "L'état c'est moi". On the daily ceremonial, see Béatrix Saule, *La journée du Roi au château de Versailles*, Paris 1986.

17 Reception of the Doge of Genoa in 1685, reception of the Siamese Embassy in 1686, and of the Persian Ambassador in 1715. On the diplomatic receptions, see Stéphane Castelluccio, "La galerie des Glaces : les réceptions d'ambassadeurs", in: *Versalia* 9, 2006, 24–52.

18 Félibien op. cit. (see note 10), p. 47.

19 Rainssant, *Explication des tableaux de la galerie de Versailles, et de ses deux salons*, Paris 1687, pp. ii and iii.

20 We can get an impression from the extant model, in addition to the descriptions and the replica in Herrenchiemsee.

21 As Claude Nivelon writes in the biography of his teacher Charles Le Bruns: "Le tout est revêtu de marbre et les balustres de bronze ciselé et doré au feu et on peut regarder ce lieu comme un des plus magnifiques qui se puisse voir dans l'Europe."

22 On the history of the major construction site, the fittings and furnishings, and the programme, in particular of the Grand Appartement, see: Nicolas Milovanovic, *Les Grands Appartements de Versailles sous Louis XIV*, Paris 2005.

23 On Le Brun's importance for the designs of the decor too, see Alain Mérot, "Le rôle de l'ornement : Le Brun, ensemblier de génie", in: Antoine Amarger/Arnauld Amelot et al., *La Galerie des Glaces. Histoire et restauration*, Dijon 2007, 154–171.

24 Here a consideration of the scientific significance of gold would be in order. The competition between alchemistical research at different courts, which at the court of Augustus II of Saxony led to the invention of porcelain while searching for gold, and from there to the success of the Meissen factory, was an integral part of self-presentation in the seventeenth century.

25 The moons of Saturn, discovered by Cassini in 1671, named by him (after his patron) the *Sidera Ludovicea,* were promptly integrated into Noël Coypel's project for the ceiling painting of the *Salon de Saturne*. The reference to the course of the stars allows the deeds of the mythological or classical heroes to be presented as operating across the ages, for example in the *Salon d'Apollon* the motif "Auguste fait bâtir le port de Misène" is transferred to the founding of harbour at Rochefort (1666), as Félibien also hints in his description.

26 Cf. Christian Baulez/ Alexandre Maral, "Les trophées en bronze doré de Pierre Ladoyreau", in: *La Revue des Musées de France. Revue du Louvre* 2007-3, Paris 2007, 65–73.

27 Colbert promoted the manufacture of mirrors against competition from Venice in Tourlaville and in 1665 with a factory in the Faubourg Saint-Antoine in Paris.

28 On the history of the *Galerie des Glaces* see: Bruno Cortequisse, *La Galerie des Glaces de Louis XIV à nos jours*, Paris 1999, and *La Galerie des Glaces. Histoire et restauration*, Dijon 2007, as well as N. Milovanovic/A. Maral (eds.), *La Galerie des Glaces : Charles Le Brun maître d'œuvre*, Versailles 2007.

29 The crisis undergone by the early modern post-Cartesian consciousness can also be seen as a crisis of perception and its conventions, formulated for example in the *Querelle des Anciens et des Modernes*. It is comprehensive in nature and tended to lead at the centre of power to a turning away from the recently rediscovered and much promoted Antiquity in the sense of a Golden Age. It looks as though the classically symbolic capital was drying up towards the end of the century. A process that started around 1680 and led at court from 1700 to a new religiosity, whose end in turn was longed for by many contemporaries.

30 On the transformation of the numerous mythological reference figures, cf. Simone Hoog, *Iconographie des Bourbons—de Henri IV à Louis XVI, d'après la collection de gravures de Louis-Philippe*, Paris 1985.

31 Cf. Arnaud Amelot, "L'ordre français à la galerie des Glaces de Versailles : un modèle resté sans postérité", in: *La Revue des Musées de France. Revue du Louvre* 2007-3, Paris 2007, 55–64.

32 Cf. Pierre Laurens/Florence Vuilleumier, "Les inscriptions de la galerie : une épigraphie de la gloire", in: *La Galerie des Glaces. Histoire et restauration*, Dijon 2007, 194–211.

33 Sieur de Combes, *Explication historique de ce qu'il y a de plus remarquable dans la maison royale de Versailles*, Paris 1681.

34 The ceiling painting by Claude-François Vignon, *Mars avec le Capricorne*, has been destroyed.

35 Cf. Jules Guiffrey (ed.), *Inventaire général du mobilier de la couronne sous Louis XIV (1663–1715) publié pour la première fois sous les auspices de la Société d'encouragement pour la propagation des livres d'art (1885)*, Paris 1885, e.g. pp. 5–13: "Vases et autres pièces d'orfèvrerie d'or".

36 On the silverwork and its history cf. Catherine Arminjon (ed.), *Quand Versailles était meublé d'argent*, (exhib. cat. Musée National du Château de Versailles et du Trianon, Versailles), Paris 2007.

37 Cf. on the examples of works by Charles Le Brun, Henri Testelin, Sébastien Bourdon, Nicolas Loir, Claude Audran the Younger, and especially Eustache Le Sueur highlighted in gold or painted on a gold ground in the lectures at the Academy: Jacqueline Lichtenstein and Christian Michel, *Les Conférences de l'Académie Royale de Peinture et de Sculpture*, Tome 1: *Les conférences au temps d'Henry Testelin, 1648-1681,* Paris 2006.

38 On this see e.g. Jean-Christian Petitfils, *Louis XIV*, Paris 2008, pp. 251f.

39 Cf. Pascal Bertrand et al., *L'intimité retrouvée—La restauration de la salle de bains de Louis XV à Versailles*, Paris 2004.

Vienna Gold Guide

Bodo Hell

widely scattered

gold and rolled gold: both seem to be indestructible and imperishable in their effect, capable of radiating their sheen for a long time, and thus attracting naive attention, indeed in the form of 'gilding' and in subtle gradations of colour and meaning gold can coat not only other metals, for example silver (the 'guilder' was a silver coin after all), but also imperial palaces (Schönbrunn yellow) and national dishes (cf. the proletarian golden-yellow breading of the Wiener schnitzel: the aristocrats decorated their food with genuine gold leaf), not to mention Ottakringer Goldfassl (a brand of Viennese beer) and the dubious Goldflitter (a brand of sparkling wine) for trivial celebrations

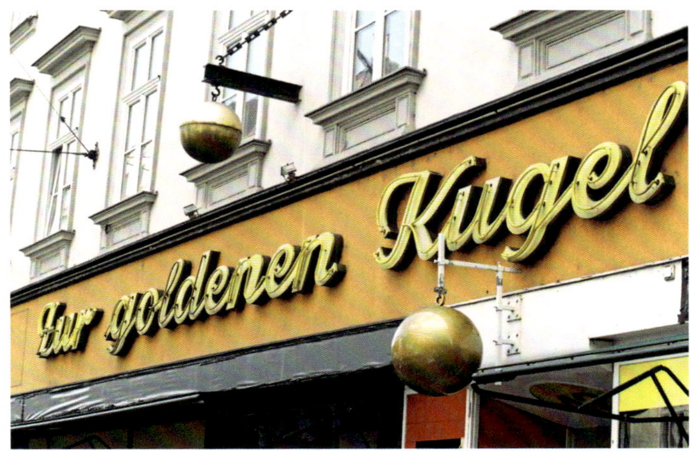

Fig. 1 Ironmongery "Zur Goldenen Kugel"
Wiedner Hauptstrasse 42, Vienna

there seems to be hardly any city that can do without a luminous **set of golden domes** visible from afar (thus there is little point in getting worked up about the *vieux riche* gold lustre of St Petersburg or the *nouveau riche* gold lustre of Orthodox Moscow), first and foremost, and furthest west in Vienna, radiating far and wide, the dome of the church of St Leopold on the Steinhof, a key building of the much-invoked turn-of-the-century, whereby Otto Wagner made the scales that cover this dome of die-cut and hammered copper plates, folded and thus moderately mobile, as a protection against the hail (the meteorological hail, not the hail of contemporary protest) and then gilded them, so that the result is a kind of floating gold balloon, which inspired the Viennese, who after all were quite at home on Lake Garda, to create associations with lemons, just as the gilded laurel-leaf dome of the perforated Secession dome soon came to be known as the 'Krauthappl' (bunch of weeds) in popular parlance (what is probably less well known is that its renovation was partly financed by the cosmetics magnate and respected former American ambassador to Austria Ronald Lauder), the Otto Wagner Villas I and II on Hüttelbergstrasse (whose further opulent decoration was, as is well known, taken on by an old-masterly artist, who, together with his colleagues can now call a Museum of Fantastical Realism in the Palais Pálffy their own) and the Art Nouveau buildings of the Linke Wienzeile are mentioned merely *en passant* (although each of these objects deserves a mono-graph of its own)

golden globes will doubtless catch the eye of the stroller of city streets here and there, for example the ones on the shoulders of the Atlases on the roof of the venerable National Library, to the right and left of the central dome (Josefsplatz), the blue-and-gold globe on the Military-Geographical Institute on the Zweierlinie, the pylon-flanked gold bulge over the redesigned entrance to the Kunstforum on the Freyung, the cupolas of the Russian Orthodox church on the Rechte Bahngasse of the S-Bahn (that just slipped in by mistake), then the two golden globes (one of them recognizably imitation as it swings in the wind of the diplomatic quarter) on the eponymous hardware shop on the Wiedner Hauptstrasse (the GOLDENE KUGEL (*anglice* GOLDEN GLOBE) was formerly a large shop, occupying several buildings, selling ironmongery and kitchenware), not to mention the many little globes, often surmounted by a cross, which the baby Jesus, as ruler of worlds, holds in his hand in the sculptural depictions of this

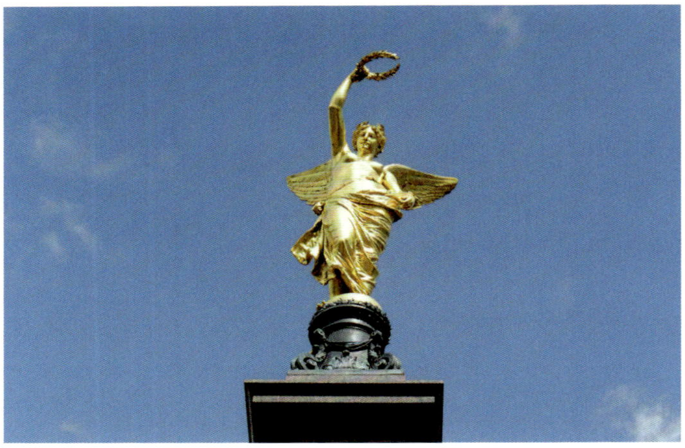

Fig. 2 Johann Silbernagel, *Nike*, on the Liebenberg Memorial, Vienna

primal scene (mother with sonny boy on her arm), and furthermore the great multitude, which no man can number, of heraldic **eagles, single and double-headed** alike, with and without a hammer and sickle, as well as broken chains (although in this case there would hardly be any example of the gilding of revolutionary emblems to be found), but also gold **portrait medallions** and radiant **goddesses of victory** like the one on the Liebenberg Memorial opposite the main building of the university on the Dr. Karl Lueger Ring

Karlsplatz

from afar (that is to say while walking down Lastenstrasse to the laurel-leaf dome of the Secession) there appears in the background the high tambour dome of the Karlskirche, standing out between the treetops of Karlsplatz and, when the position of the sun is right, the globes on the dome and the smaller ones on the balustrade on the balcony of the dome glow gold, otherwise it is just the **eagles**, with their wings interlocked, perching on, or taking off from, the platforms above the two twisted columns which reflect the imperial gold down on to passers-by, quite apart from the **gilt inscription** (in Roman capitals), whose forcefulness is (allegedly) enhanced by the use of precious metal: VOTA MEA REDDAM IN CONSPECTV TIMENTIVM DEVM PS XXI (I will pay my vows before them that fear God, psalm XXI), both of those involved with the church were incidentally called Charles: the founder (Emperor Charles VI) and the patron saint (Charles Borromeo, arch-bishop of Milan, implementer of the Tridentine reforms and one of the many plague-saints), and the programme of this building by Fischer von Erlach is said to have been thought up in part by the philosopher Gottfried Wilhelm Leibniz (who lived in Vienna in 1713/14), in the left-hand tower pavilion (a cycle-path passes between) two music students could just about stand, singing two-part folk-songs of the homeland, a hat with a few exemplary coins at their feet, Ulrike Truger's Empress Elisabeth, seeming to liberate herself with her fan from the surrounding marble block, has disappeared from her place, the presbytery behind has, over its entrance, the white stucco figure of St Charles Borromeo, a red cross and a six-pointed star on a black ground (the symbol of the Canons Regular of the Order of the Holy Cross, or Crosiers, in the eponymous Kreuzherrengasse) and on two chandeliers to the side gilded heads of angels, a statue commemorating Christoph Willibald Gluck (holding an operatic mask by its hair) must make do with an out-of-the-way position on Argentinierstrasse, and as the access to the Panorama Lift and the Museo Borromeo is by way of the right-hand tower pavilion, the central temple staircase and columned vestibule can be used by the public as a place to rest (the body of Anton Bruckner was doubtless borne through this central doorway after the funeral and before being transported by rail to St. Florian), the two angels on the flight of steps (the one on the right with a crushed serpent and the one on the left with a brass snake around a cross) symbolize not only the two testaments (New and Old), but also seem to address passers-by directly with their palpable presence, on the left-hand wall in the passage of the Wienmuseum to Symphonikerstrasse there can be seen in gold spray the big A for Anarchy in its circle (otherwise executed in red) and likewise in **gold spray** the sentence: "The wall is open"

Schwarzenbergplatz

passing the island of the French embassy with its creepers around its wrought-iron balconies and the two golden statues of female figures on the main traffic-plagued façade, one comes, when the traffic lights give you permission, to the back of Schwarzenbergplatz (now raised at the edges and around the trees, and also generally renovated and paved) and immediately the eye is struck by the black perforated Outer Space sculpture, penetrating the ear with its sound effects in the form of commissioned compositions as background noise: time and again, one would like to step into this perforated sound-sculpture called *the morning line* by Matthew Ritchie, which has the best sound technology currently available (Francesca Habsburg), and traipse through it in all directions, even though it seems to end as soon as it's begun on the front and rear sides, and in any case will soon be only virtually accessible, and the Hi-jet Fountain projects its cooling spray to different sides depending on the direction of the wind, while the shield, steel helmet and flagpole of the victorious Russian soldiers with shrouded machine-gun up there gleam golden in the sunlight (the shield, seen directly edge on, looks more like a chocolate coin wrapped in gold paper) and the capital Cyrillic letters in the semi-circle of columns (26 of them) reflecting the words: ETERNAL GLORY TO THE SOLDIERS OF THE RED ARMY WHO FELL IN THE STRUGGLE AGAINST THE GERMAN FASCIST OCCUPATION FOR THE FREEDOM AND INDEPENDENCE OF THE PEOPLE OF EUROPE, and (with the expression 'occupation') Austria is indirectly (and in the Moscow Declaration explicitly) presented as the first victim of Hitler's aggression, and that the **Russian Memorial** (allegedly largely unloved from the outset) is today preserved as the last Soviet edifice in a former Western country, seems more to do with stamina than with convictions, and as a place for the local employees to sit down

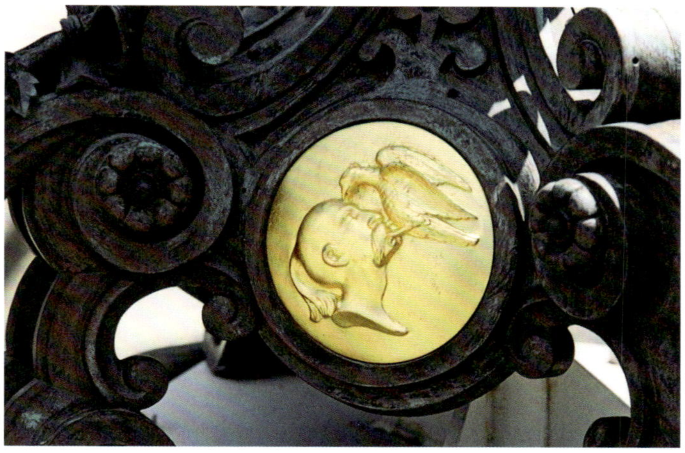

Fig. 3 Plaque with Turk's head and eagle on the Palais Schwarzenberg, Vienna

during their lunch breaks it may well even have become extremely popular, the view through to the semi-circular wall behind has remained open (the memory of a murder committed some time ago now in the shrubbery has largely faded), LEO HERE and also KEN C. have immortalized themselves, temporarily at least, in the supporting wall, while architecture students from Vienna's technical university have been surveying the whole monument photogrammetrically now for days on end, which is why we see, everywhere, little pieces of paper with black bull's-eyes in the sights sticking on the sweeping curve of the monument and a couple of students looking busy with a theodolite, the old filling station on the left has long since disappeared, while the signs advertising CROY DESIGN and DUPOUX painting restorers have remained, NO ACCESS TO THE BELVEDERE we are warned on several occasions, while someone has hung a notice in English *DON'T DISTURB PLEASE* on a French window, but parked on Vienna's most secret parking lot (entry and exit possible round the clock) in front of the closed Hotel Schwarzenberg are a sightseeing land-train as an empty prop and the mobile health-screening van belonging to PROVINCE OF BURGENLAND, while a company called Newport based in Tulln sees to security (Attention: guard dogs on patrol); time and again it attracts lost tourists trying to get to the museums of the Upper and Lower Belvedere who approach one for information, instead of orienting themselves by the city map; the now empty access ramp of the former luxury restaurant and the cast-iron lamp posts are restored, while on the medallions there appear a number of **armorial motifs of the Schwarzenbergs** in gold, above all that Turkish baldpate with a single pigtail, which a bird of prey, holding fast to a Turkish beard, is pecking in the middle of his forehead, also the Serbian embassy on Rennweg opposite has had bits of its balcony gilded, and the entrance to the museum has been moved from the path to the garden to the first access courtyard, or else the second entrance has been opened from the garden side (in spring the horticultural staff have their work cut out with the beds and the shrubberies, the strips of lawn and the gravel areas, the pools and the rows of statues in the open palace gardens, and it is almost impossible to imagine a time when no gardening, restoration, rebuilding or replanting was going on here)

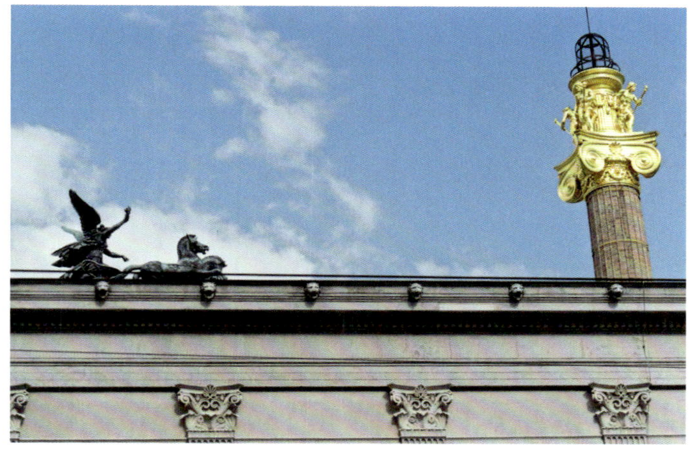

Fig. 4 Theophil Hansen, *Flue Column* on the Parliament building, Vienna

Parliament

if the fancy takes you, while the representatives of the people are in session, to circumambulate this extensive, pseudo-classical, pseudo-Palladian edifice, and immerse yourself in its details, it may well happen that one of the police officers on guard will keep a discreet eye on what you are so urgently intending at the rear of this precarious complex, and maybe it will occur to this young guardian of the public order for the first time how gleamingly gilt the two **cagelike chimney-hoods** at the rear, flanked by dark quadrigas with goddesses of victory, rise up into the summer sky, taking up once more, in their turn, as a decorative motif the functionless Ionic scroll capitals of the façade frieze in front of them, and that the colour decoration and gilt ornamentation should, according to the ideas of the naturalized Danish architect (Theophil Hansen) have been more opulent, can be seen on the front left-hand corner of the building, where there is some sparse **trial gilding**, but then you're already in the stream of visitors and photographers on the steps and ramps at the front of the building, and you just about have time to dodge to the left or right behind the imposing figure of Minerva, in other words behind the **Athene Fountain** with its riverine allegories of the Danube, Inn, Vltava and Elbe, whereby the goddess's helmet

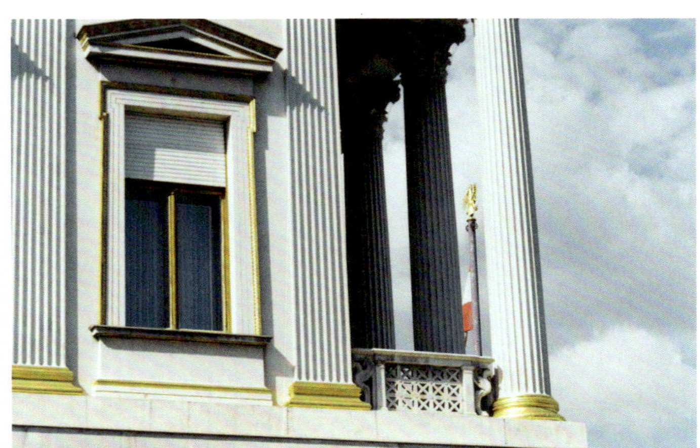

Fig. 5 Theophil Hansen, Trial gilding on the south-east corner of the Parliament building, Vienna

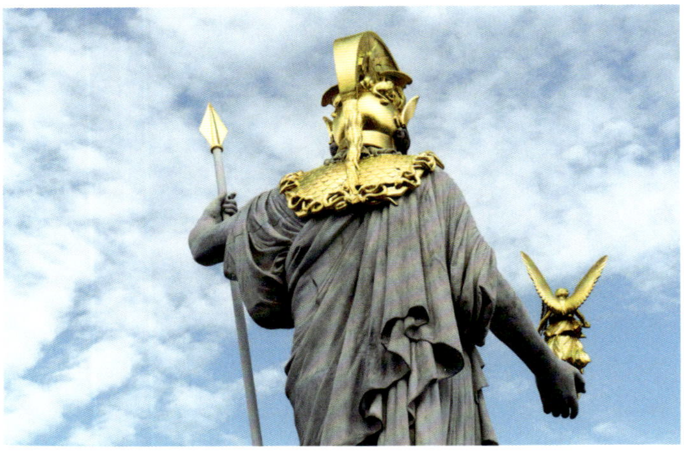

Fig. 6 Theophil Hansen (design), Carl Kundmann (execution) *Pallas Athene,* in front of the Parliament building, Vienna

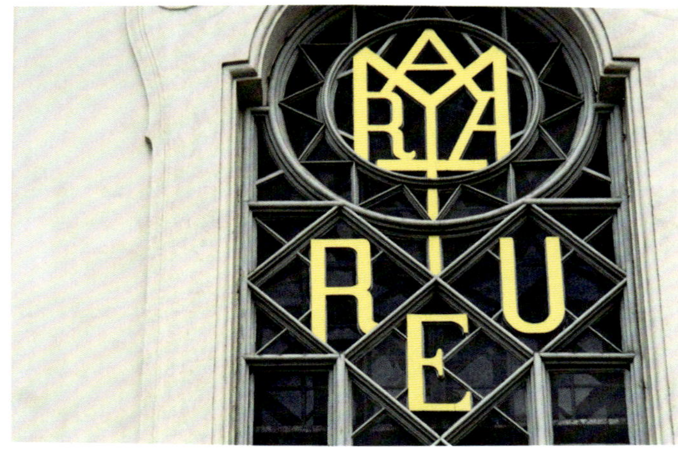

Fig. 7 Marianogram on the church of Maria Treu, Vienna

and hairpiece and corselet and the little figure of Nike on the globe in Athene's right hand are also gilded (which the ancient historian-figures on the ramp hardly notice), the gold-mosaic-backed colourful scenes of the inner picture fields in the upper vestibule (personified AUSTRIA with the crown lands doing homage) provide barely tolerable depictions (a naked stable-lad and a half-naked female carrying a wheatsheaf, bent women with false amphorae and excessively heavy and hence unstable jewel boxes etc.), and for completeness' sake we should also mention that the two flagpoles with their gilt upright republican armorial eagles only actually bear a flag when the parliament is actually in session; from the other side of the Ring the golden portrait of Julius Raab, retro-fitted into the garden railings, could shine across and its (today only semi-comprehensible) sentence: I ASK YOU FERVENTLY TO HOLD HIGH THE RED WHITE AND RED FLAG could only be read from the path at the back of the railings

Maria Treu

if we should be walking through Vienna's Quartier Latin (for example on our way to the city's most committed baker, namely Dieter Smolle vulgo Kornradl in the passage at Lerchenfelder-strasse 13), and if we should, not far from the start of Lange-gasse, raise our eyes to the house fronts on the left-hand side (those with the even numbers), on the right the Women's Café (which has long since lost the Women's Bookshop which it had in the golden years of the feminist upheaval) reminds us that it is in the 33rd year of its existence, then we should soon be struck by a restored Art Nouveau façade (which we probably would not have expected here), on which, in a number of wavy stripes which snake drily around the façade, a quantity of applied round gold platelets are to be seen, each one a decent distance from the next: this is doubtless not how we imagined the **Rhinegold**, rather they are evenly distributed **lucky tokens**, still waiting for their anglers and divers, and indeed our gaze falls, to the left of the entrance, on the name GLÜCK (*anglice* LUCK), that famous housing-estate architect who in recent decades has placed extensive residential complexes of the new trough-greened variety with swimming-pool on the communal roof, cellars on the second floor, and yellow handrails everywhere, fairly cost-consciously on the brown-field sites of Favoriten, Alterlaa and Fünfhaus, and who has his office here near the Zweierlinie, just as opposite, in the Seventh District, in the short Döblergasse, a generously proportioned Otto Wagner house has ventured into the late nineteenth-century surroundings, which one would also not expect, and if we stroll back to the little square on Kellermanngasse, where good old Augustin can be seen, and where bus route 13A reaches its lowest point (after a 1950s mosaic has familiarized us with the appearance of Kara Mustafa's tent), then our eye may be struck by a little gilded equestrian statue on the corner, with railings around: the vanquished (or at least he was sent packing) seventeenth-century Turkish general has been portrayed so twee, but at least he has been honoured by the former enemy in the form of an albeit miniaturized statue

now we shall not, repeat not, take the no. 13 to the former garden palace of Prince Eugen in order to take in the freshly restored white bodies and conspicuously Viennese-feminine charms of the crouching sphinxes in the open air, let alone to the rooms upstairs with the good view to examine Klimt's gold grounds more closely, but rather have ourselves driven two stops in the other direction to the theatre in Josefstadt, and then make our way on foot towards the doubly occupied Maria Treu Platz (coffee-house and pizzeria chairs) in front of the basilica of Maria Treu (if we had a picture frame we wanted gilded, there would be an opportunity in the workshop of a young craftswoman opposite), nor shall we venture into the church interior (in order to listen to Bruckner on the organ), nor through the passage of the Piaristengymnasium, a venerable high school named after and formerly staffed by the Ordo Clericorum Regularium Pauperum Matris Dei Scholarum Piarum, in order to view the unrendered brick rear of the splendiferous edifice, because what we're actually interested in is the gilt façade decoration of the Baroque church by Lucas von Hildebrandt: first of all on the square in front the interstellar Mother of God (Immaculata) is standing on a column decorated with clouds and angels, while sun and crescent moon glitter golden at her feet, maybe also the wreath of her son's stars around her head, her parents Joachim and Anna and foster-father Joseph are in attendance, but above all it is the heraldic pieces on the sweeping façade of the church, which above the VIRGO FIDELIS AVE COELESTIS MATER AMORIS transmit their golden lustre down to us: extraordinary decorations on the verdigris bell-garlanded tower roofs, the cross and the seven-pointed star in the middle, an oval medallion with mother and child (like an over-size gold necklet), mitre, broad-brimmed hat and bishop's crozier, and the five rows of cardinals' tassels and an armorial escutcheon in gold (cross with two bars), the large figures of women in the round, and the actual *clou* of this façade animation: these two exuberant ladies FIDES and SPES (one holds a medium-size golden cross, the other a stately gold anchor) have, as a narrow support for their backsides, nothing more than the none-too-broad cornice, and if the pair were to wiggle even more brazenly with their allegorical attributes, one might fear that perhaps they would slip off, and could, while falling, no longer manage to unfold their safety wings and float to earth, angel-like, between the crowds of Piarist school kids, while the centrally placed golden Marianogram (MARIATREU or MARIAREU: *anglice* MARIA TRUE or MARIA RUE) remains firmly anchored to the central window and poses its Marian letter puzzle in the manner of visual poetry

Order of the Golden Fleece

immediately to the left of the famous Neo-classical Canova tomb together with the sculpted cortège for Marie Christine (as is well known, wife of the founder of the Albertina and favourite daughter of Empress Maria Theresia) in the interior of **St Augustine's church** there is, above an unassuming side altar, the painted portrait of the Blessed Emperor Karl I (stylistically it comes across more like the image of a Christian-Social politician of the inter-war years, namely in hair-style and physiognomy, than as the likeness of the sovereign of the Order from 1916 to 1922), set up here in April 2005 (a brass plate on the kneeler in front tells you all this), and on the blue breast of his uniform (framed in the bright red lining of his cloak) we see, among other insignia, the **Burgundian Lamb** (the Habsburgs' domestic order since Philip the Good in 1430), hanging from a retaining ring fore and aft with skin and wool and head, and since there would be a great deal to say about the Golden Fleece (Maria Callas as Medea for P. P. Pasolini) and its order of knights (accompanying malefactors to the place of execution and drawing up a book of the dead, or would by any chance the Order of St George have dedicated itself to this task?), but here only this: in Colchis in Georgia and elsewhere too, rams' fleeces were used to sift gold out of rivers, in view of the fact that the gold-bearing sand easily got stuck in the wool of sheep, but it only remains here to say (in no sense in conclusion) that the Herzgrüfterl where the hearts of the Habsburgs are preserved in this same St Augustine's church (where Abraham a Sancta Clara also lies buried) is, contrary to expectation, not open at all times

1 URL: <http://www.TBA21.org> [read on 15 Nov. 2011]

Gold Rush

Linda Pelzmann

"Towards gold throng all
to gold cling all, yes, all!
Alas, we poor!"

These famous words are spoken by Gretchen in Goethe's *Faust*, and they could not be more up-to-date. Since 2000 the demand for gold has increased tenfold; the price has risen by 140 percent since 2005. Gold is the precious metal that is supposed to offer investors a safe haven in time of trouble. They buy gold because they believe in the constancy of its value compared with other assets. But in another sense, gold is a poor investment; it yields no interest.

And yet gold is sought after, as protection against the economic uncertainty that increases the drive to avoid liquidity crises on a personal level. Those thinking of escape also lay up a reserve of gold. It is rumoured that the late Libyan dictator Muammar al-Gaddafi, foreseeing the political turbulence that was to topple him, stashed away no less than 150 tons of the stuff. But even in small quantities, gold has proved its worth as an escape-aid. Felix Rohatyn, who saved New York City from bankruptcy in the 1970s by collecting private money, recounts in his memoirs how he managed, with the help of gold coins hidden in toothpaste tubes, to flee the Nazis and make his way to the United States via Brazil.[1]

The Magic of Gold

"Fifty thousand years ago man confronted the dangers and opportunities of his environment in a way which some 'pre-historians', sociologists and ethnologists agree was roughly equivalent to the attitude of modern primitives. Two elements of this attitude are particularly important for us: the 'collective' and 'affective' nature of the primitive mental process and, partly overlapping, the role of what, not quite correctly, I shall here call magic."[2]

Joseph A. Schumpeter

Gold has to say something that is not rational. The perception is permeated by primitive, magic conclusions. People associate gold with solid constancy of value, security, order, discipline and stability; in contrast to arbitrariness, chaos, movement and openness, vitality and change. Whatever we call 'golden' we associate with durability, equilibrium, closed ranks. The rise of order out of chaos, the ordering principle of the universe, is one of the most powerful archetypes, nourished by feelings of insecurity, fear, and the need for shelter. Our spirit always desires security. We seek for hidden analogies that link the unknown with the familiar and place the uncertain in a new light.

The rise of order out of chaos is a leitmotif of all mythologies. Every credible explanation, whether correct or not, has a calming effect. This is why we feel a sense of relief when puzzling and threatening events can be explained as manifestations of a divine order or a general law of nature. For the opposite of order is chaos—and that means unforeseeable events, shock, lack of shelter, exposure to the caprices of fate and the whims of unpredictable powers.

Even pain and suffering are mitigated by a feeling of relief as soon as they can be categorized. Even the bloodthirstiest goddesses of the Hindus and the wrathful deities of the Greek pantheon had a calming effect on people—for were they too not subject to human passions, passions that could be understood, so that there was an explanation for everything that might happen to one. The idea that 'mere chance' could decide our fate is intolerable. We seek protection from this unpredictability and this uncertainty, and we seek it in gold.

Gold Fetters

" You have to choose between trusting to the natural stability of gold and the natural stability of the honesty and intelligence of the members of the Government. And, with due respect for these gentlemen, I advise you, as long as the Capitalist system lasts, to vote for gold."[3]

George Bernard Shaw

The USA had solid reasons to abandon the Bretton Woods system with its gold standard in 1971. America was in the process of losing the Vietnam War, and its debt had risen to (for the time) immense sums. When the budget deficit became unmanageable, vast quantities of gold flowed to the creditor nations. As the dollar was pegged to gold, the Federal Reserve could not simply start printing money. When the 'gold fetters' were smashed, the era of freely floating currencies began. The value of paper money varied with the daily ups and downs of the markets. Overnight, budget deficits became invisible. The Western world followed America's example, and began allowing its deficits to rise unhindered. Gold by contrast is in short supply, its quantity cannot be increased at will, and so the world accepts it as the ultimate standard of stable value.

In the Gold Rush

"Most, probably, of our decisions to do something positive [...] can only be taken as a result of animal spirits—of a spontaneous urge to action rather than inaction [...] Thus if the animal spirits are dimmed and the spontaneous optimism falters, leaving us to depend on nothing but a mathematical expectation, enterprise will fade and die;—though fears of loss may have a basis no more reasonable than hopes of profit had before."[4]

John Maynard Keynes

The less transparent our environment and developments become, the greater is the probability that the market will not be 'moved' by the maximization of utility, but that another guiding principle will intervene and become dominant, namely the herd instinct. Mass-psychological dynamics are controlled by psychological contagion. I shall describe this process by recounting a story told by the legendary American investor Warren Buffett about oil prospectors. Here, I have substituted 'gold' for 'oil'.[5]

Gold Discovered in Hell

A gold-digger, having shuffled off this mortal coil, knocks at heaven's gate and desires entry. St Peter receives him in a friendly manner, saying, "Well, I checked you out, and you meet all of the qualifications. But there's one problem." He said, "We have some tough zoning laws up here, and we keep all of the gold diggers over in that pen. And as you can see, it is absolutely chock-full. There is no room for you." And the prospector said, "Do you mind if I just say four words?" St. Peter said, "No harm in that." So the prospector cupped his hands and yells out, "Gold discovered in hell!" And of course, the lock comes off the cage and all of the gold prospectors start heading right straight down. St. Peter said, "That's a pretty slick trick. So," he says, "go on in, make yourself at home. All the room in the world." The prospector paused for a minute, then said, "No, I think I'll go along with the rest of the boys. There might be some truth to that rumor after all."

King Midas—The Curse of Gold

Since time immemorial there have been legends concerning the great appetite for gold and its unexpected consequences. In the second half of the eighth century BC, King Midas ruled the Phrygian empire, which occupied much of Anatolia. He is said to have rescued a drunken companion of the youthful god Dionysus from the scorn of the peasants, and Dionysus, the god of the vine, wished to show his appreciation: "I will grant you a wish! For I would like to show my gratitude." King Midas did not take long: "Immortal god, if I may choose, then let everything I touch be turned to gold."

Dionysus fulfilled the wish reluctantly. Midas could hardly wait to try out the gift of the god. He started touching things. The twig he broke from the tree turned to glittering gold; the stone he picked up became a lump of gold, the grain and the corn that he plucked shone golden in his hands. Even the water that his hands touched turned to gold! Intoxicated by this miraculous metamorphosis, the king sat down to eat, reached for bread and meat, and found his hands full of sparkling gold. In a fright, he lifted the goblet to his lips: the dark wine had solidified into gold.

The dubious gift of having everything he touched turn into the precious metal also led Midas to turn his own daughter into gold. It would inexorably lead to starvation. Only then did the king see where his blindness had led him. He could not assuage his hunger and thirst, and his death seemed certain. He raised his hands and implored Dionysus to take back the deadly gift.

The god looked with pity on the penitent fool who had been seduced by his avarice. "Go to the River Pactolus, to the place where it springs from the rock. There at the source plunge your head into the cool water, and along with the gold, wash away your guilt!" Gratefully, Midas followed the instruction and freed himself from the baleful magic power. This was transferred to the water of the river, which thenceforth was panned for gold.

The Gold of the Kings

"It is the aura of the precious and the magic that we associate with gold."[6]

Peter Baum

At the pinnacle of his economic success, the industrialist and museum-founder Herbert W. Liaunig was gripped by a new collecting passion: gold. The gold of the Akan had whetted his appetite. He was overwhelmed by their charisma, their authentic rarity, and their craft.

On his visits to Zurich, Liaunig had bought 40 gold objects before acquiring the entire gold collection of René and Denise David. It comprised 400 items which René David, who had spent most of his time in Ghana since 1957, had assembled over four decades. The successful transfer of the property was favoured by the time, the meeting, fortunate circumstances and personal sympathy. Royal gold jewellery, insignia, and ceremonial objects of the Akan, who are indigenous to Ghana and the Ivory Coast, now arrived in the Liaunig Museum in Neuhaus, southern Carinthia. The precious works date from across the last 160 years, while some are much older, going back 400 years. Experts compare Liaunig's gold collection to the world's most famous museum collections of Ghanaian cultural items and insignia: the Museum of Fine Arts in Houston and the Gold of Africa Museum in Cape Town; the collection in the British Museum in London is smaller.

The catalogue conveys some idea of the magic of the collection, and tells of the magical relationships between animals and people. Imperishable archetypes cast their spell on visitors. The golden chief's ring with the crocodile holding a catfish in its jaws embodies the proverb: "When the catfish swallows something, it does it for its master, the crocodile." In other words, chiefs profit from the successes of their subjects.

The chief's ring with the golden scorpion is a metaphor of the perfect warrior: "The scorpion bites not with his mouth, but with his tail. If the scorpion stings you mercilessly, then you must kill him in the same spirit."

The gold ring with the Gabon viper and the hornbill symbolizes the relationship between debtor and creditor, which according to ancient tradition was also the relationship between the two creatures: "The hornbill borrowed money from the viper; and when the viper sought to call in the debt, the hornbill flew to the top of a tree and challenged the viper to climb the tree and recover the money. As the viper was unable to do this, he hid beneath a *dofokyee*, a decayed tree-trunk close to a pool. After a while, the hornbill forgot the incident and drank water from the pool. Whereupon the viper attacked him."[7]

By dint of the sheer mass of gold, the insignia convey a clear message: gold makes one immortal, the equal of the gods. René David however exchanged this unique collection of gold for a full life and a flourishing love for the beautiful Enam, the young daughter of a chief.

May every hoard of gold experience such a life-giving dedication.

1 Felix G. Rohatyn, *Dealings: A Political and Financial Life*, New York 2010.
2 Joseph A. Schumpeter, *Capitalism, Socialism and Democracy*, trans. unknown, London & New York 1943, quoted here from http://sergioberumen.files.wordpress.com/2010/08/schumpeter-joseph-a-capitalism-socialism-and-democracy.pdf, p. 121.
3 George Bernard Shaw, *The Intelligent Woman's Guide to Socialism and Capitalism* [1928] New Jersey 1984, p. 263
4 John Maynard Keynes, *General Theory of Employment, Interest and Money*, ch. 12, section VII. Quoted from http://www.marxists.org/reference/subject/economics/keynes/general-theory/ch12.htm
5 The story is told in Alice Schroeder, *The Snowball: Warren Buffett and the Business of Life*, London 2009, p. 12
6 Peter Baum, "Gold aus Afrika—Magie und modernes Design", in: *Das Gold der Akan* (exhib. cat. Museum Liaunig, Neuhaus), Vienna 2008, p. 7.
7 Doran H. Ross, quoted from *Das Gold der Akan* 2008 (see note 7), pp. 11, 77, 102.

"All that glitters..." Of Orpiment, Chrysotypes and Super-black

Thomas Zaunschirm

Gold exhibitions tend to take a liberal attitude regarding the material nature of the noble metal. Neither the pioneering, albeit encyclopaedically overloaded exhibition *Das Goldene Zeitalter. Die Geschichte des Goldes vom Mittelalter zur Gegenwart*[1] nor the most recent show *Aurum. Gold in der Zeitgenössischen Kunst*[2], which faded history out in favour of the present, confined themselves to gold the metal. The latter exhibition in Biel had examples of glazed ceramics, gold-coloured acrylic paint, gold-coloured ink, bronze pigments, gold spray, vinyl, various gold-leaf imitations, photographs, and videos, as well as 'mixed techniques'. Here there is no common theme, nor are there any limits to whatever interpretations anyone may care to apply.

Numerous works on art history content themselves with the golden sheen. This was already true in the nineteenth century, when the modern sense of space and corporeality clashed with the desire to integrate a medieval gold ground.

As one example among many we might point to a relief by Adolf von Hildebrand, an influential sculptor and theoretician. The piece is to be seen as an experiment by the artist: the background and the hair on the terracotta *Elisabeth von Herzogenberg as St Cecilia* (1897) were coloured gold, but without the use of gold leaf (fig. 1).

Quite generally, inventories do not note the differences in the materials. It can be seen from a painting by Erich Buchholz that he 'intended' a gold ground: does it then really matter whether he used gold leaf or gold paint?

The character of gold is intrinsic to many paintings, for example those by Hans Bischhoffshausen. But the 'gold' is gold-coloured paint, or (as one says euphemistically when the actual technique is unknown) some 'mixed technique' or other. Some authors say that the artist would have liked to use gold, but could not afford to. And indeed, when asked, many artists say precisely that. Gor Chahal writes: "Using

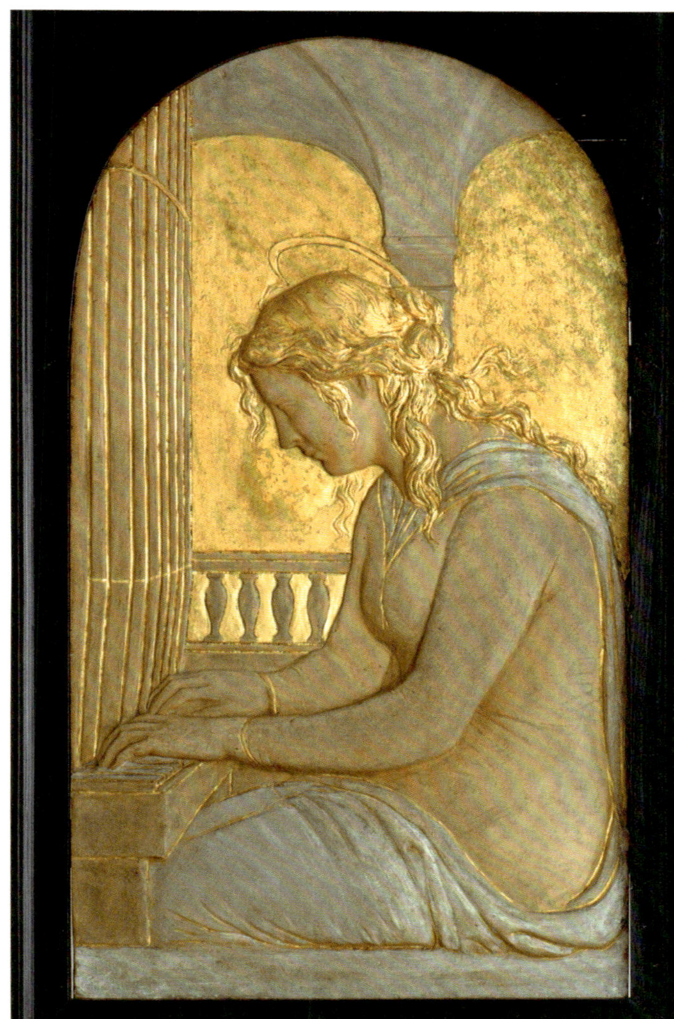

Fig. 1 Adolf von Hildebrand, *Elisabeth von Herzogenberg as St Cecilia*, 1897, Marburg, Universitätsmuseum

oracal or other materials looks like gold instead real gold for me was only economical question" [sic]. In such a case, might there not be the possibility of looking loftily in the other direction, as the beholder is not in any case capable of telling the difference? Especially when artists themselves use the word 'gold' in the titles of their pictures.

Often enough, gold is not a statement of the material used, but a poetic paraphrase. In *Venezia era tutta d'oro* (1961) Lucio Fontana uses gold-coloured oil paint. He does not use gold in paintings, but only to gild sculptures. *Zlata Praha (Golden Prague)* is the title Jiří David gives to his collage on acrylic (1988). What else might glitter with gold in the title, but actually simply look yellow?[3]

In titles of graphic works, Joseph Beuys uses the word 'gold', but investigations by conservators reveal it to be 'gold bronze'. The difference is clear: there is no element of gold in the mix to correspond to the golden impression. The beholder of such works is warned in the title in other ways against taking anything literally, because what we have in the aquatint from the series *Schwurhand* (1950–1970) is neither a sculpture, nor gold, but gold bronze. Then one remembers that many bronze or brass sculptures cannot be distinguished from gold ones. It becomes more complicated when bronzes are patinated with gold leaf, or when fire-gilding is involved. Patination with gold leaf often gives rise to the paradoxical impression that the smooth appearance of golden surfaces is disturbed, making them (intentionally) look old and used. Thus gold, the material, is employed to achieve precisely the opposite of what it has been seen as since time immemorial, namely indestructible and everlasting. Examples are known from such very different sculptors as Constantin Brâncuşi (*Danaide,* 1913) and Antoine Bourdelle (*Head of Apollo,* 1900–1909).

The intention behind the title of a work is not always evident. Paul McCarthy's multiple, *Brancusi Tree* (ed. 75), collapses when the integrated fan is switched off. The reference to Brâncuşi's *Infinite Column* can only be understood as ironic.

Many surfaces come across as golden, but they consist of *ersatz* materials ranging from shiny coloured paper to vinyl. Sometimes art-lovers are seriously disappointed, such as when at last they are standing in the MoMA in front of Warhol's *Gold Marilyn* (1962) only to find it's made of vinyl. Warhol at least makes up for it in the form of numerous early works with gold leaf. A golden field by Félix Gonzáles-Torres may glitter, but it consists of nothing but wrapped sweets. Nor would it go very well with Thomas Hirschhorn's garbage aesthetic if suddenly real gold leaf were to be used instead of gold-coloured gift-wrapping paper (although from his point of view the difference must be irrelevant). A stimulating confusion could also be caused by contrasting materials (papier mâché, plaster + gold leaf) as in the works of Franz West.

For the collector, the difference is in any case secondary. At an auction in the summer of 2010, Hirschhorn's necklet made $ 5000, in other words no less than if it had really been made of gold. What was evident from Warhol's dollar drawings turns out to be true time and again. A *Dollar* from his studio costs, in the nature of things, more than a dollar printed by the Fed.

Sigmar Polke certainly did work with gold leaf, for example in *Blaue Periode*, an ironic reference to Picasso's Blue Period, but when he calls a picture *Goldklumpen* ('Lump of Gold'), then in the alchemistical reaction typical of him, by using a different material, he is playing with the idea of metamorphosis. *Goldklumpen* (1981) uses highly toxic paints that are now no longer on the market, such as Paris green (which contains arsenic) and orpiment. Also known as auripigment and arsenic yellow, this latter likewise contains arsenic. Chemically, it is arsenic trisulphide (As_2S_3), and forms translucent golden yellow crystals insoluble in water or acids. Orpiment was used in Antiquity as a not exactly wholesome cosmetic. In the sense of Alberti, Polke extends the problem of depiction by the use of a material iconography. Alberti rejected the use of gold, because he regarded the use of paint to simulate the metal as superior. While Polke does not revise the use of paint to imitate gold, he replaces the appearance of gold in painting by the material appearance of gold in nature, an appearance which he introduces into his painting by using orpiment. His orpiment also depicts gold, not mimetically, but by reference to the history of deception through the use of this natural product.

Polke's means create a playful impression, but also disillusionment and disappointment. He does not cast off the compulsion to depict (gold, in Alberti's sense), but implies a pre-existing plane of deception in nature. However it is not nature that disappoints; rather our human reaction results from our own expectation. The disappointment is quite different from that provoked by Polke's 1982 black-and-white photograph of a *Lump of Gold*. Here it is merely the design potential of gold that is taken (uncontrollably) ad absurdum. It is reduced to being an unsightly lump of any old metal.

If photographs and videos are included in exhibitions about gold, then the specific properties of gold fly out of the window. A photo of a gold bar is no different from the photo of a bronze or brass copy. And it is no less of a misunderstanding to see gold-coloured substitute materials (from vinyl to gold bronze) as equally exhibition-worthy objects. These can never be more than indirect explanations. Reference is made to light phenomena such as the way sunshine falls on gold leaf, to symbolic reminiscences, to historical quotations, but gold itself plays no direct role. The reference to gold in these various media then corresponds to the depiction of gold in Alberti's sense. The surprising thing, though, is the emphasis on gold's material nature, which can certainly be iconologically charged, as a reference to the sacred or merely to the costly, the opulent or the ornamental.

On the other hand, insistence on the genuineness of the material raises the question of whether there can be any appeal in exhibiting invisible gold, or to be more precise, material gold which, by dint of chemical properties, cannot be recognized as such because, amazingly, it exhibits different colours. It should not occasion any surprise, in an age when the hunt for sub-atomic components in the elementary-particle zoo is in full swing, that 'materiality' is subjected to a transformation,

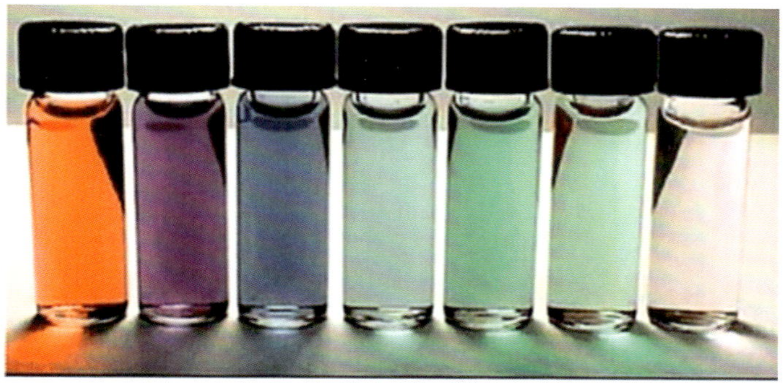

Fig. 2 Nano-gold colloides

just as the alchemists dreamed of. In the case of chrysotype, a high-quality printing technique using colloidal gold, and of nano-gold, it is perfectly possible to talk of the deconstruction of the gold phenomenon.

Chrysotype (literally: gold print) is a photographic process developed by the astronomer John Herschel in 1842. The chemist and photographic historian Mike Ware[4] has revived this form of printing with pure gold. The size of the dissolved gold particles gives rise to different colours. As the smallest particles absorb the short wavelengths of the visible spectrum (blue), the result is the complementary colour pink or red. The larger gold particles absorb long wavelengths, giving rise to blue. By mixing particles of different sizes, or also by changing their shapes, other colours (purple, violet, green) can be produced. The same material, gold, thus appears in different colours which have nothing to do with the traditional view of the sun-like sheen of gold. Amazing effects using tiny gold and silver particles were already achieved by the ancients in coloured glass. The famous Lycurgus Cup looks green in daylight, but under artificial light it is red.

The effects produced by gold particles were also used in Gothic stained-glass windows. In 1685, the Hamburg physician Andreas Cassius published a discovery which continues to bear his name: 'purple of Cassius'. The reaction of a gold compound with tin chloride gives rise to a very fine suspension of gold particles which, together with the molten glass, produce this purple. The gold content is no more than 0.08 percent.[5]

The traditional identification of gold with the sun is however taken ad absurdum in the international race to discover super-black, a surface that absorbs practically all light waves. In the dwarf world of nanotechnology (one nanometre = a billionth of a metre = 0.000001 mm) gold behaves quite otherwise than expected. With nano-gold, tiny gold-coated tubelets, the blackest black of all time was created a few years ago. Since other materials in this literally obscure race now achieve better results, gold as a particularly sun-like and thus 'symbolic' material will remain an also-ran. The mere fact that nano-gold colloids (fig. 2) take on different colours depending on the size and concentration of the particles contradicts the traditionally simple identification, popular in the history of culture, with the sacred and metaphysical.

For solid-state physicists, "gold is not a desirable noble metal, but a troublemaker that impairs the electronic characteristics of their materials."[6]

1 Tilman Osterwold (ed.): *Das Goldene Zeitalter. Die Geschichte des Goldes vom Mittelalter zur Gegenwart* (exhib. cat. Württembergischer Kunstverein Stuttgart). Ostfildern 1991.
2 *Aurum. Gold in der Zeitgenössischen Kunst* (exhib. cat. CentrePasquArt, Biel). Nuremberg 2008.
3 Cf.: Georg Salner, *Das Goldalphabet*, Vienna 2003.
4 Mike Ware, Gold in Photography: The History and Art of Chrysotype, URL: <http://www.siderotype.com/publications.html#books> [read on 8 Nov. 2011].
5 See Hans-Gert Bachmann, *Mythos Gold. 6000 Jahre Kulturgeschichte*, Munich 2006, p. 224.
6 *Nanopfeiler leuchten besser ohne Gold*, URL: <http://www.nanowerk.com/news/newsid=21884.php> [read on 8 Nov. 2011].
Cf. also: S. Breuer et al., *Suitability of Au- and Self-Assisted GaAs Nanowires for Optoelectronic Applications*, URL: <http://www.ncbi.nlm.nih.gov/pubmed/21319838> [read on 8 Nov. 2011].

Gold and Gilding Techniques

Tatjana Bayerova, Lisa Gräber, Judith Kern, Regina Knaller, Gabriela Krist, Veronika Loiskandl, Marija Milcin, Eva Putzgruber, Agnes Szökrön-Michl

The chemical element gold with the symbol *Au*, from the Latin word *aurum*, is one of the precious metals. Its colour, which does not tarnish, and its shiny nature have caused it to have been put to diverse uses in art. For that reason gold plays a significant role in almost all areas of conservation and restoration. Paintings, stone artworks, textiles and metal were usually gilded to look like real gold and also for decorative purposes. The present contribution takes the form of a discussion among restorers about gold and gilding techniques and provides an insight into the diverse uses of the precious material. The colloquy was held by the specialist staff at the Institute of Conservation and Restoration at the University of Applied Arts in Vienna.[1]

Gabriela Krist: Gold always says that the material has to fulfil the highest demands and in the past, in every era and every culture, it was reserved for the ruling houses and the senior clergy. For that reason I wish to present the following hypothesis at the start of our discussion, namely that the artists and artisans of the past used various gilding techniques on artworks in order to imitate solid gold. To get going, let us start with the gilding techniques used on paintings and on polychrome wood sculptures.

Veronika Loiskandl: The classic gilding technique is bole gilding. The first step is to apply the glue substrate, followed by several coats of chalk. This foundation, known as gesso, is warmed and applied with a brush. After it is dry, the gesso substrate is polished, after which the bole is applied. It is standard to use red bole, as well as sometimes yellow or white bole, bound by animal glue or egg white. Since the Baroque period, red bole has been used for burnished gilding and yellow bole for matt gilding. Before the gold is applied, the bole has to be smoothed by rubbing and brushing. The area to be gilded is moistened with a water and alcohol mixture, then the gold leaf was applied to the bole with an a flat brush known as a gilder's tip. Subsequently the areas covered with burnished gilding could be polished with agate or animal teeth and the matt gilded areas could be coated with more size so that a differentiated appearance with shiny and matt areas is created.

Gabriela Krist: In bole gilding and many other gilding techniques, gold in the form of gold leaf is used. That means that the gold first has to be beaten into extremely thin sheets.

Agnes Szökrön: In Europe, the craft of gold-beating was practised in monasteries up until the fourteenth century; after that secular craftsmen took over gold-leaf production. Vienna, Nuremberg and Prague were important centres until the nineteenth century. In order to manufacture gold leaf, the gold is usually alloyed with other metals such as platinum, silver and copper. The gold alloys vary in their colour nuances depending on the composition of the alloy. The amount of gold in the alloy is stated in carats. Red gold is purest, at 23 carats. The alloys are melted at around 1200 °C and cast into a bar, which is subsequently rolled out. The resulting gold sheet, cut into squares, is arranged into books or booklets and beaten under a spring hammer. The gold is subsequently beaten with special rounded hammers, until the leaves have an area of 8 cm by 8 cm and a thickness of 1/800 to 1/10,000 mm. In the past the leaf sizes varied, depending on where they were made, between 5 and 10 cm. In addition the gold leaves were significantly thicker.

Judith Kern: I would like to add something to bole gilding and the subject of material imitation. The gesso substrate is often processed further before the gold leaf is applied. In the pastiglia technique for example, the gesso is applied in a relief-like manner in order to produce a pattern on the surface. The gesso can also be removed in parts, before the gold leaf is applied, as is the case in engraving (fig. 1). Often the gilded surfaces received further treatment too, such as chasing and repoussage. These were used to create structured surfaces that created the effect of a worked metal surface. Textiles were often imitated using the pressbrokat technique on medieval panels and polychrome sculptures. The effect of

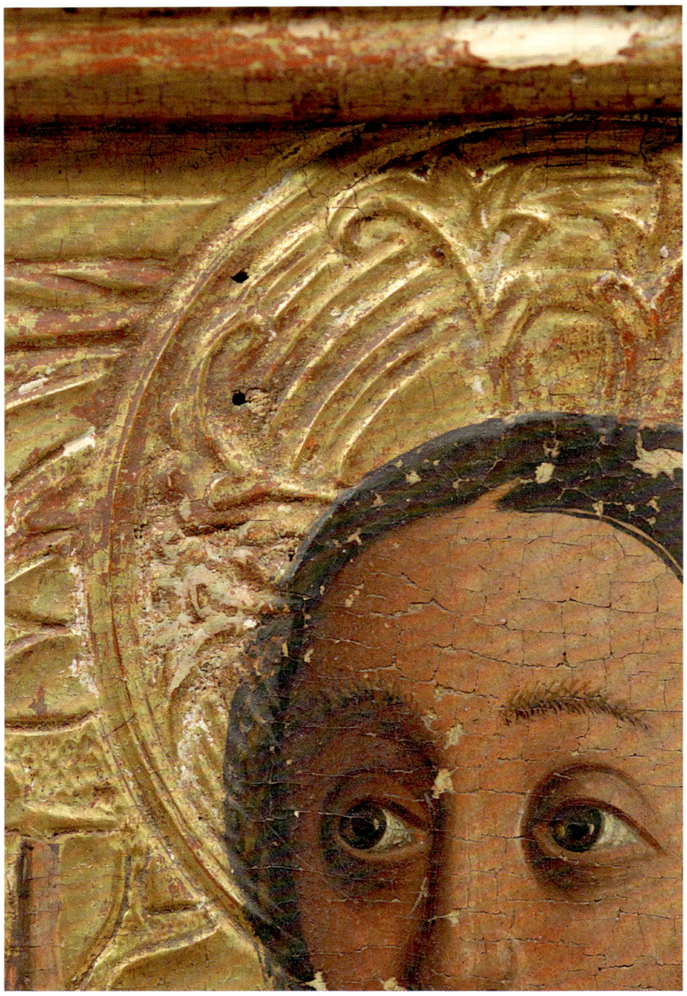

Fig. 1 *Apostles Altar,* c. 1490/95, Detail of a bole gilding with engraved chalk ground

brocades was imitated by laying a tin foil on a metal stamp and pressing both into a malleable mass. The foil served as a separating layer and could be removed from the pressed leaf after the stamp was lifted. In a different technique, ornaments were cast in moulds that were subsequently applied to the substrate.

Gabriela Krist: With that we have introduced traditional bole gilding, which was used all the way back to Antiquity. I would now like to discuss other material groups, because gilding is not just used for traditional painting backgrounds or for covering sculptures; it is also used on stone, for example.

Marija Milcin: Since bole gilding cannot be used on exterior stone surfaces because of its sensitivity to moisture, the most common gilding technique is oil-gilding. A drying oil is used as an adhesive agent between the gold leaf and the substrate, which is known as oil size. Preparing the substrate is an important step that can vary depending on the type of stone used. Non-porous stones require polishing, whereas porous stones need to have a glaze applied. The oil size is subsequently applied in a thin layer. After a period of time, when the oil is only just tacky, the gold leaf is applied (fig. 2). It

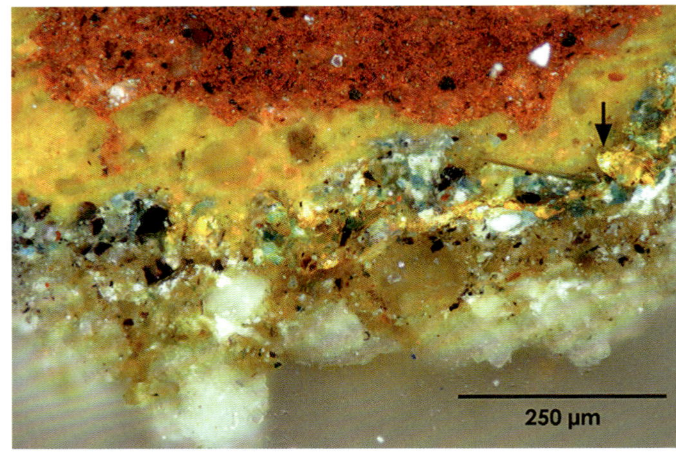

Fig. 2 *Mount of Olives relief*, 1511, Cross-section of a partial gilding on painted calcareous sandstone

should be said that oil-gilding is a matt gilding technique. Historical stone objects are largely only gilded in part, for example to accentuate the hair, the angel's wings or the edges of gowns.

Gabriela Krist: Gilding artworks fulfils several functions. Firstly it is an imitation of solid gold, and secondly, important areas are emphasized through the use of gilding. How was gold leaf used on textiles? Historical textiles were often adorned with gold wire.[2]

Regina Knaller: Hammering gold between leather skins was already mentioned in the Old Testament. To use it for gold thread, narrow strips of gold leaf, approximately 8 cm long, were cut and wound into a thread. Pliny already mentioned the technique of spinning gold around thin thread. This thread was made of linen, wool or silk and the thin strips of gold were wrapped around it. In Antiquity works produced by tablet weaving, embroidery and knitting, but not woven textiles, were decorated with gold thread. Later, the gold strip is applied to a support material, such as intestines, leather or paper (fig. 3). The carrier material allows the gold to be woven as a long strip. Textiles whose gold threads ran from selvedge to selvedge are evidence of the lavish use of gold. There is evidence of textiles produced in Europe and adorned with gold thread only with the advent of the silk industry in the northern Italian towns of Lucca, Venice and Florence from the start of the thirteenth century onwards. Animal intestines were used as a support in this region. Applying the gold to a support material has the advantage that longer gold strips can be used. In these Italian silk weaving centres the gold threads were used sparingly, only for patterns.

Gabriela Krist: When dealing with textiles the aim was once again preparing the precious metal as sparingly as possible for manufacturing processes, while maintaining the visual effects peculiar to the gold. If we move on to the gilding of metal objects, we move away from the standard gilding techniques using gold leaf.

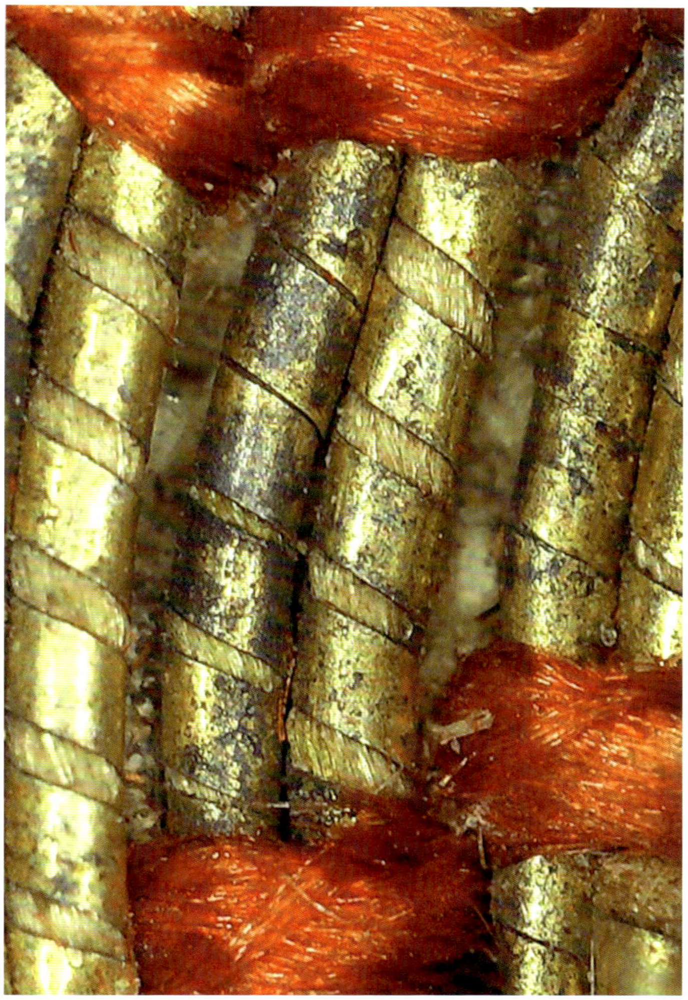

Fig. 3 *Relief embroidery*, c. 1500, Detail of a gold thread with gilt wire wound around a silk core

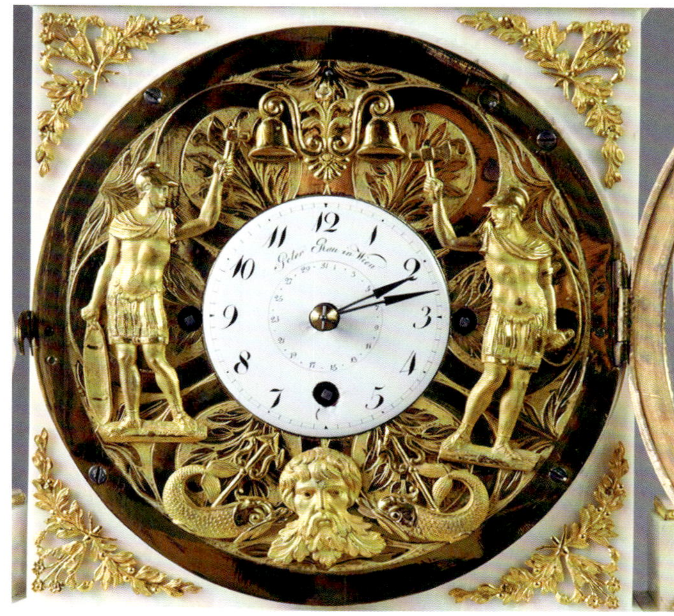

Fig. 4 *Alabaster clock*, pre-1850, Detail of the fire-gilded bronze clock-face

Eva Putzgruber: From Antiquity to the mid-nineteenth century fire gilding was the traditional method of gilding metal surfaces.[3] An alloy of gold and mercury was produced, whose melting point is much lower than that of pure gold. The resulting gold amalgam had a dough-like consistency and was kneaded under water in a porous leather sack to remove the non-amalgamated mercury. After that the metal surface to be gilded had to be cleaned and de-greased in order to allow a good bond. Sometimes the surface was also amalgamated. The gold amalgam could be applied using a spatula and distributed with a rabbit's paw. After that the object was sublimated over a weakly glowing charcoal fire. The heat caused the amalgam to melt and the majority of the mercury evaporated, while the gold and the underlying metal formed a close bond. The gilded surface was then roughened with a brass brush and finally polished with haematite (fig. 4). In fire gilding too, different alloys resulted in different shades of gold, producing visual diversity. Apart from the item's material upgrading, the underlying metal was also protected from the effects of oxidation and corrosion.

Gabriela Krist: Thus gilding has a protective function for long-term preservation here.
In addition to fire gilding, another method used on metal is galvanic gilding.

Eva Putzgruber: Yes, however, the development of galvanic gilding occurred very much later. This technique only became a serious competition to fire gilding from the mid-nineteenth century on. Galvanic gilding uses an electric current and an electrolyte containing dissolved gold. Together with the object, which acts as the cathode, a stainless-steel anode is suspended in the electrolyte, then a current is applied. Metallic gold precipitates on the surface of the object. The gilding layer thus produced is very much thinner than that of fire gilding. The two gilding techniques can be distinguished by the fact that a fire gilded object will not have its back, underside and inside gilded. For that reason the edges tend to display characteristic gilding edges. In galvanic gilding the gold can be found on all of the object's surfaces.

Gabriela Krist: Maybe this is a good time to talk about how the materials are identified. Restorers have their eyes and their experience. There are, however, analytical techniques that can be used to identify materials.

Lisa Gräber: There are several methods available. Initial conjectures about whether the material in question is sold gold or gilded can be made by observing the object's surface under an optical microscope. Often the next step is taking a material sample, which is embedded in synthetic resin before being ground and polished. Subsequently the cross-section can be examined under incident light, providing insight into the layers of the gilding. In addition it is possible to determine how thick the gold leaf is. Certain proof of gold can be obtained using instrumental analytical methods such as energy-dispersive X-ray spectroscopy (EDS). A qualitative elementary analysis will do more than just establish the presence of gold, it will also provide information about the gilding techniques. In the case of fire gilding, for example, mercury can also be found in the gold layer. Quantitative analyses also allow conclusions to be drawn about the composition of the gold alloy.

Gabriela Krist: Gold and gilding were always imitated by more inexpensive materials; during the Baroque and Rococo periods they were also deliberately contrasted with non-precious, non-ferrous metals.[4] In one technique, silver leaf was given yellow, transparent coatings. Gold-coloured metal alloys such as tombac or brass were processed into imitation gold. Often these alloys were woven as wires into textiles or ground into powders and bound and used for bronzing. These gold imitations raise technological questions, most of them complex, for conservation and restoration, which can often only be answered with the help of specific materials-science analyses. Gold and gilding techniques are thus a broad and exciting area of research for the conservation sciences.

(Glossar on p. 364)

1 Further reading on the subject of gold leaf and leaf metals: Filippo Baldinucci, *Vocabolario dell'arte del disegno*, Firenze 1681.- Gerhard Baumann, *Die Kunst des Vergoldens*, Leipzig 1988. - Diana Bigelow (ed.), *Gilded Wood. Conservation and History* (Symposium Philadelphia 1988), Madison 1991. - Thomas Brachert, "Die Techniken der polychromen Holzskulptur", in: *Maltechnik restauro,* no. 4, 1972. - Thomas Brachert, *Lexikon historischer Maltechniken, Quellen-Handwerk-Technologie-Alchemie*, Munich 2001. - Fritz Buchenrieder (ed.), *Gefasste Bildwerke*, Munich 1990. - Cennino Cennini/Mario Serchi (ed.), *Il libro dell'arte (c. 1400)*, Florence 1991. - Denis Diderot/Jean Baptiste le Rond d'Alembert, *Encyclopédie, ou dictionnaire raisonné des sciences, des arts et des metiers* ..., Paris 1751-1765. Recueil de planches, sur les sciences, les arts libéraux, et les arts méchaniques, avec leur explication, Paris 1763 (Reprint New York 1959), vol. II, pl. 418, 421-424. - Homepage of the. Dungl company, URL: < http://www.dungl.com/deutsch/dungl/herstellung.htm > [read on 11 Nov. 2011] - Oskar Emmenegger, "Metallauflagen und Applikationen an Wandmalereien", in: *Zeitschrift für Kunsttechnologie und Konservierung,* no. 3, 1989, 149-164. - Heidelinde Fell (ed.), *Vom Anschiessen und Durchreiben. Die neuen Leiden des alten Goldes* (report of the symposium on 6 February 2009 in Diözesanmuseum Graz), Graz 2009. - Alois Firlei, *Die Poliment-Glanzvergoldung und die antike Polychromie für Maler, Vergolder und Bildhauer*, Munich 1926 (Sammlung maltechnischer Schriften 12). - Cornelius Hebing, *Vergolden und Bronzieren*, Munich 1907, 15th ed.. 1990. - Brigitte Hecht, "Betrachtungen über Pressbrokate", in: *Restauro*, vol. 1, 1980, 22-49. - *Hermeneia tes zographikes. Malerhandbuch des Malermönchs Dionysios vom Berge Athos*, Munich 1960. - Aldo Hug, "Vergolder-Kasten", in: *Maltechnik,* no. 77, 1971, 6-10. - Hans Kellner, *Vergolden. Das Arbeiten mit Blattgold*, Munich 1992.- Johannes Klinger/Roland Thomas, *Die Kunst zu Vergolden. Beispiele-Techniken-Geschichte*, Munich 1989. - Manfred Koller, "Die Lüsterfarben—zu ihrer Geschichte und Konservierung", in: *Maltechnik,* no. 1, 1975, 20-34. - Manfred Koller, "Barockaltäre in Österreich: Technik, Fassung, Konservierung", in: *Restauratorenblätter,* vol. 2, 1976, 17-71. - Manfred Koller/ Hubert Paschinger/Helmut Richard, "Gefasste Metallobjekte—Aktuelle Befunde und Restaurierprobleme", in: *Restauratorenblätter,* vol. 11, 1990, 155-162. - Manfred Koller, "Die Kreispolitur—eine vergessene Vergoldertechnik der Spätgotik", in: *Die Kunst und ihre Erhaltung. Rolf E. Straub zum 70. Geburtstag*, Worms 1990, 47-56. - Manfred Koller, "Polieren und Firnissen historischer Metallauflagen", in: *Restauratorenblätter,* vol. 11, 1990, 127-134. - Manfred Koller, "Altar- und Skulpturfassungen des Barock in Österreich—Renovieren oder Konservieren?", in: *Restauratorenblätter,* vol. 20, 1999, 41-52. - Manfred Koller, "Damastfassungen. Quellenlage und Beispiele von Skulpturen des 17. Jahrhunderts", in: *Restauro,* no. 2, 2001, 114-123. - Johann Georg Krünitz, *Ökonomisch-technologische Enzyklopädie* ..., Berlin 1773-1858, vol. 208, Berlin 1852, 91-193, URL: <http://www.kruenitz.uni-trier.de/> [read on 11 Nov. 2011]. - Hermann Kühn/Heinz Roosen-Runge/ Rolf Straub/Manfred Koller, *Reclams Handbuch der künstlerischen Techniken*, vol. 1, Stuttgart 1984. - Johannes Kunckel, *Ars Vitraria oder vollkommene Glasmacher-Kunst*, Nürnberg 1679. - Nenna Merhardt/Traudl Zulehner, *DuMont's Handbuch Vergolden und Fassen*, Köln 1987. - Hans Portsteffen, "Ein neuer Beleg zur 'praunierten Vergoldung'", in: *Zeitschrift für Kunsttechnologie und Konservierung*, vol. 6, 1992, 331-334. - Frank Reid/William Goldie, *Gold als Oberfläche. Technisches und dekoratives Vergolden, Verfahren, Anwendung, Bearbeiten der*

Schichten, Eigenschaften und Prüfung, Saulgau 1982. – Otto Rentzsch, *Gesamtgebiet der Vergolderei*, Vienna 1908 (Hartleben's Technische Bibliothek vol. XX). – Ulrich Schiessl, *Rokokofassung und Materialillusion*, Mittenwald 1979. – Ulrich Schiessl, *Techniken der Fassmalerei in Barock und Rokoko*, Worms 1983. – Kurt Schönberg, *Historische Beschichtungstechniken. Erhalten und Bewahren*, Berlin 2002. – Erling Skaug, "Cenniniana: Notes on Cennino Cennini and his treatise", in: *Arte christiana,* no. 754, vol. LXXXI, 1993, pp. 15–22. – Heinrich Friedrich August Stöckl, *Praktisches Handbuch für Künstler, Lackirliebhaber, Vergolder und Anstreicher aller Ölfarben, dieselben unverändert zu erhalten. Nebst noch vielen nützlichen Arbeiten in diesem Fache,* Vierter Theil, Munich 1820. – Christian Friedrich Gottlieb Thon, *Die Staffirmalerei und Vergoldungskunst. Ein praktisches Handbuch für Maler, Baumeister, Lackirer, Ebenisten* ... Weimar 1836 (Neuer Schauplatz der Künste und Handwerke, vol. 87). – Jean Felix Watin, *L'art du peintre, doreur vernisseur*, Paris 1772 [German trans. of the 2nd edition, Paris 1773: *Der Staffirmaler, oder die Kunst anzustreichen, zu vergolden und zu lackiren* ..., Leipzig 1774]. – Kurt Wehlte, "Blattgold und sein Werdegang", in: *Maltechnik,* no. 4, 1974, 101–111. – Christoph Weigel, *Abbildung und Beschreibung der gemein-nützlichen Hauptstände*, Regensburg 1698 (Repr. Nördlingen 1987). – Württembergisches Landesmuseum Stuttgart (pub.), *Graviert, gemalt, gepresst*, Stuttgart 1996.

2 Further reading on gold wire and gold thread: Ernst Foltz, "Die Herstellung von Golddraht im Frühen Mittelalter", in: *Arbeitsblätter für Restauratoren*, no. 2, Gruppe 3, Edelmetalle, 1989, 99–106. – Sophie Fürnkranz, "Metallstickerei im außereuropäischen Raum—Beispiele aus der Sammlung des Museums für Völkerkunde Wien", in: *Technologische Studien*, ed. Wilfried Seipl, vol. 2, 2005, 10–49. – Márta Járó, "Gold Embroidery and Fabrics in Europe: XI – XIV Centuries", in: *Gold Bulletin & Gold Patent Digest*, vol. 23, no. 2, 1990, 40–57. – Márta Járó, "Catalogue of Metal Threads in Medieval Woven Textiles in the German National Museum, Nuremberg", in: Sabine Martius/ Sybille Russ (eds.), *Historische Textilien—Beiträge zu ihrer Erhaltung und Erforschung*, Nuremberg 2002, pp. 51–58 (Veröffentlichung des Instituts für Kunsttechnik und Konservierung im Germanischen Nationalmuseum, vol. 6). – Jane Lemon, *Metal Thread Embroidery*, London 2004. – William A. Oddy, "The production of gold wire in antiquity: hand making methods before the introduction of the draw-plate", in: *Gold Bulletin,* vol. 10, no. 3, 1977, 79–87.

3 Further reading on gilding on metal surfaces: Kilian Anheuser, *Im Feuer vergoldet. Geschichte und Technik der Feuervergoldung und der Amalgamversilberung*, Stuttgart 1999 (AdR Schriftenreihe zur Restaurierung und Grabungstechnik, vol. 4). – Filippo Baldinucci, *Vocabolario dell'arte del disegno*, Florence 1681. – Erhard Brepohl, *Teophilus Presbyter und die mittelalterliche Goldschmiedekunst*, vol. 2, Vienna 1999. – Erhard Brepohl (ed.), *Benvenuto Cellini. Traktate über die Goldschmiedekunst und die Bildhauerei*, Cologne 2005. – Erhard Brepohl, *Theorie und Praxis des Goldschmieds*, München 2008. – Benvenuto Cellini, *Due trattati. Uno intorno alle otto principali arti dell'oreficeria. L'altro in material dell'arte della scultura*, Florence 1568. Thomas Jelinek, *Praktische Galvanotechnik*, Saulgau 2005. – Oskar Krämer/Robert Weiner/ Max Fett/Robert Weiner (eds.), *Die Geschichte der Galvanotechnik und die Entwicklung der galvanischen Metallüberzüge bis zur Neuzeit*, Saulgau 1959 (Schriftenreihe Galvanotechnik). – C. Plinius Secundus/Roderich König (ed.), *Naturkunde*, Buch XXXIII, Munich 1984. – Terry Drayman-Weisser (ed.), *Gilded Metals—History, Technology and Conservation*, London 2000.

4 Further reading on mosaic gold, aurum musivum and bronzing: F. Ames-Lewis, "Matteo de' Pasti and the use of powdered gold", in: *Römisches Jahrbuch für Kunstgeschichte*, vol. 29, 1984, 351–359. – Louis E. Andés, *Blattmetalle, Bronzen und Metallpapiere*, Wien/Pest/Leipzig 1902. – Ekkehard Hans, "Musivgold", in: *Restauro,* H. 6, 2005, 420–425. – Manfred Koller, "Zur Bronzierung von Skulpturen des 18. Jahrhunderts in Österreich", in: Ingeborg Schemper-Sparholz (ed.), *Georg Raphael Donner. Einflüsse und Auswirkungen seiner Kunst* (Symposium in der Österreichischen Galerie, Juli 1993) Vienna 1996, 127–137 (Jahrbuch der kunsthistorischen Sammlungen in Wien, vol. 92). – Janet L. Ross, "A note on the use of mosaic gold", in: *Studies in Conservation*, no. 18, 1973, 174–176. – Ulrich Schiessl, "Musivgold", in: *Maltechnik restauro,* no. 4, 1981, 219–229.

The Themes of the Exhibition

Thomas Zaunschirm

The rich and extensive use of gold in art is thematically structured in the first few rooms of the exhibition tour. This section of the catalogue gives an overview of these various aspects of the way gold has been interpreted and used by artists. An occasional excursus will allow the topics to be looked at in greater depth and breadth. The catalogue of works on display follows the arrangement in the individual rooms.

A Short Artistic History of Gold

The GOLD exhibition at the Belvedere concentrates on modern and contemporary artists. However in the first room, the earlier history of the gold ground will be sketched out with the help of a few examples. The first known gold ground is found in mummy portraits dating from the second or third century AD, most of them from the Egyptian oasis of Fayoum. The example now kept in Berlin was discovered by the French archaeologist Albert Gayet in Antinoupolis in 1904/05. These portraits were already hung in the houses of the sitters during their lifetime. On their death, and after the mummy had been prepared, the portraits were subjected to a final process: their surfaces were gilded. These mummy portraits did not come into use until after the abandonment of the gold masks which had previously been widely used, and which are still what springs to mind when we think of mummies and the sarcophaguses in which they were placed—one need only think of Tutanchamun. When the masks were discontinued, the gold ground was invented. At the same time, though, the show of expense took a back seat, because there are in fact far more mummy portraits in which colours other than gold were used. The gold therefore comes across as an addition ordered specifically by wealthy clients.[1]

Since this time, the gold ground has not had an unbroken development. The icons and mosaics, the gold leaf of altar-pieces and illuminated books are generally seen as phenomena of Late Antiquity and the Middle Ages. That may not be entirely accurate, for Albrecht Dürer and Raphael still used the gold ground from time to time, and the tradition of icon painting is alive and well even today, but post-medieval use of the gold ground is certainly uncommon. In the fifteenth century gold gradually retreated from paintings. Masterpieces dating from the past five centuries demonstrate paradigmatically the various possibilities of the use of gold.

The gold ground as the characteristic of medieval art, as can be seen in the Prunkstall of the Belvedere, generally disappeared after the onset of the Renaissance with its modern perspective. Among the rare exceptions is a *Madonna and Child* on a gold mosaic ground by Marcello Provenzale in the Galleria Borghese.[2] The work was created to mark the Holy Year 1600, which was also called a 'golden year'.[3]

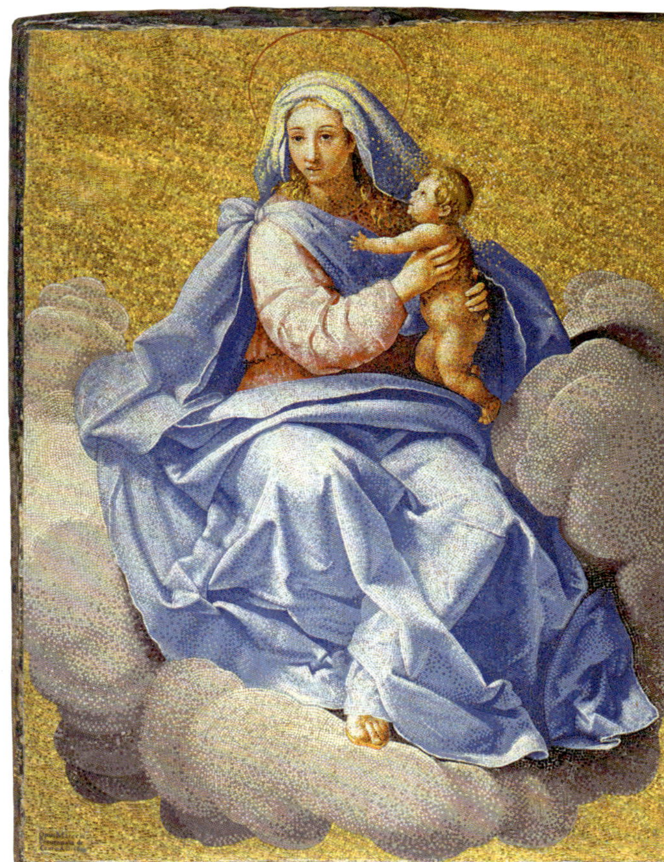

Fig. 1 Marcello Provençale, *Madonna and Child,* 1600
Galeria Borghese, Rome

Seventeenth century

The chronological start of post-medieval art at the exhibition is marked by a seventeenth-century work: *The Doctor's Visit* by Frans van Mieris the Elder (1657). The doctor is laying a finger on the patient's pulse to measure the heart rate. Raising an index finger, he is telling us that he is tracking the sick man's constitution which is hidden beneath the skin.[4] If we apply this clue to the technique of the picture, we shall discover that the gold, while not visible on the surface, is present beneath the oil paints. The artist painted this work, dating from the 'golden age' of Dutch painting, on copper, to which he had applied a ground of gold leaf.

Excursus: Gold grounds

The history of art has, for reasons that can be readily understood, not given great attention to this circumstance. The gold ground was invisible beneath the paint, and thus played no particular role. It needed the expertise of restorers to discover this technical background in the first place, and to this day no one knows how many paintings actually have a gold-leaf ground in this sense. After the well-known and variously interpreted disappearance of the gold ground as used in medieval painting, at the latest by the start of the sixteenth century, the noble metal went underground. Technical examinations have always been good for a surprise. Thus Tintoretto, for example, used gold leaf in his *Presentation of the Virgin in the Temple* (1553, Madonna dell'Orto, Venice).[5] Very little attention has been given to the history of the 'hidden gold ground'. In his inventory of 1801 Saint Saphorin wrote in respect of the early Rembrandt, *Old Woman Praying* (c. 1630), now in the Residenzgalerie in Salzburg: "[...] it is painted on gilded copper, in order to enhance the transparency."[6] The technique of applying a gold ground can be traced all the way to the twentieth century, an age in which the picture support no longer had to be of copper. Thus it is known that Giovanni Segantini used silver and gold leaf. In 1921 Walter Gramatté applied oils to a canvas grounded with gold. As late as 1936 Alexej von Jawlensky was inspired to paint five *Meditations* on drawing paper to which gold leaf had been applied. In these examples, the gold ground would only be apparent to the observant expert. It can be taken as certain that future research will come up with further examples of this largely unknown and almost invisible phenomenon, with which artists hope to create unusual light effects. Whether Saint Saphorin's opinion concerning Rembrandt was shared by anyone else is not at all certain. Of four early works painted on copper, only two are known for certain to have a gold ground, the one in Salzburg and the *Laughing Man* (c. 1628, Royal Picture Gallery Mauritshuis, The Hague). The intention was to use a protective layer to prevent a chemical reaction between the copper and the oils in the paint.[7] This was also possible by using silver. Adam Elsheimer, who preferred copper plates, painted a few pictures after 1603 on silvered copper.

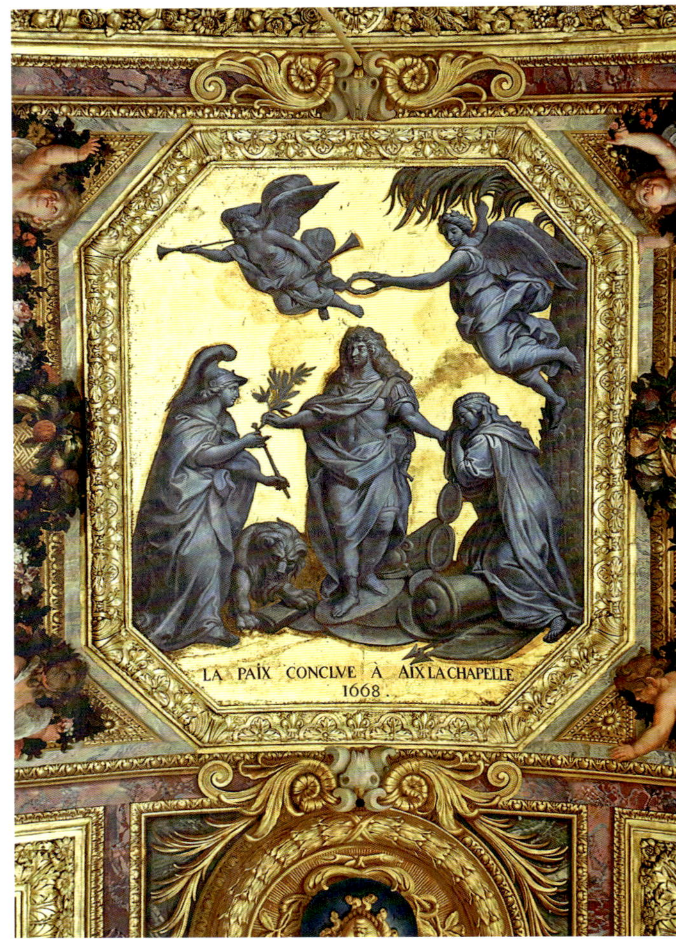

Fig. 2 Charles le Brun, *Peace of Aix la Chapelle,* ceiling of the Hall of Mirrors, Versailles, 1662

Philipp Otto Runge, in his picture *The Day* (1803), was also striving for light effects. To this day, this work is seen as the first post-medieval attempt to revive the gold ground.[8]

However the gold ground, having disappeared from paintings, maintained a continued existence not just on concealed gilded copper plates. The actual opulent use of gold was transferred to frames and, not least, ceilings, in the context of the luxuriant decorative system of regal display. Some inkling of this golden opulence can be obtained from a glimpse of the court of King Louis XIV of France, and quite generally in the palaces of the aristocracy of the Ancien Régime.[9] In the grisailles on a gold ground, such as can be seen in the Hall of Mirrors in Versailles, the gold ground is there to glorify the effulgence of the monarchy, while at the same time being the preserve of an art genre on the fringe of painting, namely the grisaille.

Extravagant as the use of gold was, its charisma did not match that of paintings distinguished by their use of colour, nor were grisailles on a gold ground ever seen as pictures in their own right. A history of this medium is yet to be written. In the vast wealth of decorative variety found in Baroque palaces, variations of this technique are constantly turning up. It is difficult to estimate how often, and for what reason, a grisaille on a gold ground was seen as an autonomous work.

Eighteenth century

In May 2009 the extension of the Arp Museum Bahnhof Rolandseck by Richard Meier was inaugurated with the exhibition *Tiepolo und das Antlitz Italiens*. Visitors were without warning confronted in the first of the darkened rooms on the top floor with six monumental pictures (1757) by Giovanni Domenico Tiepolo.[10] No history of art mentions this sudden appearance of the use of gold after the medieval period but before the beginning of the Romantic movement c. 1800, which finally rediscovered the Middle Ages.[11]

The six frescoes, mounted on canvas, from the Palazzo Porto in Vicenza by Andrea Palladio, two of which are to be seen in the present exhibition, relate the history of the family from 1022 to 1661, when Giovanni Battista Porto was appointed Generalissimo of the Venetian forces. The use of grisaille as a technique is no less striking than the gold ground. Was it chosen on the model of Versailles, where likewise the same techniques were use to point to a king's historic feats? Of course these frescoes do not relate to the medieval gold ground. There is nothing reminiscent of icons or altarpieces with saints. Rather, the monochrome painting evokes yet another art genre. The figures cast their shadows on the gold ground, and accordingly are to be perceived as statues or reliefs. The gold ground is not there to recall antiquarian painting, but to suggest sculptural monuments, which were often gilded or bronzed. The most closely comparable objects are gilt reliefs such as those on the choir stalls of the Frauenkirche in Munich by Ignaz Günther (1774). As the gold ground points to a different artistic genre, it can only be understood to a limited extent using the categories appropriate to painting.[12]

The renunciation of colours and the allusion to sculptural techniques must have seemed to the artist to be an appropriate means of representing the subject matter of these frescoes. Against the background of the *paragone* debate of his age, this is understandable.[13] To sculpture is assigned truth (in contrast to painting, which is put in the category of dream), because touch (rather than sight) had advanced to the status of most important sense (Johann Gottfried Herder, *Plastik*, 1778). In our context it is also significant that art theory (Johann Joachim Winckelmann, 1764, François Hemsterhuis, 1765) assigned sculpture a greater age and venerability. Körner summarizes the paradoxically complex view of art at this period as follows: "The art-lover apostrophized by Herder is not the blind man, but the sighted man, who through seeing becomes the blind, feeling man. For Herder, touching means above all tactile seeing, a form of seeing that has learnt to replicate real tactile experiences, in a sublimated fashion, in the fictive tactile gaze."[14]

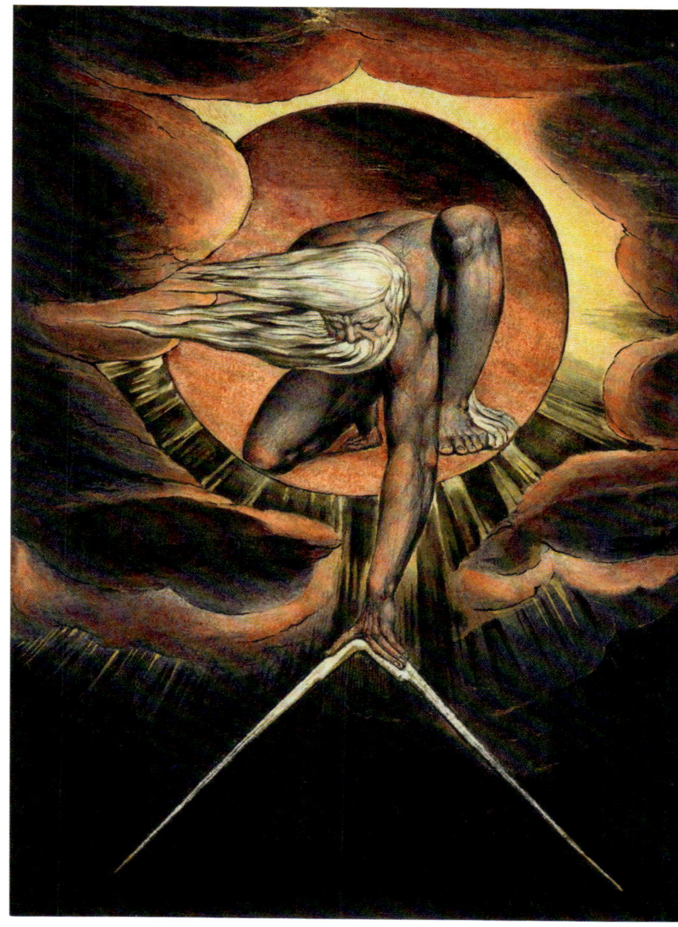

Fig. 3 William Blake, *The Ancient of Days,* 1827
The Whitworth Art Gallery, University of Manchester

The ban on gold pronounced by Gian Battista Alberti in 1435 at the dawn of post-medieval painting affected only painting, not relief or sculpture.[15] Gilded materials such as bronze, stucco, terracotta or wood and even papier mâché[16] were used by numerous artists. The disappearance of true colours, and the flickering shadows in many places recall, in addition, the popularity during this period of sculptures that were viewed by torchlight at night.[17] The idea was to give the statues life, to which Tiepolo may have reacted by seeking for a metamorphosis of painting into sculpture.
The claim to durability or even eternity, in other words the abolition of temporal factors, also motivated this work by Tiepolo. It is an astonishing work, without any parallel. By contrast, the decorative systems of the ceilings and walls of Venetian palaces are still waiting for an investigation of their own. Giovanni Battista Tiepolo, the father of Giovanni Domenico, worked in the Palazzo Papadopoli. Here, grisaille paintings on a gold ground in stucco-framed medallions have been preserved; these may also have served as models for the Porto cycle.

The grisaille-on-gold frescoes by Giovanni Domenico Tiepolo remained a one-off, and did not lead to the rediscovery of the gold ground for paintings. The use of gold leaf was epoch-making nonetheless, as it ran counter to Alberti's pronouncement that the effect of gold was to be achieved illusorily by the use of paints, and not by the use of the material itself. A renaissance of the gold ground demanded a contact with the chiaroscuro colour space of painting.

As maybe the first example of this interest we can look at an early work by William Blake. He used gold in various ways, to mix with paints or to highlight pictures. He only used gilt stars at the start of his career. As an eighteen-year-old apprentice to the engraver James Basire, he supplied copies of images in Westminster Abbey from which prints were made for publication. In 1775 the monument to King Sæberht of the East Saxons, the purported founder of the Abbey, was uncovered. Blake was commissioned to document this, and also to prepare pictures of the kings standing in the choir stalls, probably Henry III and Edward I. Blake painted in watercolour with pen-strokes of black ink and mosaic gold (stannic sulphide) on graphite or on plain paper. Blake only used golden motifs in such copies. In the next few years he seems to have shown no interest in the use of gold leaf.

It was only some years later that he was to use gold again, albeit not as a ground but to highlight tempera pictures on canvas, as in *The Spiritual Form of Pitt Guiding Behemoth* (c. 1805). The gold covers the paints and not vice versa, just as Runge was doing at almost the same time. During the next few years too he made effective use of gold. Perhaps the most famous sheet, the creator god Urizen wielding his dividers, an image influenced by medieval forbears, formed the cover of his book *Europe A Prophecy*. Blake reworked the relief etching of 1794 with gouache and gold only in later editions published in the 1820s.

Blake's early use of the gold material was based in history, but it did not motivate him to use a gold ground later for his own work. The gold ground underwent a metamorphosis, which replaced the gold-leaf with paints into which gold had been mixed, with gold bronze, and with glossy mixtures of paint. These did not lead to a renaissance of the gold ground, but fulfil an historical function of reminiscent quotation.

Nineteenth century

Already in the eighteenth century silhouettes had become a fashionable phenomenon. Named after their inventor, the French finance minister Étienne de Silhouette (1709–1767), and realized both as cut-outs and in Indian ink, they retained their popularity throughout the first half of the nineteenth century. In Runge's early silhouettes, such as the *Temple of Happiness*[18], gold-leaf also makes an appearance. The 22-year-old Ferdinand Georg Waldmüller presumably gave his *Self Portrait* (1814) on a gold ground to his bride. By dint of the technique, this early work by a painter who was to discover the Biedermeier landscape looks ahead to the theatrical effects of Hans Makart, who used gold to close the background of his *Moderne Amouretten* (1868). However Makart mostly used gold paint and only rarely actual gold. He marks an end-point in the post-medieval development of space, because he reduced the view of the landscape and contented himself with theatrical pomp. The pale golden ground prevents any distant view. While he used gold for the lunettes of the Kunsthistorisches Museum, Vienna had to wait for Gustav Klimt to take up and pursue this lustrous legacy.

In the nineteenth century the gold ground was the hallmark of a reawakened interest in the Middle Ages. The Nazarenes painted it at first (without mosaic gold), without however renouncing the achievements of the post-medieval development of space and corporeality. It is characteristic of this century that the medieval quotations collide with the spatiality of post-medieval painting. The nineteenth century. however, relativized not only the gold ground, but also other medieval techniques such as the golden lustre of mosaic. As early as 1727, the Studio del Mosaico della Reverenda Fabbrica die San Pietro in Vaticano was founded under Pope Benedict XIII.[19] For reasons of restoration, numerous medieval mosaics were reproduced. The tradition (above all in Venice) of mosaics glowing in constantly varying hues continued until about 1900. In 1889 in Berlin the Puhl & Wagner factory was set up. to be followed by the Mosaik-Werkstätte in Vienna in 1906.

In addition, the nineteenth century witnessed a triumphal march of the gold ground through the interiors of churches and prestige buildings (stately homes, opera houses, museums, theatres). Only in such places can it be properly admired. On the one hand, reference was made to medieval models by alluding to the gold ground, while on the other, gold leaf was used only rarely. This use of gold in the nineteenth century evoked no religious images, but, rather, costly glitz. Only from mid-century onwards, among the Pre-Raphaelites and the Symbolists, can we find gold leaf and occasional gold grounds in autonomous pictures.

Surprisingly, the Rome-based German artist Gustav Seyfferth placed his eclectic *Autumn Festival in Rome* (c. 1875) on a gold ground. With the Romantic rediscovery of the Middle Ages, of fairy-tales and poetry, there was a particular predilection in the nineteenth century for Dante. One favourite motif from the *Divine Comedy* was the story of Paolo and Francesca, who fell in love while reading *Lancelot*: "And all the while one spirit uttered this, The other one did weep so, that, for pity, I swooned away as if I had been dying, And fell, even as a dead body falls."[20] In his watercolour *Paolo & Francesca* (1877) the Lombard painter Mosè Bianchi uses the non-spatial gold ground to emphasize the weightlessness of the souls of the amorous couple floating past Dante and Virgil.

It was not until the end of the nineteenth century that the more or less correct historicism led to a renaissance of gold in pictures, particular in the aestheticism of Art Nouveau. The influence of the Japanese gold-ground tradition also played an important role here. Emil Orlik, for example, was inspired to create his garden in golden surroundings by a visit to Japan in 1900. The much-travelled artist from Prague was also a member of the Viennese Secession (1899–1905).

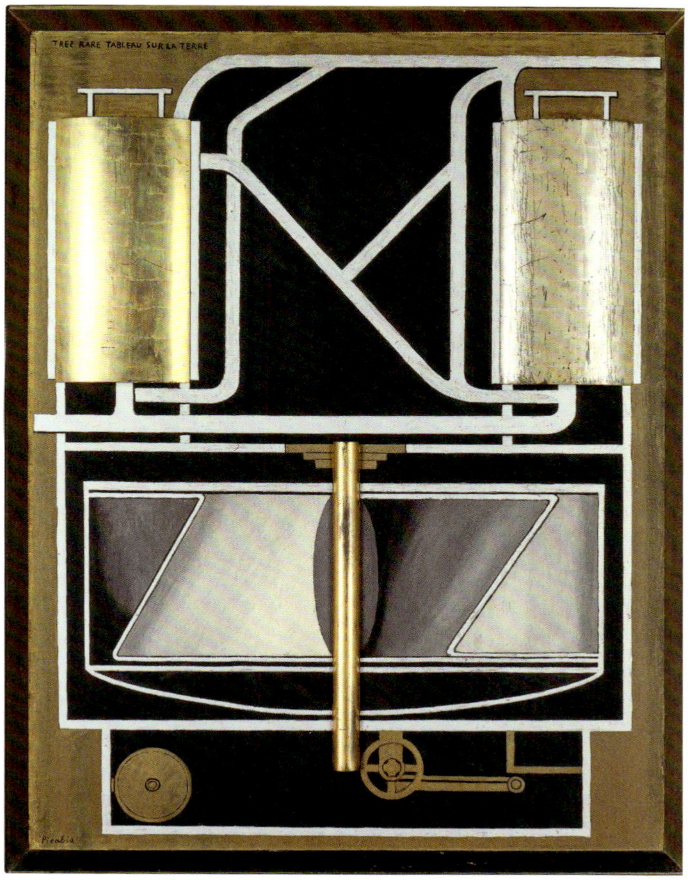

Fig. 4 Francis Picabia, *Very Rare Picture on the Earth,* 1915
Peggy Guggenheim Collection, Venice
(Solomon R. Guggenheim Foundation N.Y.)

Twentieth century

The avantgarde by contrast, like all of post-medieval painting, was either indifferent or hostile to gold. Hints of a gold ground using paint rather than gold leaf (here we mention by way of example works by Erich Buchholz) were, as in the first half of the nineteenth century, rare. When individual pictorial elements consist of gold leaf, this is sometimes traceable to the consequences of Cubism. As in the collage, a uniform composition is secondary.

Francis Picabia accurately called his gold-leaf-enriched picture *Very Rare Picture on the Earth* (1915). A further example would be *Er kommt* (1919) by Johannes Molzahn (Staatsgalerie Stuttgart). There is also a *Kinderbild* (1921 - 1923) by Johannes Itten, which can only with difficulty be explained in terms of his colour theory. A later famous work would be *I Saw the Figure 5 in Gold* (1928) by Charles Demuth.

In the first half of the twentieth century the use of gold was the exception. Willi Baumeister, in his series of *Mauerbilder* created figural compositions collage-like with geometric basic elements. He was thinking of an "as yet non-existent new architecture as support"[21] for these pictures. Baumeister was thus anticipating the post-war fashion for art in the public space.[22] However the *Mauerbild mit Metallen* (1923) is an exception even for this artist. The occasional pictures with a gold-leaf ground are due to the artist's individual aims in each case.

The picture *To the Victims of Fascism* painted after the Second World War (1946/47) by Hans Grundig, a co-founder of the Leipzig School, occupies a place all of its own. The gold ground in his autobiographically motivated painting—it makes reference to his own number 18061 in the Sachsenhausen concentration camp—fills the broad space of the area within the barbed wire and lifts the dead body out of the inhuman reality.[23] In principle, though, it was apparently less the committed realistic works than abstract ones that allowed the use of gold.

Only in the 1950s was gold leaf rediscovered (for example by Andy Warhol and Robert Rauschenberg). The polarization of abstraction and realism had played only a peripheral role in the art-critical discourses since the 1960s. The museums were the neutral ground for the reconciliation of opposite aesthetic positions. Richard Hamilton, the European father of Pop Art, had, since 1965, obsessively created variations on the façade of the New York Guggenheim Museum by Frank Lloyd Wright, opened in 1959. He drew the popular building, published a copper-coloured screen-print in 1965, and began that same year on a series of technically demanding synthetic reliefs in rainbow colours (1966), and from 1970 in black (edition of 750), in white, and in chrome. Hamilton himself wrote: "It was an attempt to mirror the whole activity of architecture in the confines of a four-foot-square panel."[24] The version ennobled with gold leaf (1965/66) stresses the artistic aspiration to appreciate the museum as an iconic work and the artwork as museum-worthy.

As a counterpart to the works of Tiepolo that marked the post-medieval revival of the use of gold, the large-format *Mennigebild 37/3* ('Red Lead Picture', 1976/2010) by Imi Knoebel represents the present day. The Suprematist utopia of Kazimir Malevich, on whose *Black Square on White Ground* (1915) Knoebel oriented himself, motivated him here no more than did the pathos of his teacher Joseph Beuys. The minimalism of the hardboard rectangles and frames pushed over one another depicts nothing. In 2010 Knoebel took up once more the Red Lead series of 1976, in which he had used cheap rustproofing paint (hence the name) and industrially produced hardboard. In view of this sprawling heptagon, one critic wrote of a 'halo for plywood'.[25] But the monumental gold ground has no referential character, it merely seals a form which is at the same time 'shaped canvas' and 'colour-field painting', recalling those principles of the 1970s from which the concept of the at once simple and complex outline derives.

Excursus: Entropy and Gold Bars

In the history of culture, information theory and social sciences, the term 'entropy', borrowed from thermodynamics, is used to mean the tendency to disorder. "The visual arts too, insofar as we encounter them on the art market, are in a state of advanced entropy [...].

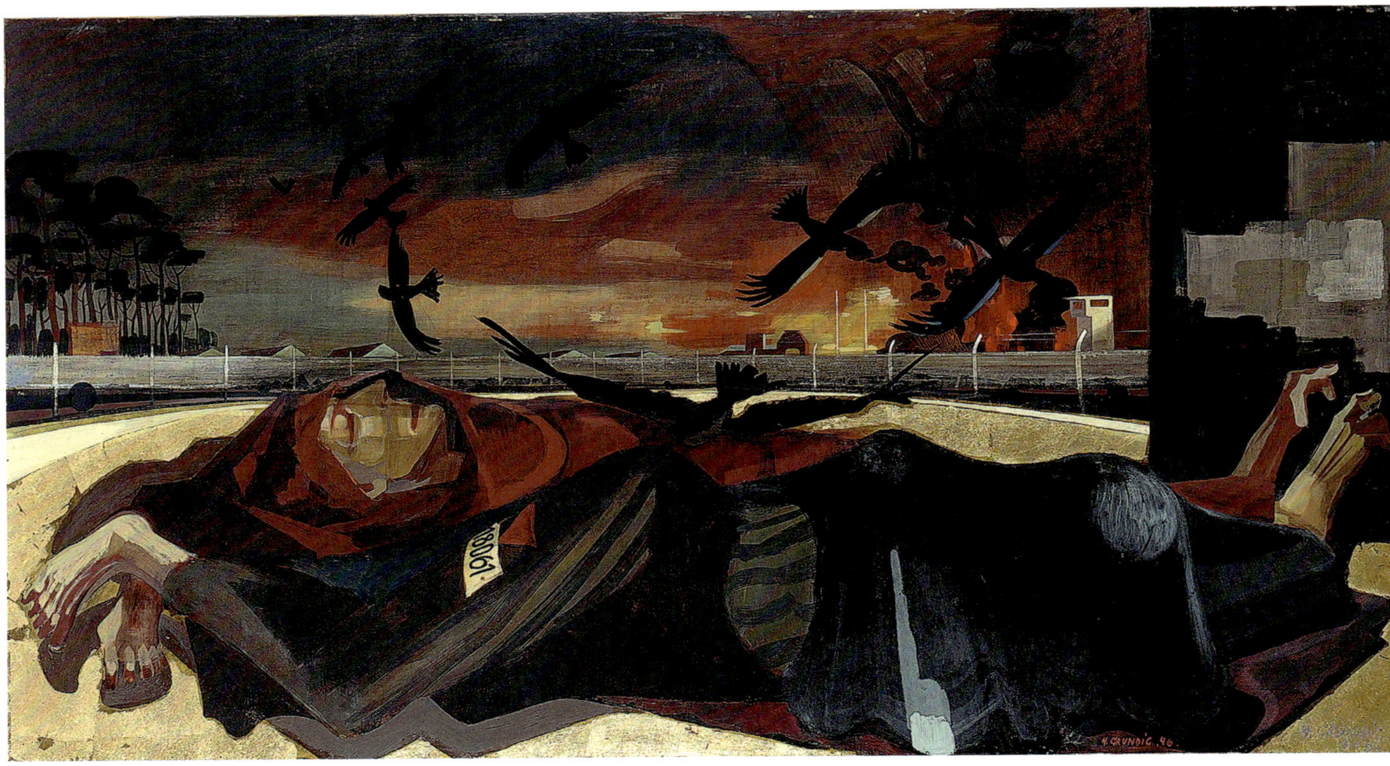

Fig. 5 Hans Grundig, *To the Victims of Fascism,* 1946, Museum der bildenden Künste, Leipzig

A modern artist actually needs to master just one segment of art, *namely the art of marketing.*"[26]

But even grand masters of the art market can come a cropper with such enterprises. For his exhibition *One Ton One Kilo* in March 2009 at the Beverly Hills branch of Larry Gagosian's gallery empire, Chris Burden needed a hundred one-kilogram gold bars. For the installation *Tower of Power—A Hundred Kilograms of Gold*, which had already been shown at the Museum Moderner Kunst Stiftung Ludwig in Vienna in 2002, the bars had been borrowed. Burden set the price for a possible sale at 15% above the cost of the material. He had already, at another exhibition, employed a guard to watch over four gold bars. At the time, the public believed the bars to be fake, simply because he *had* employed a guard. "The point was to make people think you were faking when it was actually real gold."[27] Here, reality really had caught up with art. The real gold bar is thought to be its own mimetic imitation. If, in this situation, it is no longer possible to distinguish between art and reality, a high degree of entropy (mixing of two systems) has been reached. But one can also think in the opposite direction. In 2008, Tom Früchtl painted, in *hybrid*, a genuine gold bar with gold paint. Cheap-looking artifice conceals the costly reality.

Gagosian decided not to borrow the bars, but to buy them. In times of gold speculation, one might think that nothing could go awry. He ordered the gold from Stanford Coins & Bullion in Houston and transferred the money, about 3.3 million dollars, in February. Just two weeks later, before the bars had been delivered to Gagosian, R. Allen Stanford, the owner of the company, had been arrested on suspicion of "massive ongoing fraud", and the US Securities and Exchange Commission had his assets confiscated. Stanford is still being held on remand, and has since sued the SEC for 7.2 billion dollars in damages. Investment in the gold tower has, in a sense, fallen victim to the entropy of fraudulent financial speculations or, to look at it from a different angle, has led, because of the rising gold price, to virtual profits in the event that Gagosian gets his gold after all.

Damien Hirst has been luckier in his dealings with gold bars. His company 'murderme' bought *100 gold bullion bars,* a pallet of just what the title says, from Paul Insect; the work differs from Burden's ziggurat only in that one bar appears to have been nibbled at. Hirst has reduced gold from auratic material to consumer commodity.

Gold bars as the material for artworks are used by numerous artists. In *Frustrum* (2007) Gary Hill submerged a 12.5 kilogram gold bar in crude oil. Two years later Michael Sailstorfer buried 28 gold bars in Pulheim near Cologne, and invited the public to dig them up. Other gold-bar works have been created by Evariste Richer (*Lingot mort*, 2007), Rebecca Horn (*Gold Rush*, 1986), Thomas Lehnerer (*Stadt aus Gold*, 1986/91)

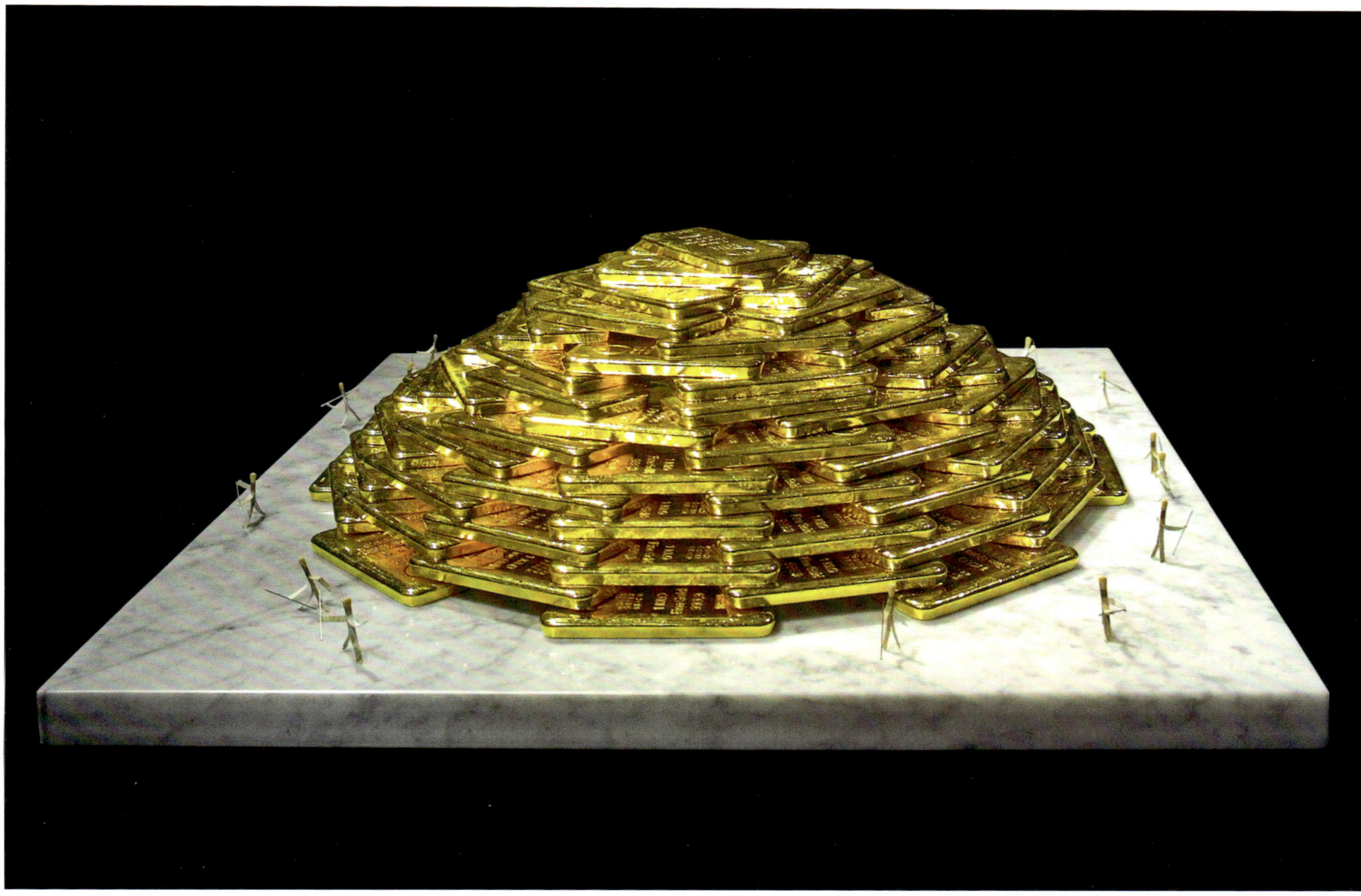

Fig. 6 Chris Burden, *Tower of Power—A Hundred Kilograms of Gold,* 2002, Exhibition view, MUMOK Vienna 2002

and Jonathan Monk (*The Weight of my Hand in Gold,* 2005). Carmen Perrin (*Flac*, 2008) has exhibited a molten bar. There are also 'gold bars' that manage without gold, such as Pavel Braila's *Baron's Hill* (2005) and Thomas Hirschhorn's *Gold Mic-Mac* (1998).

Crucial to all the works mentioned is the relationship between the material value and the artistic value. It is not useful to use the word 'entropy' in the sense of an alleged decay of art, but rather in the relationship between the hardware and soft-ware values. In the physical sense, entropy refers to the transition from two separated gases to an indistinguishable mixture; differentiated order becomes disorder.

Gold bars were used for the first time in the history of art by Yves Klein. In a group exhibition in Antwerp in 1959, he fixed the price for the declared *Zone of immaterial painterly sensitivity* as a one-kilogram bar of gold. He himself seemed to have been surprised to have demanded nothing for this gold bar, and above all to have got it all the same. Regardless of whether one agrees with Yves-Alain Bois in calling this 'psychic frippery', 'idealistic junk' and 'ostentatious fraud',[28] the decisive point is that the artist, while integrating gold bars into his work (*Ex-Voto, dedicated to Saint Rita of Cascia*, 1961), he was still paid in them by the buyers. It was part of his concept to throw half of the gold he had received into the Seine. From this point of view, one cannot really agree with Benjamin Buchloh in calling him the 'artist of advanced capitalism'.[29]

In 1971 Marcel Broodthaers offered a gold bar with his logo, a stamped eagle, for twice the gold price fixed on any given day. As a result of the contractual link with the gold price, *Lingot d'or* is not subject to a revaluation by the art market, but rather to the volatile price of the metal. Even so, his gold bar differs from commercial gold bars because of the eagle, which also explains the difference between the two values. Entropy sets in when the art market offers no more for gold bars as art than for gold bars *per se*. Why should the artistic market value fluctuate just like the gold price?

Things are different when the gold cannot be piled up in the form of bars. On principle the price of the artwork can be higher than the market value of the gold used. Subodh Gupta calls his work in 24-carat gold *1 kg of War* (2007). The question concerning the quantity, "How much gold does one want?" is linked to the question of the war which has been accepted along with the gold.

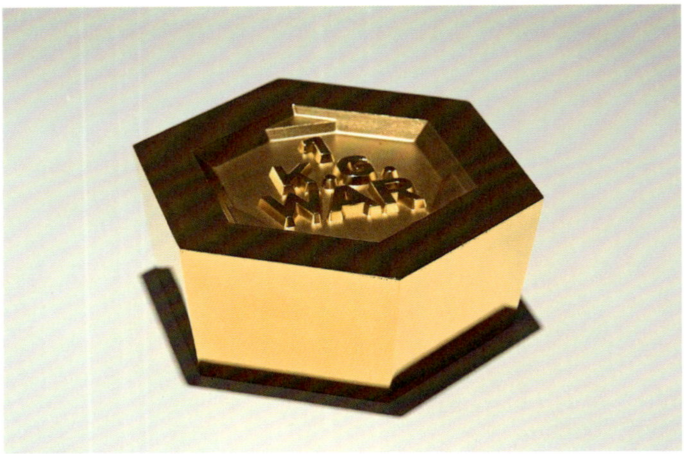

Fig. 7 Subodh Gupta, *1 Kg of War*, 2007

Thus in view of these works definable by the weight of gold used, two questions arise: Is property a form of violence?[30] Is the aesthetic dimension a *quantité négligeable* for the art-market price? For Chris Burden, gold is a symbol of power. That was the reason the Spaniards went to America. What happened to the shiploads of gold which the sailors brought for the Spanish crown? His *Gold Bullets* (2003) are cartridges cast in 22-carat gold. He calls them, depending on their shape, *Pointies* and *Roundies*. From the point of view of these cartridges, gold is not just the subject of wars, but also a weapon in its own right. In recent months, the gilded weapons of dictators have pointed to the connexion between violence and gold.

While modernist art, by dint of the Readymade principle, has not shied away from using real objects, one can still ask whether art achieves a premium over the material used by virtue of what it depicts, or of the fame of the artist. The miracle of transformation and metamorphosis continues to be a strategy. Alicja Kwade vouches for this with *Kohle (1T Rekord)*, a bronze cast of a coal briquette gilded with real 24-carat gold (2006).

In all the examples, the material, its lustre and its value are glorified like no other material in art. Not Vital glorifies the classic form of this idolatry with his *Golden Calf* (2001) in the Orangery.

1 Hilde Zaloscer, *Vom Mumienbildnis zur Ikone*, Wiesbaden 1969.
2 I am grateful to Michael Brunner, Überlingen, for this information.
3 Quoted from: Arndt Brendecke, *Die Jahrhundertwenden. Eine Geschichte ihrer Wahrnehmung und Wirkung*, Frankfurt/New York 1999, p. 374, note 4.
4 Dorothee Weber-Weigand, *Herzminutenvolumenbestimmung nach dem Fickschen Prinzip. Vergleich von zwei Methoden in Ruhe und unter Belastung.* diss. Universität Würzburg, 2007, p. 3.
5 "Recent technical examination has revealed that Tintoretto even went so far as to employ dabs of gold leaf to bring out the elaborate design on the patterned risers of the steps, a technical archaism which proposes a connection with much older Venetian art." Tom Nichols, *Tintoretto. Tradition and Identity*, London 1999, p. 47.
6 Roswitha Juffinger, "Ein Rembrandt für die Czernin'sche Gemäldegalerie—Die Ankäufe Johann Rudolph Czernins aus der Gemäldesammlung des Armand François Louis de Mestral de Saint Saphorin", in: *Gesamtverzeichnisses der Gemälde der Residenzgalerie Salzburg*, Salzburg 2010, p. 467. Cf. Erika Oehring, *Grünspan & Schildlaus. Meister der Residenzgalerie Salzburg und ihre Arbeitsweisen* (mus. cat. Residenzgalerie Salzburg), Salzburg 1996, pp. 20f. (I am grateful to R. Juffinger for this reference).
7 It could take the form of other metals or a prepared oil-paint layer. Cf.: Veronika Poll-Frommel, *"Beobachtungen zur Maltechnik von Adam Elsheimers 'Flucht nach Ägypten'"* in: *Von neuen Sternen. Adam Elsheimers "Flucht nach Ägypten"*, ed. by Reinhold Baumstark (exhib. cat. Alte Pinakothek Munich), Cologne 2005, pp. 202f.
8 Cf. Markus Bertsch et al. (eds.), *Kosmos Runge. Der Morgen in der Romantik* (exhib. cat. Hamburger Kunsthalle, Hamburg/ Kunsthalle der Hypo-Kulturstiftung, Munich), Munich 2010, p. 155. Very cautiously, it is said here of the picture now known as *Mother Earth with her Children*: "Possibly this painting is one of the first, if not the first, picture to be painted on a gold ground in nineteenth-century Germany."
9 "Painting 'sur fond d'or' and 'rehaussement en or'—in all its derivatives, which always involve the Maniera of the artist—are both of course ubiquitous and not confined to the Châteaux du Roi, they were so to speak a must for the furnishings of the nobility. Hardly any of the great painters did not experiment in this way (Henri Testelin, Sébastien Bourdon, Nicolas Loir in the Appartement in the Tuileries, Claude Audran the Younger and especially Eustache Le Sueur in the Louvre, in the Hôtel Lambert and of course Le Brun)." Markus A. Castor, letter dated 2 Sept. 2011. Cf. also the catalogue essay by Markus Castor.
10 Heike Frosien-Leinz, "Sechs Tiepolo-Fresken des Palazzo Porto", in: *Tiepolo und das Antlitz Italiens. Kunstkammer Rau*, ed. by Oliver Kornhoff (exhib. cat. Arp Museum Bahnhof Rolandseck, Remagen), Cologne 2009, figs. 27-32.
11 Runge was involved in the rescue of medieval paintings in Hamburg and is said to have written a manuscript on the significance of the medieval gold ground. It has not been preserved.
12 Oliver Kornhoff (ed.), *Tiepolo und das Antlitz Italiens. Kunstkammer Rau.* (exhib. cat. Arp Museum Bahnhof Rolandseck, Remagen), Cologne 2009, 88-101.
13 Cf. Hans Körner, "Paragone der Sinne. Der Vergleich von Malerei und Skulptur im Zeitalter der Aufklärung", in: *Mehr Licht. Europa um 1770. Die bildende Kunst der Aufklärung*, ed. by Herbert Beck (exhib. cat. Städelsches Kunstinstitut und Städtische Galerie, Frankfurt am Main), Munich 1999, 365-378.
14 Körner 1999 (as note 12), p. 376.

15 There are countless examples of the tradition of gilded materials in post-medieval sculpture, from Lorenzo Ghiberti's *North Portal* (1402–1424) and *Paradise Door* (1430–1437) in the Baptistery in Florence via Donatello's Annunciation Tabernacle in Santa Croce (c. 1433), Benvenuto Cellini's *Golden Saltcellar for Francis I of France* (1540–1543) and the gilded stucco octagons in Pietro da Cortona's ceiling frescoes in the Palazzo Barberini in Rome (1633–1639) to Gian Lorenzo Bernini's *Cathedra Petri* (1657–1666) and Corneille van Clève's relief on the *High Altar* of the palace chapel at Versailles (1709) or the *Chinese Teahouse* in the park of Sanssouci palace (J. P. Benckert and M. G. Heymüller, 1754–1756) with the lifesize gilded stone figures, and Jean-Antoine Houdon's *Statuette of the Seated Voltaire* (1779).

16 In the main residence of the dukes of Mecklenburg-Schwerin, Schloss Ludwigslust, this ersatz material was used for ornaments for want of any other.

17 Oskar Bätschmann, "Pygmalion als Betrachter. Die Rezeption von Plastik und Malerei in der zweiten Hälfte des 18. Jahrhunderts" (1974), in: *Der Betrachter ist im Bild. Kunstwissenschaft und Rezeptionsästhetik*, ed. by Wolfgang Kemp, Cologne 1985, 183–224.

18 Illustrated in: Bertsch 2010 (as note 8), cat. no. 261.

19 Barbara Lorenzer, "Eine kurze Geschichte der Mosaiktechniken von der römischen Antike bis zum Jugendstil", in: *Europäische Mosaikkunst vom Mittelalter bis 1900*, ed. by Ottobrina Voccoli/ Michael Brunner (to accompany the exhibition at the Städtische Galerie Überlingen), Petersberg 2011, 5–11.

20 Dante, *The Divine Comedy. Inferno,* Canto V, Circle 2. (Sins of Love).

21 Willi Baumeister, 1934. quoted from: URL: <http://www.willi-baumeister.org/index.php?menuid=24> [read on 21 Nov. 2011].

22 Translator's note: More specifically in German: 'Kunst am Bau'– 'art on/next to buildings'

23 *Hans Grundig,* Kleine Werkmonographie Nr. 18 (Museum der bildenden Künste Leipzig), Leipzig undated.

24 URL: <http://www.moma.org/interactives/exhibitions/1999/muse/artist_pages/hamilton_guggenheim.html> [read on 21 Nov. 2011].

25 Peter Bausch, "Ein Heiligenschein für Sperrholzplatten", in: *Sindelfinger Zeitung—Böblinger Zeitung*, 26 March 2011.

26 Manfred Wöhlcke, *Das Ende der Zivilisation. Über soziale Entropie und kollektive Selbstzerstörung*, Munich 2003. URL: <http://www.kultur-fibel.de/Kultur%20Fibel%20Buch,%20W%F6hlcke,%20Das%20Ende%20der%20Zivilisation.htm> [read on 21 Nov. 2011]. In art criticism, the term need necessarily not be used as a pessimistic battle cry. Cf. Rudolf Arnheim, *Entropie und Kunst. Ein Versuch über Unordnung und Ordnung,* Cologne 1979.

27 Dana Goodyear, "Goldless", in: *The New Yorker*, 23 March 2009, URL: <http://www.newyorker.com/talk/2009/03/23/090323ta_talk_goodyear> [read on 21 Nov. 2011].

28 Cf. the splendid essay by Yve-Alain Bois, 'Yves Kleins Aktualität', in: *Yves Klein* (mus. cat. Museum Moderner Kunst Stiftung Ludwig, Vienna), Vienna/New York 2007, 102–115.

29 Quoted from Bois 2007 (as note 27), p. 105.

30 According to section 854 para. 1 of Germany's Civil Code "The possession of a thing is acquired by obtaining actual power over the same." The Swiss Civil Code also says in Article 919 para. 1: "Whoever has actual power over a thing is the possessor of the same." Translator's note: the German word 'Gewalt' means both 'power [over]' in the legal sense used here, and 'violence' as the author clearly implies in the body of the article. In connexion with Burden, he uses a different German word: 'Macht'.

Unknown artist
Mummy portrait of a young man, c. AD 120–130
Encaustic on wood with gold leaf, 41 x 20.2 cm
Ägyptisches Museum und Papyrussammlung, Staatliche Museen zu Berlin

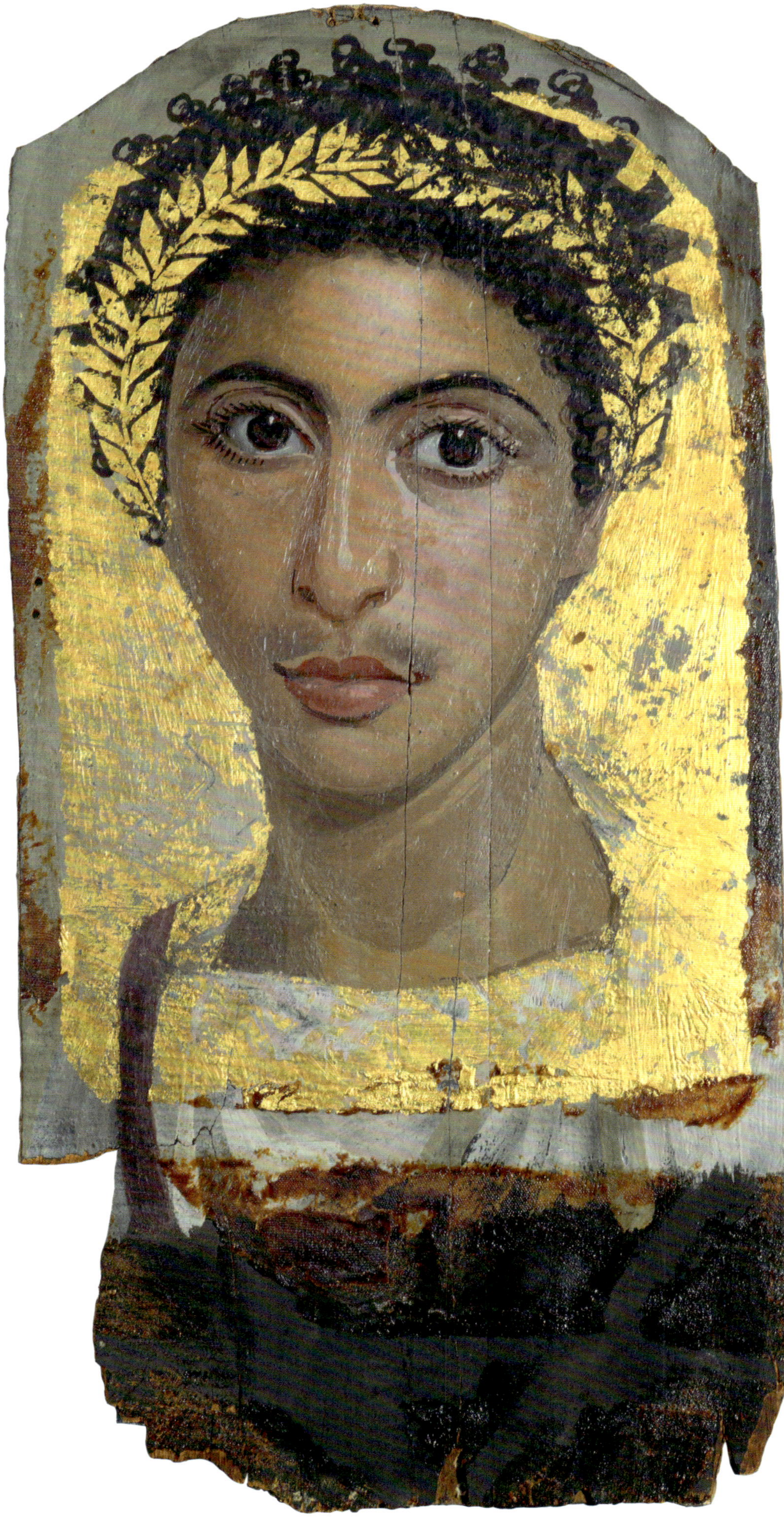

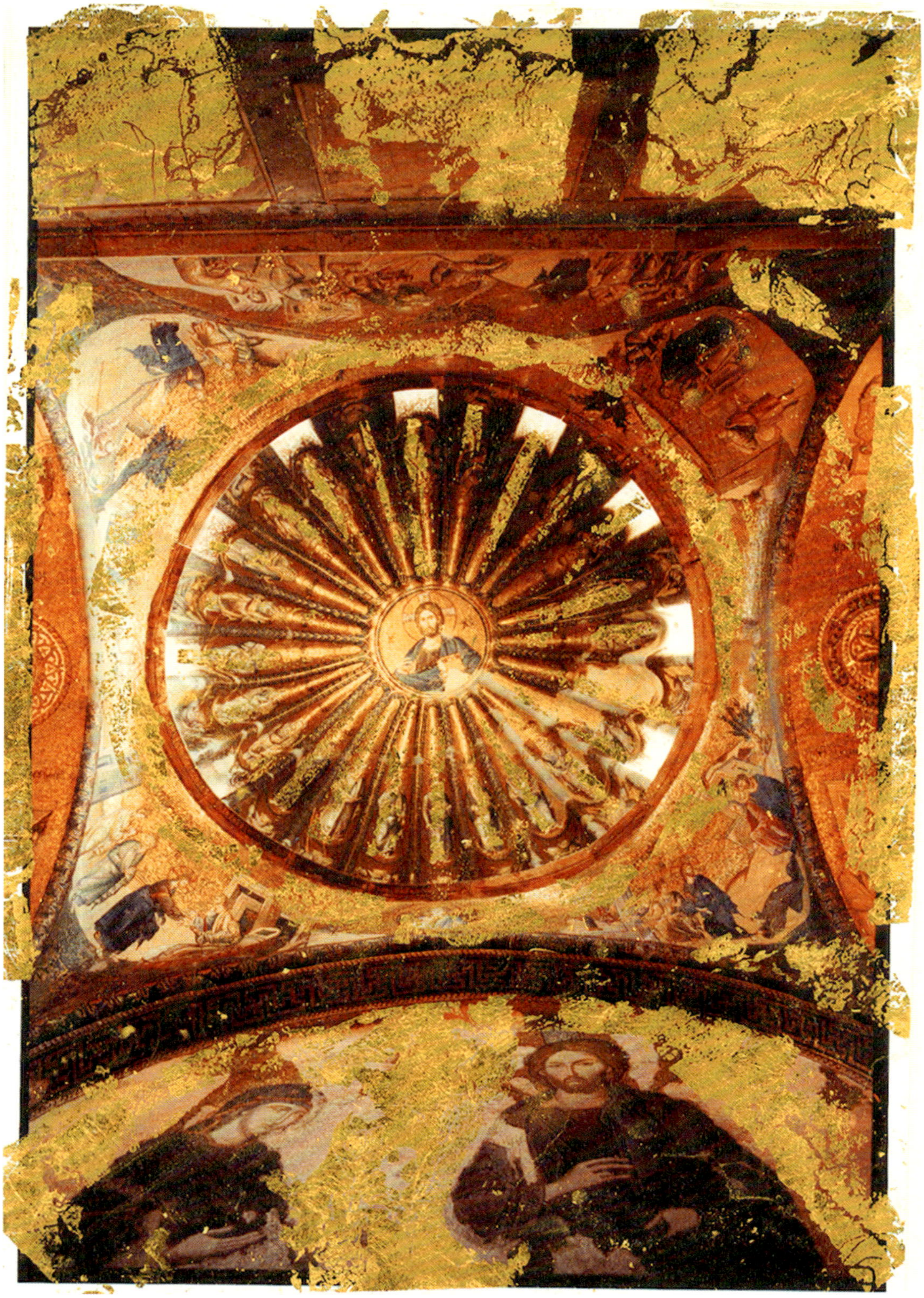

Gundula Schulze Eldowy
Das lebendige Bild, Konstantinopel, 1997
Photograph with gold leaf, c. 189 x 130 cm
Private Collection

Frans van Mieris the Elder
The Doctor's Visit, 1657
Oil on copper-gilt, 34 x 27 cm
Kunsthistorisches Museum, Vienna

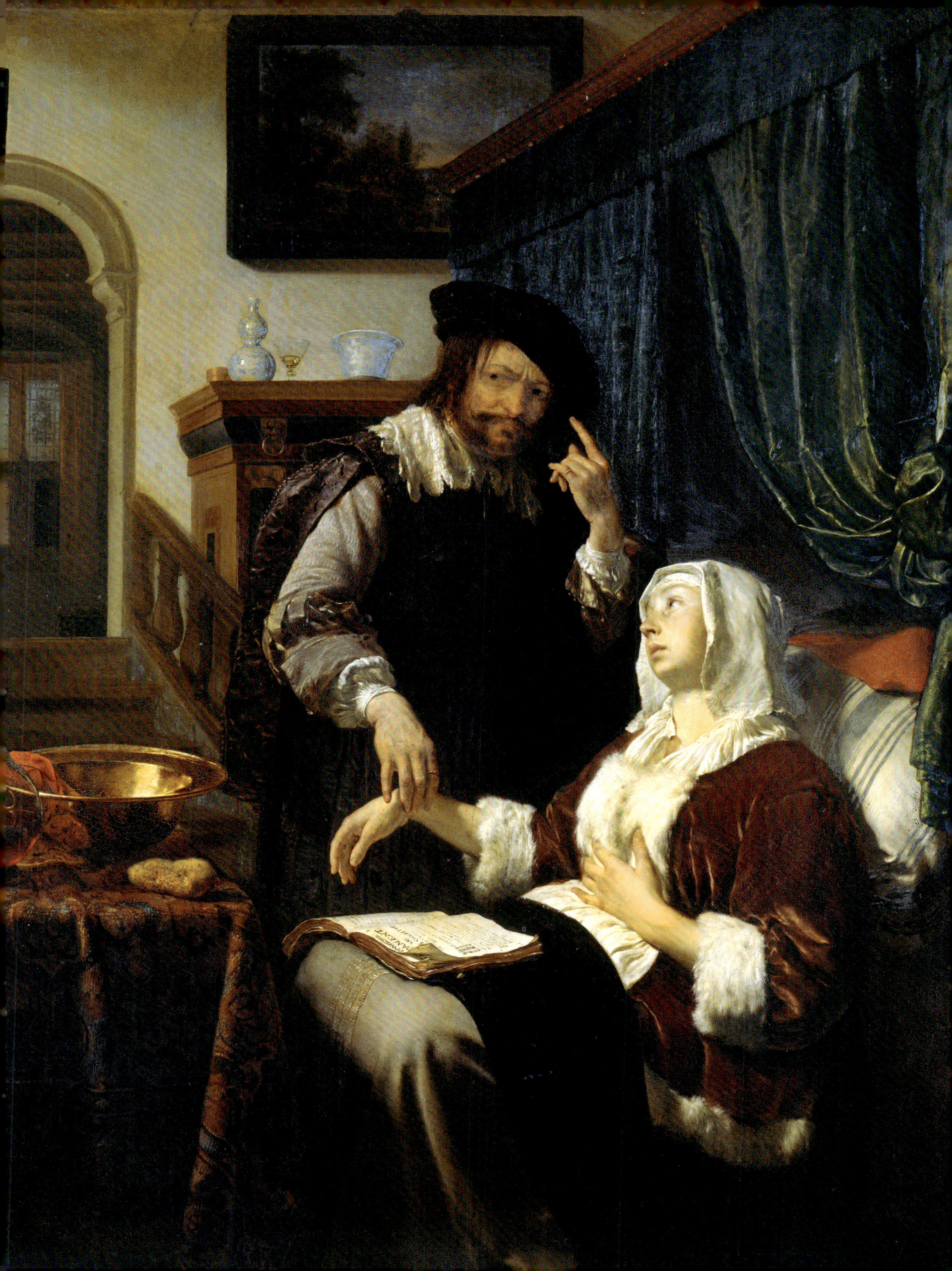

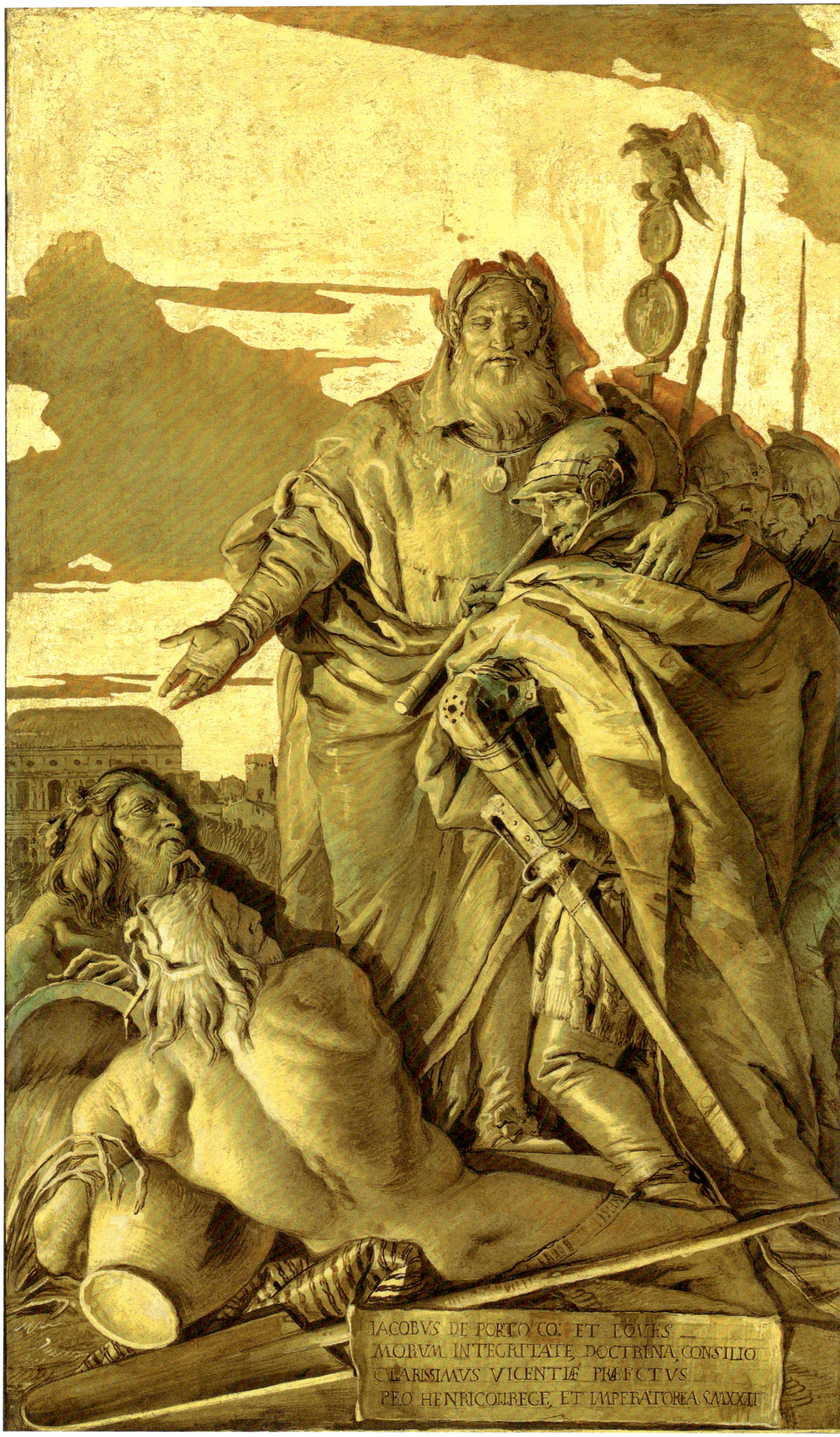

Giandomenico Tiepolo
Jacopo Porto is appointed Governor of Vicenza in 1022,
c. 1757
Fresco in grisaille and gold leaf, mounted on canvas,
262 x 183 cm
Arp Museum Bahnhof Rolandseck / Collection Rau für UNICEF

Giandomenico Tiepolo
Donato Porto Becomes a Venetian Patrician in 1379
c. 1757
Fresco in grisaille and gold leaf, mounted on canvas, 262 x 183 cm
Arp Museum Bahnhof Rolandseck / Collection Rau für UNICEF

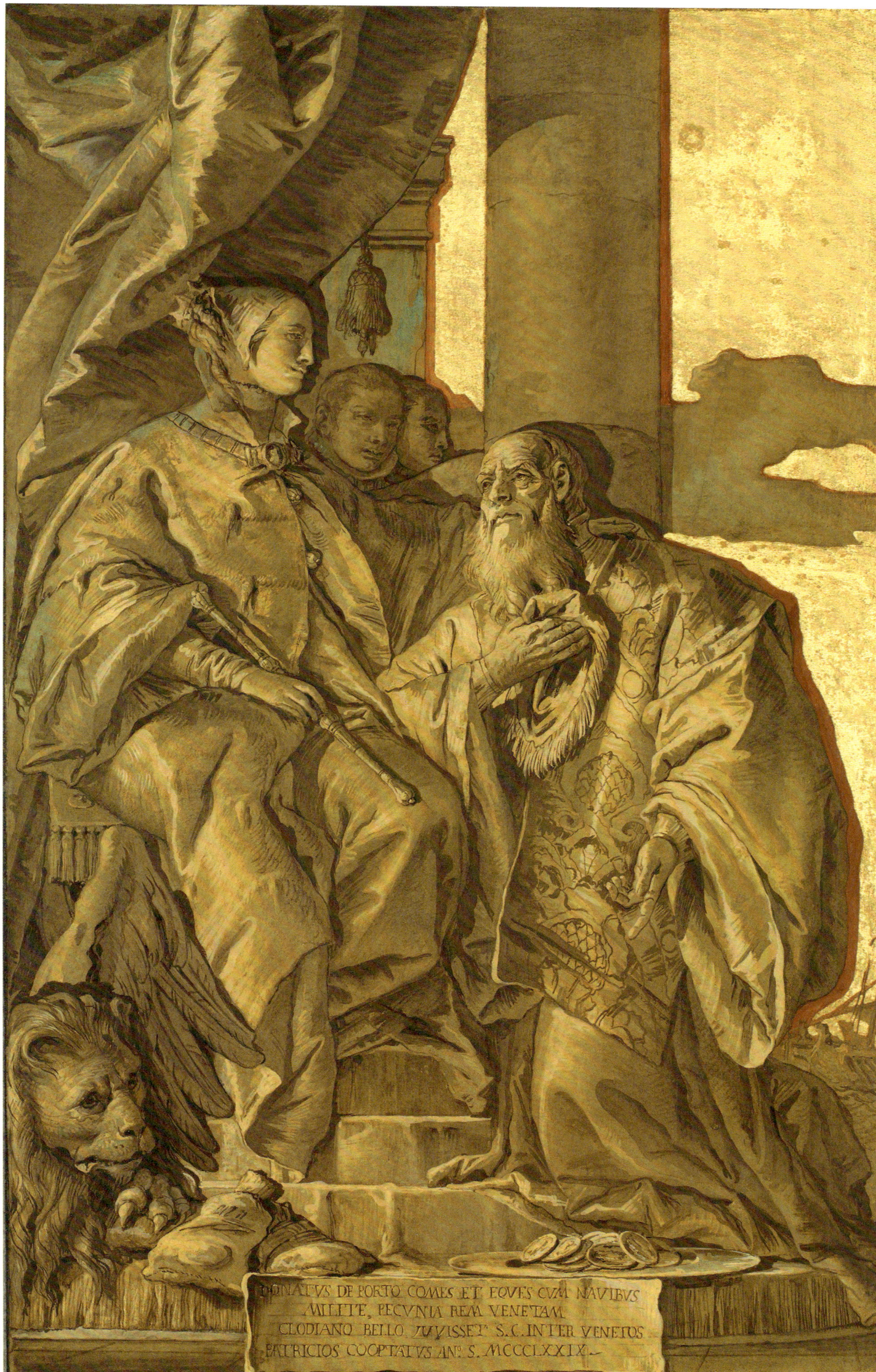
DONATUS DE PORTO COMES ET EQUES CUM NAVIBUS
MILITE, PECUNIA REM VENETAM
CLODIANO BELLO IUVISSET S.C. INTER VENETOS
PATRICIOS COOPTATUS ANº. S. MCCCLXXIX.

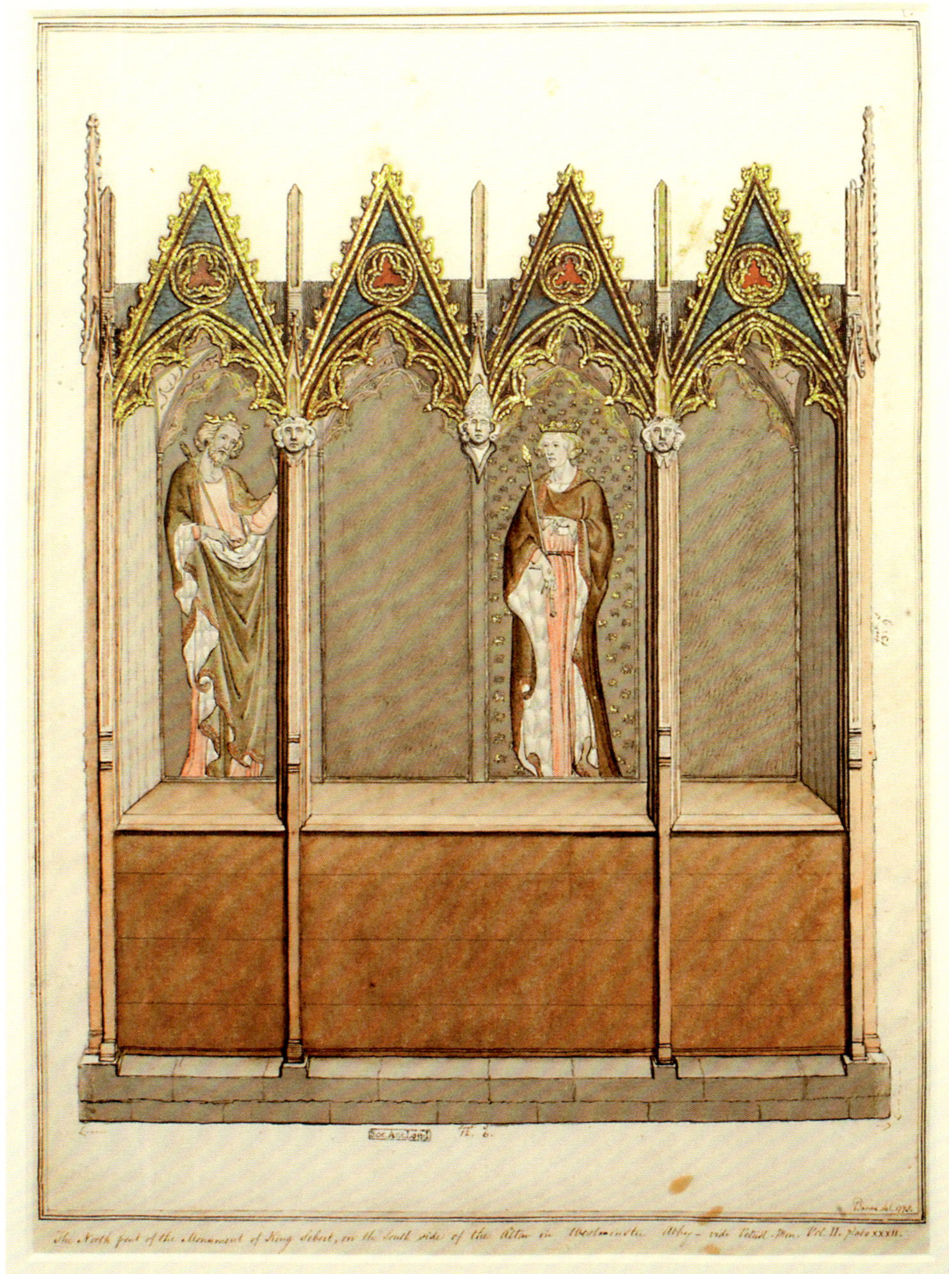

William Blake
King Sebert, the North Front of his Monument, 1775
Watercoulors, ink, mosaic gold, pencil on paper, 36.8 x 28 cm
By permission of the Society of Antiquaries of London

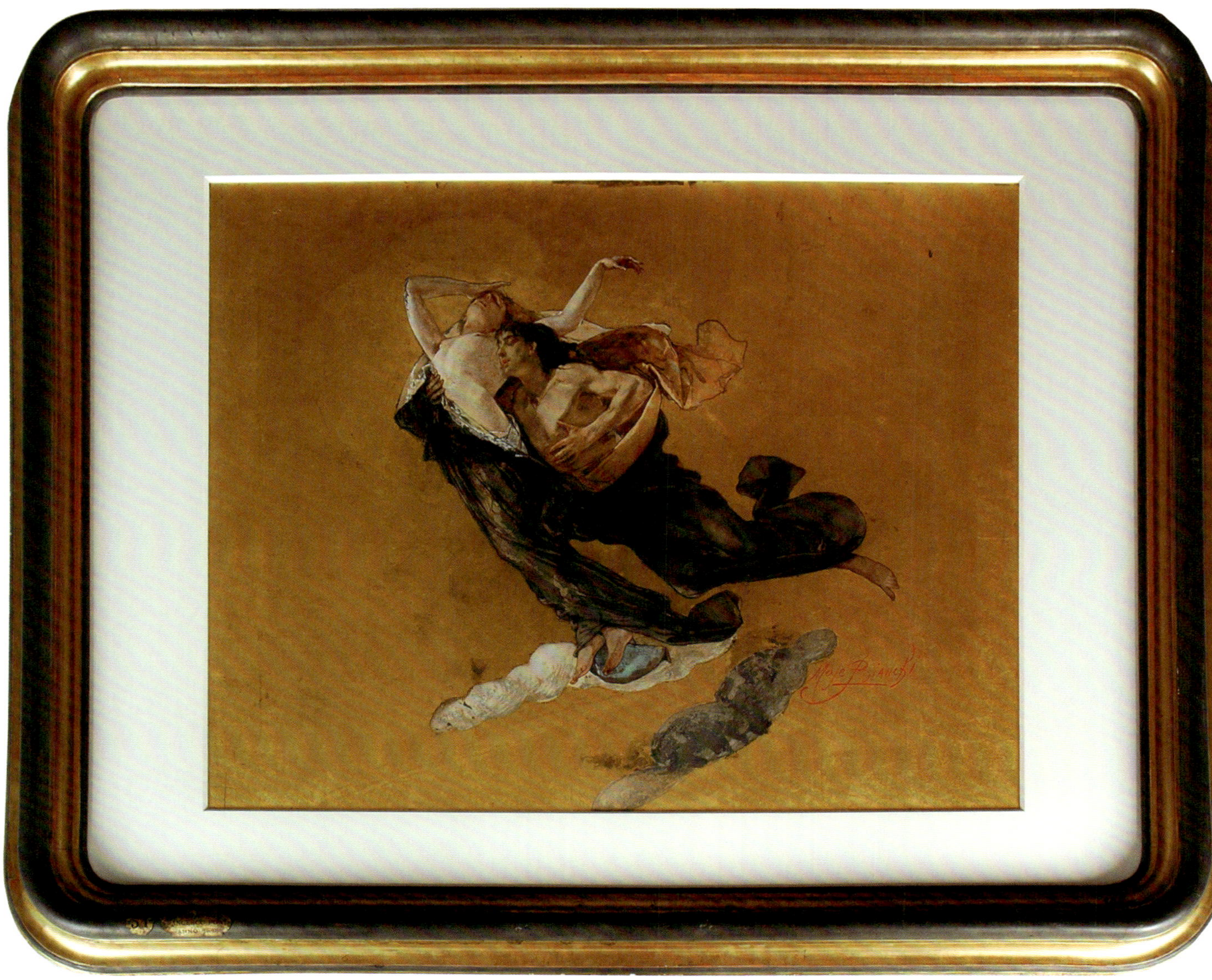

Mosè Bianchi
Paolo & Francesca, um 1877
Watercolour and gold on board, 64.5 × 85.5 cm
Comune di Milano, Galleria d'Arte Moderna, Milano

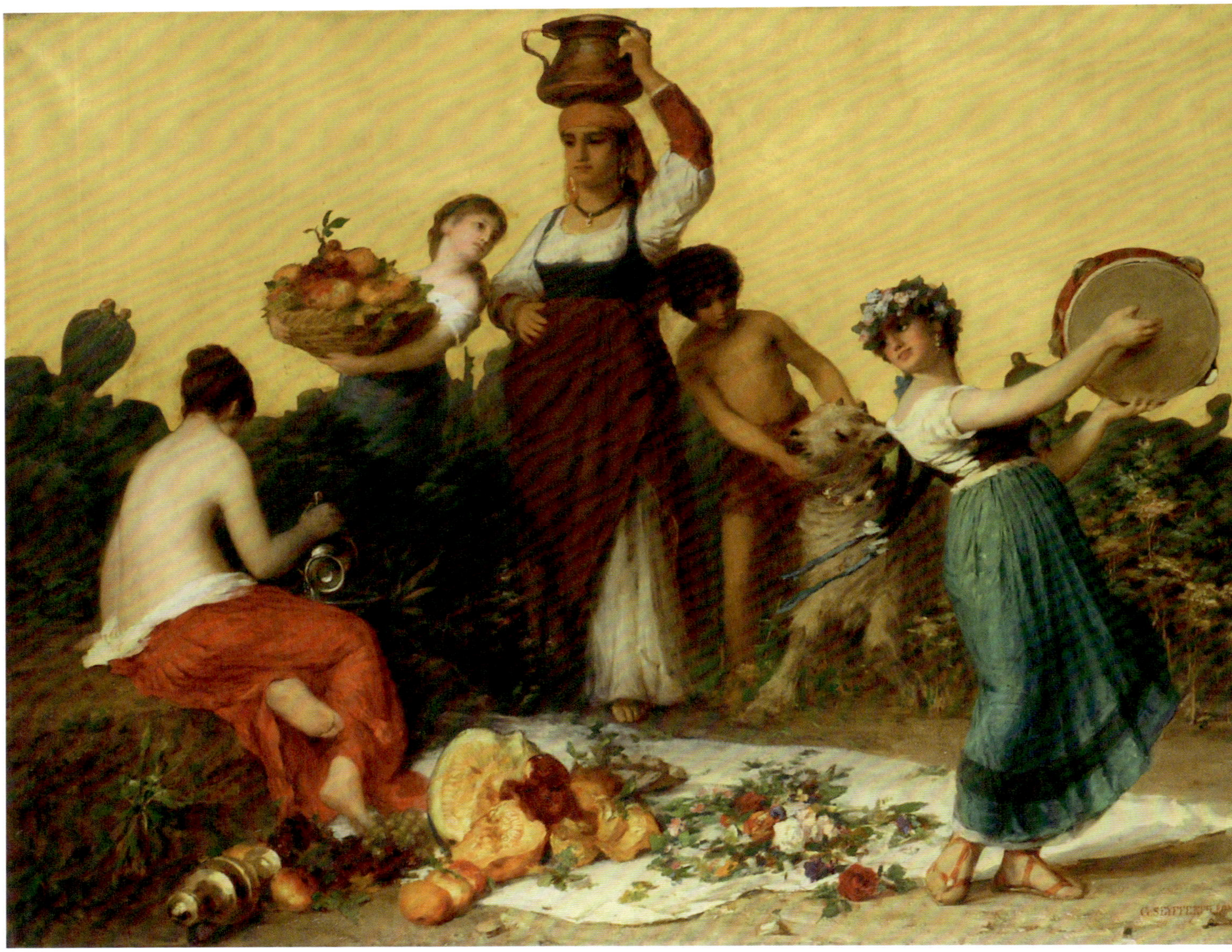

Gustav Seyfferth
Roman Autumn Festival, c. 1875/76
Oil, gold leaf on canvas, 124 x 159 cm
Frame richly decorated, c. 20 cm broad
Neue Galerie Graz at the Universalmuseum Joanneum

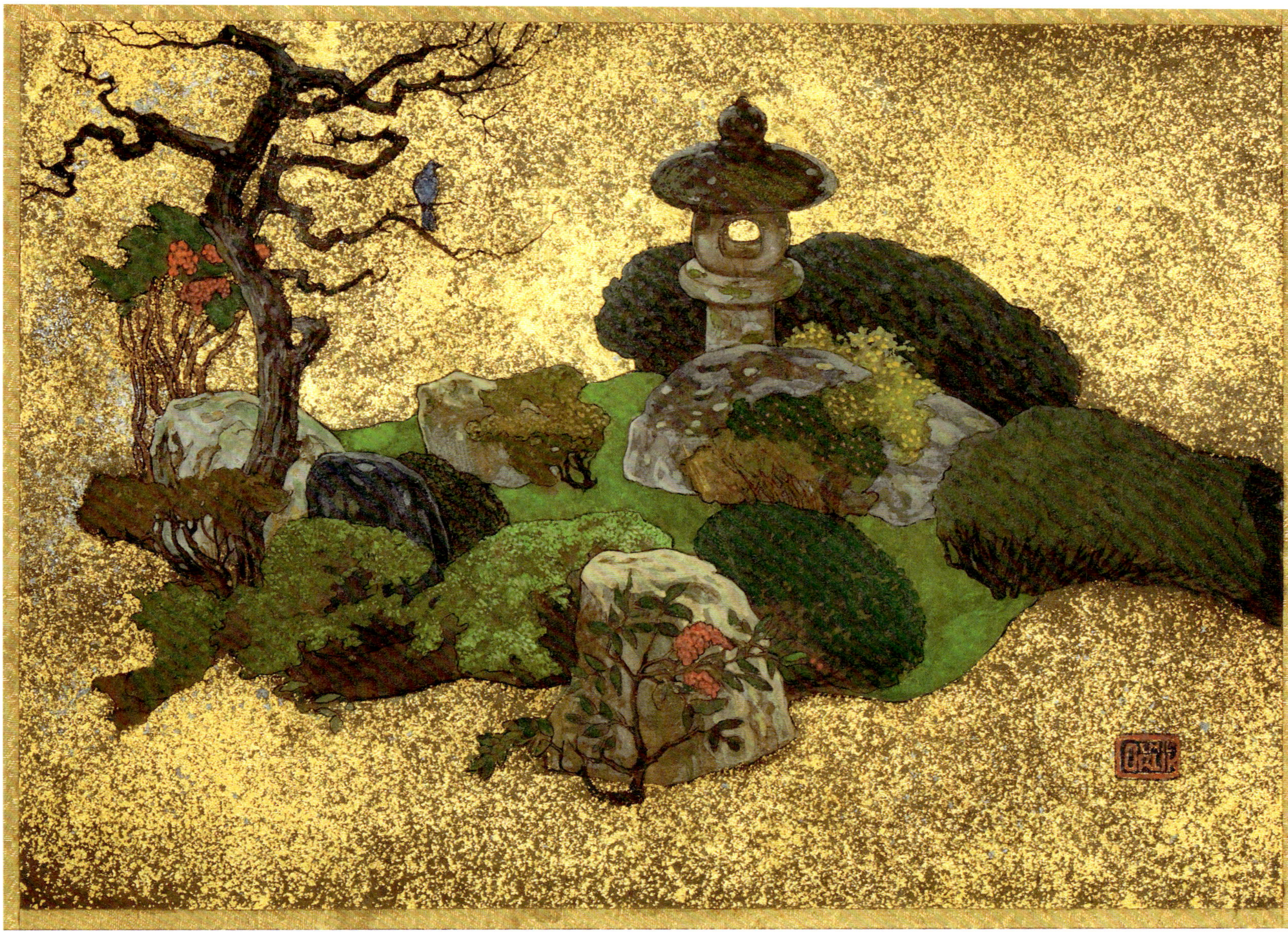

Emil Orlik
Japanese Garden, 1901/02
Tempera and bronze-gilt on paper, silk, 37.5 x 54 cm
Belvedere, Vienna

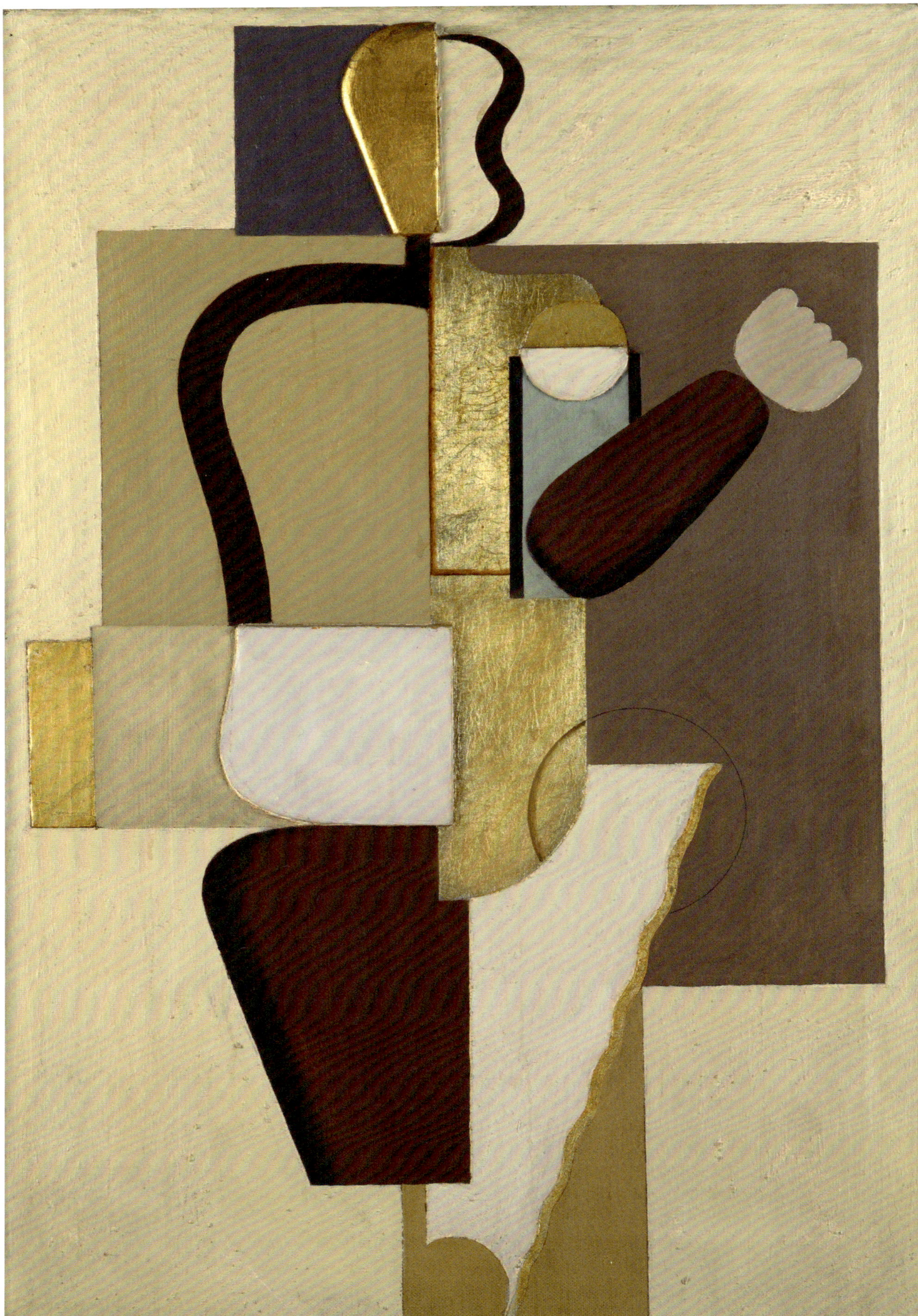

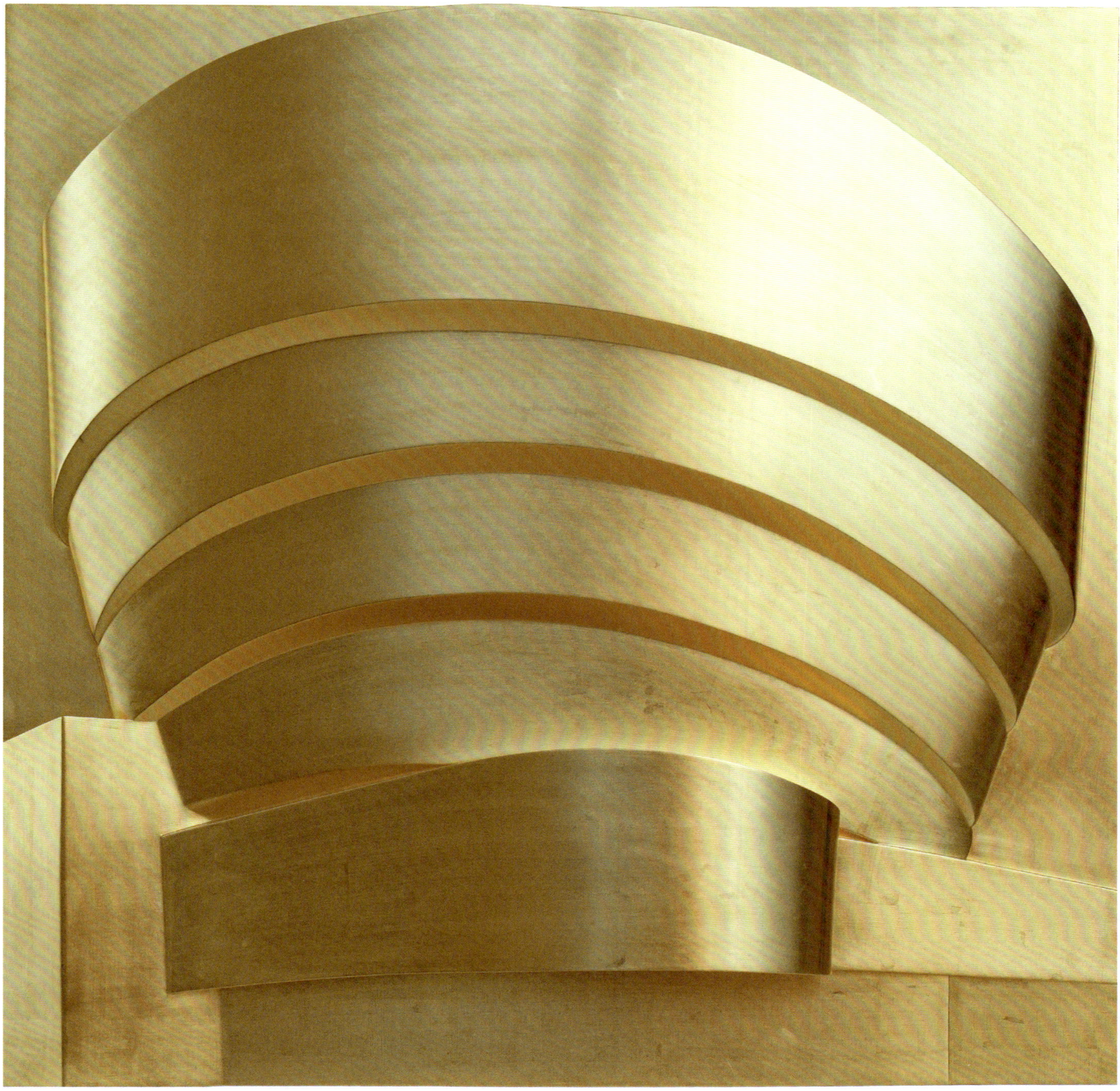

Richard Hamilton
The "Gold" Guggenheim, 1965–1966
High relief, gold leaf, fibreglass and cellulose, 121.8 x 122.4 x 20.5 cm
Louisiana Museum of Modern Art, Donation: The Museum Foundation

Willi Baumeister
Mauerbild mit Metallen, 1923
Oil, gold leaf and board on canvas, 89 x 61 cm
Stiftung Kunstsammlung Nordrhein-Westfalen, Düsseldorf

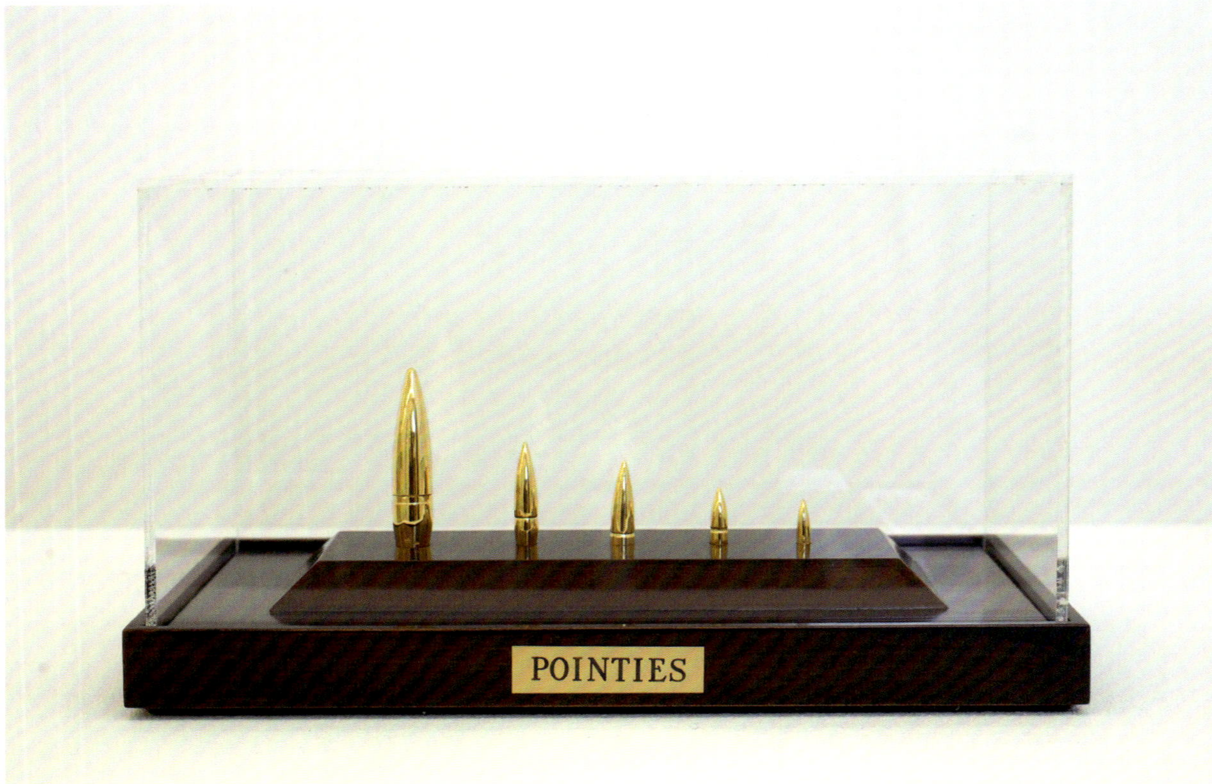

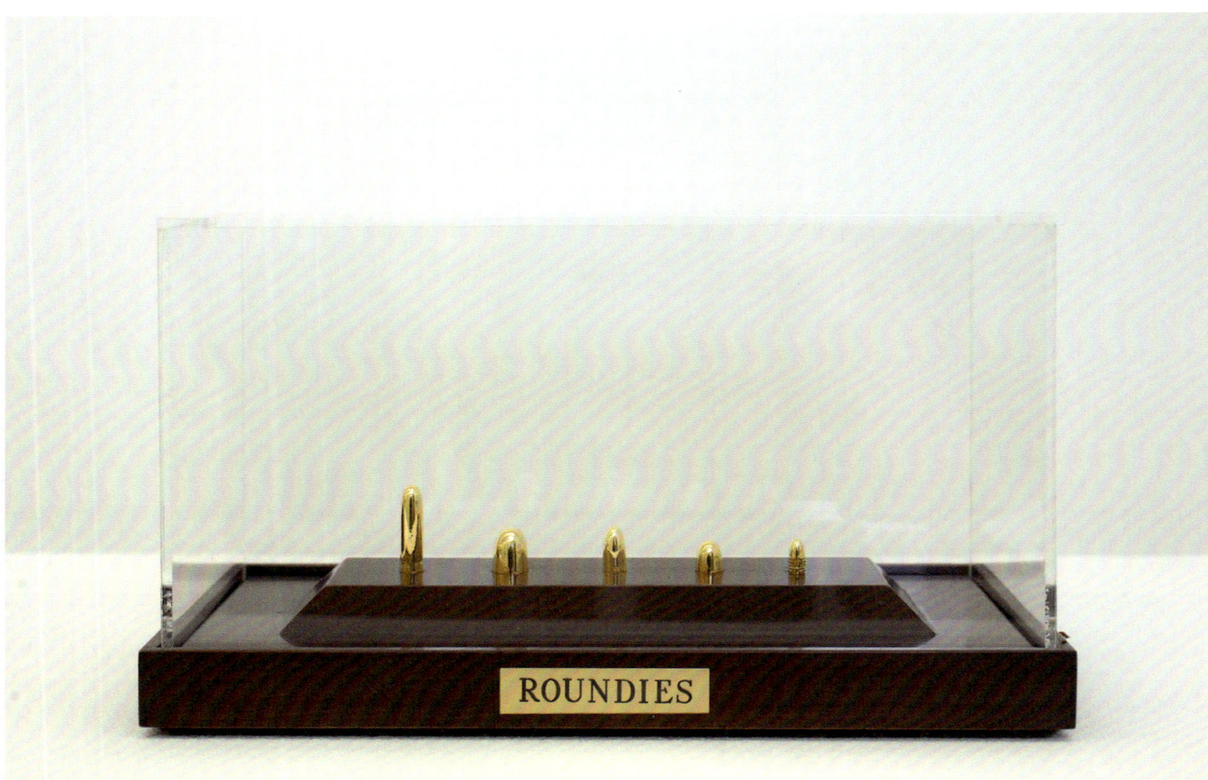

Chris Burden
Gold Bullets - Pointies and Roundies, 2003
10 22-carat gold cartridges, 2 cabinets (wood, Perspex), each 26 x 14.6 x 15.9 cm
Galerie Krinzinger, Vienna

Imi Knoebel
Menningebild 37/3, 1976/2010
Transfer gold leaf on wood, 190.7 x 351 x 9 cm
Private collection, Sindelfingen

Not Vital
The Golden Calf, 2001
18-carat gold
71 x 85 x 29 cm
Private collection

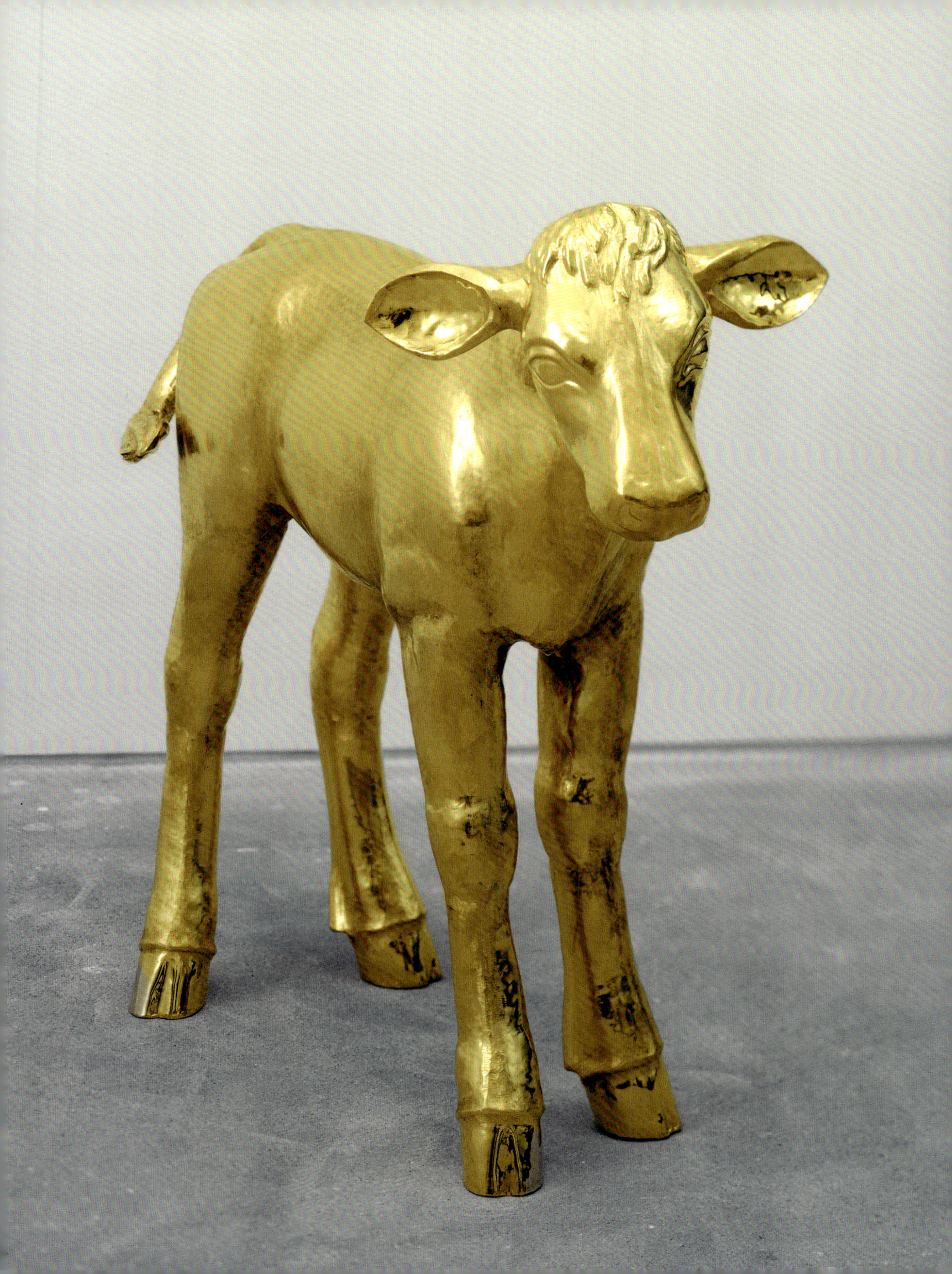

Frames

At the end of the Middle Ages, there was still a desire to preserve the gold ground in pictures. But with the perspective construction of the pictorial space, the disappearance of the gold ground from pictures was inevitable. From the Renaissance on, the gold withdrew to the decorative 'vanishing space' of the frame. In the nineteenth century, the frame took on iconographic functions, and a variety of artists (Philipp Otto Runge, Caspar David Friedrich, Anselm Feuerbach, Edgar Degas all the way to the Secession) designed their frames themselves.

Excursus: The disappearance of frames

The spread of the gold ground outside of the picture, in the frame, in the opulent furnishings of festive rooms in villas and castles all the way to the golden cabinets decorated entirely in gold, distorts our view of the actual function of the frame. To this day pictures are framed, indeed, in recent years the frame has experienced a veritable renaissance.

The avantgarde distanced itself from the gold frame at an early date, however. "The Louvre at first was only gold frames to me, gold frames which were rather glorious, and looking out of the window of the Louvre with the gold frames being all gold behind within was very glorious."[1] This was how Gertrude Stein characterized the museum as the place characterized by gold. In her shift of emphasis, she cumulates not just the gold grounds, but the gold frames, and the pictures disappear in the 'glory' of the collection. She looks out of the window and enjoys the treasures on which she has turned her back as a piece of the past: it is over. Again, as for Alberti[2] at the beginning of the post-medieval period, the window is the threshold, but this time as a liberation: away from splendid art, and into the new reality of the twentieth century. In 1906 Gertrude Stein hung her *Portrait* of Picasso programmatically without a frame to start with. The other pictures in her Paris flat were framed. But just two years later the Picasso too had a frame.[3] The frame is not just a distinguishing feature in art, but also confers dignity.

Since the Renaissance, the frame has served to raise the apparent reality of the pictures out of the empirical reality, and distinguish the one from the other. When pictures came to renounce illusionism, the function of the frame became questionable once more. It became itself a part of the work, before disappearing altogether from Modern art. Or at least, that was the idea. However, it was not so easy to change people's frame habits. When Emil Nolde chose 'serious black wood frames', the public 'could not bear it'. "My way of doing things was not what the public wanted; they wanted gold frames," he complained.[4] Kazimir Malevich called his *Black Square* (first exhibited in 1915) the 'naked unframed icon of my age'. And Piet Mondrian, self-confident but blissfully ignorant, wrote in retrospect in 1943: "As far as I know, I was the first to take the picture out of the frame instead of putting it into one."[5]

Modernism liberated the frame from its function as a picture margin and boundary. It was available or not, as the case might be, to expand the possibilities of perception. But it was not just artists who, in Modernism, renounced plinths and frames, but also artists who at the same time acted as dealers and exhibition curators allowed such interventions, and also revisions. In 1926 Marcel Duchamp bought 80 works by Francis Picabia and had them re-framed for sale. For the inaugural exhibition at the Art of This Century gallery (New York 1942) Frederick Kiesler denuded the pictures of their frames and fastened them to a movable mechanism which allowed visitors to perceive them from different perspectives.[6]

Illustrations of picture frames are only found in museum catalogues and coffee-table books when it cannot be avoided. In an opulent catalogue of the Louvre, the glorious-gold-frame Louvre in Gertrude Stein's view, there is, in the section on post-medieval art, just one (!) painting with a frame. For all the others, the graphic artists have cleanly cut away the frames. It seems as if the graphic designers were following the recommendation of Denis Diderot, who advised visitors to the 1763 Salon to look at the pictures with special glasses that would exclude the frames and the surroundings.

The printed result of illustrated frames seems all the more embarrassing insofar as the frame embodies the conditions of the creation and display of pictures, and not just in a direct sense. The history of the frame points to the history of roles and functions of paintings in their respective space. It is not surprising that artists wanted to liberate themselves from it, but it is hard to see why it was not clear to art historians that, as a result, they had faded out the historical dimension—even though as long ago as 1927 Wilhelm von Bode had staged the first exhibition of frames in Berlin.

In the seventeenth and eighteenth centuries, there was an endeavour to create coherent images for galleries, and to this end original frames were removed and all the pictures were uniformly reframed. In the nineteenth century old frames were copied, or new ones were made to designs (for example by the great architects Leo von Klenze or Karl Friedrich Schinkel) to the taste of the client with a view to selling them or giving them as presents. "Since the twentieth century most museums pay careful attention to giving the pictures stylistically, chronologically and topographically suitable frames. Only in very rare cases do the paintings still have their original frames."[7] It took some time for the cult of modern framelessness in avantgarde painting to reach the museums. In his museum 'architecture' in Venice (Museo Correr, 1957 or Pinoteca Querini Stampaglia, 1961) Carlo Scarpa cleared the frames out. They were either (shamefully) soon sold, or (luckily) found their way into the depositories. This purism was linked to the idea of democratization. "There was a belief that by getting rid of the frames, one was also getting rid of all the ballast surrounding their creation, the concentrated burden of history, of knowledge and information, that was associated with these works of art."[8]

But the pendulum has long since been swinging the other way. Museums seem slowly to have become aware of the reverse side of the circumstance, for they are reacting on their homepages. The website of the Louvre, for example, largely includes the pictures' frames once more.

In the publications, pictures are given a white frame, such as was first chosen by the Impressionists. As golden frames were obligatory in the exhibitions at the Salon, this option was not available to the Impressionists for this reason alone.

In many avantgarde pictures, in which the form is reduced, and colours, space and content all tend to zero, the question concerning the equally vanishing frame remains. Where have all the frames gone? They have taken on a life of their own and are reconquering the terrain which has been lost in the pictures. Jasper Johns, Arnulf Rainer, Daniel Buren have hung pictureless frames on the wall. In all three examples, however, the form of the frame is irrelevant. Since then, empty frames have been filled with contents, in other words with meanings or at least with formal signs. Empty frames can also stand for contents. They can demonstrate that something may not be shown, that the picture has been censored, that a dictatorship has threatened to destroy a gallery or the pictures in an exhibition. Thus the Iranian artist Parastou Forouhar decided in 2009 to remove the pictures from their frames and to open the exhibition *Blind Spot* in this way, which everyone could understand. "You can say a lot in Iran. Especially with frames that don't contain pictures."[9]

The most radical position in this respect has been taken by Christian Eckart, who chose his name after the German mystic Meister Eckhart (1260–1327). In programmatic statements such as *Picturing God: An Economy of the Zero Project* (1997)[10] the Canadian artist confronts the concept of a god who cannot be pictured. In various series of pictures, *Eidolon, Andachtsbilder, Sacra conversazione in art* he reflects on past eras and pathos formulae. The closed, broken gold frame covering the picture is itself the picture field of *Deus absconditus* (hidden God), and, through the elimination of what was ever on display in altarpieces, draws the logical conclusion from the impossibility of depicting God. The gap surrounded by the frame shows nothing outwardly, but the inward yearning for God, who in Himself corresponds to nothing. Eckart sees himself as a kind of icon painter, altogether in the Modernist tradition from Kazimir Malevich to Lucio Fontana and Yves Klein.

The gold leaf may dispense with Christian devotion, but in his opinion it relates to the highest degree of spirituality of which a material is capable. For Eckart, gold leaf may be a central argument, but it is not the only material which he employs. He has, for example, used the processing techniques applied on Bentley cars, which go back to Japanese lacquer techniques, and he also uses hand-brushed aluminium.

Many artists play with the aspects of the frame as picture-boundary, with the variations of architectural logic (Franz Josef Altenburg)[11] as perspective space for the gold, as a window for their own work: "If you sit on the bench and have the window opposite, or at the end of the room, in front of you, you see nothing but the frame and the 'moving picture' within." (Gerwald Rockenschaub)[12]

Josef Kern carves the gilded wooden frames for his pictures himself. Like a dream of the sleeping *Julia* (2006) the sixteen referential fields develop around the picture. They recall the

Ship of Fools (1494) by Sebastian Brant or other picaresque medieval pictures like *Till Eulenspiegel* (1515).

Michael Bergt also adapts the frames of his mythological and allegorical miniatures individually to the pictures. The gold ground integrates both the figures and the frames, but the boundaries of the pictures are always preserved in the process. "This is the problem with the frame: it is always present, but you don't realize."[13]

It is only in recent decades that the frame has replaced the pictures, or an intermingling has taken place. Artists are no longer prepared to accept the traditional separation of picture and frame, and they are addressing the relationship between them. To the extent that the avantgarde renunciation of depiction is retained, the gold takes on a major role as picture support. An empty frame has not necessarily lost its picture, not must the interpretation fill the empty space in a theological sense. Jan Maarten Voskuil's *Gilded Frame* (2002) has exchanged its function as picture support in favour of the new orientation. Because it has slipped from the wall to the floor, it becomes, simply and clearly, part of the installation that includes the horizontal floor. This frame corresponds to no traditional picture.

Cornelius Kolig allows the frame an extravagant life of its own. Invoking pristine childlike creativity[14] the ordering geometry of the traditional edge is lost. The frame as a means of detachment, of distancing oneself from the picture, but also as a bridge to society, obeys no regulated system. The centre of the luxuriant *Gold Frame* (1986/2011) remains an empty projection surface.

The frames by Nedko Solakov leave the craft domain in order to proclaim their autonomy as art. This is not simply postulated, but demonstratively asserted by the use of gold. The gold from the frame flows back into the picture field. Parts of the gilded frame belong to the story, just as elements of the golden ground of the picture could be carved. The frame cannot be exchanged for another, above all it shows, in relief, parts of the story. To dispense with the frame would be like cropping the picture. The gold ground and the frame no longer form a contrast, and come alive through their respective histories.

The narrative complexity within these frames can only be understood in close-up, as in medieval manuscripts. Here, as in illuminated books, the gold ground and the narrative text are united in pictures. Crucial in this development is not only the abolition of the boundary between the 'craft' edge and the 'art' field, but also the potential exchangeability of the iconographic elements and the personal activation of various traditions, from medieval models to icons. The references are to the Old Testament, to Buddhism, and to what might be called individual mythology.

The Self and its Frame

James Lee Byars found in gold the central expression of his actions and works dedicated to the artistic 'mystery'. Unforgotten is the installation consisting of a room, hung with gold leaf, *The Death of James Lee Byars* (1982/1994). His idealized *Self Portrait* consists of a golden globe. His *Autobiography* (1986) emphasizes the round golden frame of his existence, whose nucleus he illustrates as a sphere of bread lying on red silk. In the sarcophagus-like version *alla veneziana* (1986), too, a bread sphere lies on a satin cushion.[15] While 'man does not live by bread alone', for those of little faith what W.H. Auden is said to have written can at least make sense,: "Art is our chief means of breaking bread with the dead."

The ideas of the concept artist Norbert Fleischmann, who works with the means of the old masters, circle around the ego-concept and the museum. Between the self-perception and the gold there is an antagonistic relationship. To reduce it to a simple denominator, the post-medieval self has driven gold from painting. In three selected examples, Fleischmann revises this historical development in paradoxical manner.

On its way into the museum, an object loses its original context, its mobility; it is withdrawn from sale and gains security and shelter. Many artists yearn for this in respect of their own life's work. They have in principle an inclination to-wards solipsism, to narcissism, to exaggerating their own achievement.

Whether and how thought gets the ability to constitute the self is not a question we can go into here. The Cartesian 'Cogito, ergo sum', in the repetition of the 'I' in the English translation 'I think, therefore I am', sounds tautologous. Much as it is insisted upon in philosophical writings and novels, the postulate of the unknown self is ultimately empty. Therefore one reads the '*ich*' in golden letters in the centre of an excessively broad horizon in *expand II* (2008) with a degree of caution. Should one be allowed to place oneself in the centre like this, must the self appear in gold, as it were outshining the darkness of the world? Is this '*ich*' the only message of an artist to whom the world appears too unimportant? Is every sense of proportion missing from this expanse?

The 'I' appears in Norbert Fleischmann's horizontal format in lower-case: *ich*. It is set in italics, in gold, in itself a contradiction. On the one hand the gold, which since the Renaissance has been looked down upon in pictures, and on the other the italics, which, similarly, originated in the quattrocento. The design of the italic typeface is attributed to Aldus Pius Manutius (1449–1515). In some languages, for example German, italic type is known by a word cognate with 'cursive', which comes from the Latin word *currere* (to run). This lower-case '*ich*' is running around in the darkness of the wide world. Even though it is in the middle, it is not for this reason to be understood as static. The self has taken over responsibility for what used to be the religious context. If an object of depiction turns to gold itself, then, superficially, it becomes heightened and costly, but at the same time vague. The contradiction between gold the material and italic typeface characterizes the tension of an insecure self. The expanse of the framed black offers no protection, but is the distinguishing feature of all negation and ignorance. This recalls the Romantic contrast, or perhaps the transcendental philosophy of Johann Gottlieb Fichte. "Fichte put all the experiential possibilities of the world into this 'I' when he contrasted it with the not-yet-experienced, but yet experienceable as non-I."[16] The blue flower of the Romantic *summon* (2011) has become a symbol of the 'striving for the knowledge of the self'.

In Fleischmann, the self-confident Cartesian I of the 'cogito, ergo sum' has become a stuttering little '*ich*'. And yet the polarity of gold and black is accepted wordlessly and thus surprisingly conflict-free. We have a share in the ignorance, in the vanishing sacredness, in the wandering, in the inevitable awareness of being in the centre as far as oneself is concerned. Not least, it seems strange that when reading the '*ich*' we automatically think of the 'I' of the artist, who places himself in the centre, and not of ourselves or other selves as objects of the depiction, in other words we do not think of the general problem of I-ness. In the catalogue of the exhibition *Ich ist etwas Anderes. Kunst am Ende des 20. Jahrhunderts*, Slavoj Žižek writes "The self is an object which creates from the outside a minimum of substance for the subject, this void of pure negativity."[17] From Arthur Rimbaud's assertion in his famous letter of 13 May 1871, "I am someone else" to Žižek's parable, the modern I veers between megalomania and self-denial.

The self-perception of artists is documented by the signatures on their pictures. The forging of signatures is punishable under copyright law. At the same time it is easy to overlook the fact that it is not the signature but the signed work which enjoys protection as 'intellectual property'. In *call*, Fleischmann assembles initials and highlights them in gold. This museum accumulation is sufficient unto itself. They are the signs of the 'I's' of other artists, which in accordance with the title of the picture he 'calls on' or by whom he is reached.

Fleischmann unites the stages of post-medieval self-awareness which resist gold and places them at our disposal through the use of gold.

1 Gertrude Stein, "Pictures" in J.D. McClatchy (ed.) *Poets on Painters*, Berkeley, Los Angeles and London, 1988, 81–102, here p. 91.
2 Cf. the essay in this volume by Thomas Zaunschirn, *Alberti's Gold Ground.*
3 Christine Traber, 'In Perfect Harmony? Entgrenzungen in der Kunst des frühen 2o. Jahrhunderts', in Eva Mendgen (ed.) *In Perfect Harmony. Bild + Rahmen 1850–1920*, (exhib. cat. van Gogh Museum, Amsterdam—Kunstforum Wien), Zwolle 1995, p. 220.
4 Ibid., p. 234.
5 Ibid., pp..237, 248.
6 Veit Görner, *Der Betrachter als Akteur. Partizipationsmodelle in der frühen Kunst des 20. Jahrhunderts*, diss. Hochschule für Bildende Künste Braunschweig, 2005, pp. 134, 165.
7 Helge Sievert, "Geschichte der Rahmensammlung", in: *Rahmenkunst. Auf Spurensuche in der Alten Pinakothek* (exhib. cat. Bayerische Staatsgemäldesammlungen Munich), Ostfildern 2010, p. 37, fig. 27.
8 Johann Kräftner, *Halt und Zierde. Das Bild und sein Rahmen* (exhib. cat.. Liechtenstein Museum Wien) Vienna 2009, introduction, p. 22.
9 Parastou Forouhar quoted from: URL: <http://en.qantara.de/Suspect-Commemoration/6958c7027i1p450/[read on 13 Dec. 2011]
10 Cf. URL: <http://www.christianeckart.com/[read on 22 Nov. 2011]
11 "Because they are so small, frames allow you to neglect the statics. Immediacy is possible, and for me they have the spontaneity of a drawing." Quoted from: *Franz Josef Altenburg* (exhib. cat. Landesgalerie Linz), Weitra 2006, p. 29.
12 Gerwald Rockenschaub, *Variations on Classic* (exhib. cat. BAWAG Foundation, Vienna), Vienna 2001, p. 20.
13 Ernst Pöppel, *Der Rahmen. Ein Blick des Gehirns auf unser Ich* (2006), Munich 2010, p. 502.
14 Thomas Zaunschirn, "Gold", in Thomas Zaunschirn (ed.): *Wiener Diwan—Sigmund Freud heute*, (exhib. cat.. Museum des 20. Jahrhunderts, Vienna), Klagenfurt 1989, p. 6
15 On bread as an artistic material see Jürgen Raap, "Das Abendmahl in Religion und Kunst. Essen & Trinken A-Z. Brot", in: *Kunstforum International*, vol. 160, June/July 2002, 140–145.
16 Gerhard Schulz, *Romantik. Geschichte und Begriff*, Munich 1996, p. 96. The further history of the self/ego/I in Kant, Husserl, Sartre, Heidegger, Wittgenstein down to modern neuroscience has only strengthened doubts concerning the substance of a conscious self.
17 Slavoj Žižek, "Das Selbst als das Andere des Subjekts", in: *Ich ist etwas Anderes. Kunst am Ende des 20. Jahrhunderts* (exhib. cat. Kunstsammlung Nordrhein-Westfalen, Düsseldorf) Cologne 2000, p. 79.

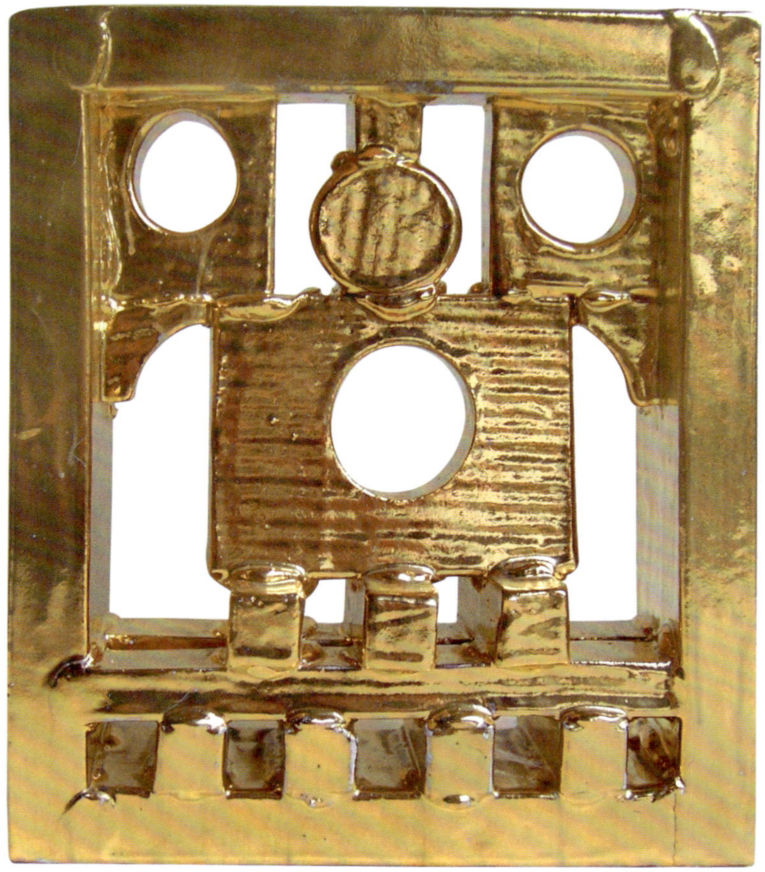

Franz Josef Altenburg
Frame, 2005
Ceramic, polished gold
26.5 x 22.5 x 6 cm
Courtesy the artist

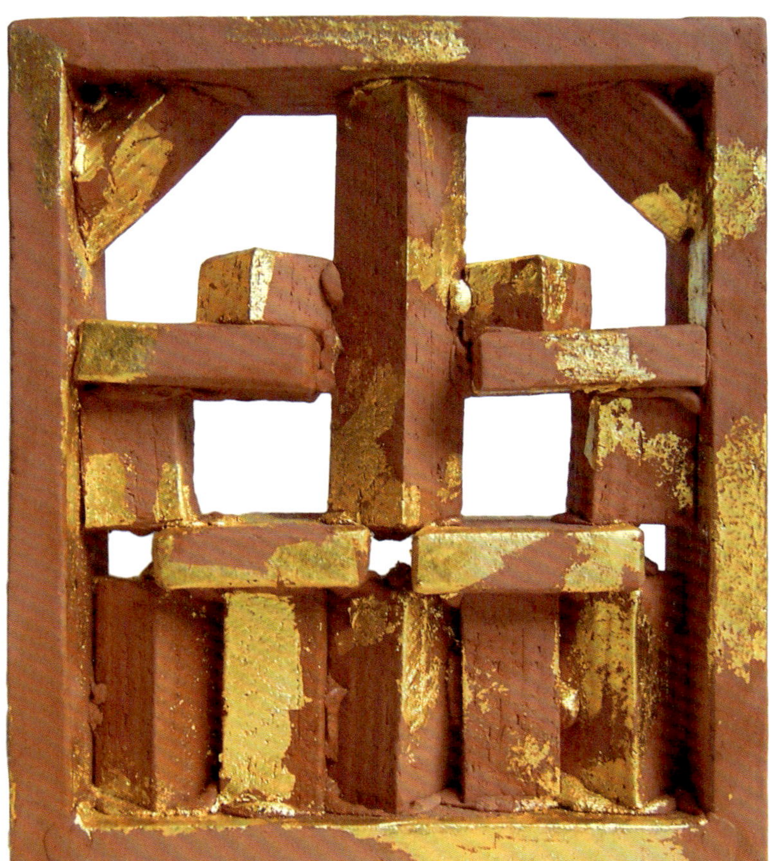

Franz Josef Altenburg
Frame, 2006
Ceramic, polished gold
26.5 x 22.5 x 6 cm
Courtesy the artist

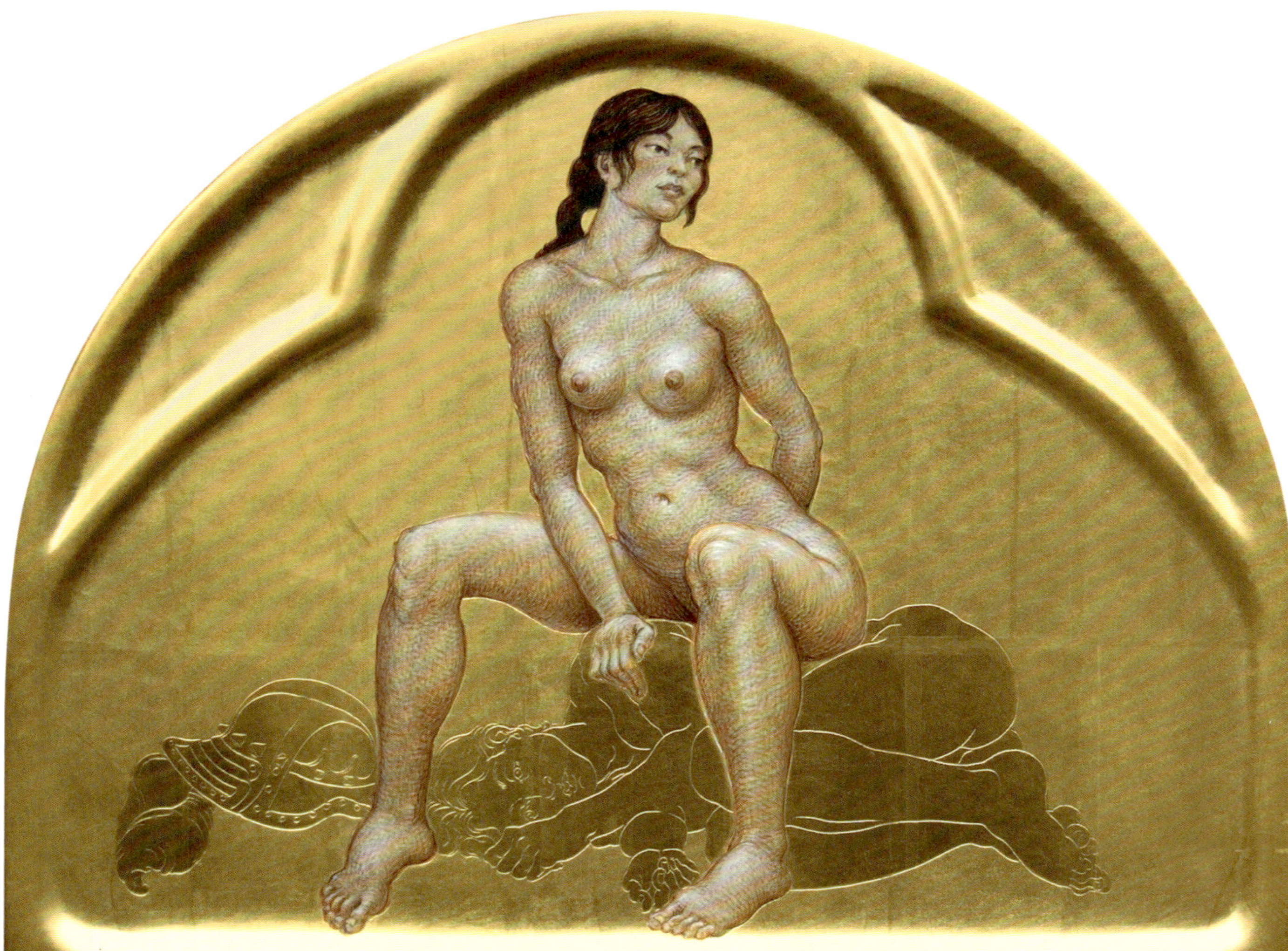

Michael Bergt
Venus and Mars, 2009
Egg tempera and gold leaf on wood, 22.9 x 29.2 cm
Courtesy the artist, via Jane Sauer Gallery, Santa Fe, New Mexico

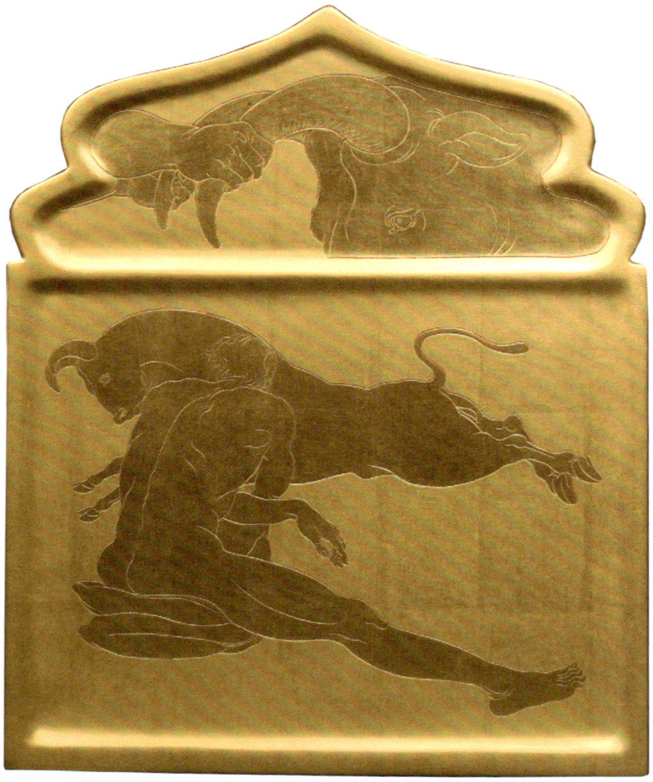

Michael Bergt
By The Horns, 2009
Gold leaf on wood, 31.1 x 25.4 cm
Peter Selig, San Antonio, Texas

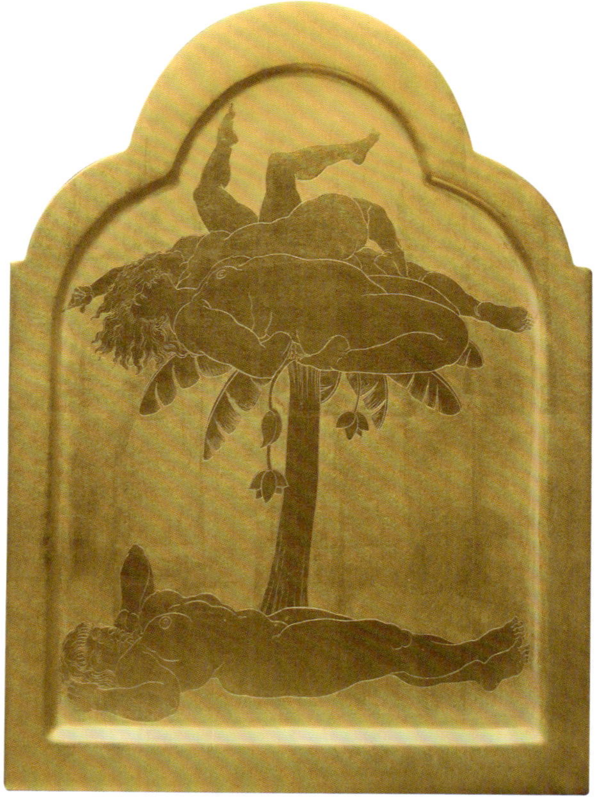

Michael Bergt
Tree of Life, 2009
Gold leaf on wood, 40.6 x 33 cm
Yolanda and Abram Eisenstein, USA

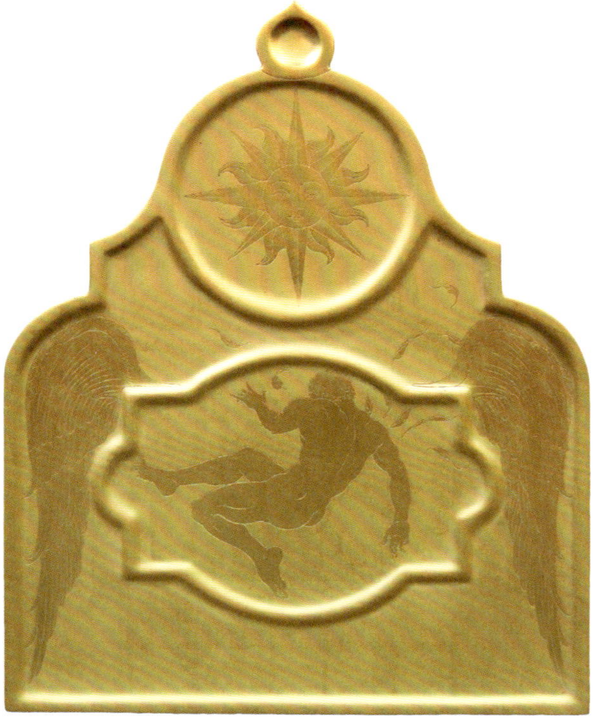

Michael Bergt
Icarus, 2009
Gold leaf on wood, 45.1 x 35.6 cm
Courtesy the artist, via Jane Sauer Gallery,
Santa Fe, New Mexico

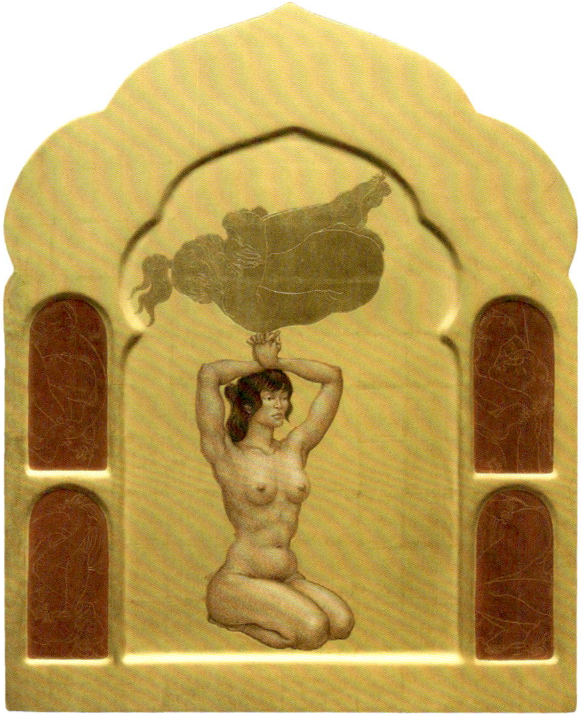

Michael Bergt
Balancing Act, 2009
Egg tempera and gold leaf on wood, 50.8 x 38.1 cm
Courtesy the artist, via Jane Sauer Gallery,
Santa Fe, New Mexico

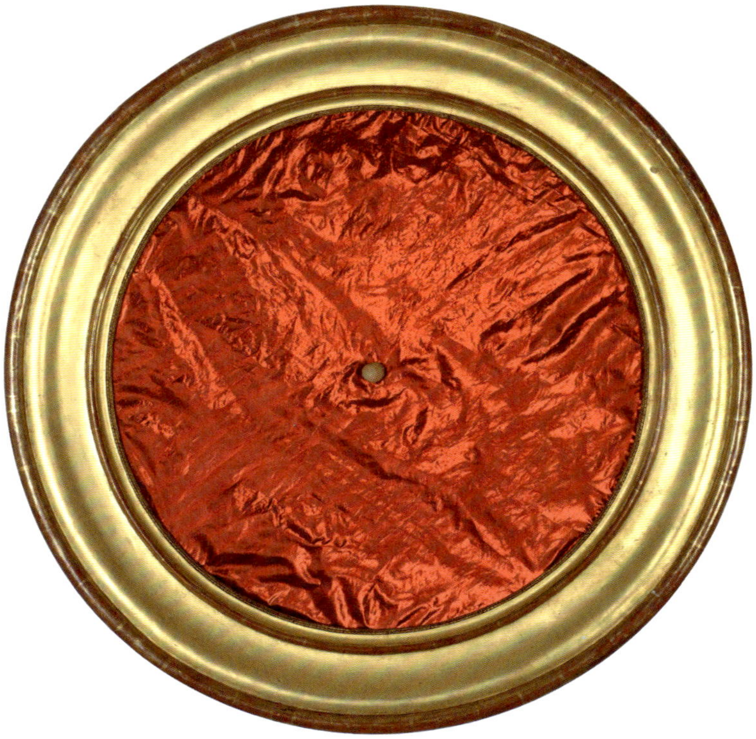

James Lee Byars
Autobiography, 1986
Bread ball on red silk in gold frame, diameter 70.5 cm
Kunstmuseum Bern, Collection Toni Gerber, Bern—purchased 1993

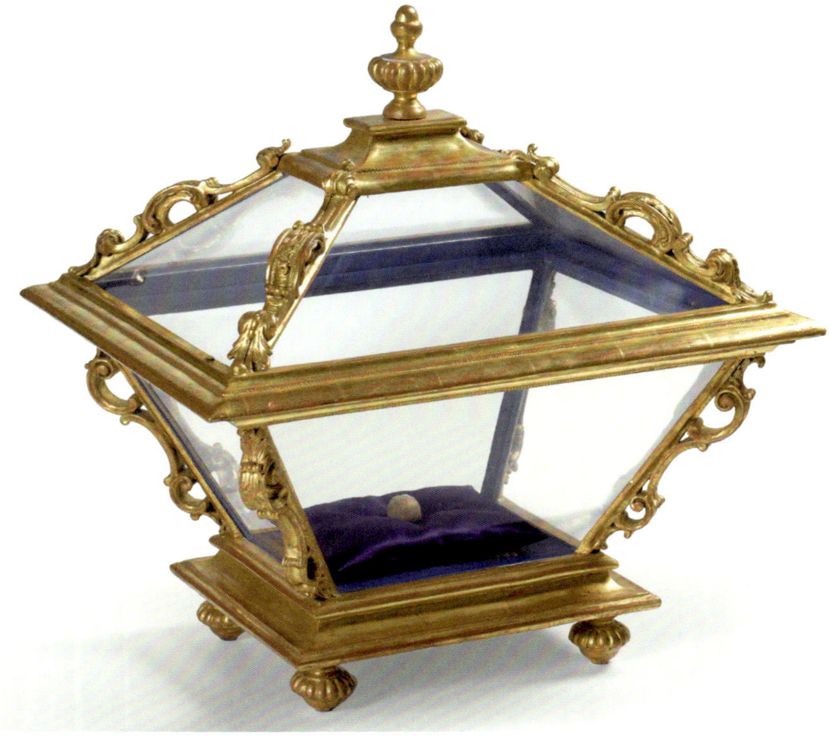

James Lee Byars
Autobiography alla veneziana, 1986
Glass and gilded wood, bread ball on satin cushion, 55 x 52 x 42.5 cm
Kunstmuseum Bern, Collection Toni Gerber, Bern—purchased 1993

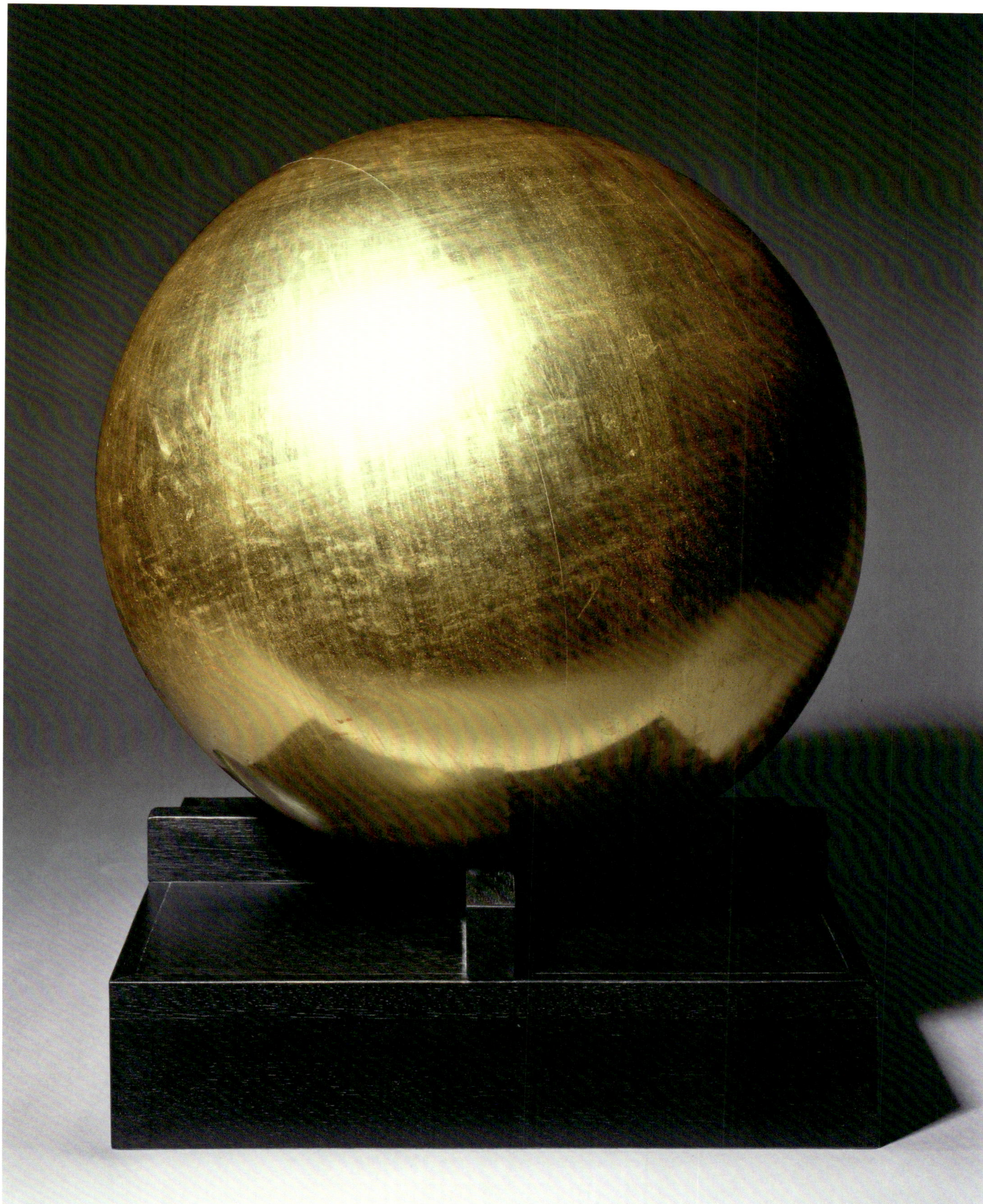

James Lee Byars
IS, 1991
Wooden sphere, gilded, diameter 33 cm
Collection Titze

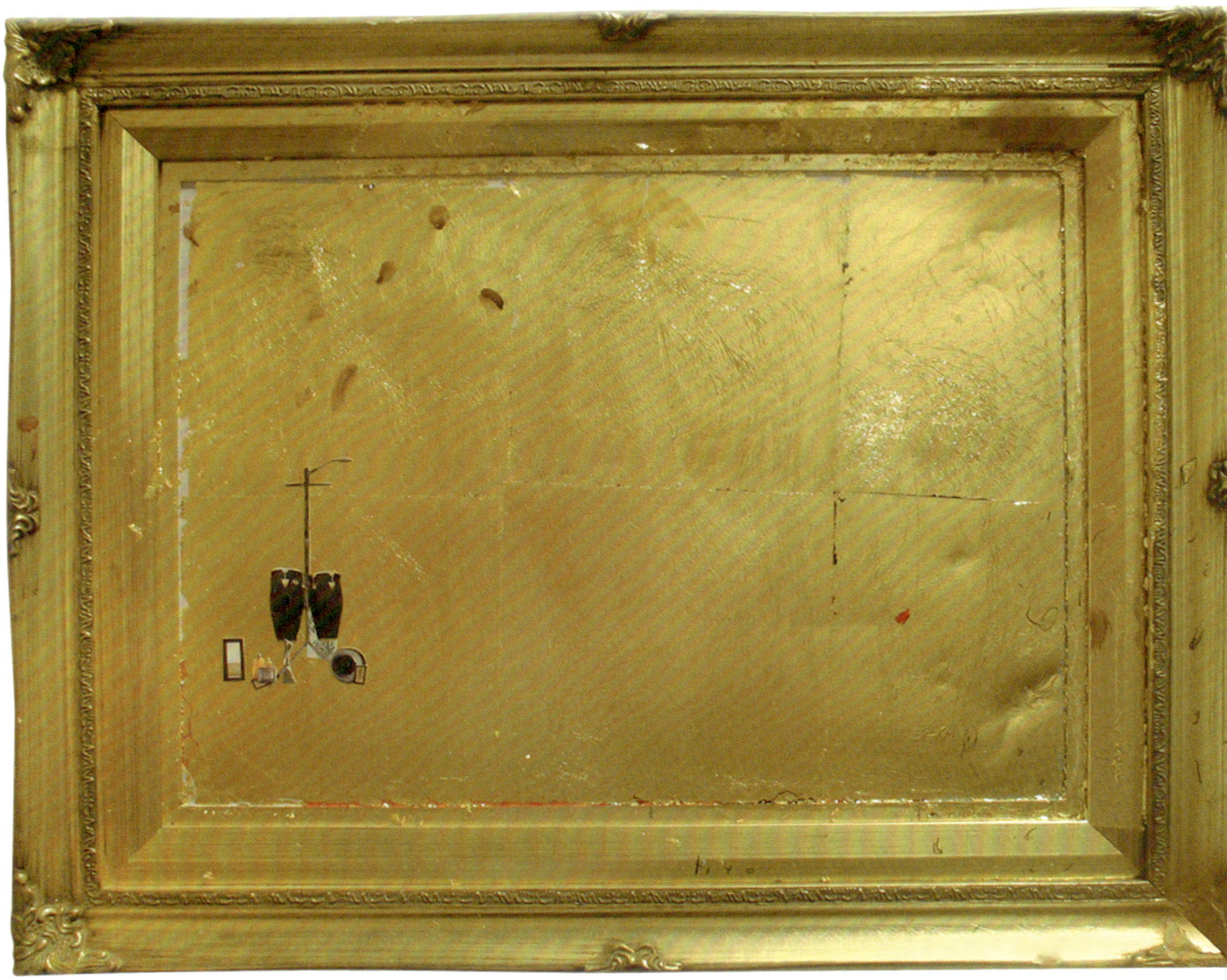

William Cordova
Untitled, 2010
Gold leaf and collage in vintage frame, 42 x 56 cm
Private Collection, Germany, Courtesy ARNDT, Berlin

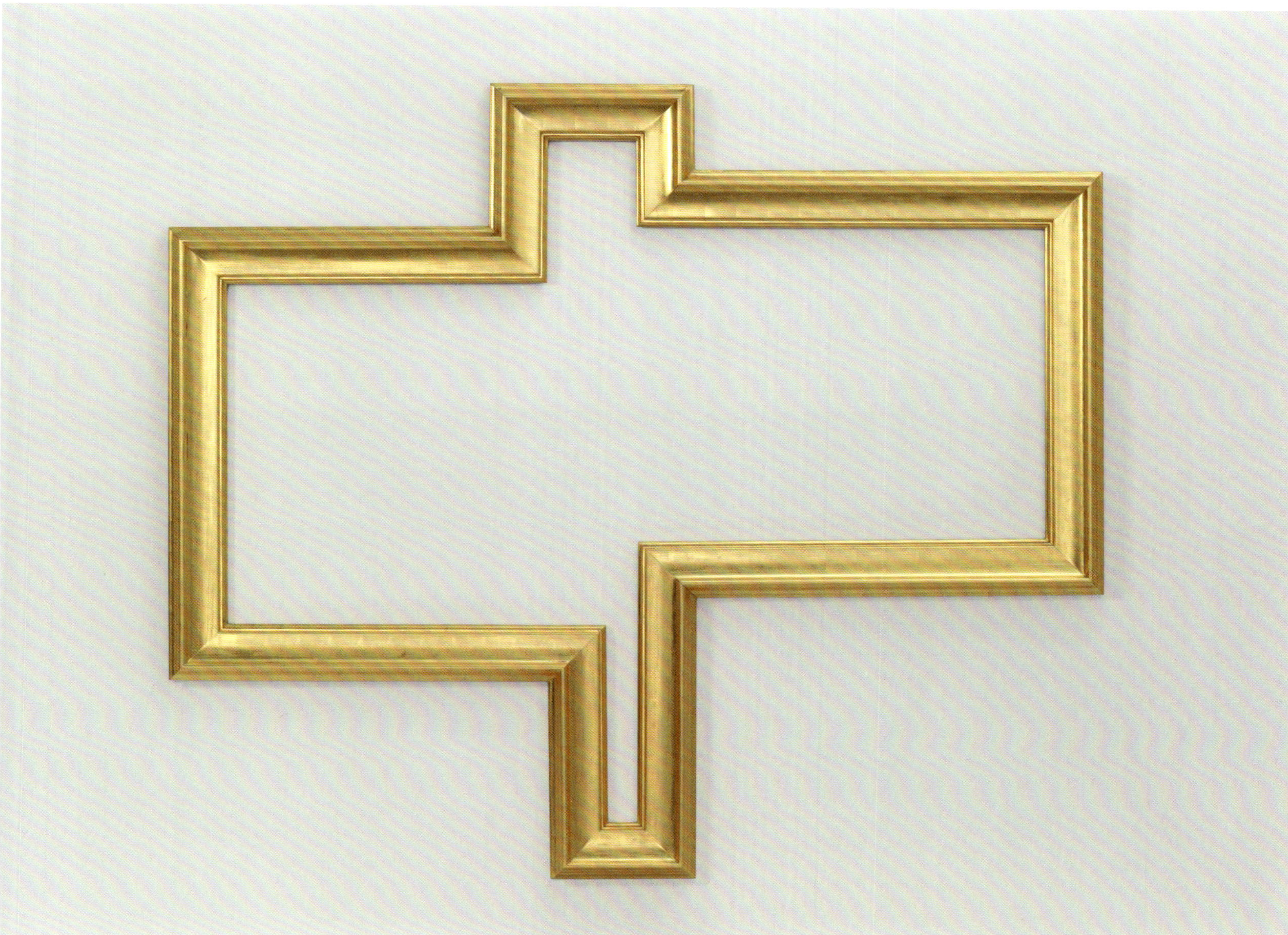

Christian Eckart
Eidolon # 1101, 1989/2000
23-carat gold leaf on poplar wood, 213.4 x 243.8 x 35.6 cm

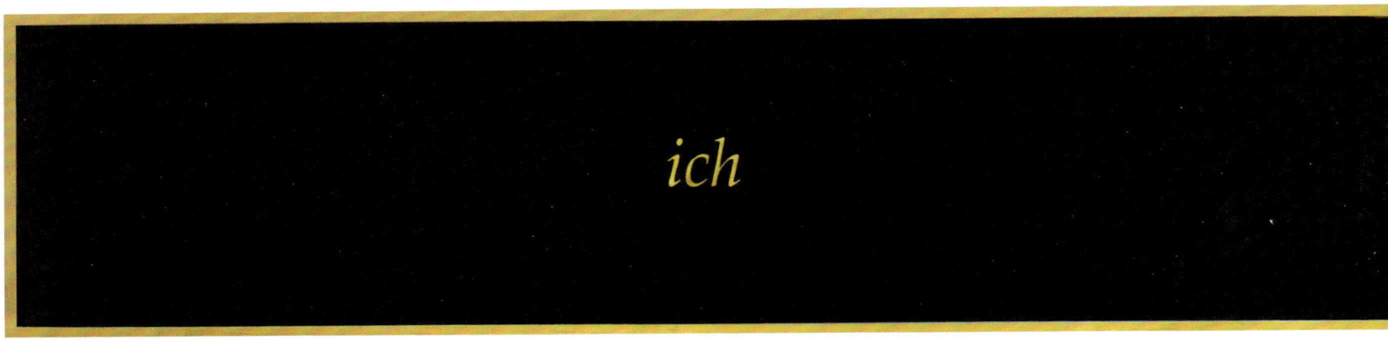

Norbert Fleischmann
expand II, 2008
Screenprint on glass, gold leaf, framed, Edition 1/3 bis 3/3, + 1e.a + 1p.p., Werkstatt Stalzer, Vienna, 37.5 x 166 cm
Norbert Fleischmann

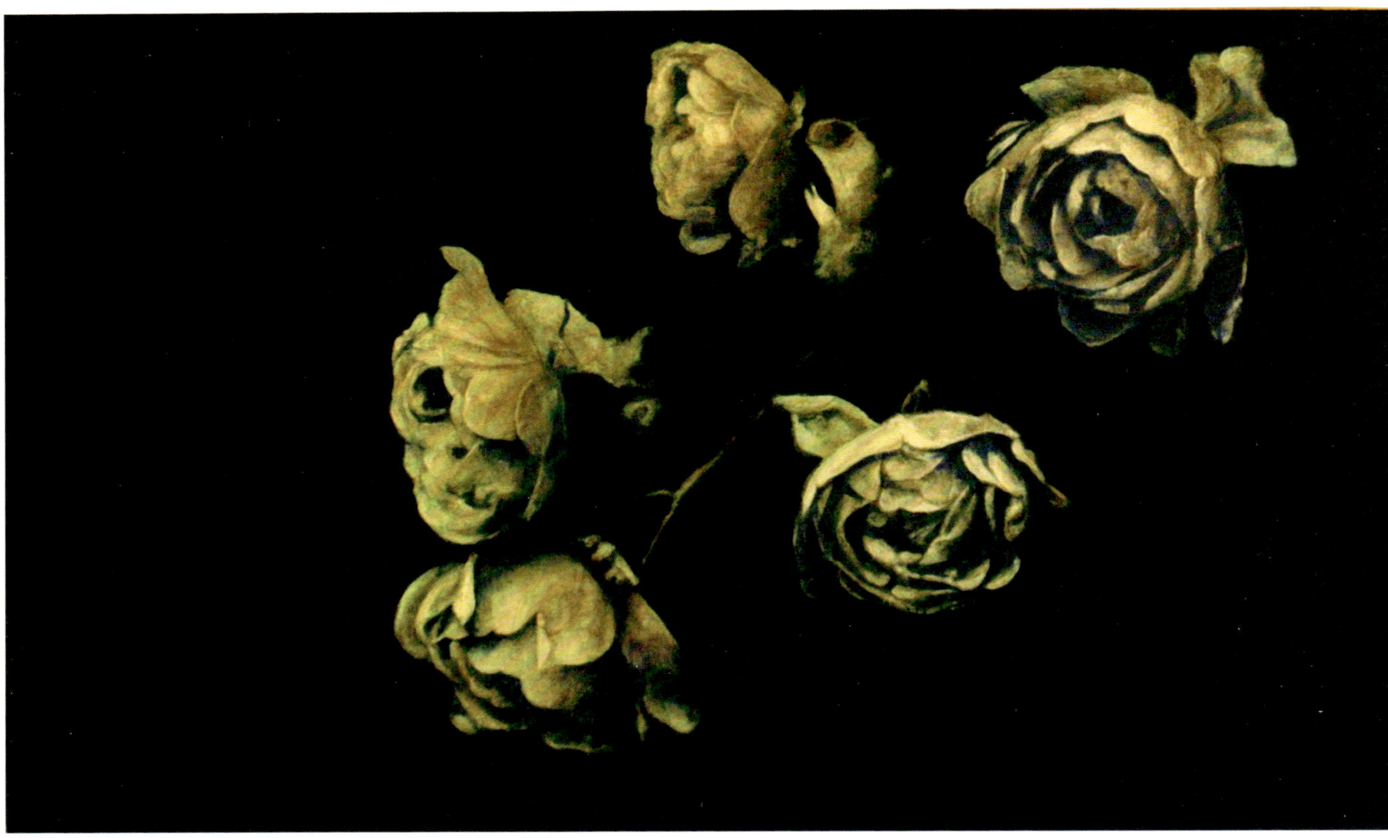

Norbert Fleischmann
summon, 2011
Resin oil on gold ground, wooden construction, 30 x 50 cm
Norbert Fleischmann

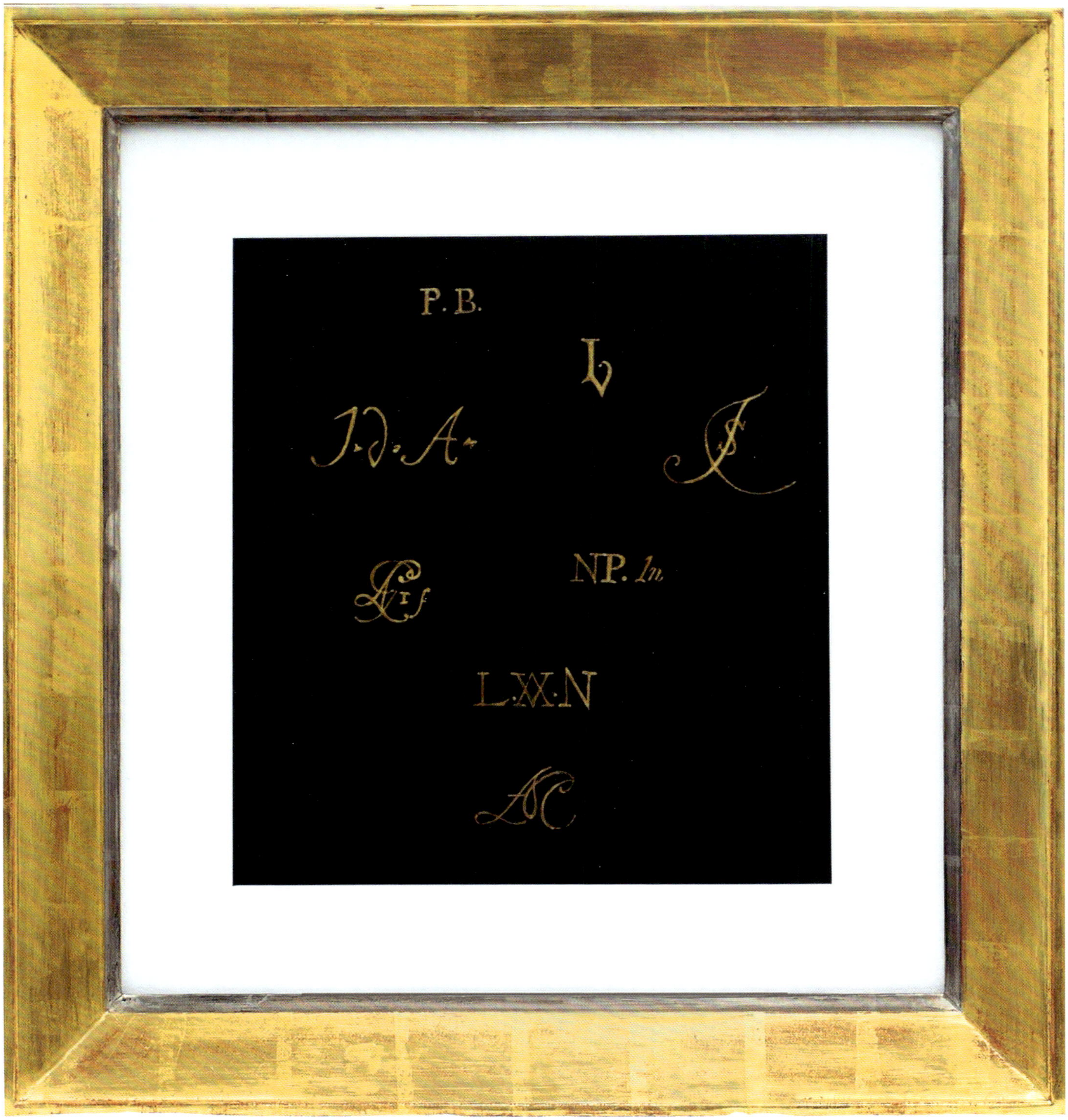

Norbert Fleischmann
call, 2011, Acrylic
Gold paint on lined canvas, framed, 53 x 49.5 cm
Norbert Fleischmann

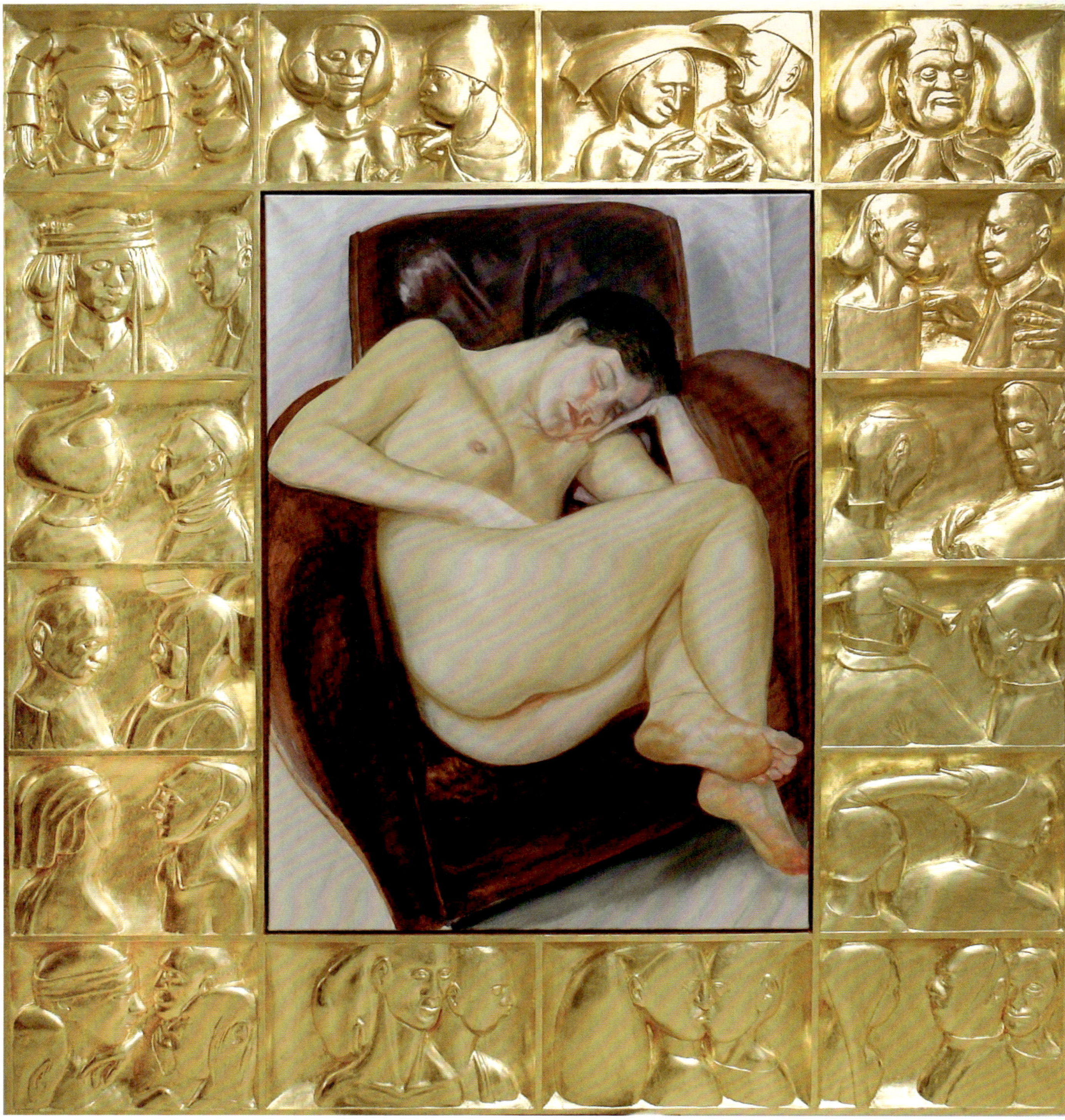

Josef Kern
Julia schläft, 2006
Wood, gold leaf, 185 x 172 cm
In the artist's possession

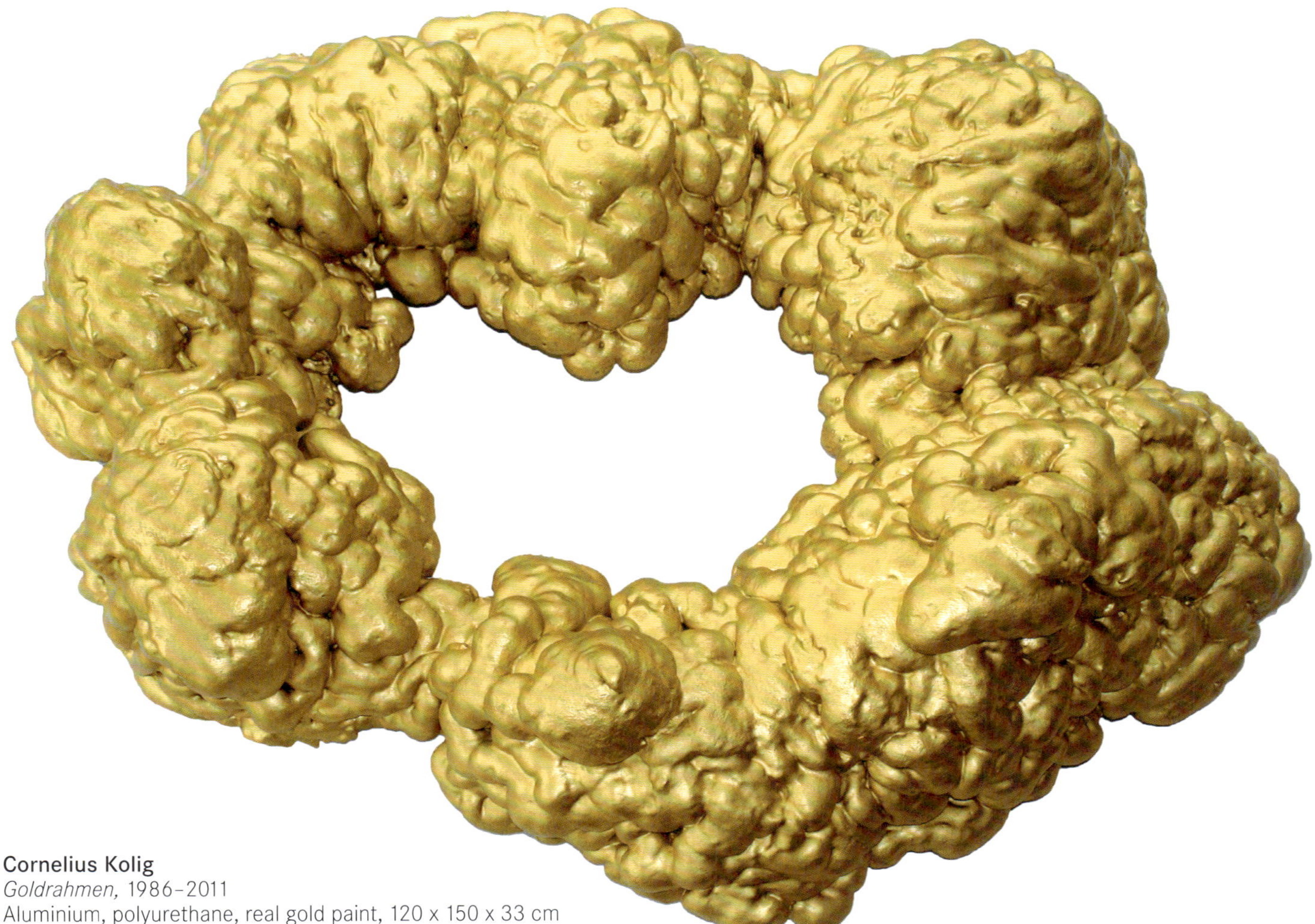

Cornelius Kolig
Goldrahmen, 1986–2011
Aluminium, polyurethane, real gold paint, 120 x 150 x 33 cm
Cornelius Kolig

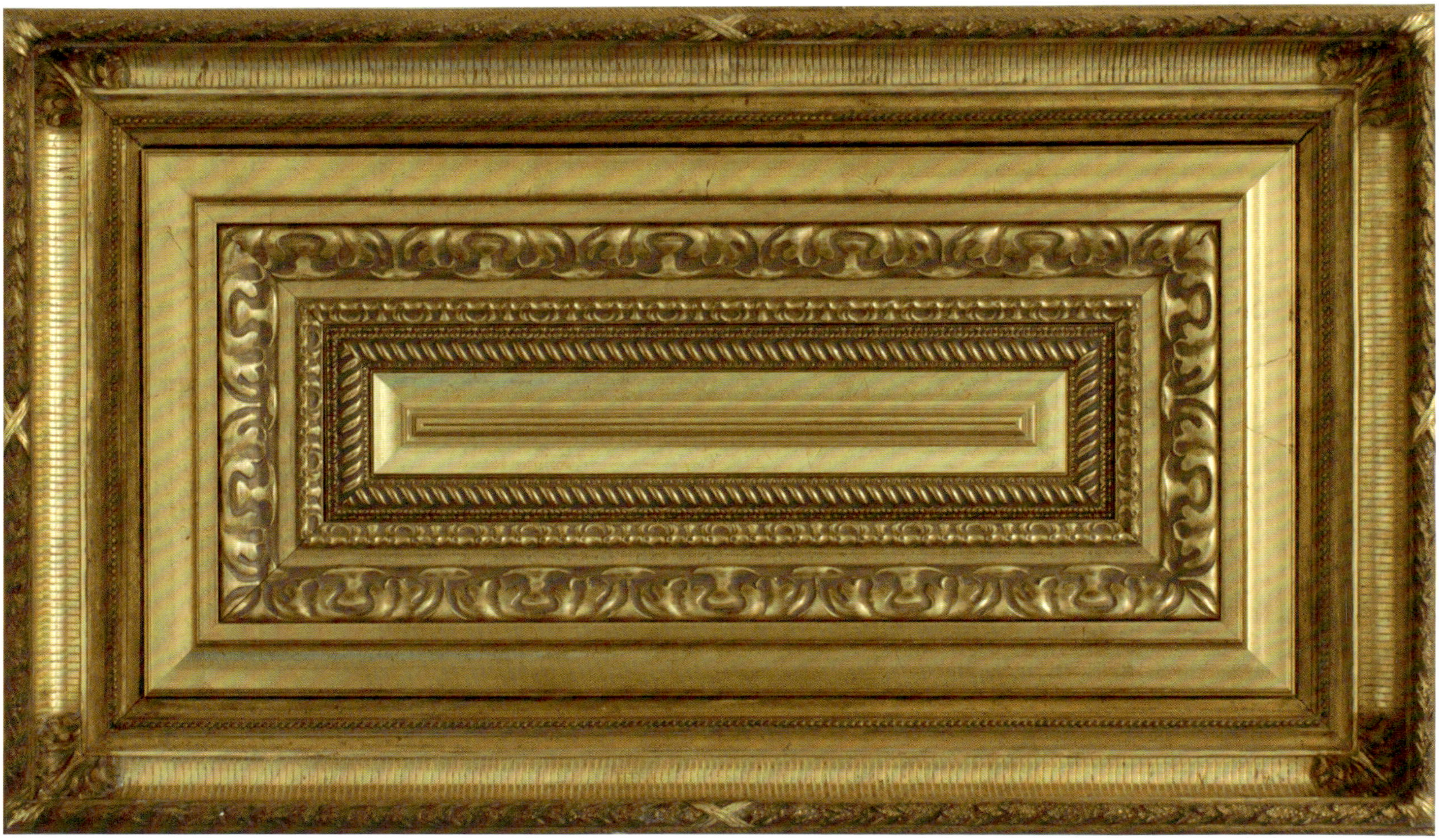

Kolkoz
Marine, 2009
Gilded frame elements, 87.5 x 147.5 x 11.5 cm
Courtesy Galerie Perrotin, Paris

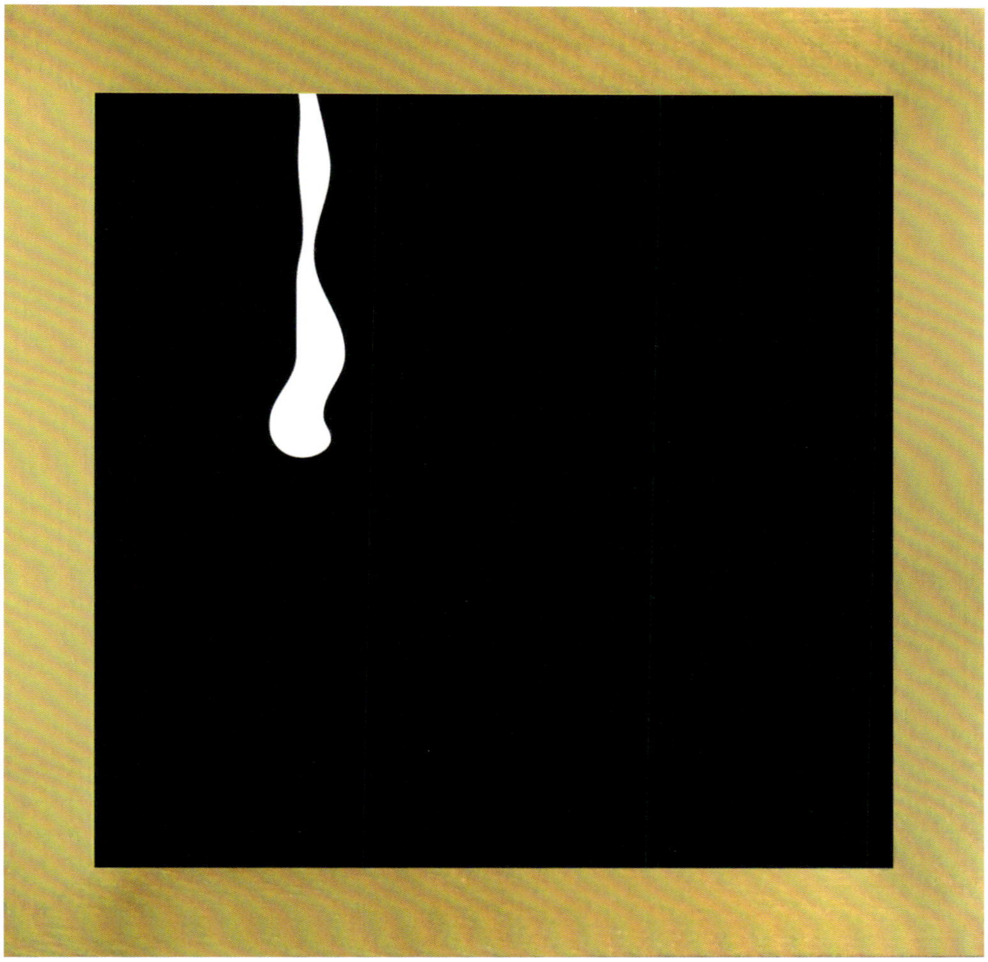

Gerwald Rockenschaub
Untitled, 2008
Gold leaf frame, acrylic glass,
40 x 40 x 5 cm
Courtesy Galerie Vera Munro

Gerwald Rockenschaub
Untitled, 2008
Gold leaf frame, acrylic glass,
40 x 40 x 5 cm
Courtesy Galerie Vera Munro

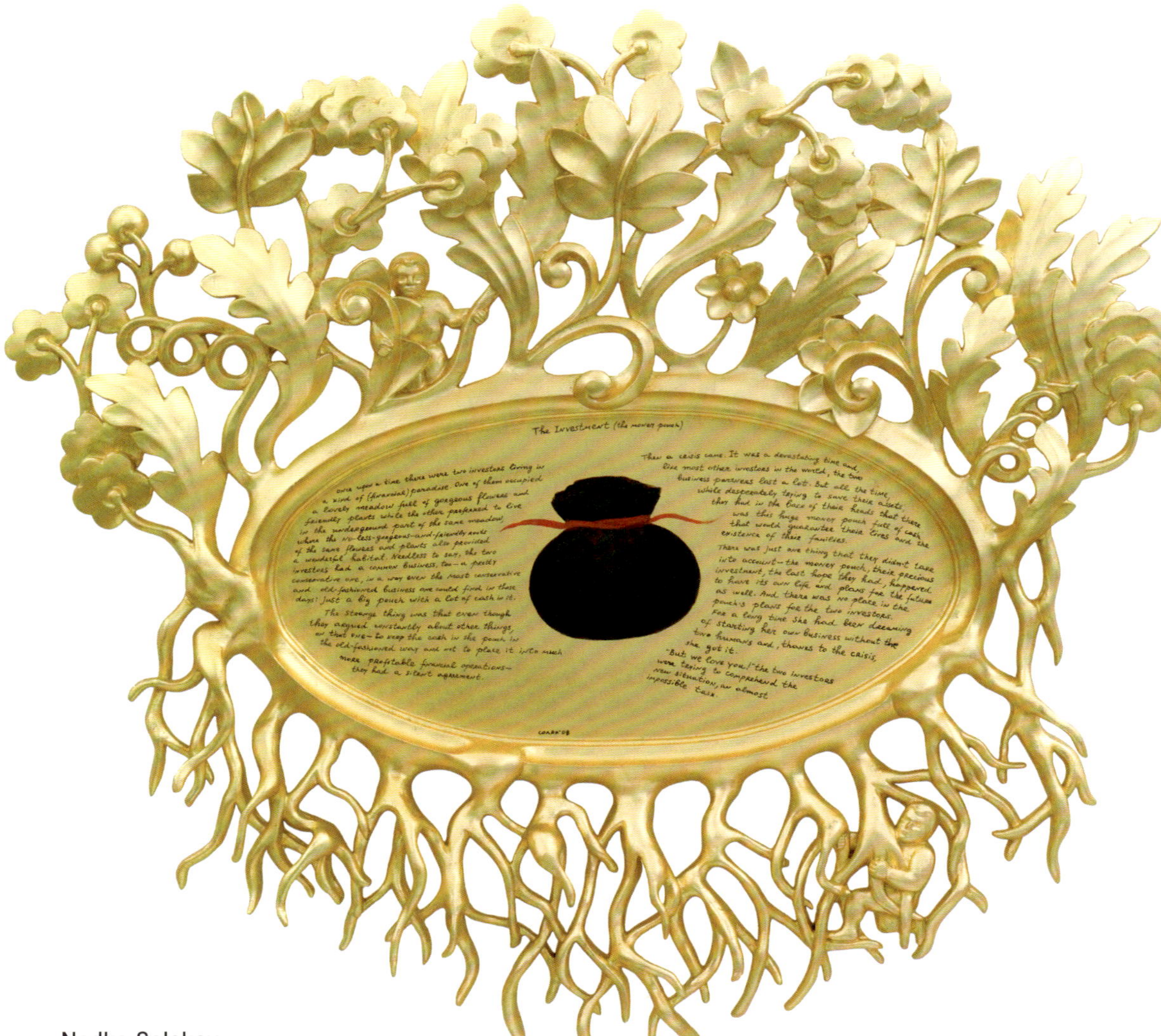

Nedko Solakov
The Investment (the money pouch), 2008
Acrylic and India ink on carved lime wood, 77 x 83 x 9.5 cm
Courtesy private collection, Australia

The Investment (the money pouch)

Once upon a time there were two investors living in a kind of (financial) paradise. One of them occupied a lovely meadow full of gorgeous flowers and friendly plants while the other preferred to live in the underground part of the same meadow, where the no-less-gorgeous-and-friendly roots of the same flowers and plants also provided a wonderful habitat. Needless to say, the two investors had a common business, too—a pretty conservative one, in a way even the most conservative and old-fashioned business one could find in those days: just a big pouch with a lot of cash in it.
The strange thing was that even though they argued constantly about other things, on that one—to keep the cash in the pouch in the old-fashioned way and not to place it into much more profitable financial operations—they had a silent agreement.

Then a crisis came. It was a devastating time and, like most other investors in the world, the two business partners lost a lot. But all the time, while desperately trying to save their assets, they had in the back of their heads that there was this huge money pouch full of cash that would guarantee their lives and the existence of their families.
There was just one thing that they didn't take into account—the money pouch, their precious investment, the last hope they had, happened to have its own life and plans for the future as well. And there was no place in the pouch's plans for the two investors. For a long time she had been dreaming of starting her own business without the two humans and, thanks to the crisis, she got it. "But we love you!" the two investors were trying to comprehend the new situation, an almost impossible task.

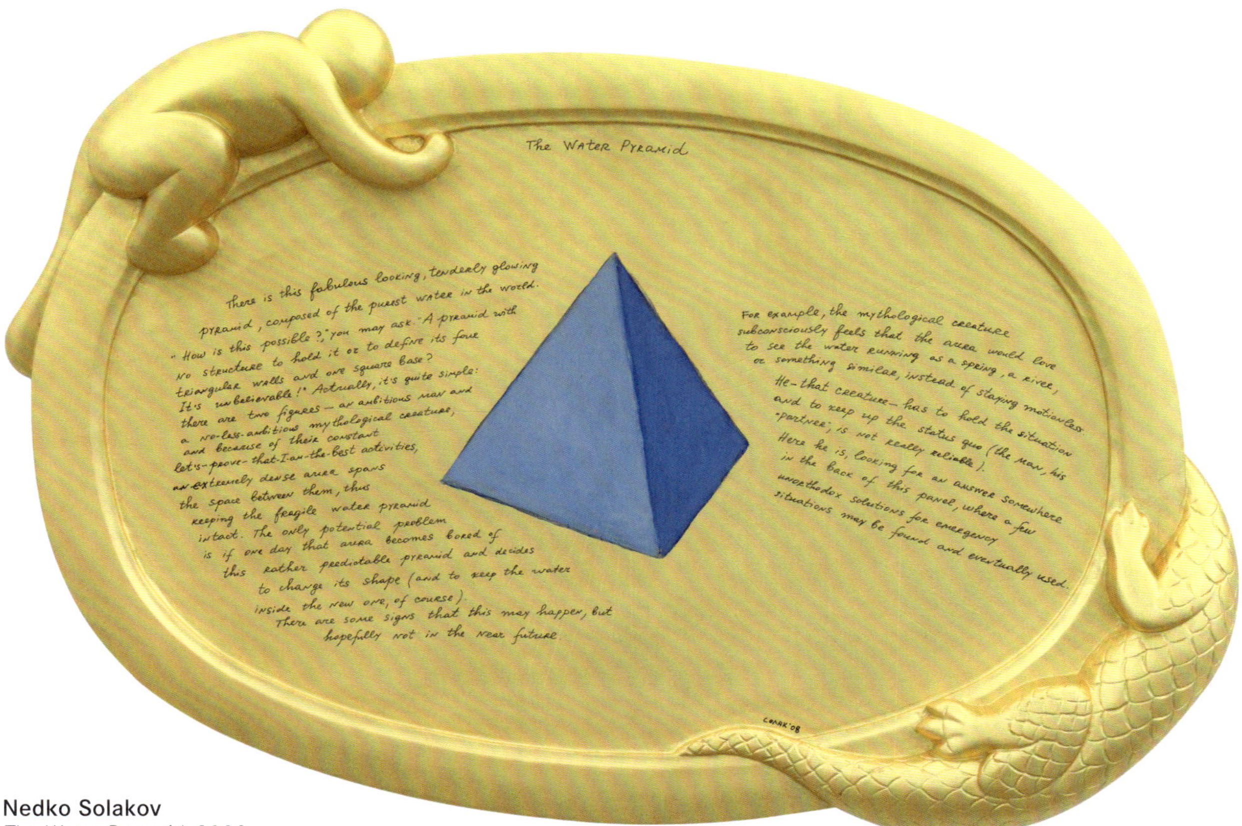

Nedko Solakov
The Water Pyramid, 2008
Acrylic and India ink on carved lime wood, 45 x 69 x 8 cm
Courtesy the artist, ARNDT, Berlin

Solakov—The Water Pyramid

There is this fabulous looking, tenderly glowing pyramid, composed of the purest water in the world. "How is this possible?" you may ask "A pyramid with no structure to hold it and to define its four triangular sides and one square base?! It's unbelievable!" Actually, it's quite simple: there are these two figures—an ambitious man and a no-less ambitious mythological creature, an extremely dense aura covers the space between them thus keeping the fragile water pyramid intact. The only potential problem would be if one day that aura becomes bored of this rather predictable pyramid and decides to change its shape (and to keep the water inside the new one, of course). There are some signs that this may happen, but hopefully not in the near future. For example, the mythological creature subconsciously feels that the aura would love to see that water running as a spring, a river or something similar, instead of staying motionless. He—that creature—has to hold the situation and to keep up the status quo (the man, his "partner", is not really reliable). Here he is: looking for an answer somewhere in the base of this panel, where a few unorthodox solutions for emergency situations may be found and eventually used.

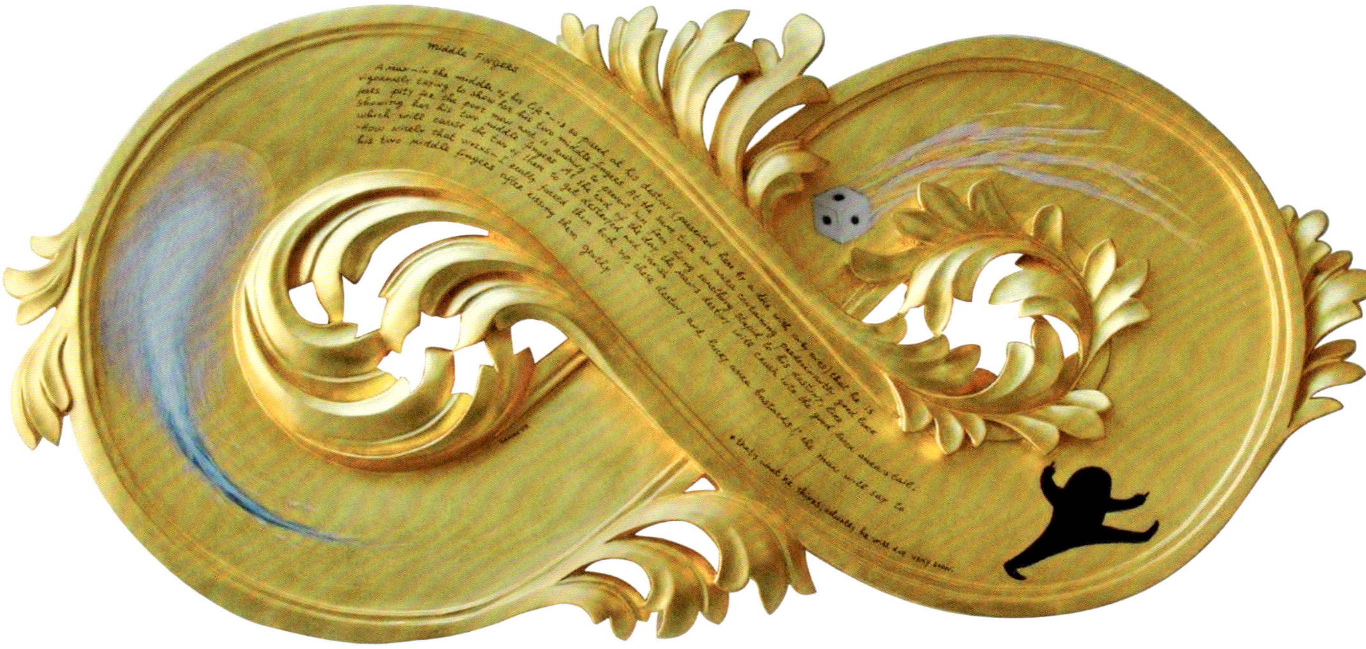

Nedko Solakov
Middle Fingers, 2009
Acrylic and India ink on gilded, carved lime wood, 43 x 88 x 7 cm
Private Collection Milan

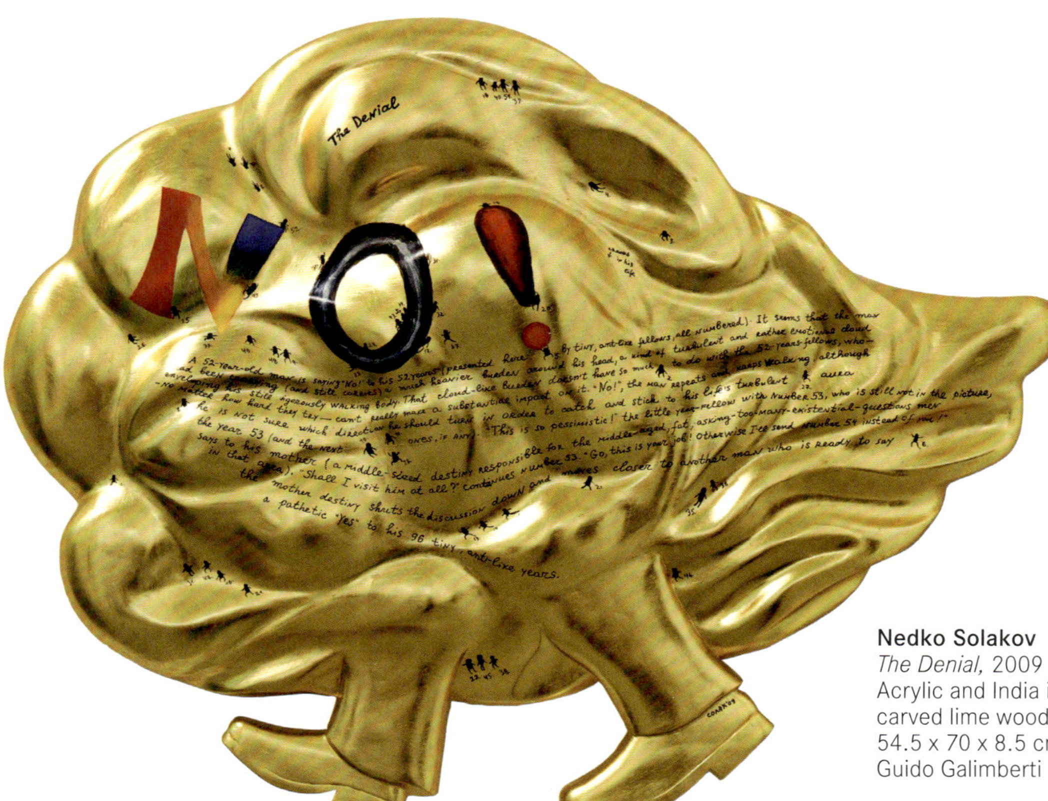

Nedko Solakov
The Denial, 2009
Acrylic and India ink on gilded, carved lime wood, 54.5 x 70 x 8.5 cm
Guido Galimberti

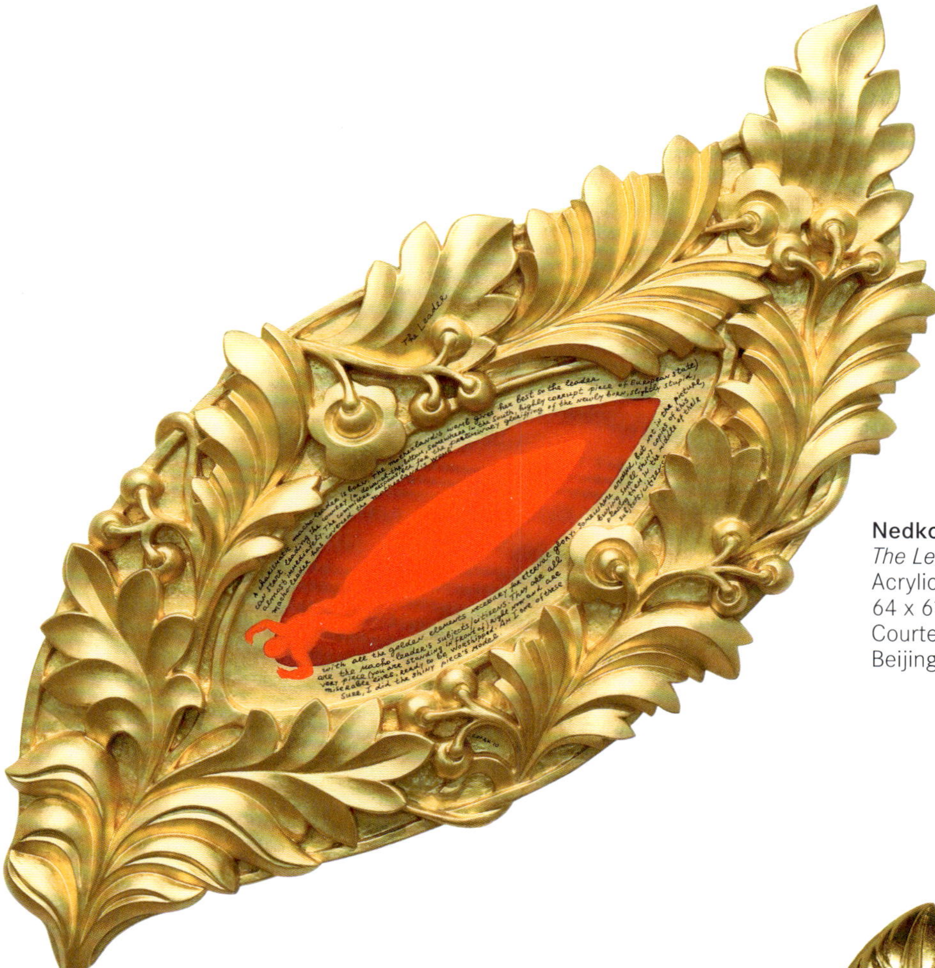

Nedko Solakov
The Leader, 2010
Acrylic and India ink on gilded, carved lime wood,
64 x 61 x 4.5 cm
Courtesy GALLERIA CONTINUA, San Gimignano/
Beijing/Le Moulin

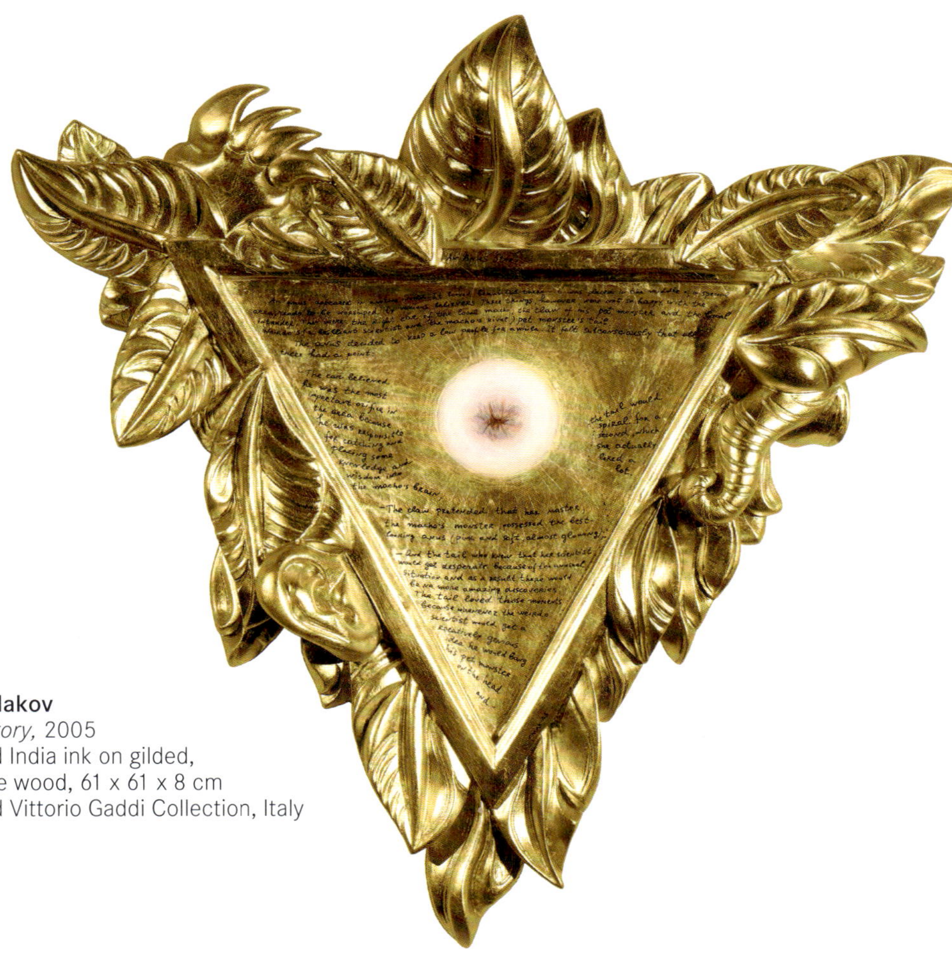

Nedko Solakov
An Anus Story, 2005
Acrylic and India ink on gilded,
carved lime wood, 61 x 61 x 8 cm
Nunzia and Vittorio Gaddi Collection, Italy

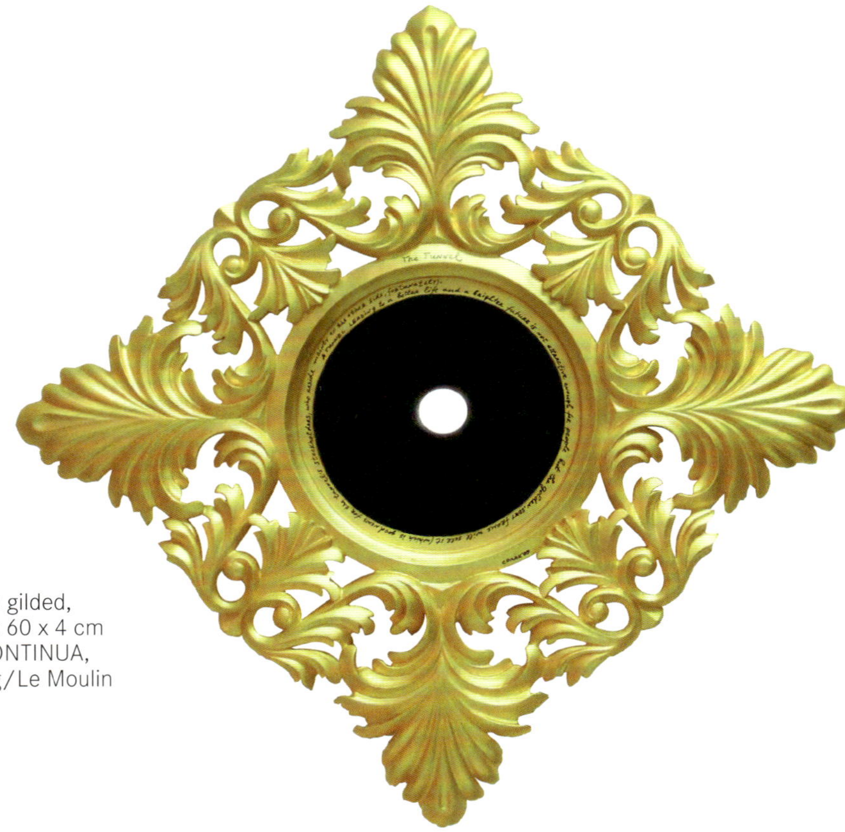

Nedko Solakov
The Tunnel, 2009
Acrylic and India ink on gilded, carved lime wood, 60 x 60 x 4 cm
Courtesy GALLERIA CONTINUA, San Gimignano / Beijing / Le Moulin

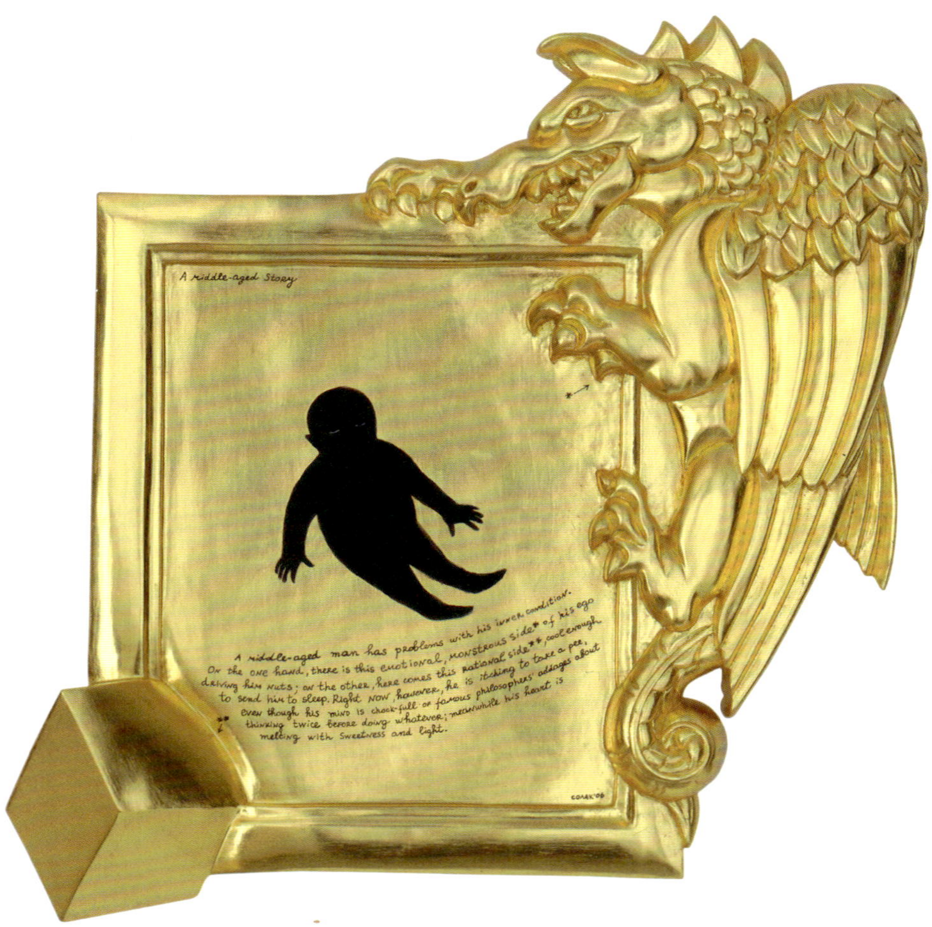

Nedko Solakov
A Middle-Aged Story, 2006
Acrylic and India ink on gilded lime wood, 61 x 60 x 8 cm
Collection of Aaron and Barbara Levine, Washington D. C., ARNDT, Berlin

Jan Maarten Voskuil
Gilded Frame, 2002
Wood, gold leaf, 145 x 115 x 60 cm
Courtesy the artist

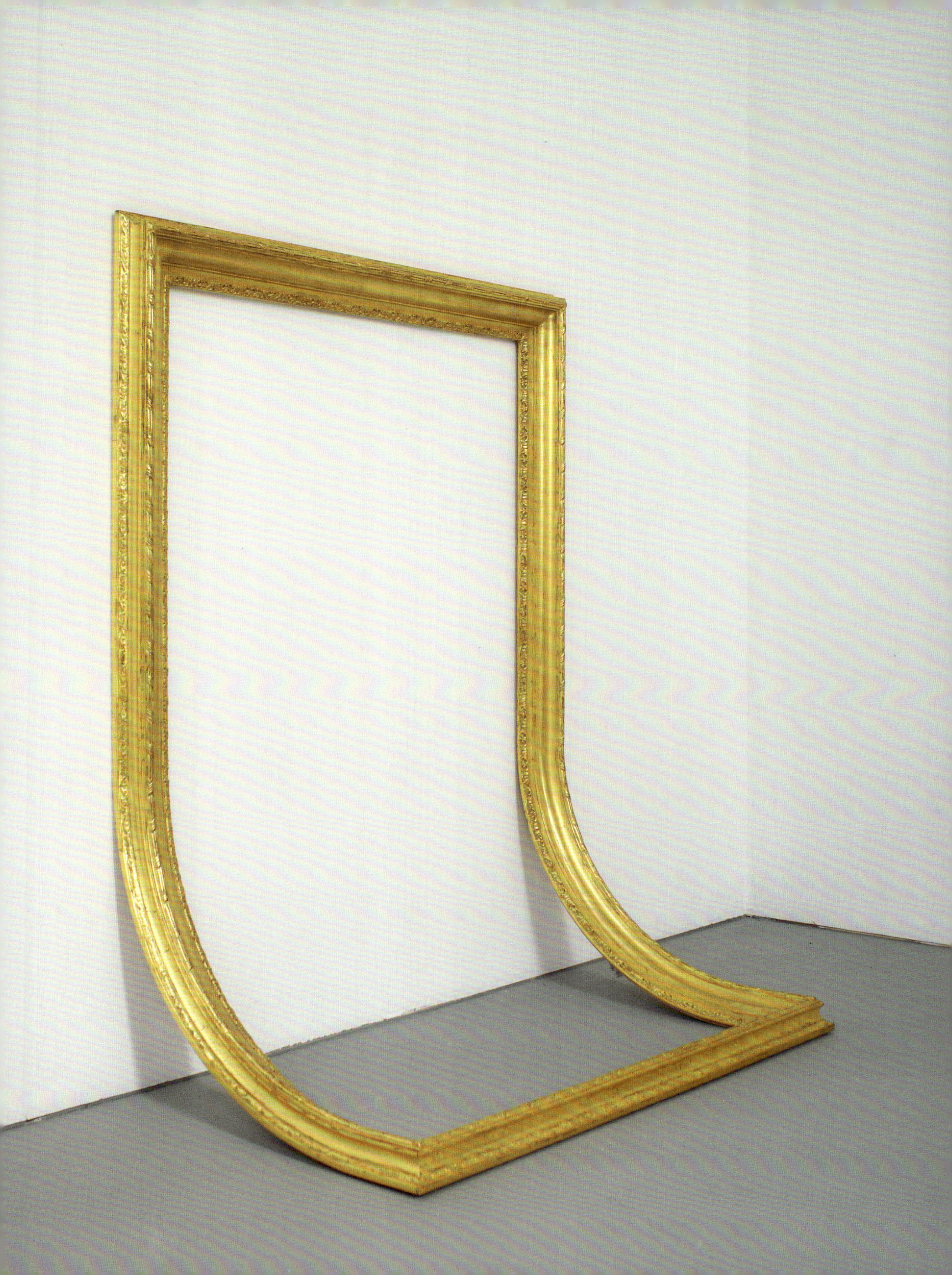

Still Lifes

Excursus: The Still Life

The still life appeared late in the history of painting. At the time of the gold ground in medieval art, there was no interest in it. As an art genre in its own right the still life, popular since the seventeenth century, not without a touch of *vanitas* symbolism, was still fighting for recognition in the nineteenth. In contrast to history painting, there was no grand theme. The still life, *nature morte*, *natura morta*, in other words dead nature as an event-free world, could only establish itself without restriction when painting came to be valued for itself. The famous statement by Maurice Denis (later much quoted on behalf of abstraction) bears witness to this: "A picture—before it becomes a war-horse, a naked woman or some anecdote or other—is essentially a flat surface covered with paint in a particular arrangement." (1890)[1]

The solemn gold ground, through its noble sheen, draws attention to the composition. Many a still life on gold owes its inspiration to one by Caravaggio (fig. 1). The Italian art-historian Roberto Longhi declared it in 1950 to be the first still life in the narrower sense.[2] Art-historical attitudes to it differ.[3] The picture has been cropped. The orange ground, which on illustrations looks like gold, is a later addition.

The still life as sculpture was very much the exception in Modernism. Earlier eras did not know this genre at all. The integration of everyday objects into sculptural arrangements by the avantgarde flirted with de-familiarization. In Alberto Magnelli's *Still Life* (1914) or in Ivan Puni's *Relief with Plate* (1915–1919) the bottle and the plate do not represent kitchen items, but, as in a collage, formally surprising effects.

The shoe-pictures made in gold leaf by Andy Warhol were first shown in 1956. Having once discovered the charms of this material, he either placed objects against a gold background, or else swapped figure and ground by reproducing the objects themselves in gold. Warhol's pioneering achievement of creating still-lifes in which things were 'depicted' in gold leaf has never been followed up or imitated. It was only with the revival of Realism in the 1960s that new forms of still life were developed, like the *snare pictures* of Daniel Spoerri. However, there was no general occasion to gild them.

After training at the Royal College of Art, in 1962 the New Zealander Barrie Bates decided to start a new life in London in London. Since then, he has called himself Billy Apple. He is mainly interested in questions of identity, and, since 1983, in gold.[4] Many years before his large *Golden Apple* (1983) he created the nibbled-at three red, three green and three gilded black apples with the title *Two minutes, 33 Seconds (Gold)* (1962). The time-related title, recalling the composition *4'33"* (1952) by John Cage (1912–1992) contradicts the very idea of the still life, but the aspect of consumption hints at the *vanitas* connotation. Not least, it is a self-portrait.[5]

Marc Quinn came to fame with his *Sphinx* (2006/2008), a golden sculpture of Kate Moss. In his skeleton *A Beautiful Continuation* (2009) the heart is formed by a gilded apple. The title of the work refers to re-birth after death in Buddhist sermons. The titles of earlier skeletons in the lotus position evoke this context: *Meditation on Illusion* (2007). The golden apple with the inscription **καλλιστι** (*kallisti* = for the most beautiful one) is part of the Sacred Chao, the symbol of the English pseudo-religion of the Discordians, and represents the Apple of Discord, with which Eris unleashed the Trojan War (Paris uses the golden apple[6] to declare Venus the most beautiful goddess, because she promises him the most beautiful woman, Helen). Blossoms and fruit were already the themes of the series *The Nurseries of El Dorado* (2007).

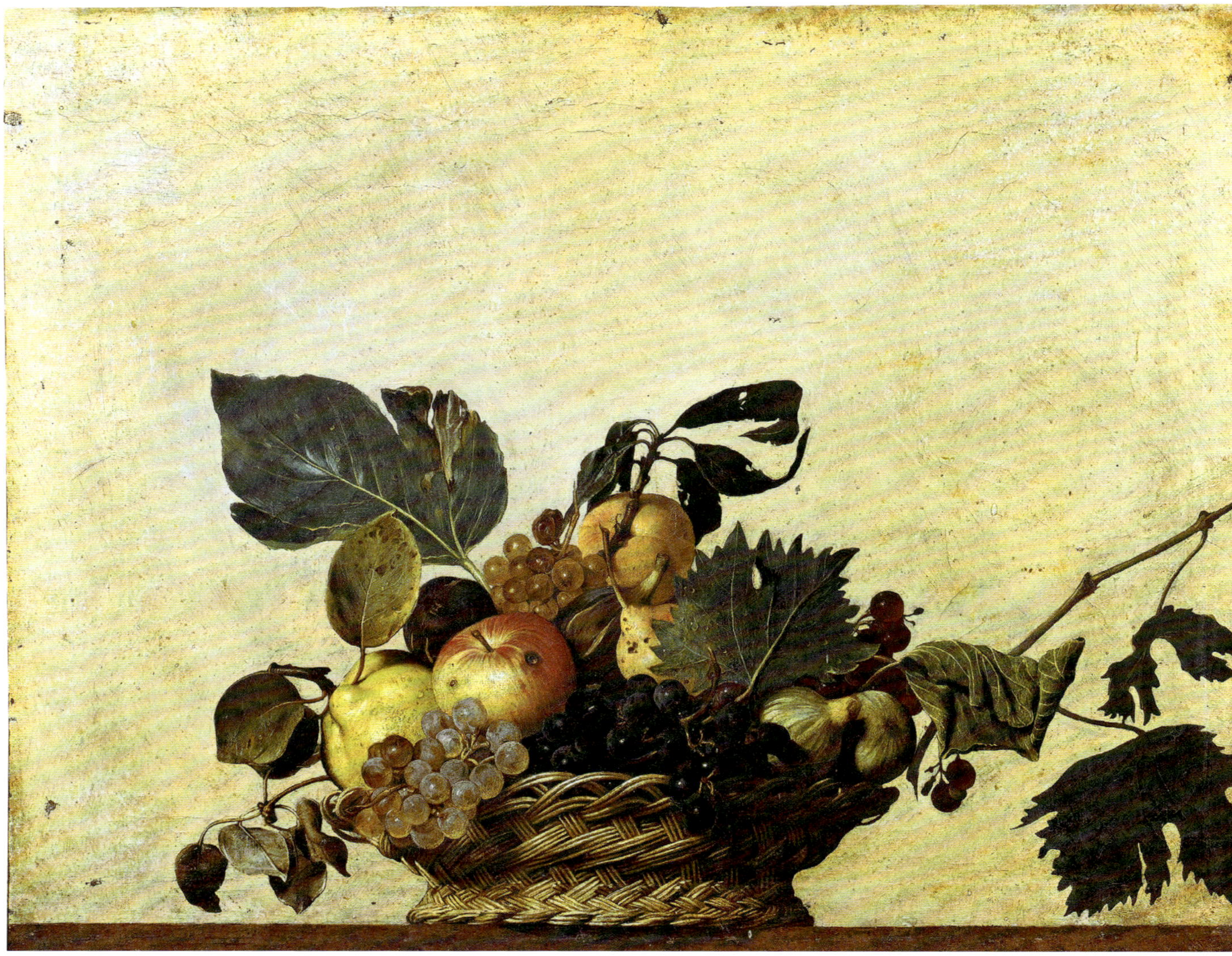

Fig. 1 Caravaggio, *Basket of Fruit,* um 1593/94, Pinacoteca Ambrosiana, Milan

Excursus: Taking a bite from the apple

The semantic range of the apple[7] has, amazingly, developed from false attributions. The biblical apple, with which Eve is said to have tempted Adam and thus brought about the Fall of Man, was plucked from the Tree of the Knowledge of Good and Evil, whose fruit is not described in greater detail. The forbidden 'apple' came from the Latin translation 'malum', which means both 'evil' and 'apple'.

That Sir Isaac Newton was prompted to think about the laws of gravity by a falling apple in his garden is a no less persistent legend. Voltaire (1694–1778), returning from England in 1728, still wrote quite generally about apples falling from trees. It was only in Leonhard Euler's (1707–1783) *Letters to a German Princess* (3 September 1760) that these became an apple that had fallen on Newton's head.[8] The book *Warum der Apfel vom Baum fällt* (Why the Apple Falls from the Tree) is naturally no longer concerned with gravitation, but with the biochemistry of the 'built-in weak link' of the apple stalk[9], in other words the answer to why fruit falls from trees in the first place.

The best-known apple from which someone has bitten out a chunk does not derive from the iconography of the Garden of Eden, but from the first advertisement for the Apple computer in May 1976. The Beatles had already founded their multimedia company in 1968. The legal dispute with the computer company was not settled until 2010. Legend has it that the apple logo with the bitten-out chunk goes back to the punning advertising slogan "Byte into an Apple". The apple is still something special, unlike other fruits. "In [Heinrich] Böll satire was the apple from the Tree of Knowledge, in his successors it is mostly only the banana," one reads, after the reference to the fact that the original meaning of 'satire' was a colourful bowl of fruit, in other words a still life.[10]

Guillaume Bruère discovered the sculptural potential of banana skins, which he spontaneously transforms into capricci. Art secures durability, not least with food.[11] Gold leaf seals and ennobles all raw materials, even chewing-gum.[12] Erwin Wurm was inspired by the shaping of the dumplings in his grandmother's kitchen to make colour spheres of his own. One, cast in silver, he gilded. Herbert Mehler oriented himself to quasi-organic forms in his pumpkin-like gilt-steel floor sculpture.

The facets of the still life range from solemn old-masterly nature studies (Pam Burnley-Schol, Robert Ginder, David Hewson and Fred Wessel) in the tradition of Caravaggio and the Dutch painters of the seventeenth century, via ecologically motivated portraits of endangered animal and plant species (Mary Frances Dondelinger), all the way to the portrait of the neighbour's cow *Ariola* (Thomas Brix).

Robert Kushner is the best-known exponent of Pattern Painting, a trend which since the 1970s has seen itself as the response to Concept Art and the allegedly bloodless, over-theoretical Minimal Art. In this opulent painting, a broad space is devoted to the decor of elegant patterns and colourful plants. The exuberantly staged aesthetic is given a solemn tone by the use of gold leaf.

Since time immemorial, Paul Morrison has concerned himself with the representation of nature. In the process, he has recourse to botany books, examples from art history, fairy-tales and comics in equal measure. The boundaries between the still-life and landscape genres are abolished at the same time. Serial patterns alternate with close-ups and distant views, processed on the computer.

Milan Kunc, the polyglot Czech exponent of a post-modern Pop Art and co-founder of the 'Gruppe Normal' (1979) places his ecologically surreal rubble-fields on a solemn gold ground. In 2005 he used a vegetable-tank (the military variety, created from tomatoes, cabbage, leeks, peas and beans) beneath thunder and lightning to caricature a *Vitamin Attack*. The assault is directed not against an enemy, but seems to mock vegetarian valetudinarianism.

1 Maurice Denis, "Définition du Néo-Traditionnisme" (1890), in: Maurice Denis: *Du Symbolisme au Classicisme: Théories, Paris* 1964, pp. 33f.
2 Roberto Longhi, "Un momento importante nella storia della natura morta", in: *Paragone*, no.1, 1950.
3 Andreas Prater, *Licht und Farbe bei Caravaggio*, Stuttgart 1992, S. 93f. - Claus Grimm, *Stillleben. Die italienischen, spanischen und französischen Meister*, Stuttgart 2011, pp.17, 81. - Sybille Ebert-Schifferer, "Caravaggios 'Früchtekorb' und sein Kontext", in: *Zeitschrift für Kunstgeschichte*, vol. 65, no. 1, 2002, 1-23.
4 *Tales of Gold—The Tale of Ray. An Installation by Billy Apple with Wystan Curnow.* (exhib. cat. ARTSPACE, Auckland), 2nd ed., Auckland 2006.
5 Zoë Gray/Nicolaus Schafhausen/Monika Szewczyk, "Introduction", in: *Billy Apple®*, Source Book 7/2009, p. 8.
6 Verbatim in *The Golden Ass* by Apuleius 'malumque bracteis inauratum'.
7 See the essay in the volume by Jürgen Borchhardt.
8 Hans Blumenberg, "Die Apfelgeschichte", in: H. Blumenberg *Die Vollzähligkeit der Sterne*, Frankfurt am Main 2011, 54-65.
9 Ranga Yogeshwar, *Ach so! Warum der Apfel vom Baum fällt und weitere Rätsel des Alltags*, 4th ed., Cologne 2010[4], p. 120.
10 Christopher Schmidt, *Schere im Kopf. Bibliothek des Humors 5*, in: *Süddeutsche Zeitung*, 4 Nov. 2011, p. 13.
11 Ralf Beil, *Künstlerküche. Lebensmittel als Kunstmaterial von Schiele bis Jason Rhoades*, Cologne 2002.
12 Cf. the *Jagdtrophäen* by Pippi Frank. The chewing-gum sculptures are gilded with gold leaf.

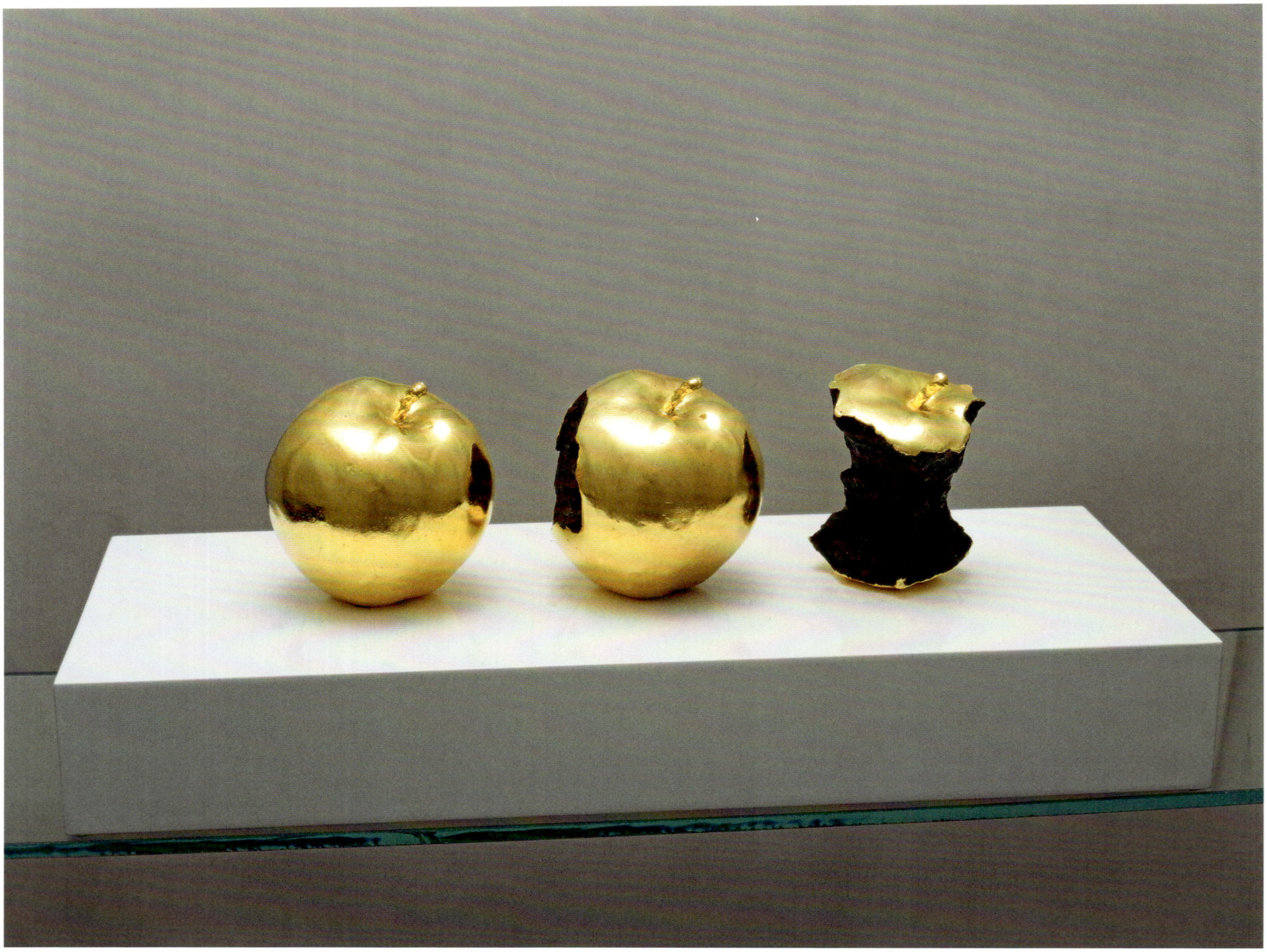

Billy Apple
2 minutes, 33 seconds, 1962
Bronze-gilt, 375 x 150 x 170 cm
Collection of Dame Jenny Gibbs

Joachim Baur
Lungot, harvest, 1992/95
Glass frame, bean tendrils, bean pods, gilded seeds, digital LCD display
100 x 30 x 14.5 cm
Kunstsammlung/Kulturvermittlung Steiermark

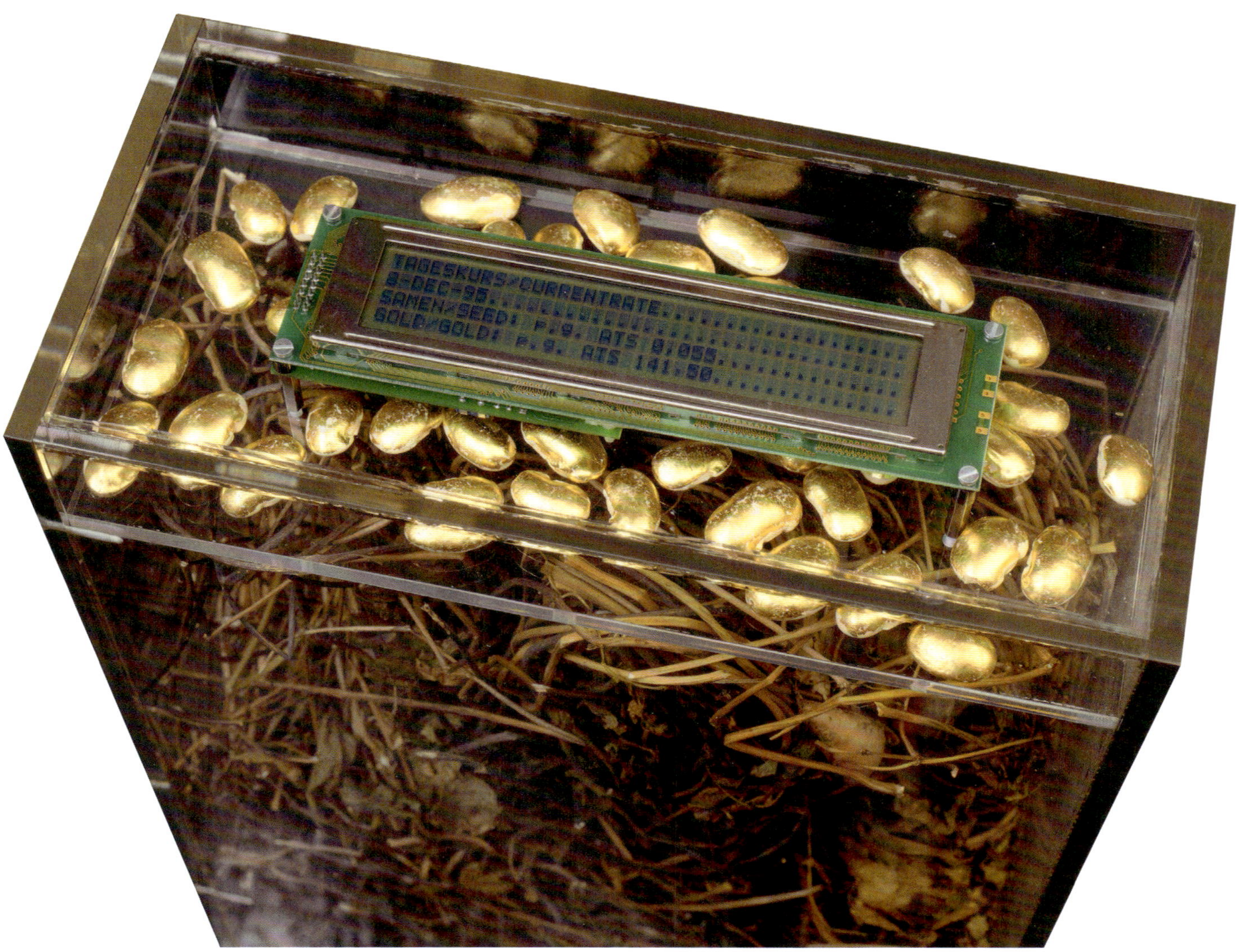
TAGESKURS/CURRENTRATE.
8-DEC-95.
SAMEN/SEED: p.g. ATS 0,055.
GOLD/GOLD: p.g. ATS 141,50.

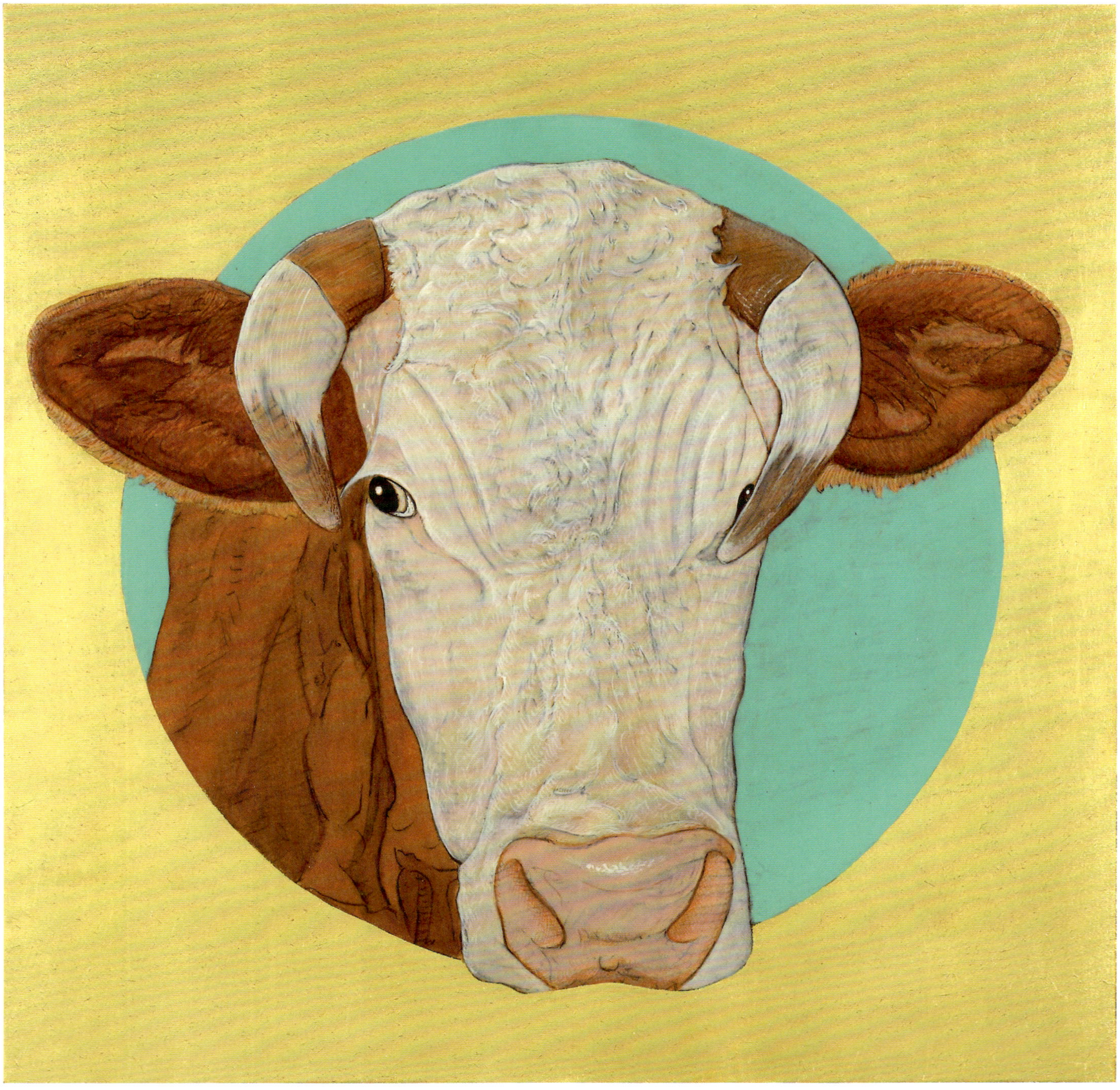

Thomas Brix
Ariola, 2002
Acrylic, gold leaf on chipboard, 60 x 60 cm
Thomas Brix

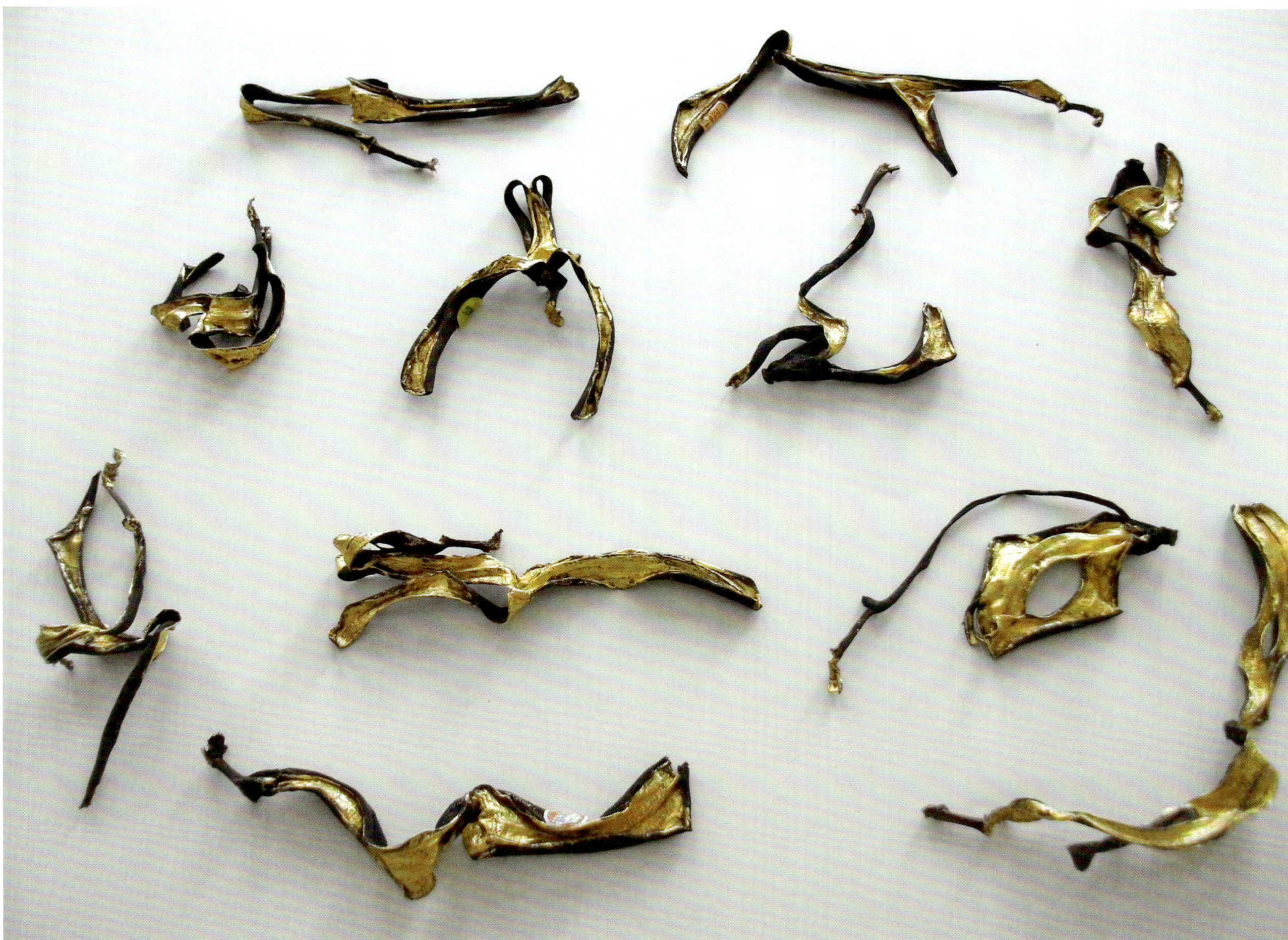

Guillaume Bruère
Bananagold, 2010
Gold leaf on banana peel, variable dimensions
Courtesy the artist and Galerie Heike Curtze, Vienna/Berlin

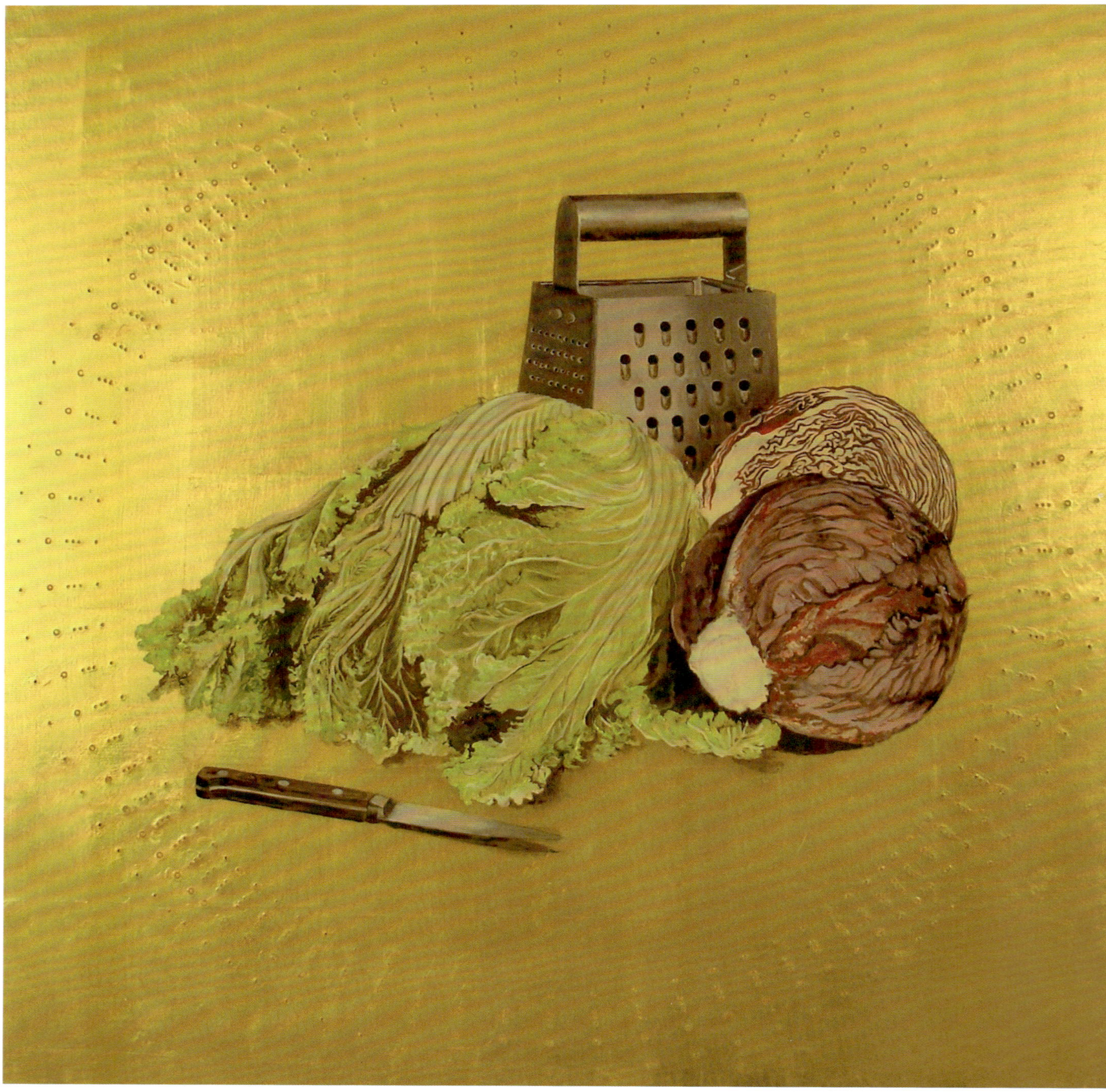

Pam Burnley-Schol
Transfiguration: Cabbage and Steel, 2010
Oil and gold leaf on wood, 61 x 61 x 5.1 cm
Pamela Blake Burnley-Schol

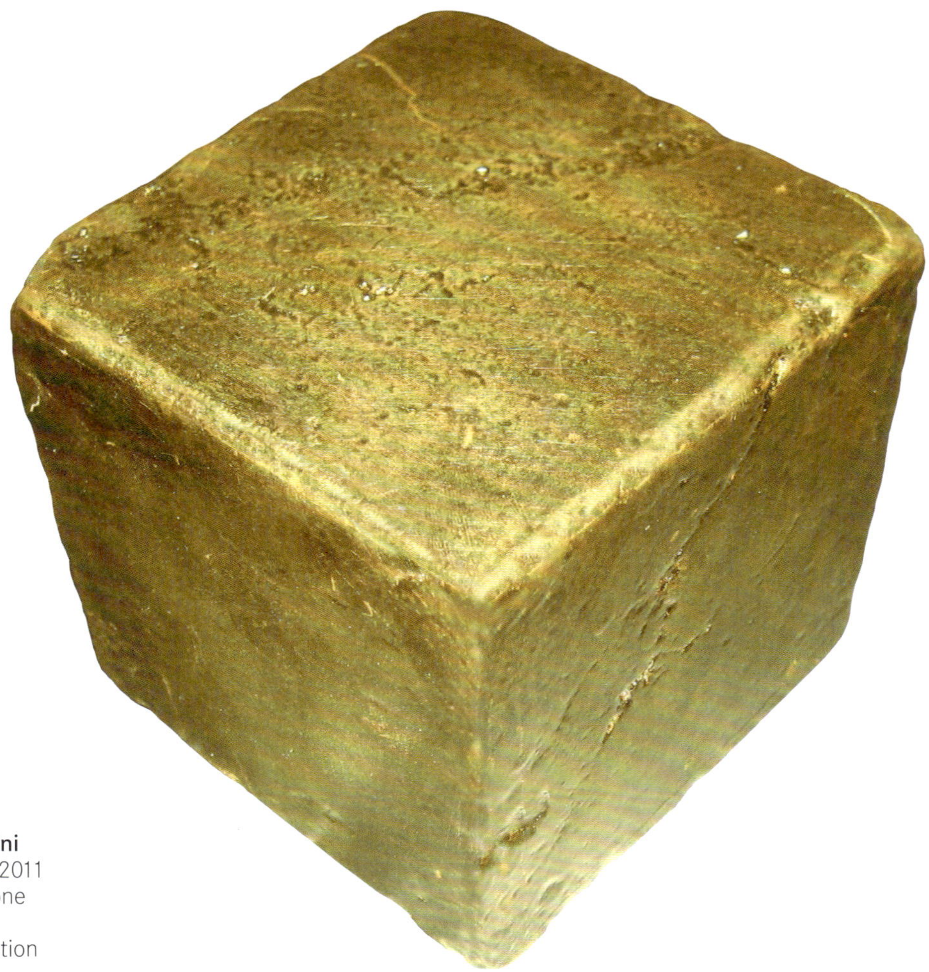

Theres Cassini
Pflasterstein, 2011
Gilt cobblestone
8 x 8 x 8 cm
Private Collection

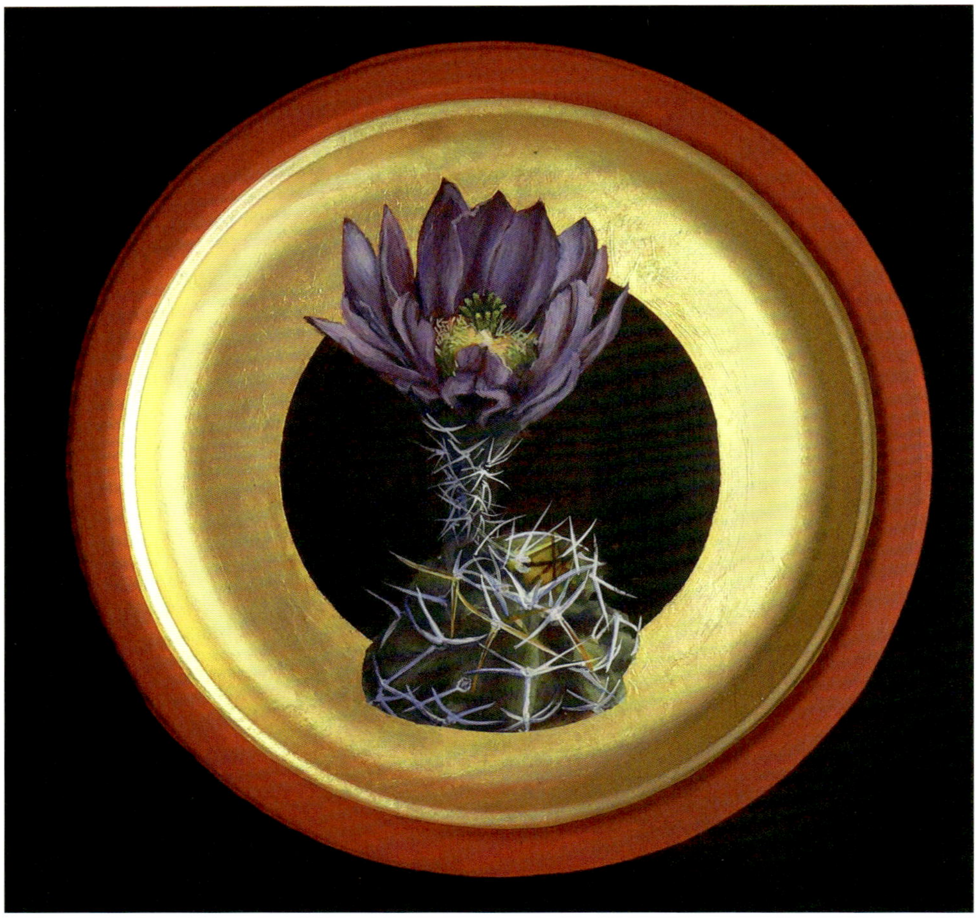

Mary Frances Dondelinger
Kuenzler Hedgehog Cactus, 2010
Egg tempera, 23-carat gold leaf,
gluten glue and acrylic on cardboard,
Diameter 25.4 cm
Courtesy the artist

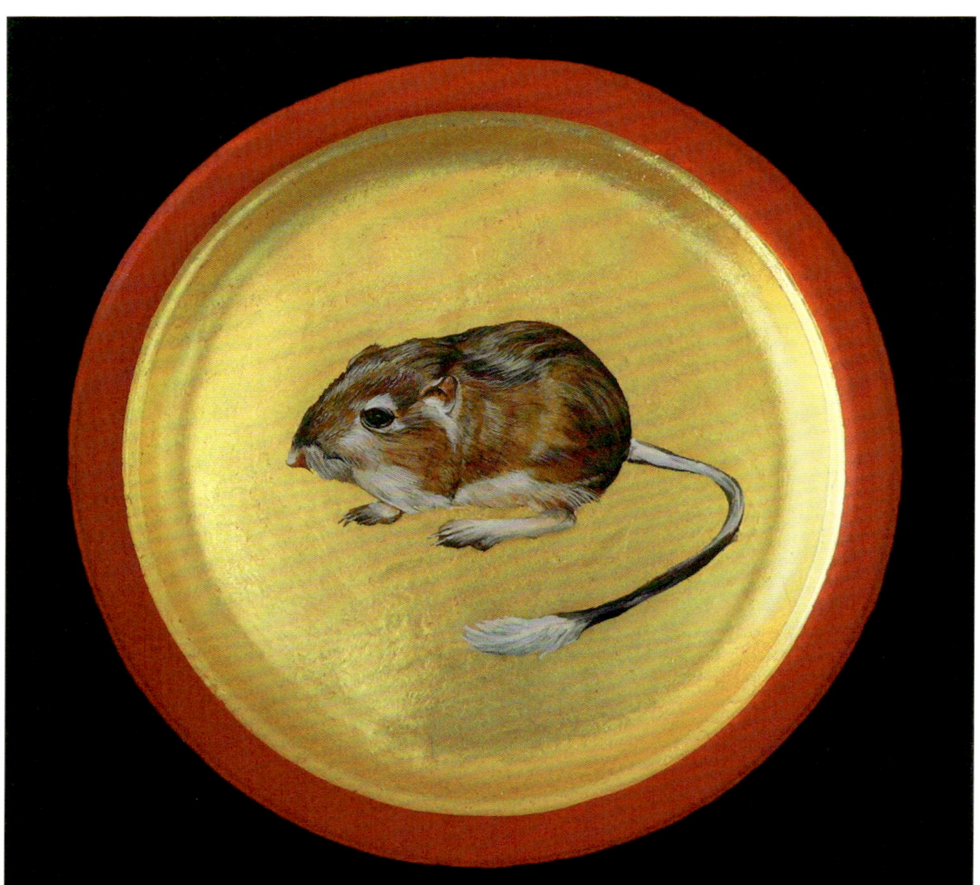

Mary Frances Dondelinger
Ords Kangaroo Rat, 2009
Egg tempera, 23-carat gold leaf,
gluten glue and acrylic on cardboard,
Diameter 25.4 cm
Courtesy the artist

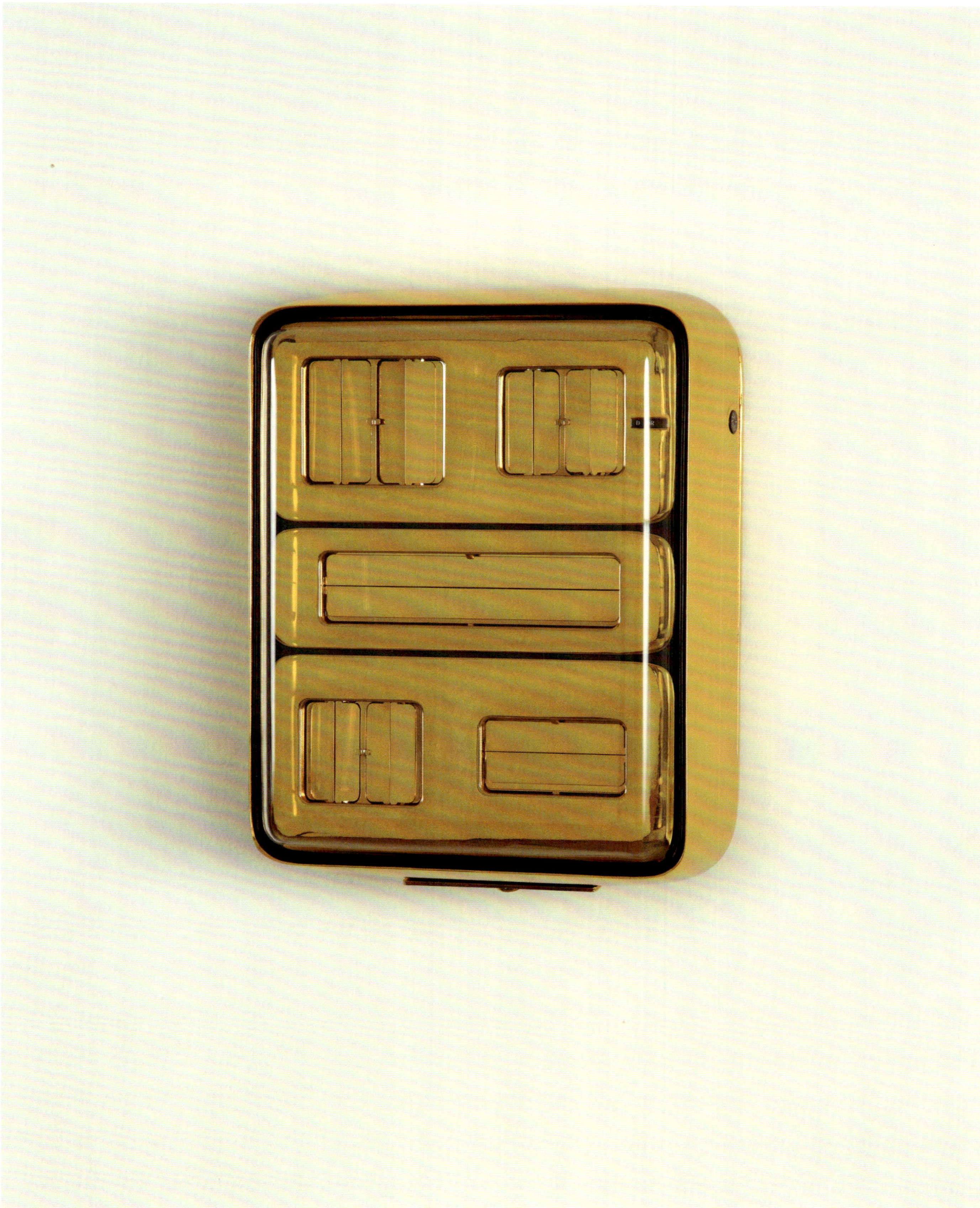

Manfred Erjautz
D OR, 1962/2012
Solari Udine flip clock, gilded and dis-figured, 34.5 x 28 x 12 cm
Courtesy the artist

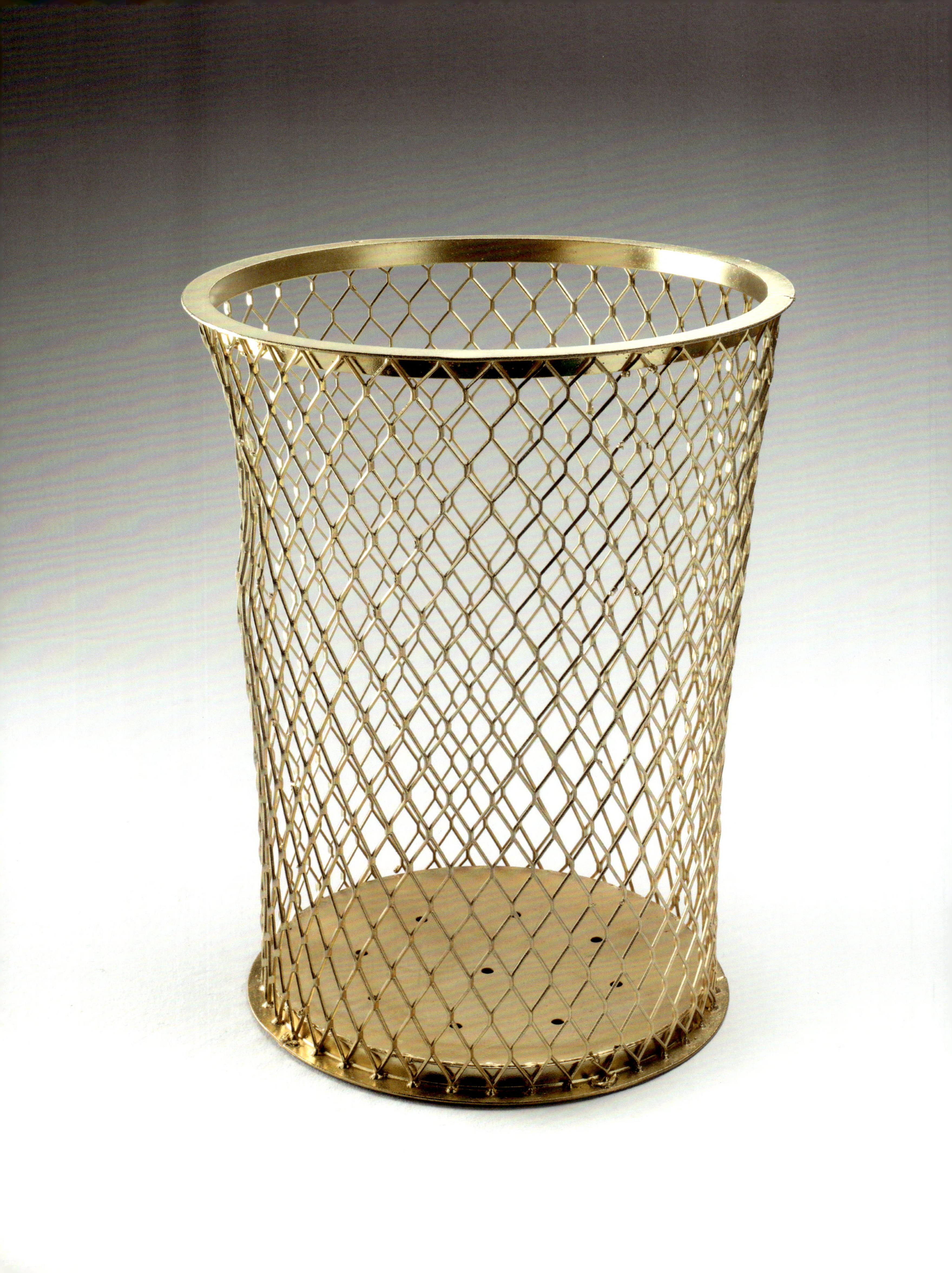

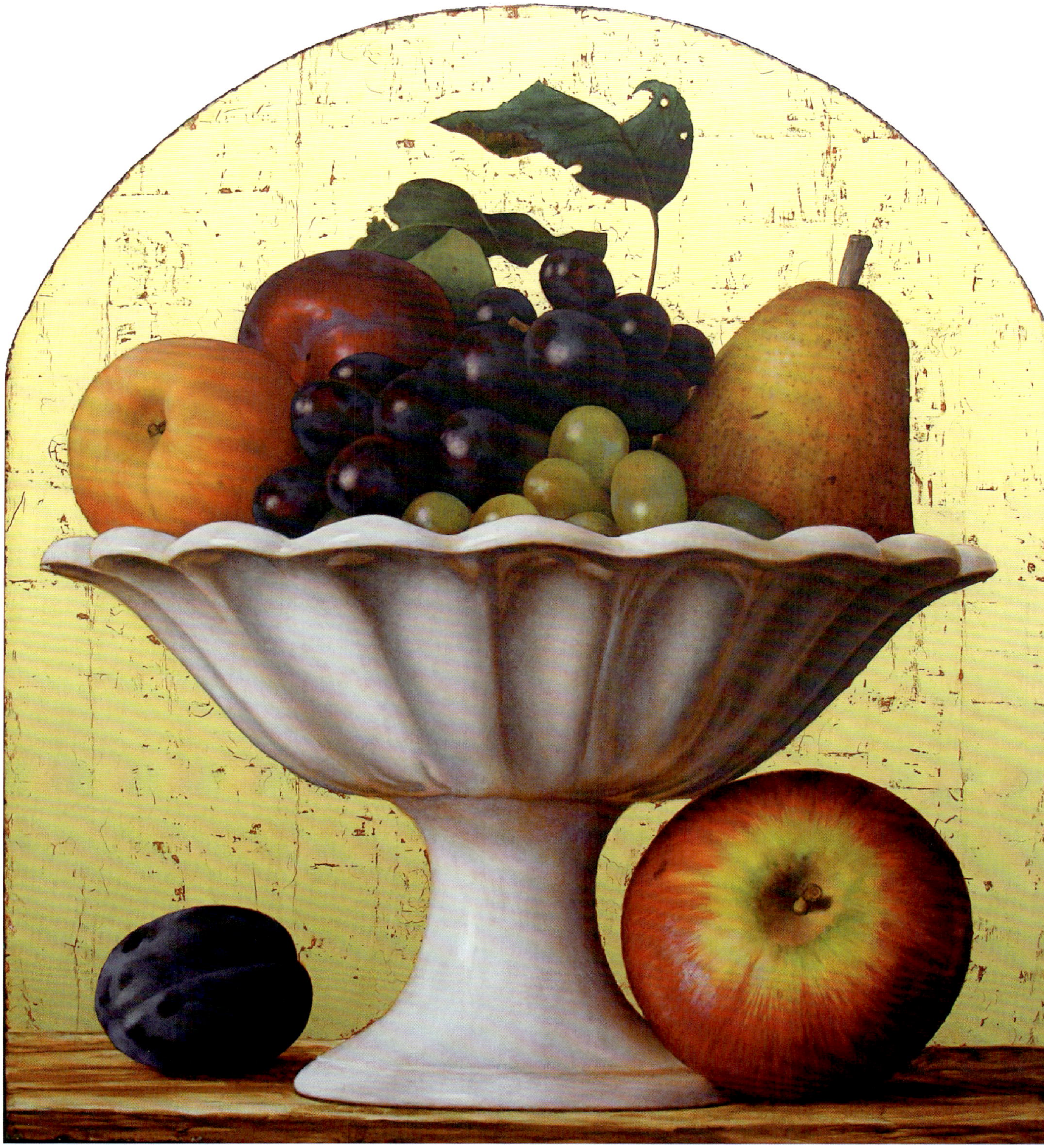

Robert Ginder
Love, 1997
Oil, gold leaf on wood, 101.6 x 91.4 cm
Lora Schlesinger Gallery, Santa Monica, L.A./New York

Sylvie Fleury
Yes to All, 2004
Steel with painted 24-carat gold plate, Edition 4/8, 71.7 x 56.5 cm
Private collection, Salzburg

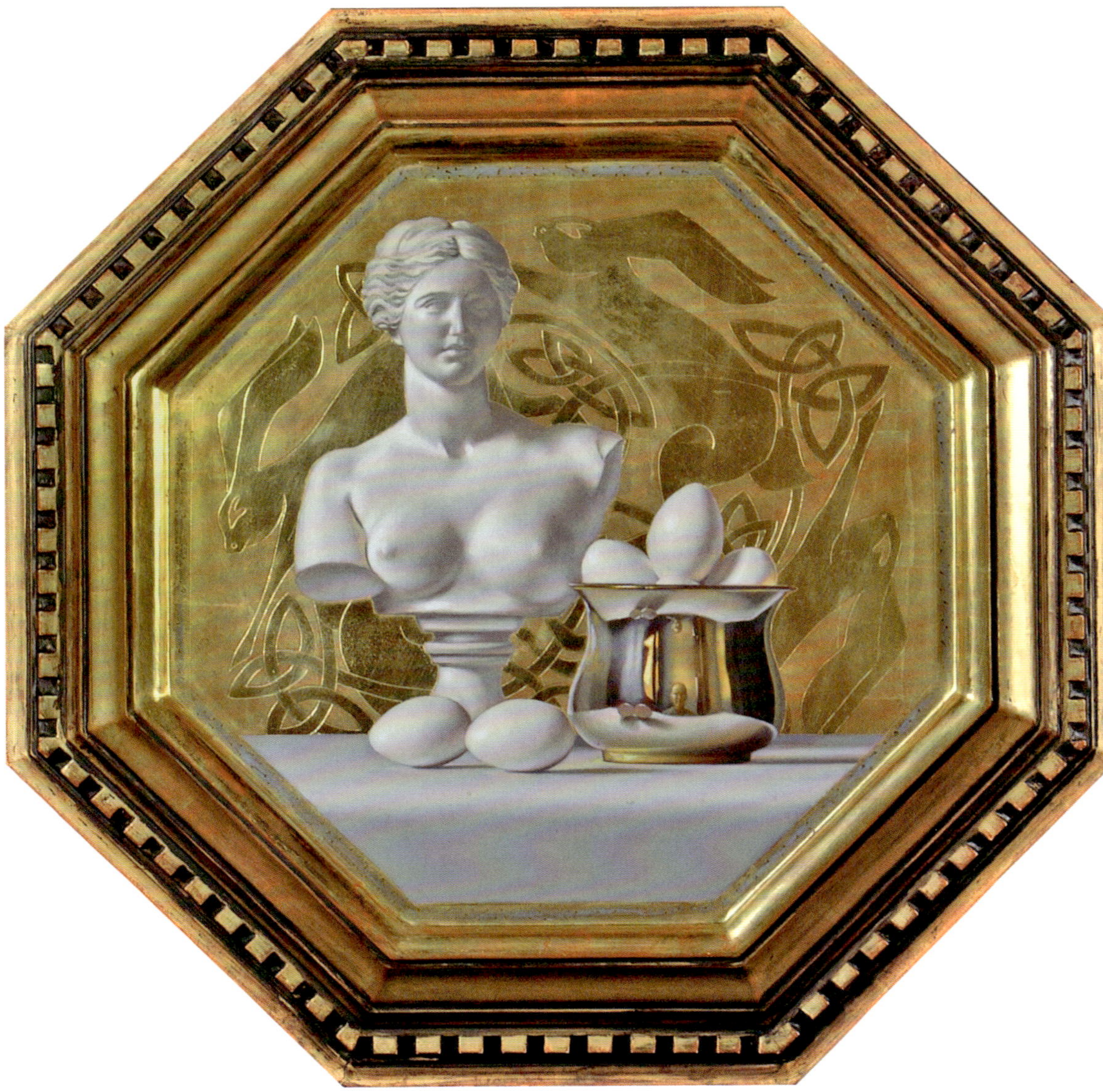

David Hewson
Moon Goddess, 2000
23-carat gold water gilt on wood, 68.6 x 68.6 cm
Susan R. Alexander

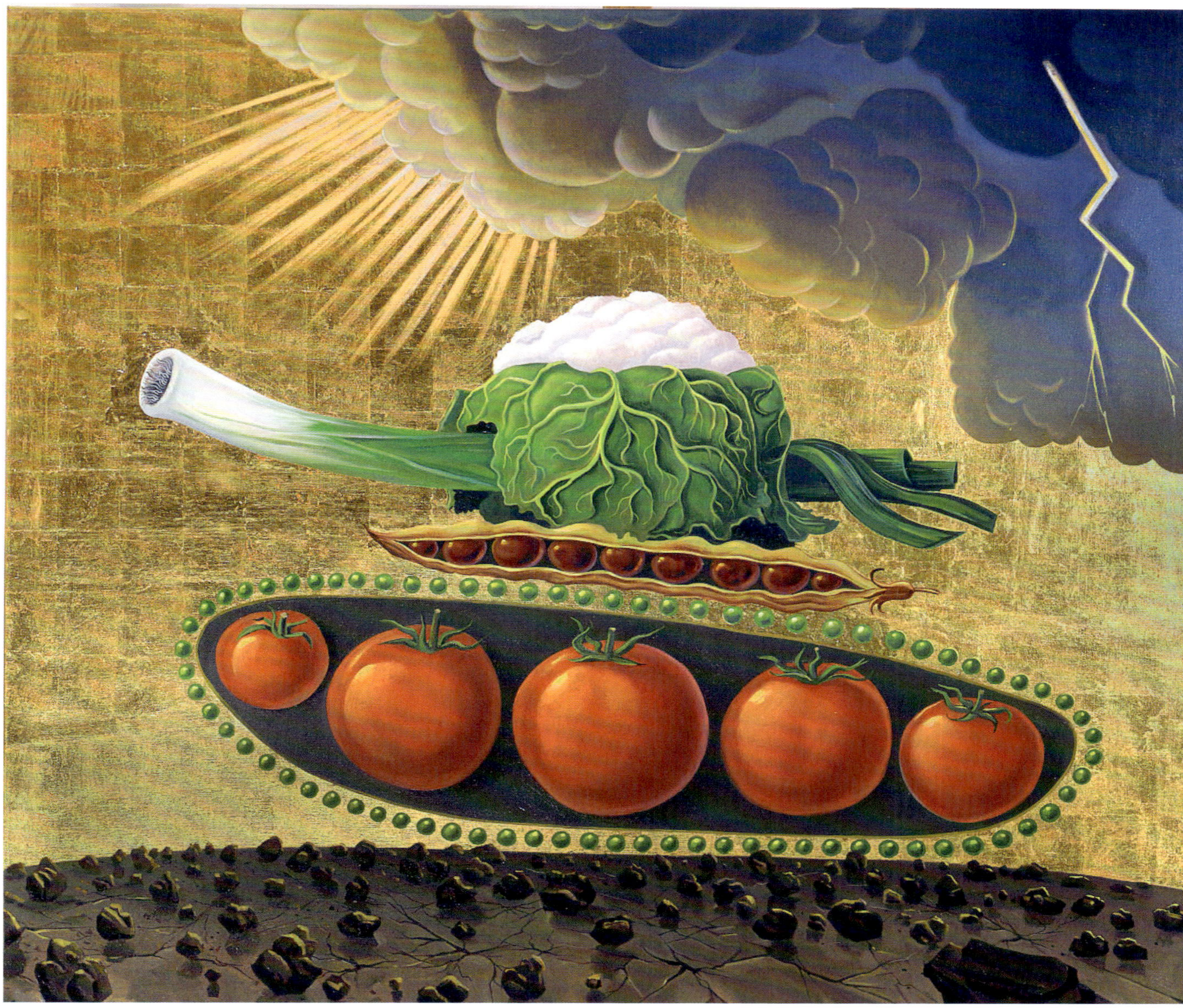

Milan Kunc
Vitamin Attack, 2005
Oil and gold leaf on canvas, 130 x 150 cm
Milan Kunc

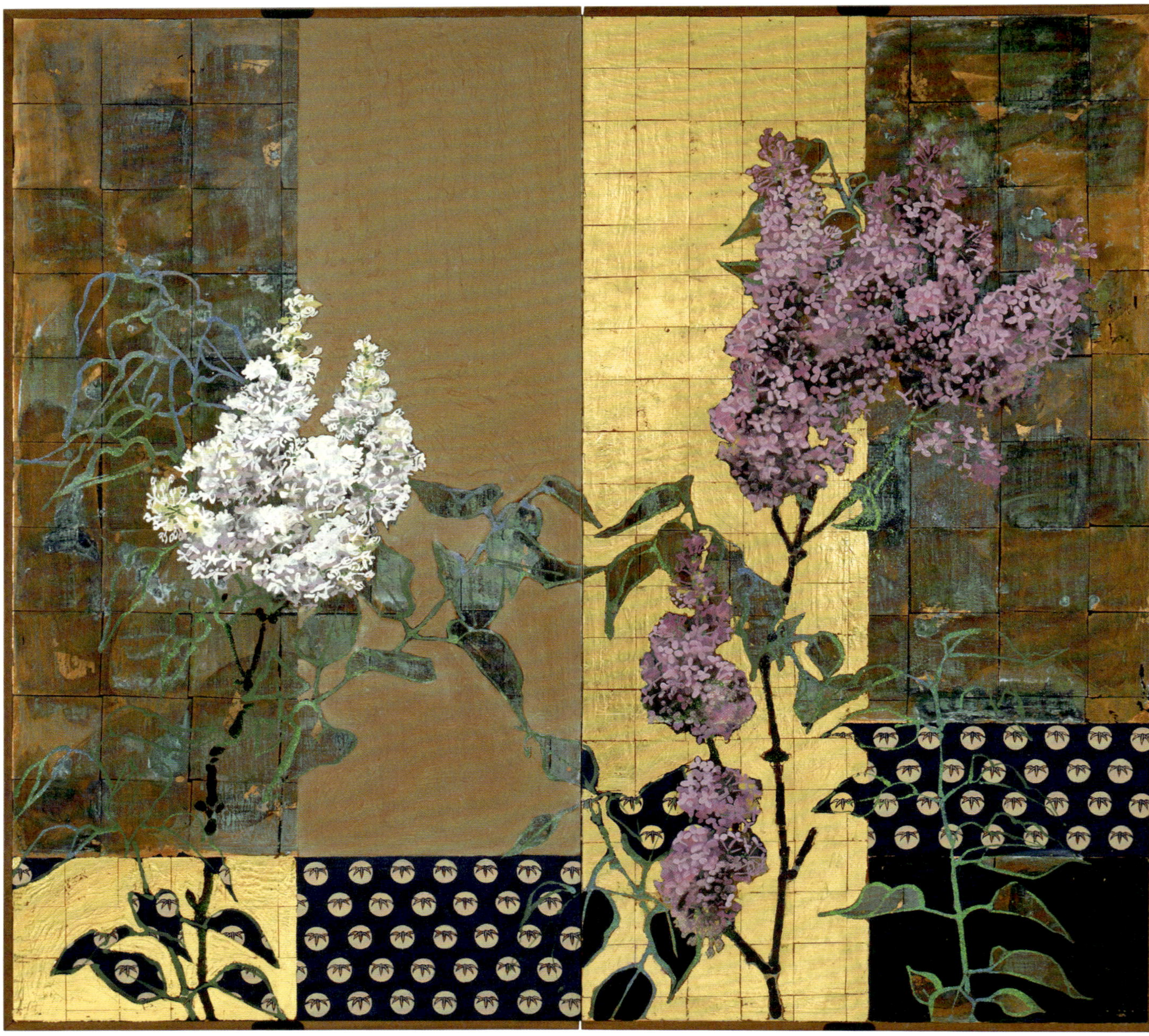

Robert Kushner
White Lilac, Lilac Lilac, 2004
Oil, acrylic, gold leaf, glitter and paper on Japanese screen, 170.8 x 187.3 cm (open)
Private collection, courtesy DC Moore Gallery, New York

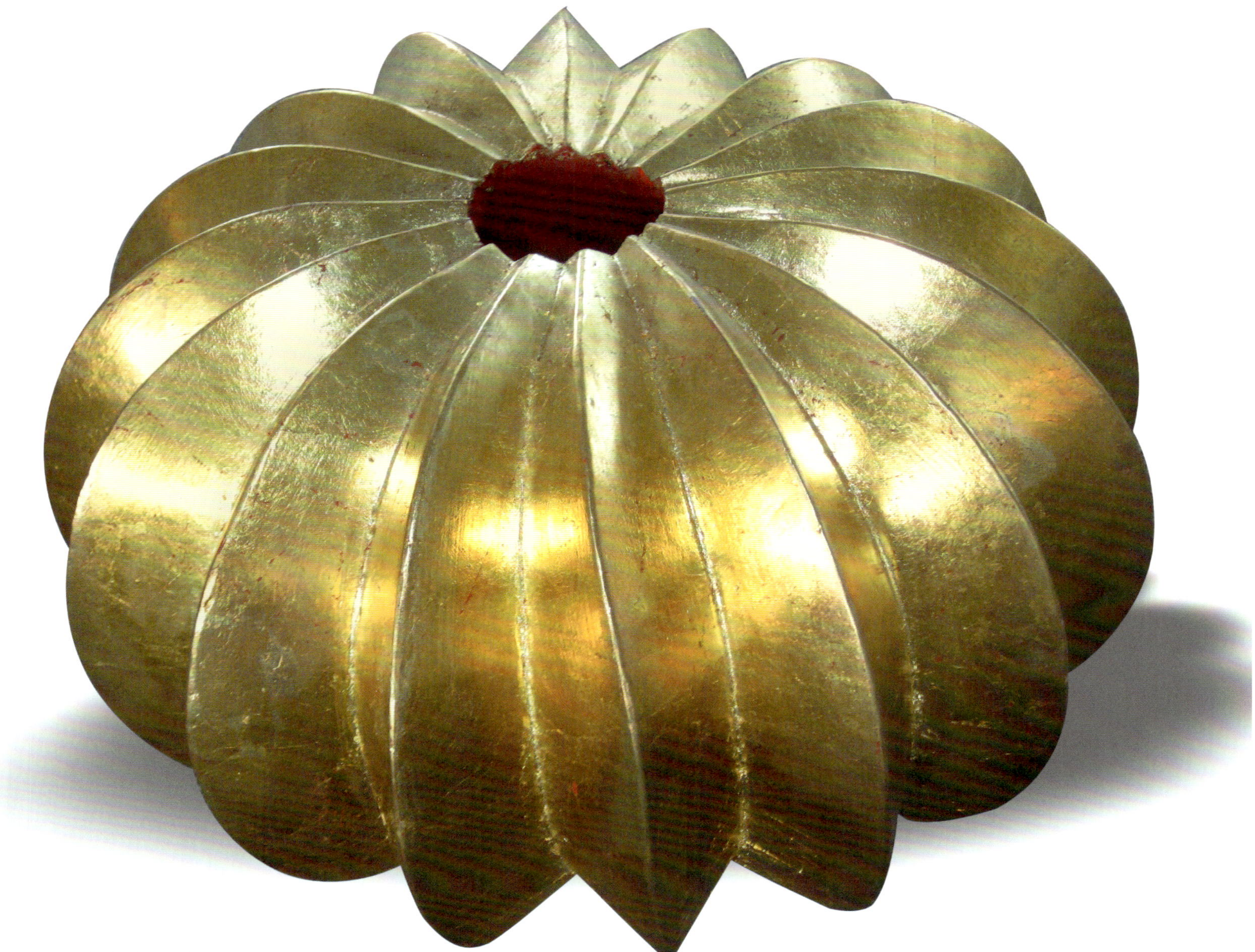

Herbert Mehler
Ondella, 2007
Corten-steel, gilded, 33 x 50 x 50 cm
LS LandskronSchneidzik Galerie und Kunstagentur

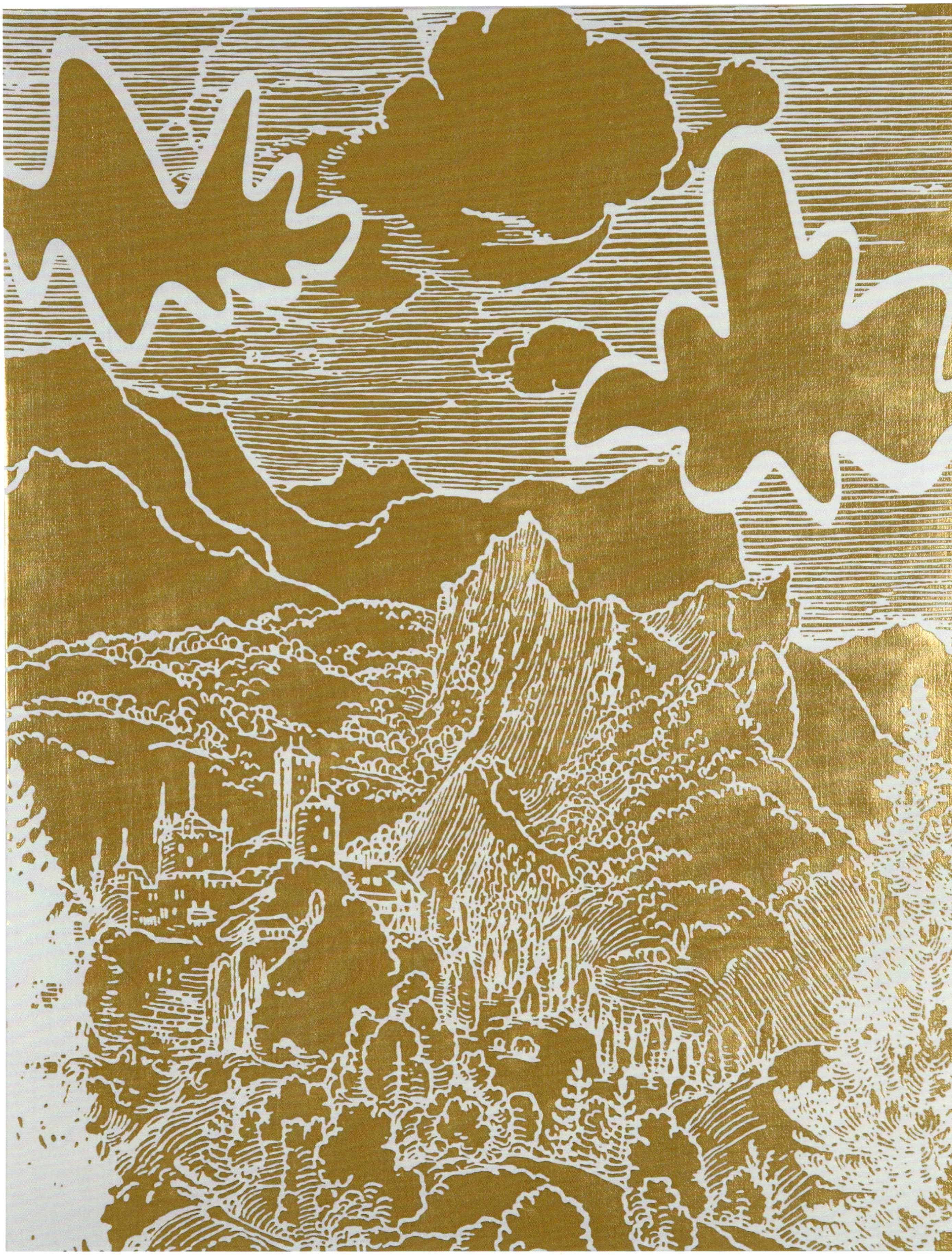

Paul Morrison
Pyxide, 2010
24-carat gold leaf and acrylic on linen, 72 x 54 x 4 cm
© the Artist/Galerie Bob van Orsouw, Zurich

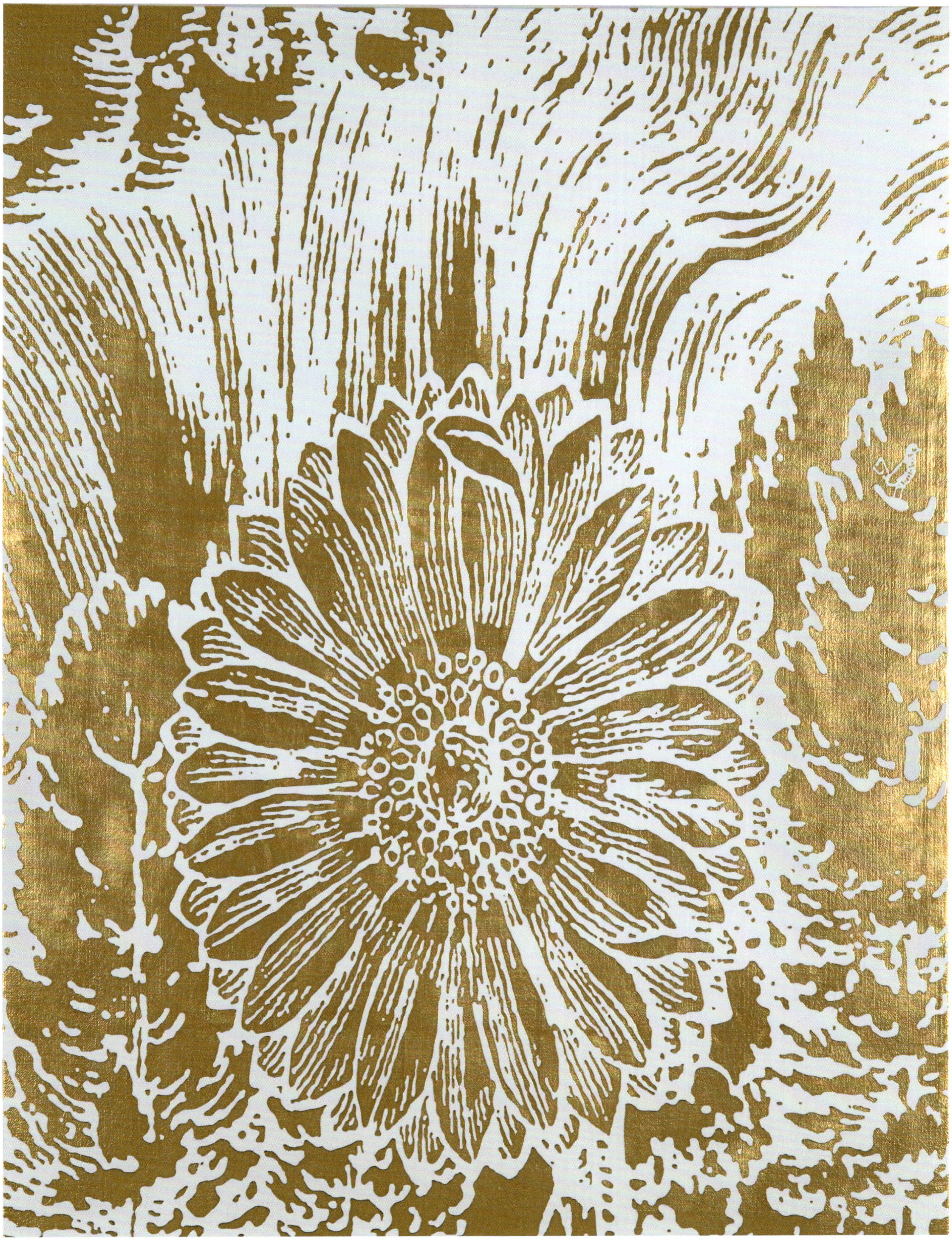

Paul Morrison
Rachilla, 2010
24-carat gold leaf and acrylic on linen, 72 x 54 x 4 cm
© the Artist/Galerie Bob van Orsouw, Zurich

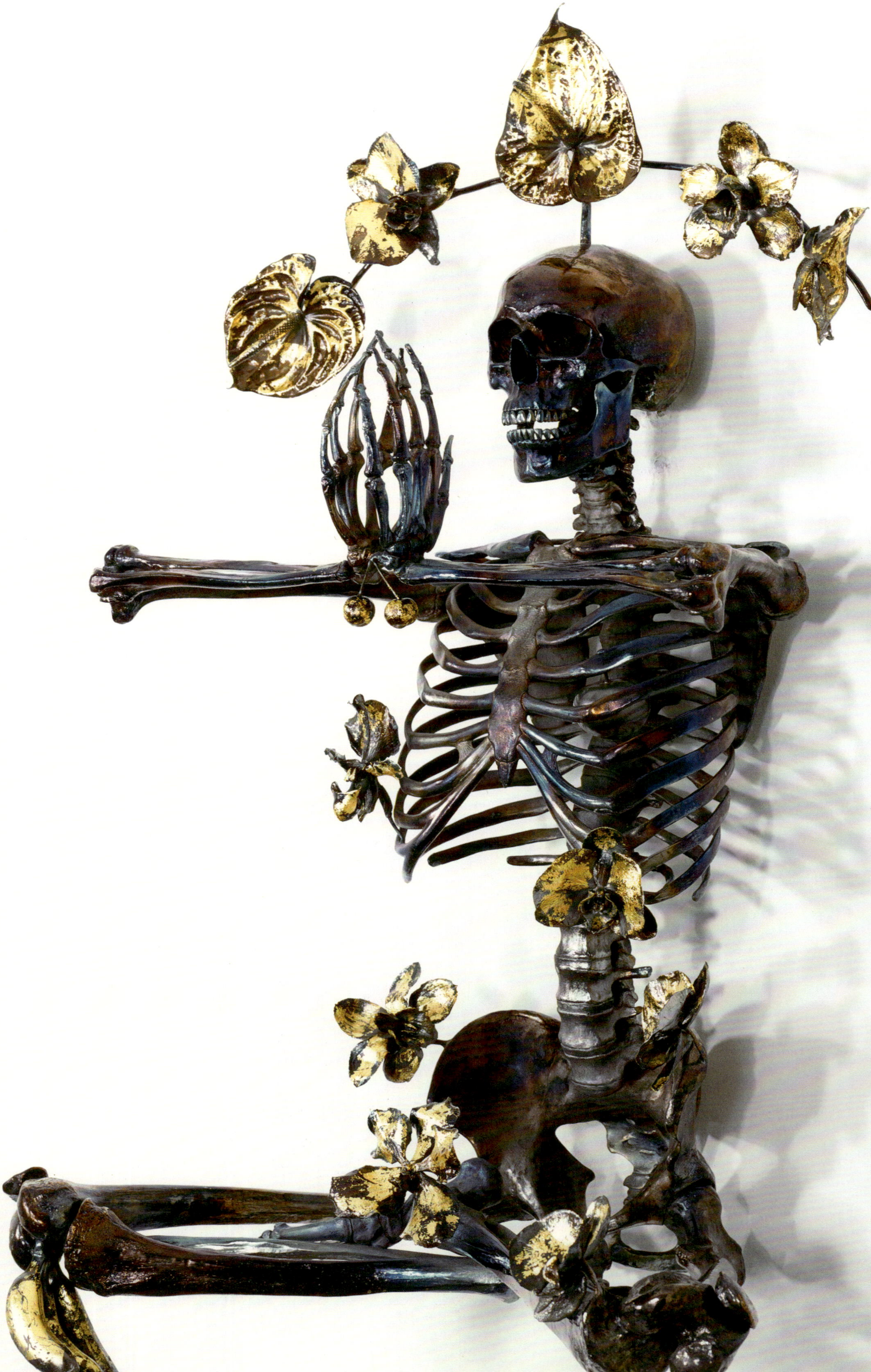

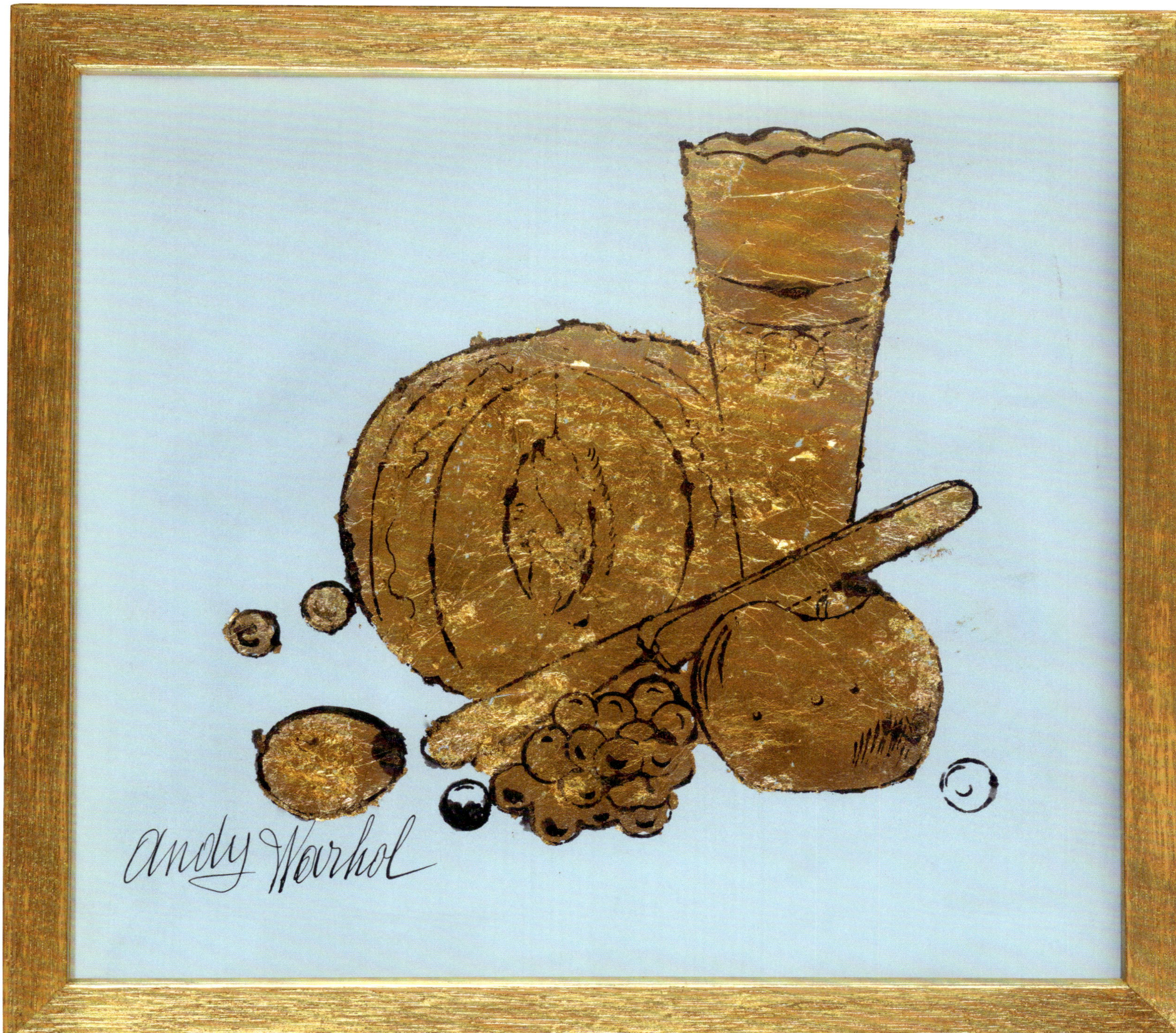

Andy Warhol
Still life, 1957/58
India ink and gold leaf on turquoise paper
22 x 24.8 cm
Private collection

Marc Quinn
A Beautiful Continuation, 2009
Chrome-plated bronze and gold leaf, 110 x 63 x 52 cm
Galerie Thaddaeus Ropac, Salzburg · Paris

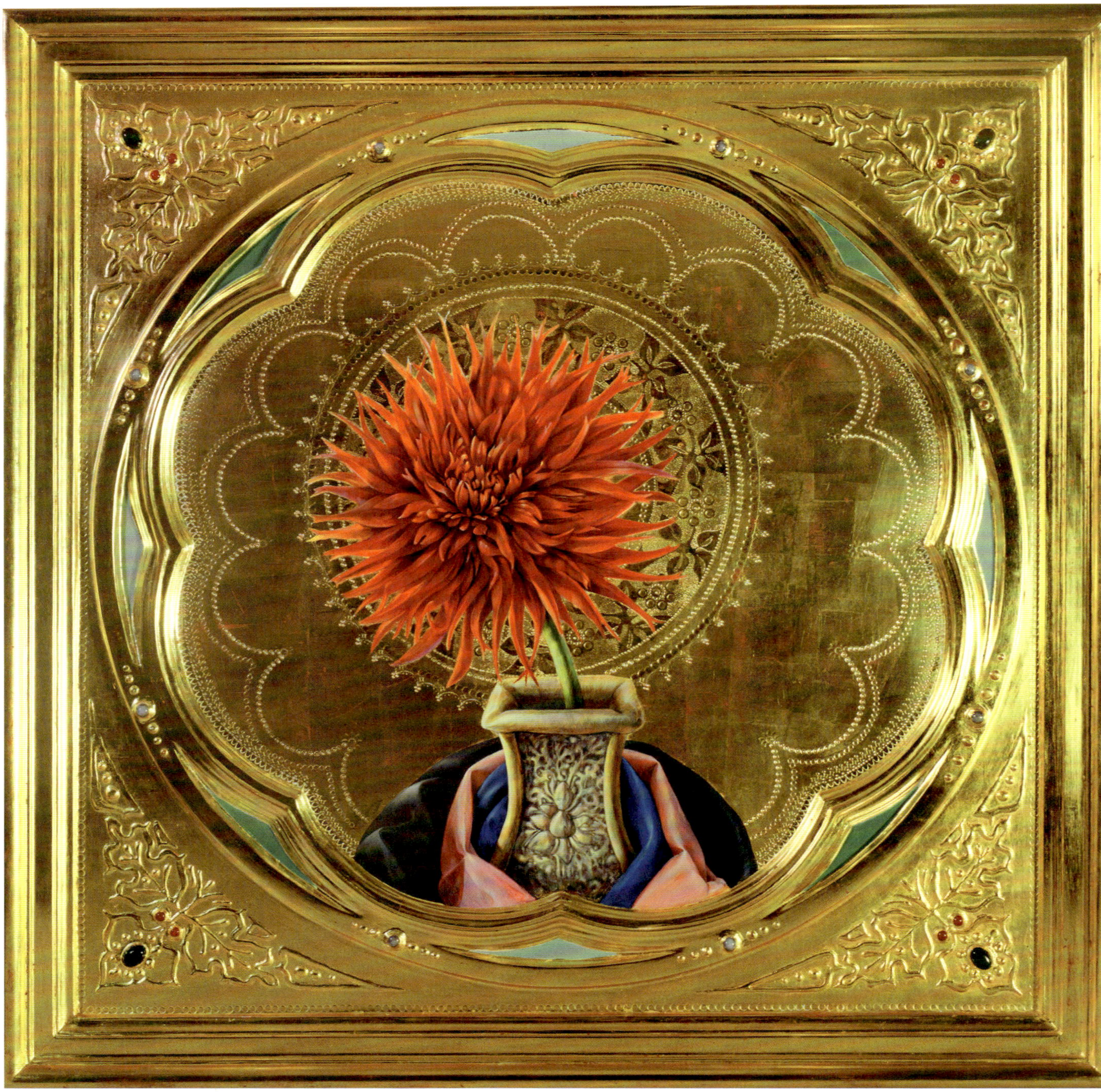

Fred Wessel
Sienese Icon (Dahlia), 1997
Tempera with gold leaf, 71.1 x 71.1 cm
Private collection

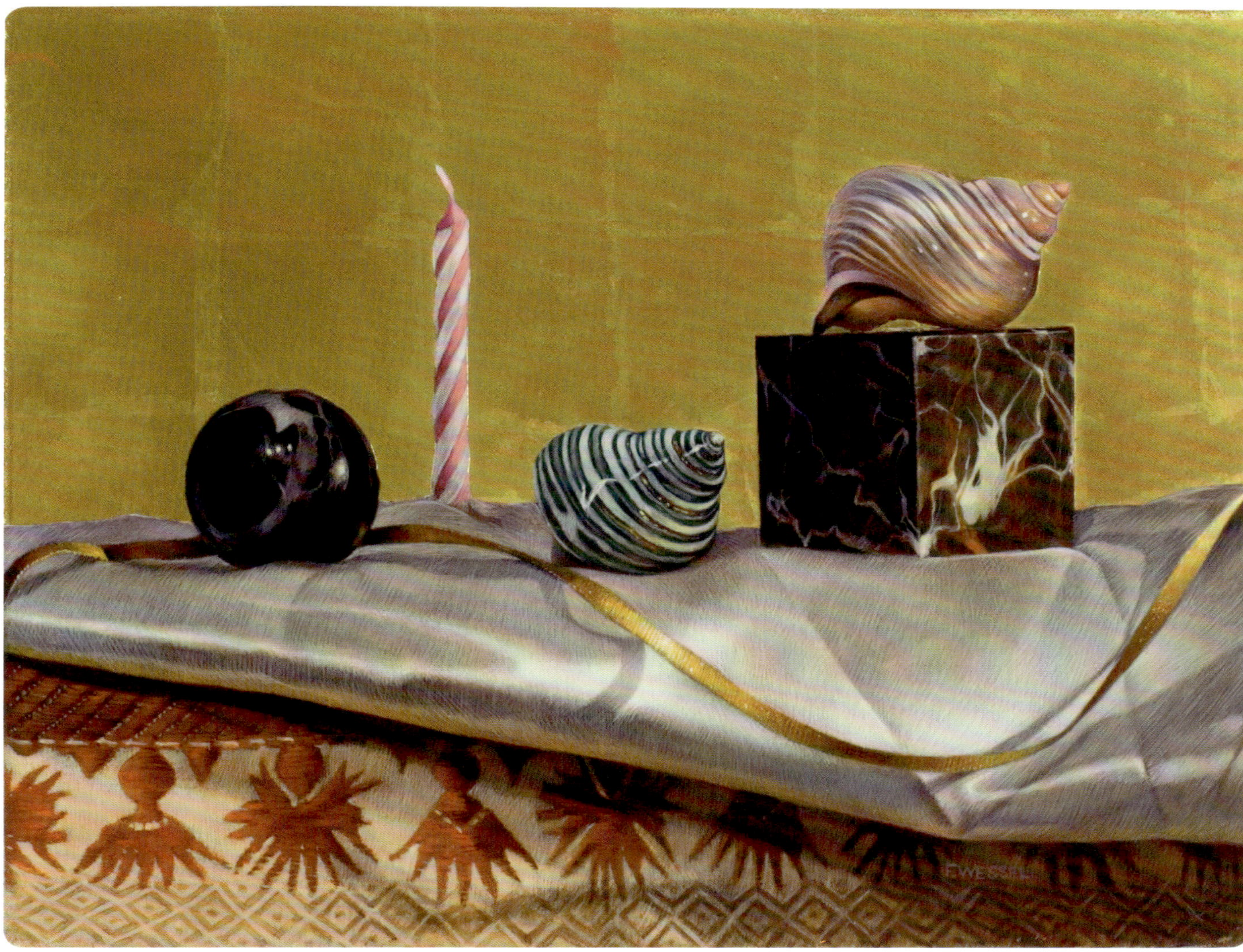

Fred Wessel
Shells, 2006
Egg tempera, gold leaf, 17.8 x 22.9 cm
Arden Gallery, Boston, Massachusetts

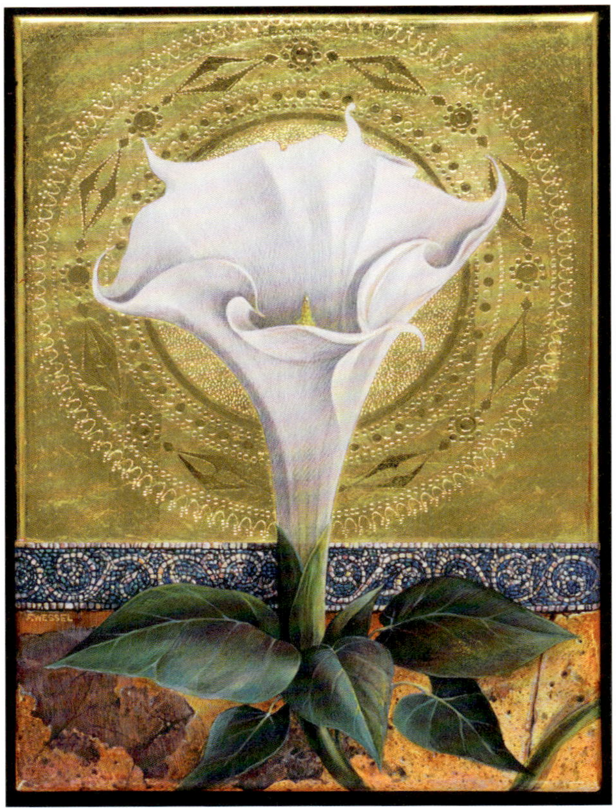

Fred Wessel
Angel's Trumpet, 1998
Egg tempera and gold leaf, 25.4 x 17.8 cm
Sheldon Bigler, Henrico, VA

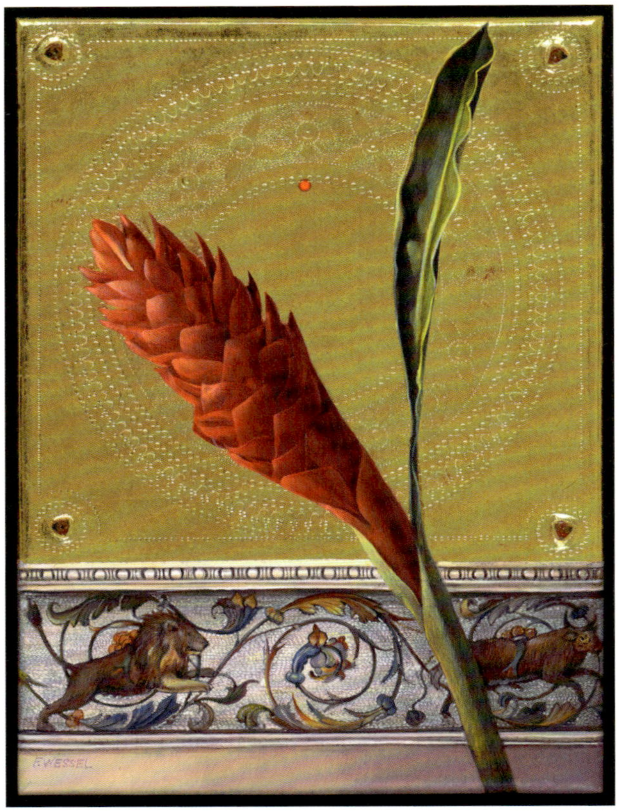

Fred Wessel
Venetian Icon (Ginger), 1998
Tempera with gold leaf, 25.4 x 17.8 cm
Linda B. Neely

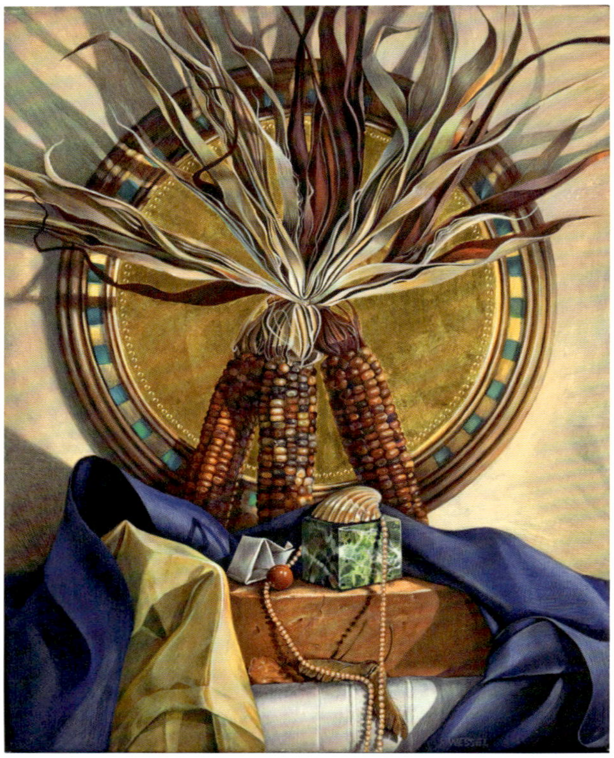

Fred Wessel
Maize, 1993
Tempera with gold leaf, 48.3 x 55.9 cm
Collection of the artist

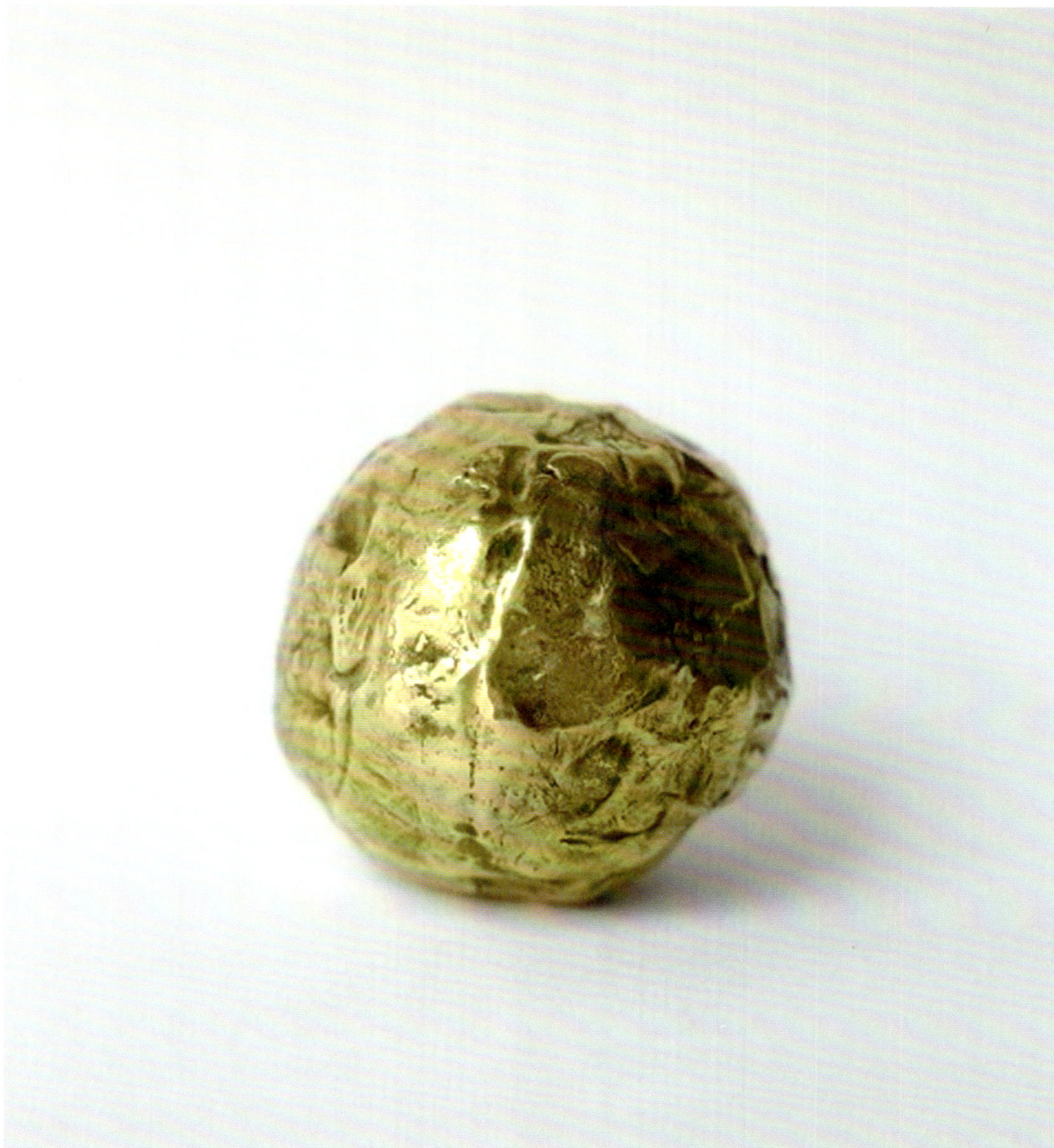

Erwin Wurm
"Goldknödel", 1989
Silver-gilt, diameter 7 cm
Private collection

Landscapes

Excursus: Gold ground and landscape

At the beginning of post-medieval painting the picture was understood as a window. The empty pictorial space was constructed by perspective, which replaced the gold ground. The space was, over the centuries, filled with landscape motifs, until the landscape became a genre in its own right. As, since Cubism, the canvas itself has been emphasized as a plane of composition, pictorial space no longer has any role. Gold leaf did not simply return to occupy the undefined space in modern painting. The combination of landscape and gold ground is in itself an illogical plan, because the space-less gold and the space-occupying view are mutually exclusive. The gold ground in the background represents a barrier to a view into the world.

What was the first modern landscape with a gold ground? Just as it is difficult in late-medieval painting to pinpoint the first appearance of landscapes, the answer in the twentieth century too depends on the definition of landscape. The criterion of topographical identifiability no longer played any role in the avant-garde. On the other hand, a tree in blossom against a gold ground and a few architectural details do not constitute a landscape. When in the first half of the twentieth century abstraction increasingly emancipated itself as the universal language, the views of external nature became inner psychological landscapes. It is difficult to say when gold first made its appearance here. The landscapes on display mostly date to between 2001 and 2010.

The landscape stages are composed of accessories that mark out the distances and spatial relationships. A boat evokes a river, a tree a grove, pictures paraphrase an interior space. In a spontaneous comparison, garden and living landscapes are created. The stranded *Golden Boat* (2002) of tulip-tree wood, blackened by fire on the inside, by Karl Manfred Rennertz, carries the inside/outside contrast within itself. It is at the same time a horizontal sculpture, a means of transport made of poisonous wood which has come to a standstill in the exhibition, the psychogram of a radiant apparition with a dark inner life. By contrast the vertical wave of gilded CORTEN steel by Herbert Mehler is a vegetal frame. He monumentalizes the *Kunstformen der Natur* ('art forms of nature'), in which Ernst Haeckel (1904) popularized the aesthetics of simple organisms, and reduces the classic *Urformen der Kunst* ('basic forms of art', 1928) by the photographer Karl Blossfeldt, which made people aware of the ornamental structures of parts of plants. "In the open air the minimalist sculptures come across as spectacles in steel—as metallic fossils or crystalline vegetation."[1]

Excursus: Artificial paradises

The most beautiful landscapes are 'artificial paradises'. The title of the book by Charles Baudelaire[2] has been used by art historians to describe greenhouses and conservatories.[3] In an age when criticism of tourism is dominated by headlines such as 'trouble in paradise' or 'paradise under threat', at the same time travellers' expectations of recuperation, relaxation but also adventure are fulfilled with artificial landscapes. This global dream is reflected most spectacularly by the 'Palm Islands' project, which has been under construction in the emirate of Dubai since 2001. The three island groups, 'Jebel Ali', 'Jumeirah', and 'Deira' are supplemented by the inclusion of 'the world'. The project is also 'high-flying' insofar as it can only be seen as a whole from the air, or in Google Earth.

In 2010 the Iranian-born Canadian artist Abbas Achavan displayed on the wall of The Third Line gallery in Dubai an aerial view (in fake gold leaf) of the coastal stretch with its artificial islands. The monumental (13.88 x 5.20 metre) *Island* (fig. 1) picture was sold during the exhibition section by section to visitors, analogously to real estate. Achavan confronted the fake gold (described as genuine in the run-up to the exhibition)

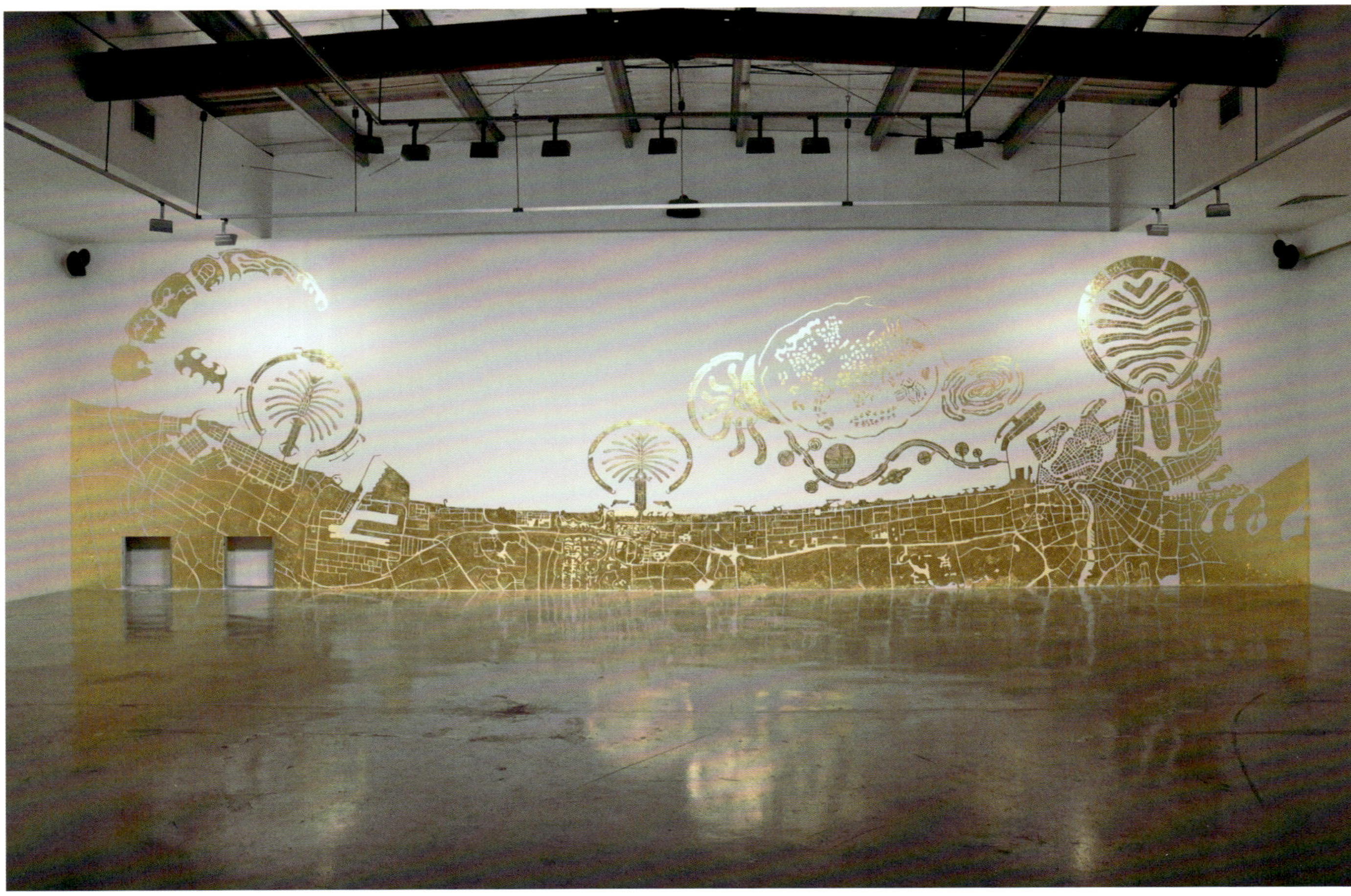

Fig. 1 Abbas Akhavan, *Island*, 2010, Exhibition view, The Third Line Gallery, Dubai

of the real-estate-bubble landscape, four sections of which are on display in the Belvedere, with less magnificent photocopies of examples of the region's original economy, such as fisheries, date harvests and animal husbandry. He emphasized the fruits of these more fundamental economic processes with genuine gold leaf.

Architectural motifs play an important role in landscapes. In the early post-medieval period, the gold ground gave way to the topographical exploration of the city. The first townscapes (like portraits too) still followed stereotypical models. Of the 68 cities depicted in the *Schedelsche Weltchronic* (1492), only 30 were topographical likenesses. However, architecture against a gold ground does not necessarily evoke the Middle Ages. Jürgen Kadow gave the Golden Pavilion (Kinkaku-ji) a gold-leaf ground. The upper section of the Buddhist temple (originally 'Rokuon-ji' or 'deer-park temple') in the north-west of the Japanese city of Kyoto is clad in gold leaf. Patricia Miranda constructs the space of her *Siena Landscapes Series* (2005) using geometric lines that integrate the modellike buildings into the pictorial space. Robert Ginder demonstrates the value placed by the middle classes on their unsullied dreamscapes. His family homes, portrayed on a gold ground, are decorated only with flowers and plants, no people or cars contaminate the view of this world where all is as it should be.

The immaculate façades of these homes are contrasted by Milan Kunc with the horror vision of a spent environment. The idyll of peaceful family life is maintained in the ecological disaster of the surreal landscape *Recycled World* (2001), compiled from the relics of civilization. Kunc loves the gold ground out of coloration considerations. In the gold ground, objects and light sources in the surroundings are reflected light or dark depending on the beholder's position. The badlands are crossed (*Cross the Badland*, 1995) in a grotesque 'bachelor machine'. According to Michel Carrouges[4] a 'bachelor machine' is a fantastical image which transforms male desires into a death mechanism. It consists of two pictorial regions, a mechanical-sexual zone and an inaccessible female zone. The vehicle signals with the death lamp in the wrong direction (to the right), in contrast to which the destination aimed at along the road gets lost in the background of the non-perspective space dimension of the gold leaf.

The gold in the picture has, in principle, two possibilities. It can spread out or it can place highlights. The gold streams into the whole picture surface like light in photography. The photographer and gilder Hermann Staudinger unites both talents and experiments with ways of bringing the two techniques closer together. "I use photographs which depict light that outshines the photographic emulsion, so that the photographic paper appears white as a result. The moment no colours can be seen any more, only pure transparency, the white, and in this printing process, the gold lustre. [...] the angle of viewing is crucial: seen in the right light, the backgrounds radiate, open up the space and take the picture on to a different, immaterial plane." Another technique he has developed is gold embossing. "Large copies of newspaper photographs are laid over gilded surfaces, the raster points and black areas of the original are traced in pencil; the pressure of the pencil produces shallow dents in the surface of the gold, and, in this hollow, reflected light is refracted when the copy of the original is removed [...]."[5]

In a traditionally composed landscape Gary Milec has the saturated tone of the fields, trees and autumnal clouded sky shine out. In the Japanese tradition, in turn, the ancient Rimpa painting can be taken to new horizons through the experience of Western Modernist abstraction. Makoto Fujimura, in an obituary for his teacher Matazo Kayama (1927–2004), recalled how great the latter's problems with the 'third colour' (silver or gold) had been. "I considered it, foolishly enough, pre-modern and inappropriate for our present time."[6] Astonishingly, he shared this attitude with many Western artists, although there has been an unbroken tradition in Japan of using gold. In his landscapes Fujimura unites Western abstraction (he reveres Willem de Kooning) and Japanese Nihonga painting with historical and religious themes.

In the meditative landscapes of the Catalan artist Carme Sanglas, gold is used to represent all of the Aristotelian elements: it is the air above the sea, it flows, it is fiery, and it forms the terra firma for the *Paths and Roads* (2003). In the *Factory* (2005) with a panoramic view of the blue waves of animal carcasses, the gold streams out of holes in the walls down into a gold pool. The reflection of the *Solstice* (2008) is seen on the grey surface of the sea from a bird's eye view. Like a boat the golden light travels along its ecliptic. The strong summer sunlight grows weaker in the autumn below until the winter solstice is reached.

1 Hansdieter Erbsmehl, "Kontinuität und Neubeginn", in: *Herbert Mehler—cavex Sculpturen*, (exhib. cat. Tammen Galerie, Berlin/ Museum am Dom, Würzburg), Berlin 2007, p. 16.
2 Charles Baudelaire, *Les paradis artificiels*, 1860. The original title has an extension—*opium et haschisch*—which points to inner paradises.
3 Stefan Koppelcamm, *Künstliche Paradiese. Gewächshäuser und Wintergärten des 19. Jahrhunderts*, Berlin 1988.
4 Jean Clair/Harald Szeemann (eds.), *Junggesellenmaschinen / Les machines Célibataires*, Venice 1975, pp. 21 ff.
5 Hermann Staudinger, quoted from: URL: <http://www.hermannstaudinger.at/html/gg-photos-ues.html> and URL: <http://www.hermannstaudinger.at/html/gg-praegung/19.html> [read on 14 Dec. 2011].
6 Cf. URL: <http://www.macotofujimura.com/worcs/images-of-grace1/>

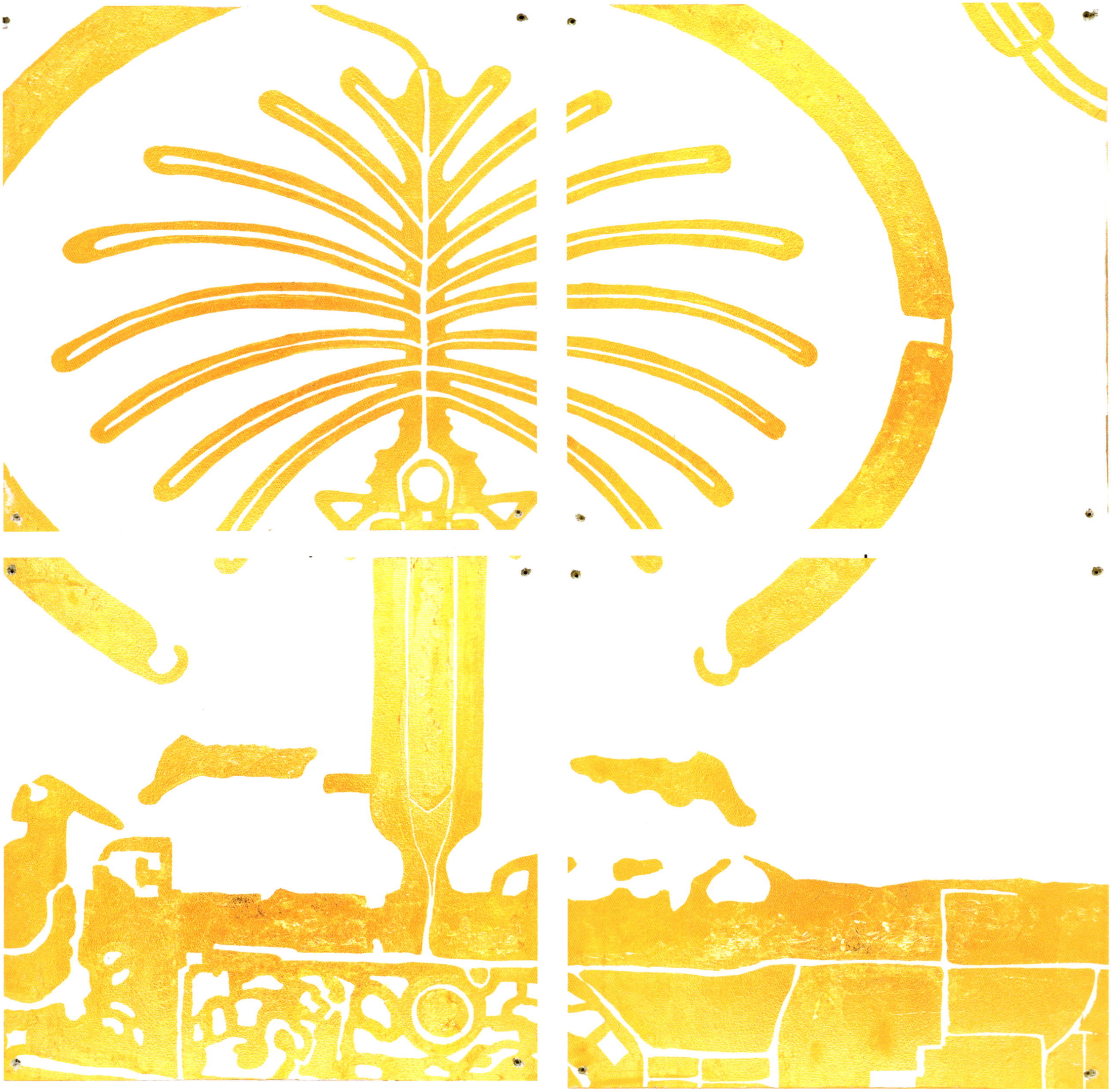

Abbas Akhavan
Islands: Units E 11 [zone 2], E 12 [zone 2], D 11 [zone 3], D 12 [zone 3]
Gold leaf on drybuild board, each 60 x 60 cm
The Third Line, Dubai

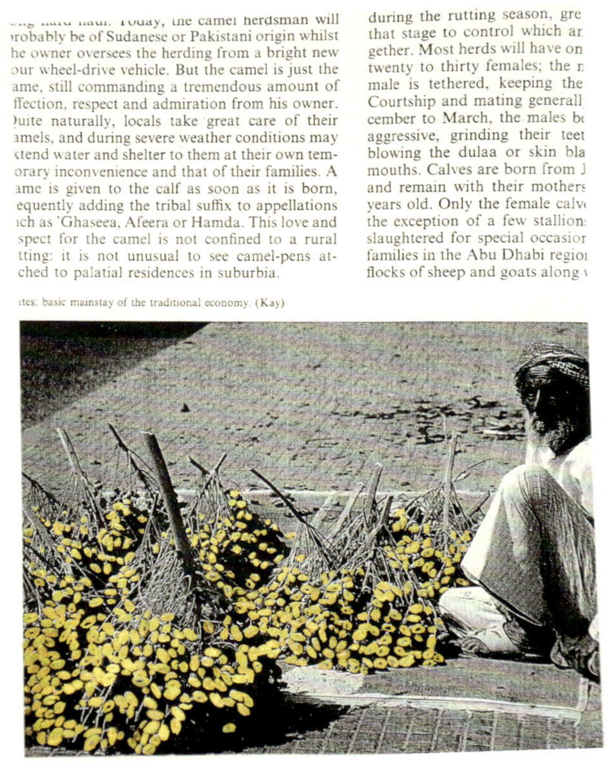

Abbas Akhavan
Dates, 2010
22-carat gold on Xerox copy, 29.7 x 21 cm
The Third Line, Dubai

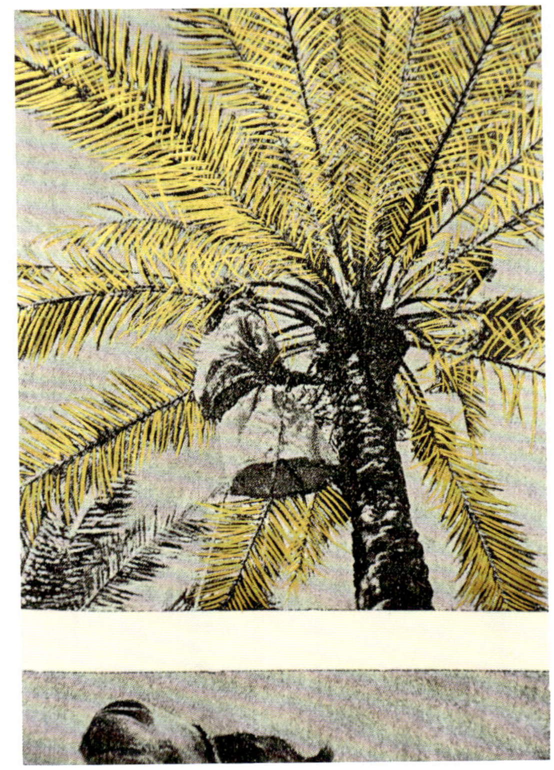

Abbas Akhavan
Palm, 2010
22-carat gold on Xerox copy, 29.7 x 21 cm
The Third Line, Dubai

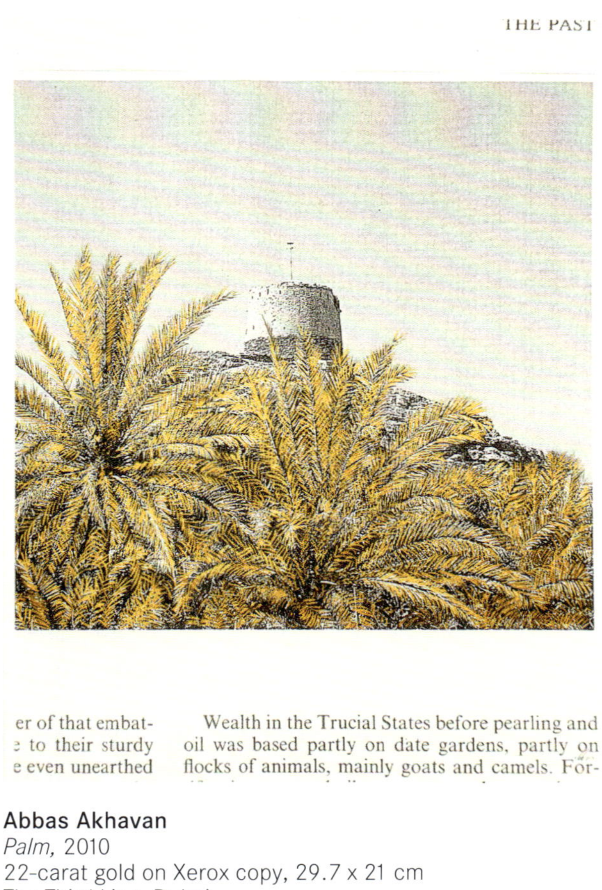

Abbas Akhavan
Palm, 2010
22-carat gold on Xerox copy, 29.7 x 21 cm
The Third Line, Dubai

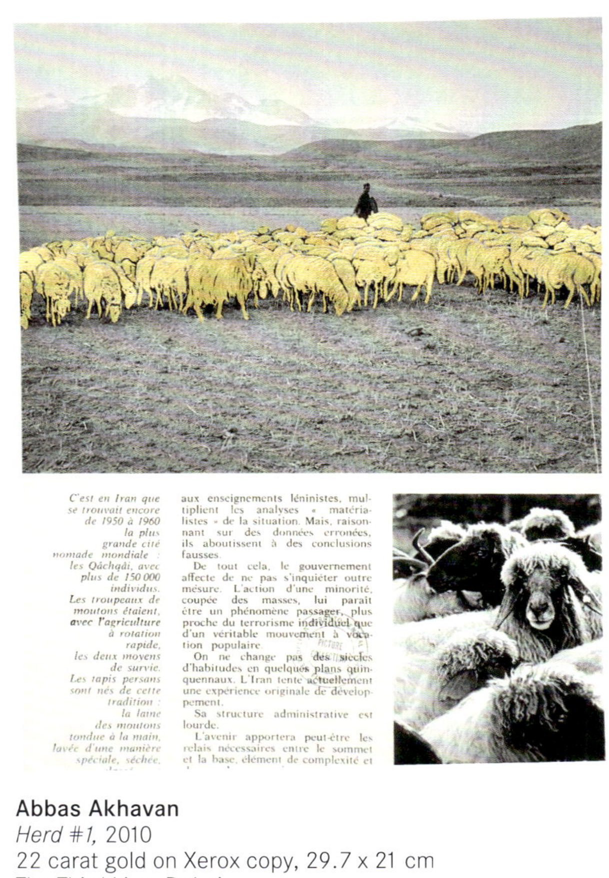

Abbas Akhavan
Herd #1, 2010
22 carat gold on Xerox copy, 29.7 x 21 cm
The Third Line, Dubai

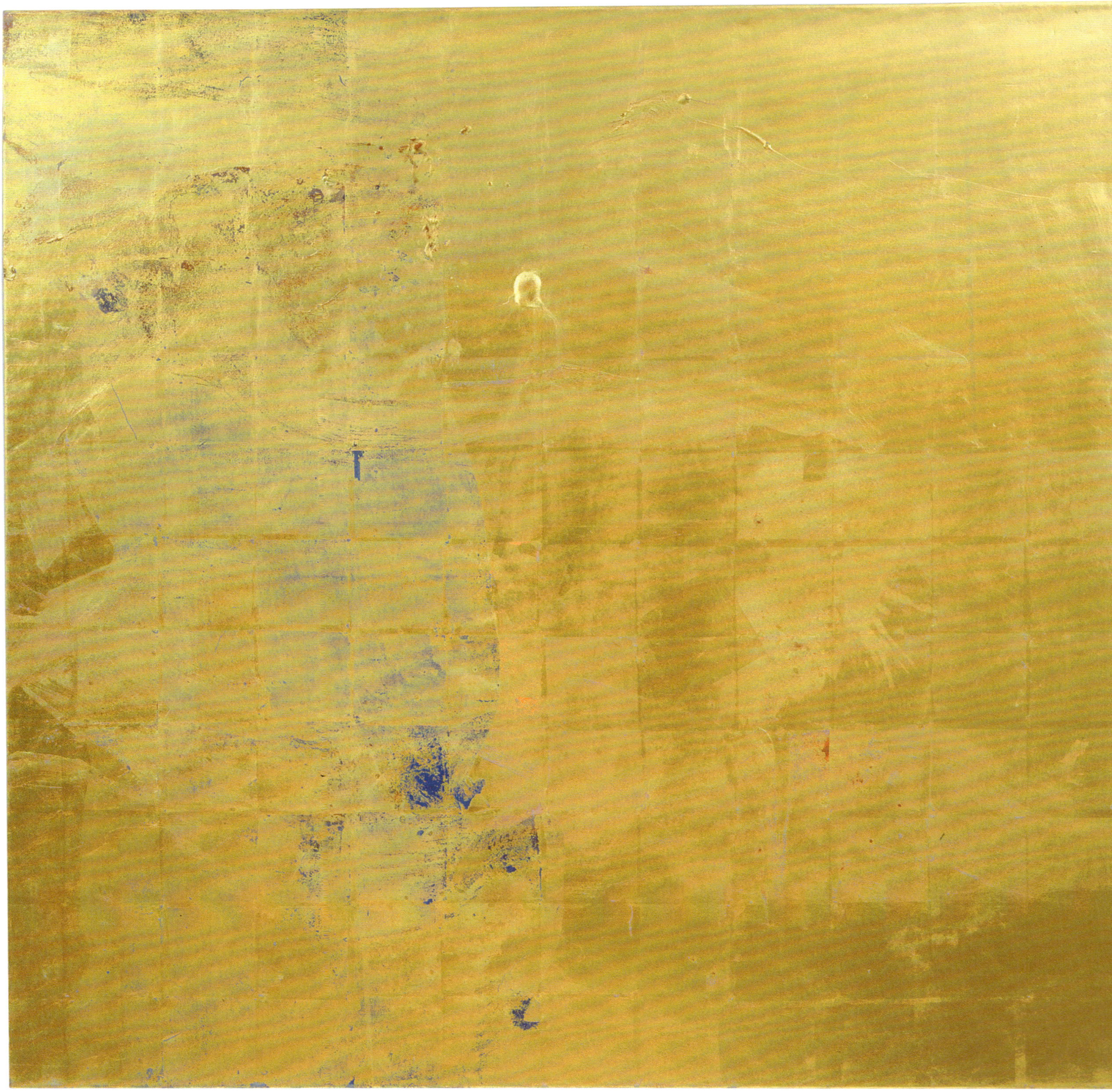

Makoto Fujimara
Interior Castles, 2008
Japanese gold leaf and mineral pigment on koudama paper
mounted on canvas, 121.9 x 121.9 cm
Courtesy of Dillon Gallery New York

Jakob Gasteiger
Klondike, 2011
Gilded aluminium, Height 62 cm
In the artist's possession

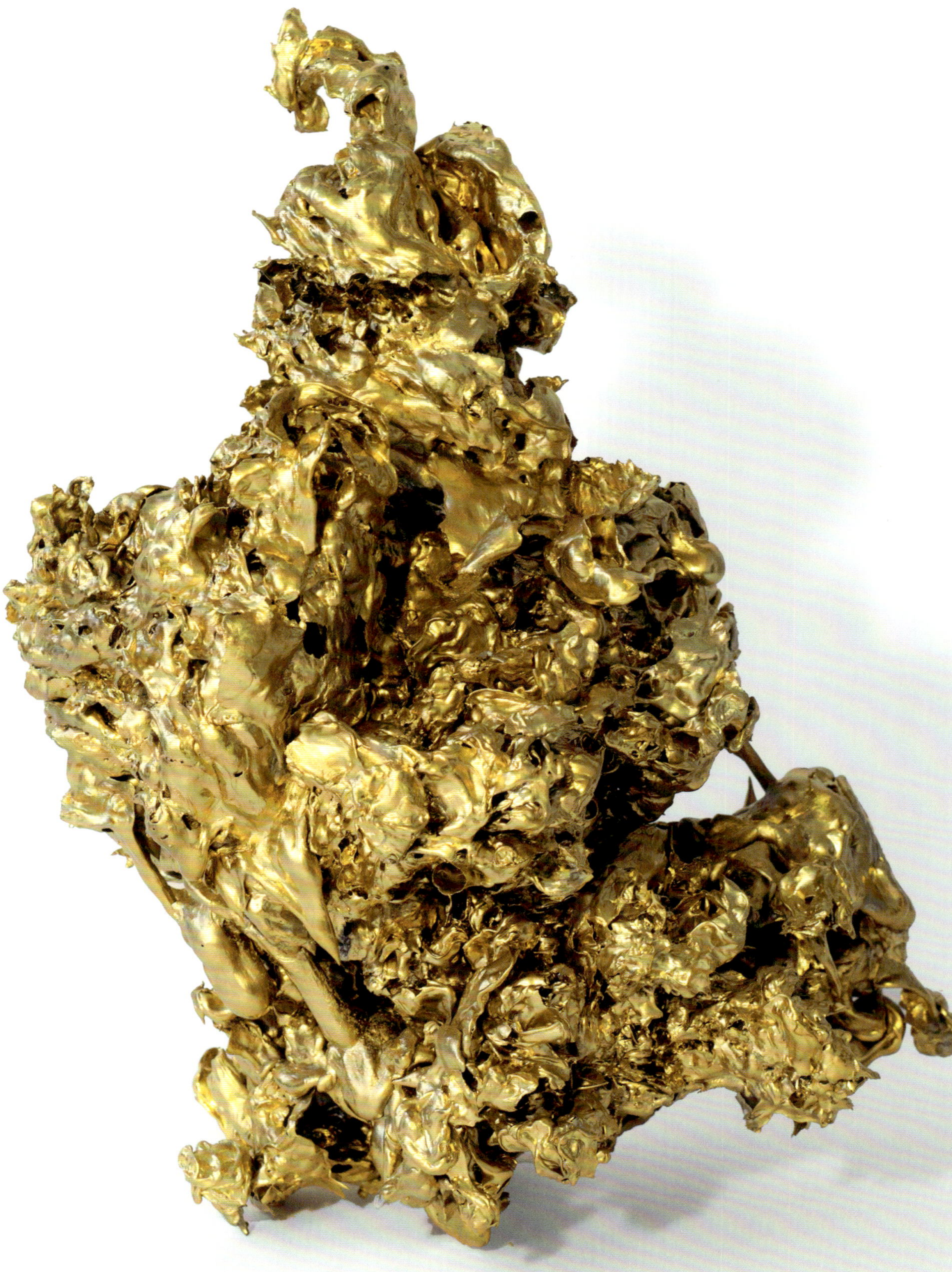

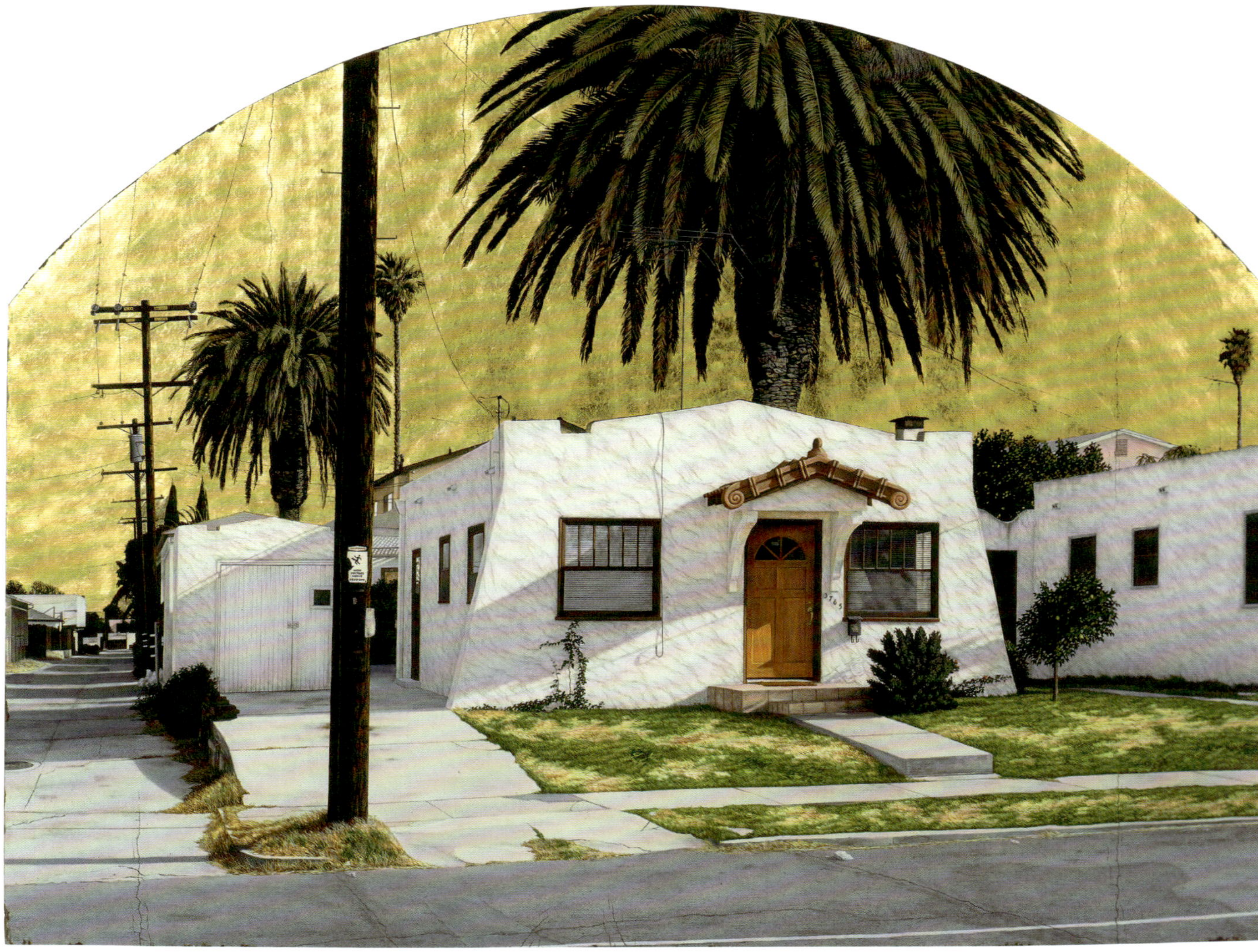

Robert Ginder
Haven, 1991
Transfer gold leaf on wood, 102.9 x 134.6 cm
Lora Schlesinger Gallery, Santa Monica, L.A./New York

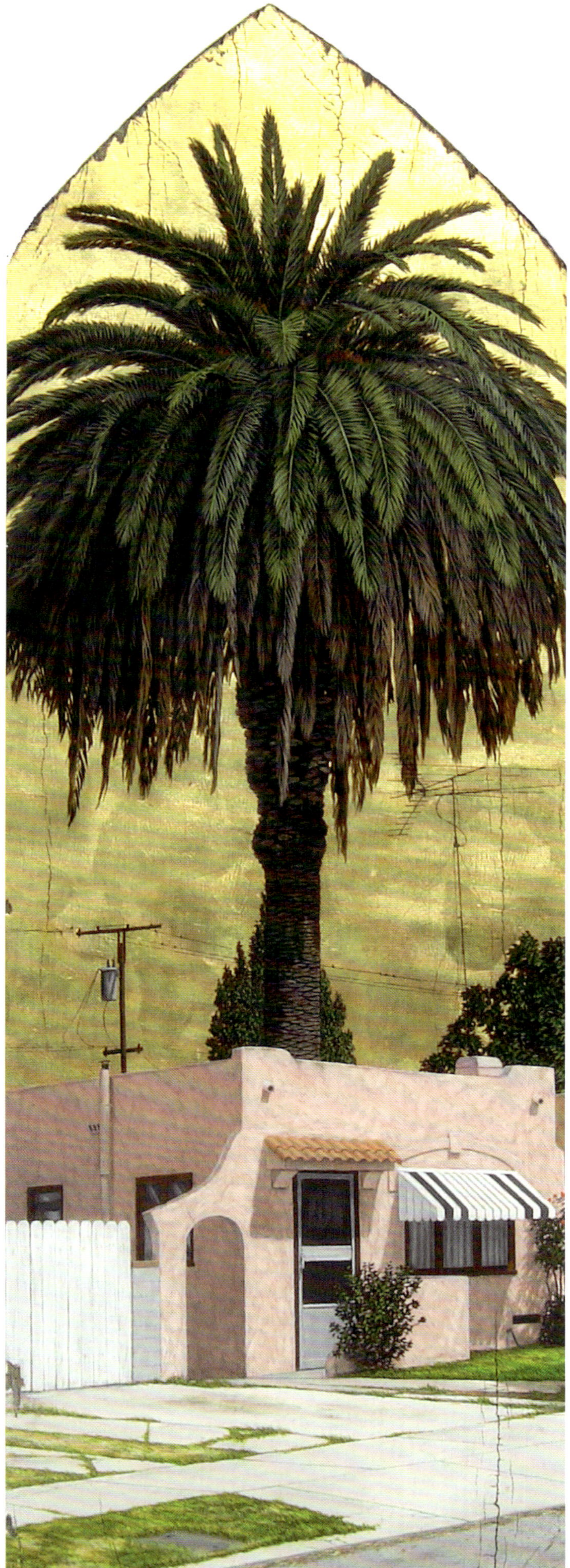

Robert Ginder
June, 1993
Oil, gold leaf on wood,
83.9 x 30.5 cm
Lora Schlesinger Gallery,
Santa Monica, L.A./New York

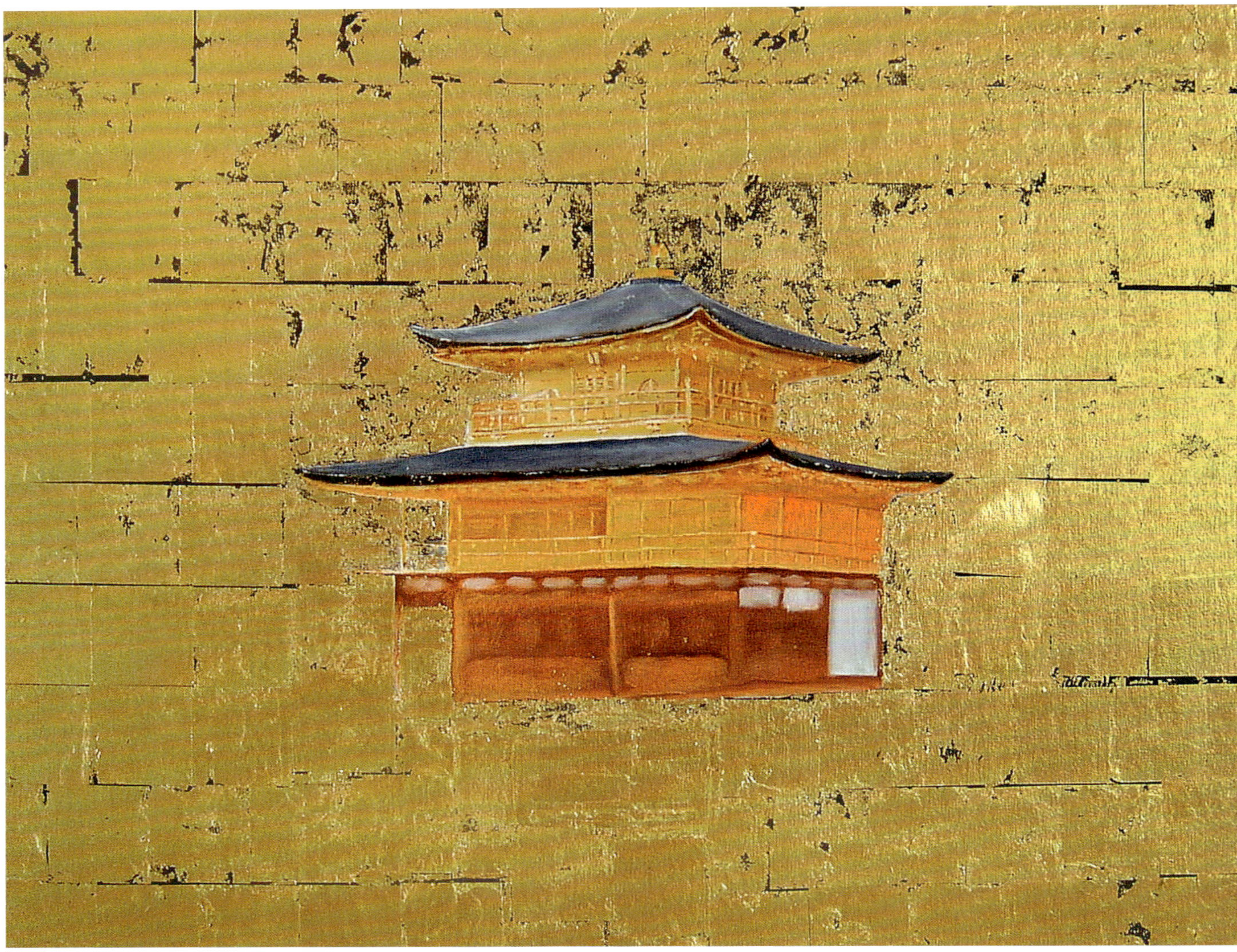

Jürgen Kadow
Untitled, 2005
Oil, gold leaf on canvas, 80 x 100 cm
Private collection, Germany

Milan Kunc
Recycled World, 2001
Oil and gold leaf on canvas, 200 x 240 cm
Milan Kunc

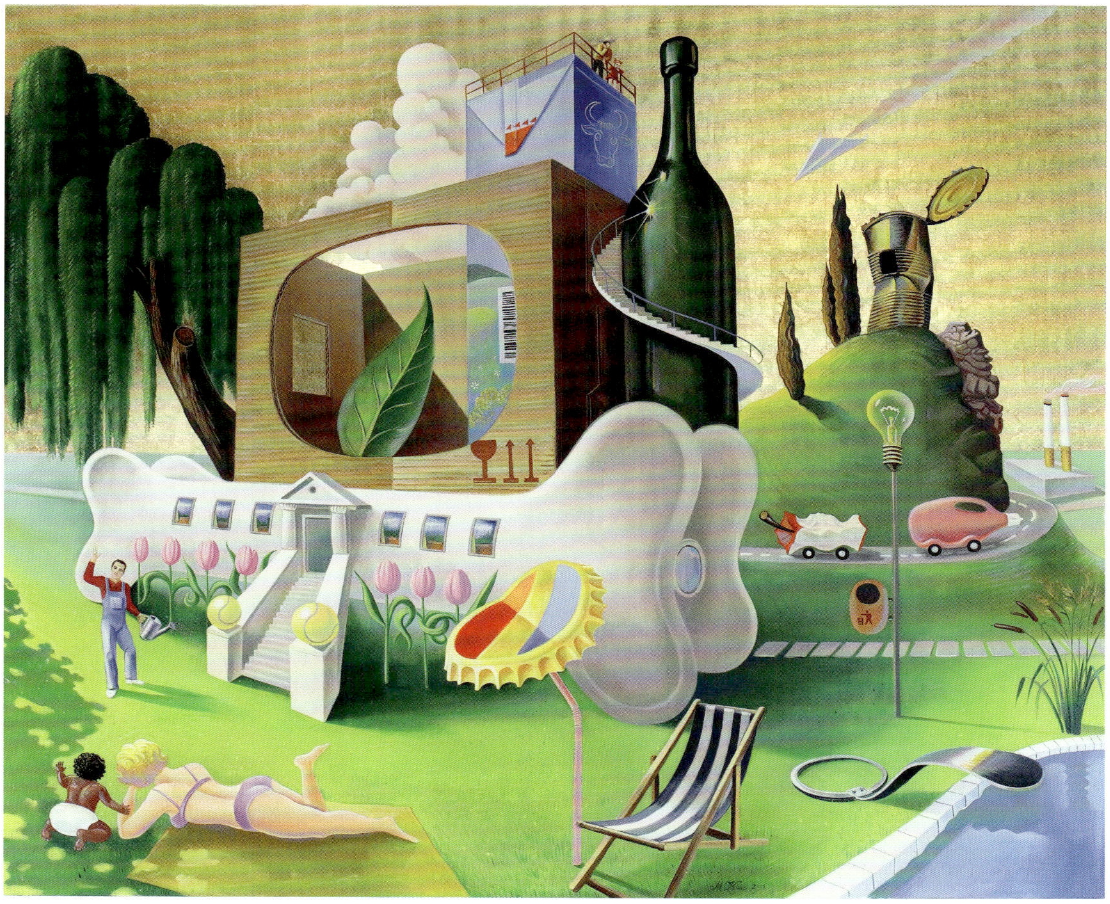

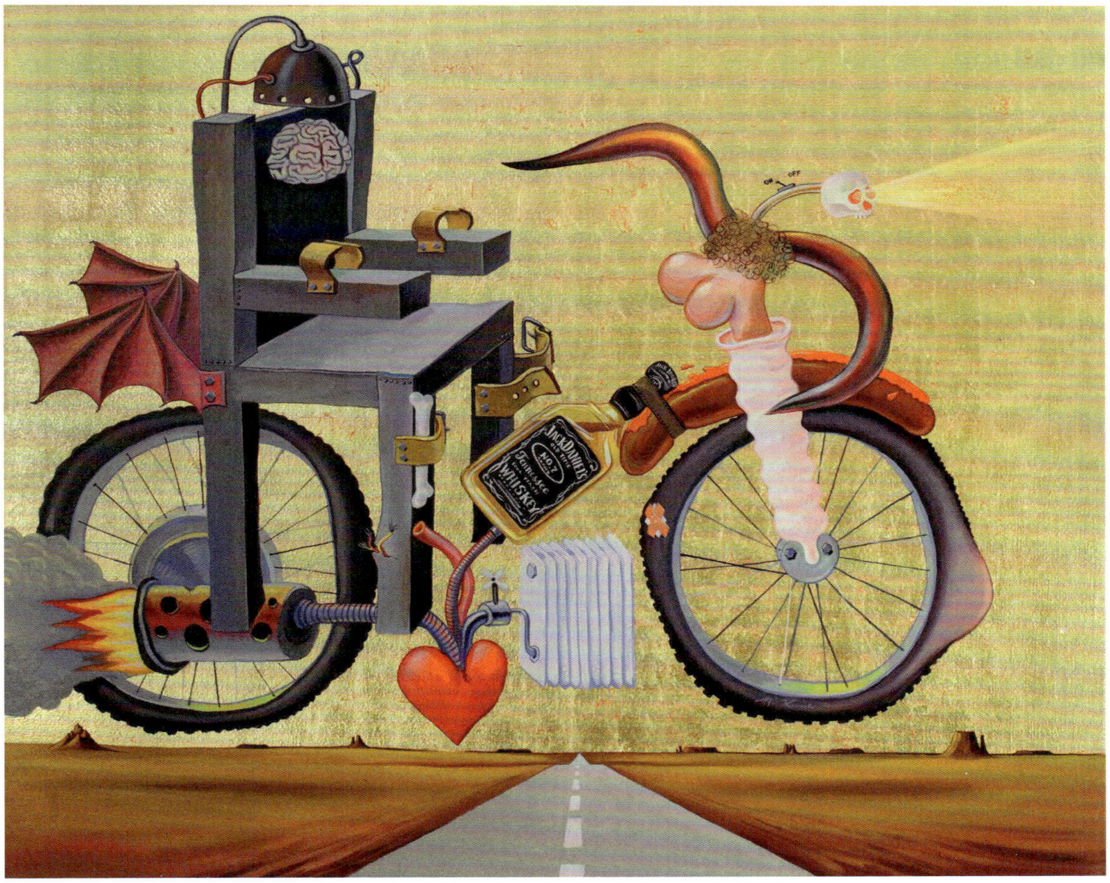

Milan Kunc
Cross the Badland, 1994
Oil and gold leaf on canvas, 190 x 220 cm
Milan Kunc

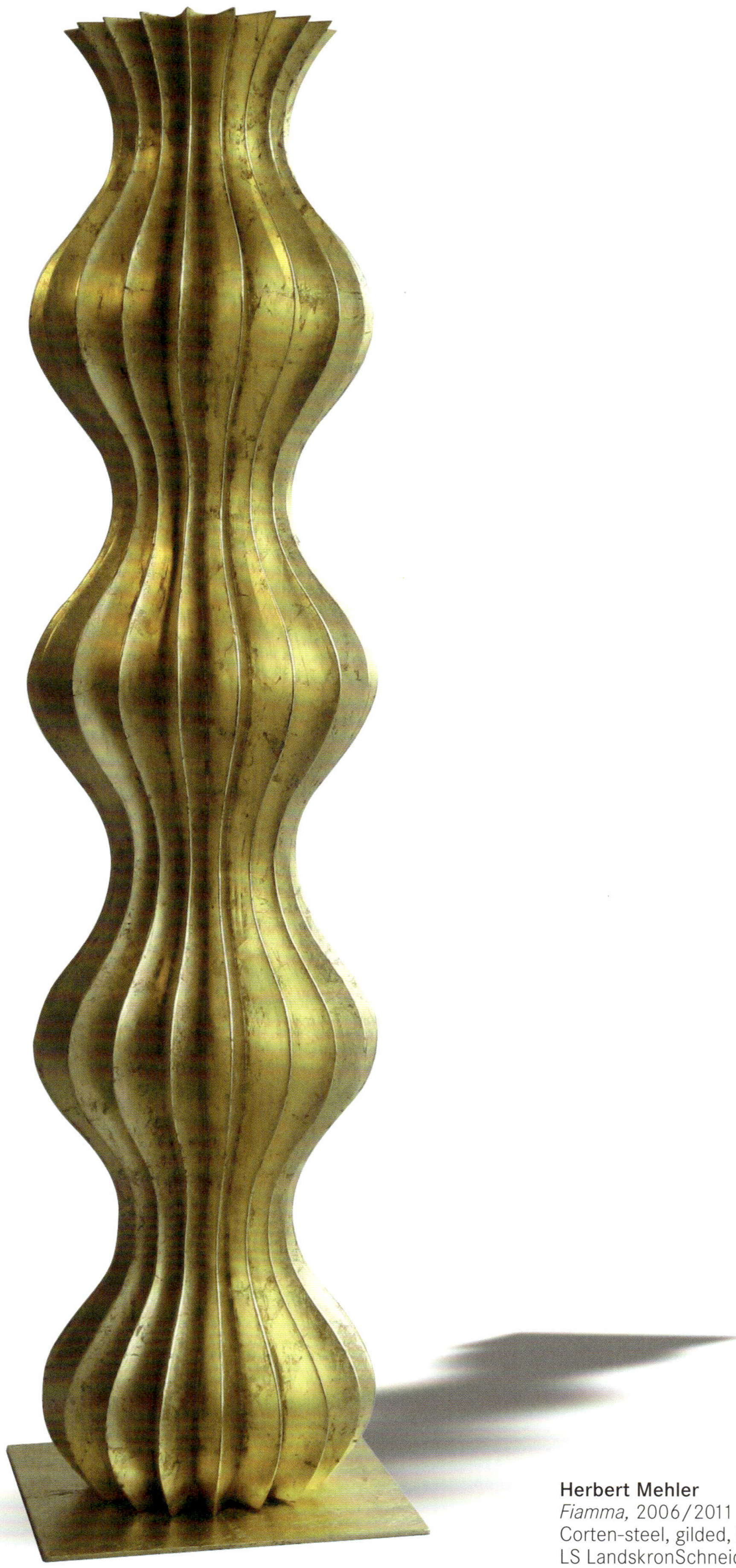

Herbert Mehler
Fiamma, 2006/2011
Corten-steel, gilded, height 225 cm
LS LandskronSchneidzik Galerie und Kunstagentur

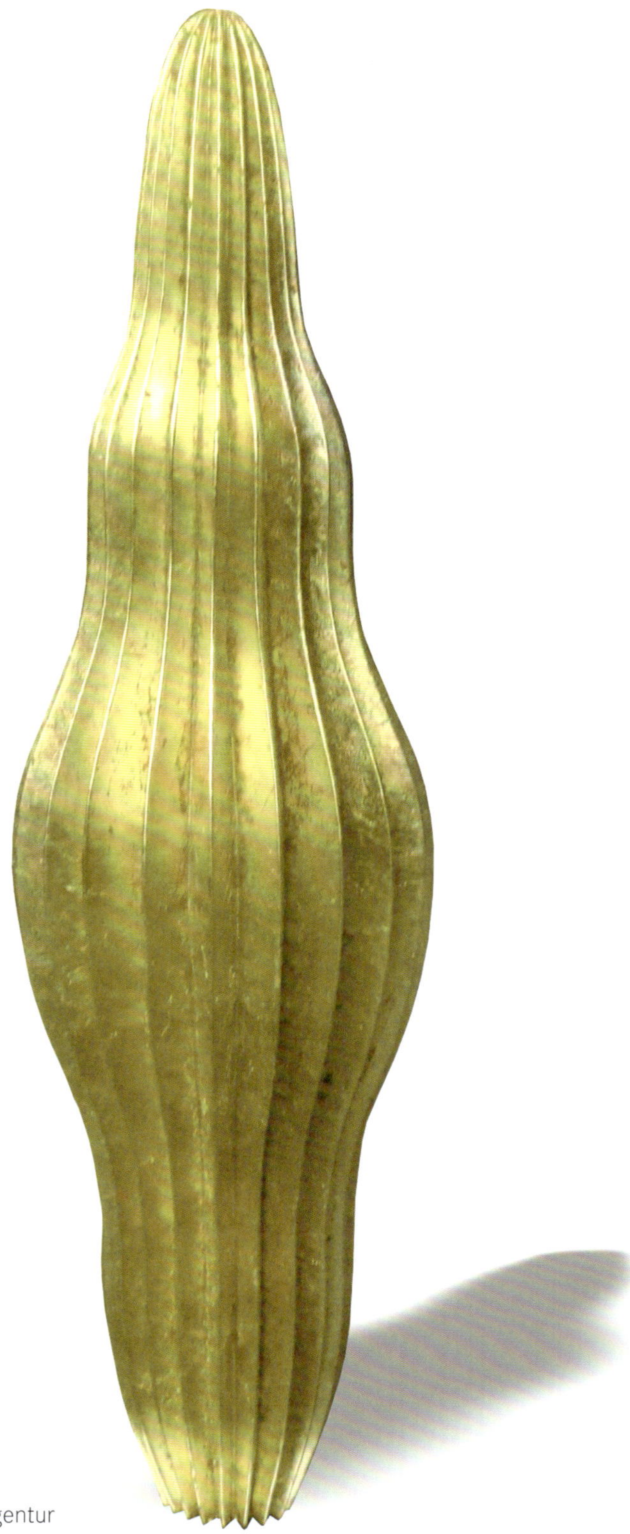

Herbert Mehler
Stella della Zucca, 2008
Corten-steel, gilded, 150 x 43 x 43 cm
LS LandskronSchneidzik Galerie und Kunstagentur

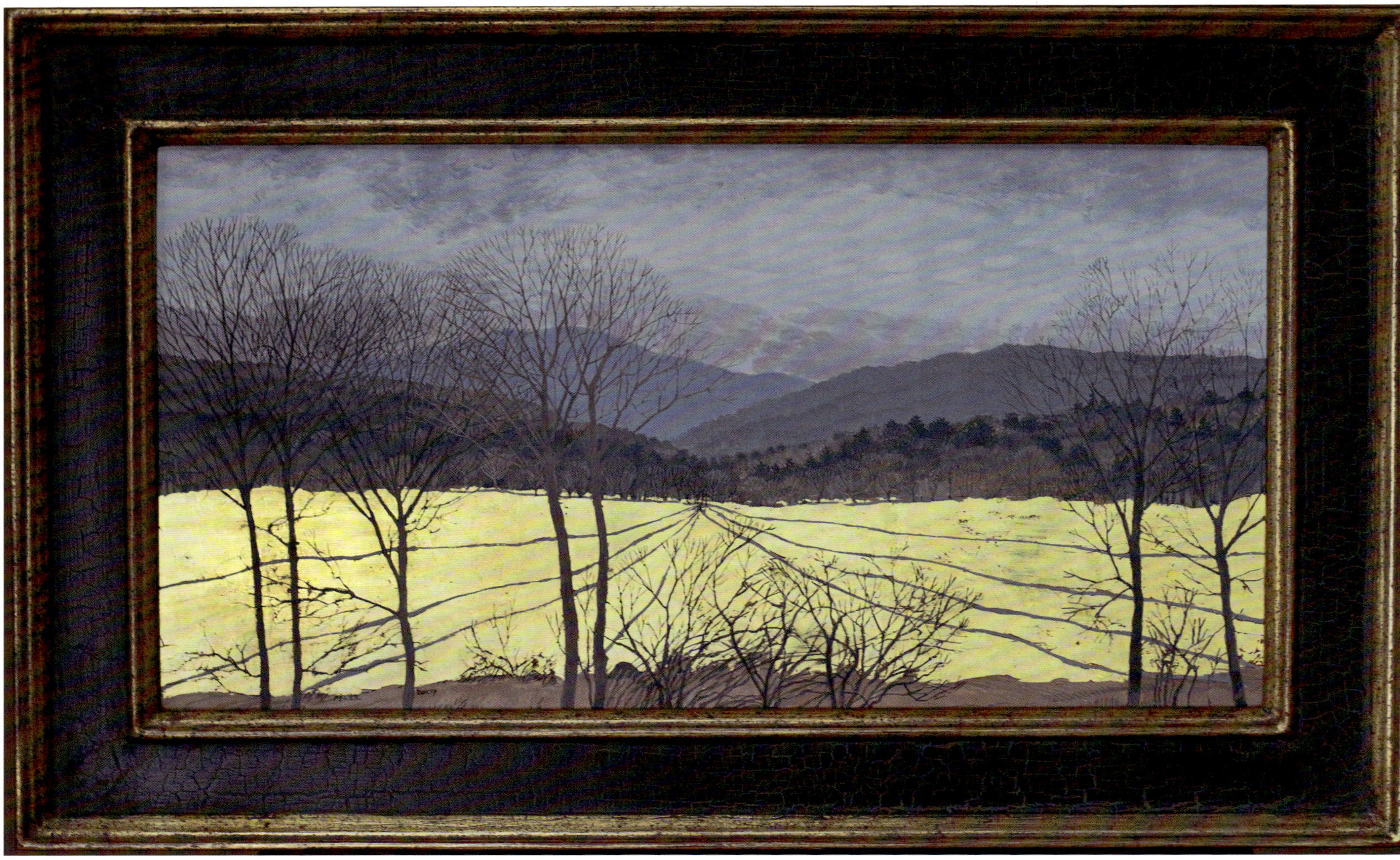

Gary Milek
November Study #2, undated
Egg tempera and gold leaf on wood, 33 x 53.3 cm
Painting on loan from Janet and Stephen Flanders of Norwich, Vermont

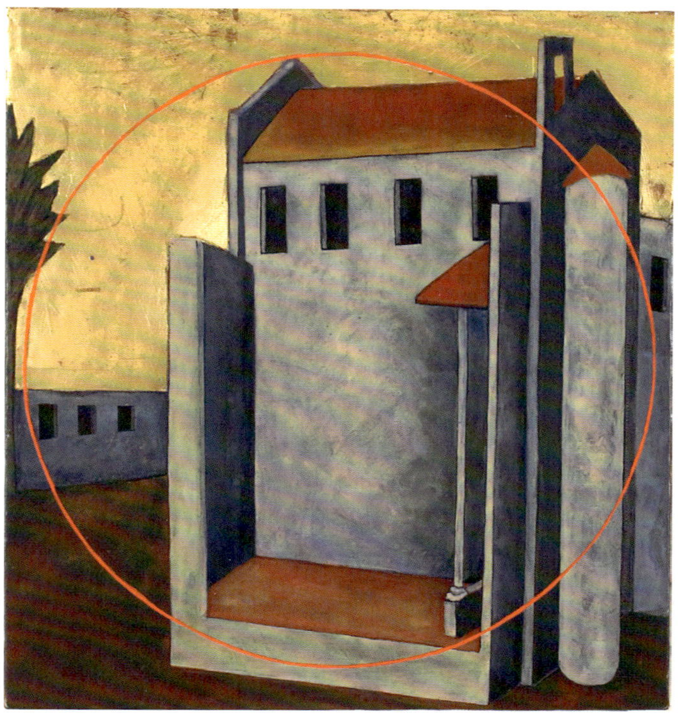

Patricia Miranda
Siena Landscape Series #3, 2005
Egg tempera and 22-carat gold leaf on panel, 30.5 x 24.4 cm
Patricia Miranda

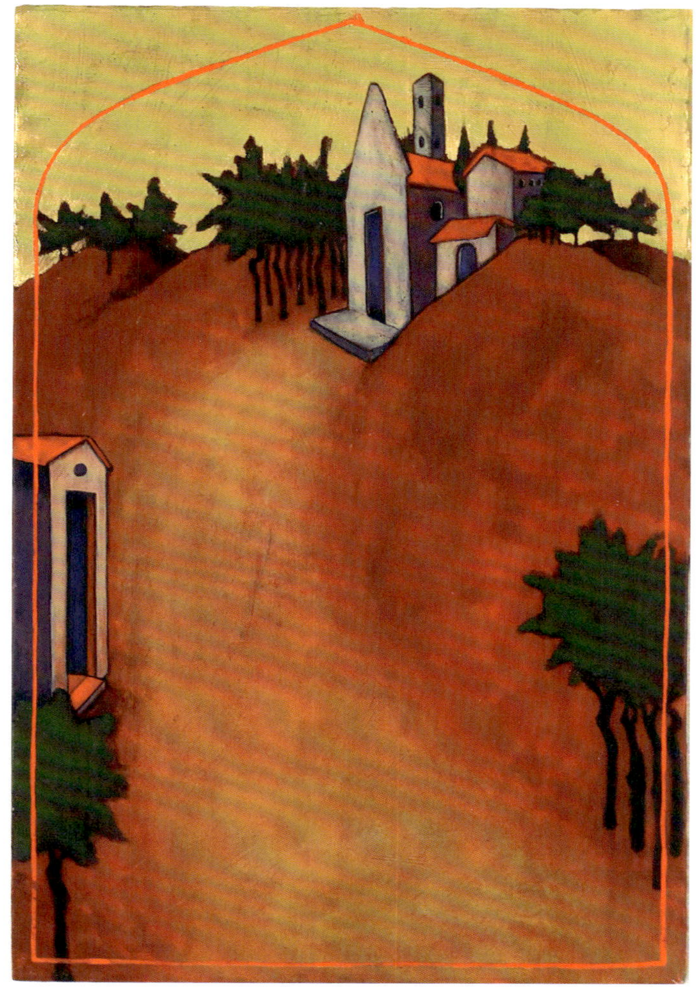

Patricia Miranda
Siena Landscape Series #1, 2005
Egg tempera and 22-carat gold leaf on panel, 22.9 x 15.2 cm
Patricia Miranda

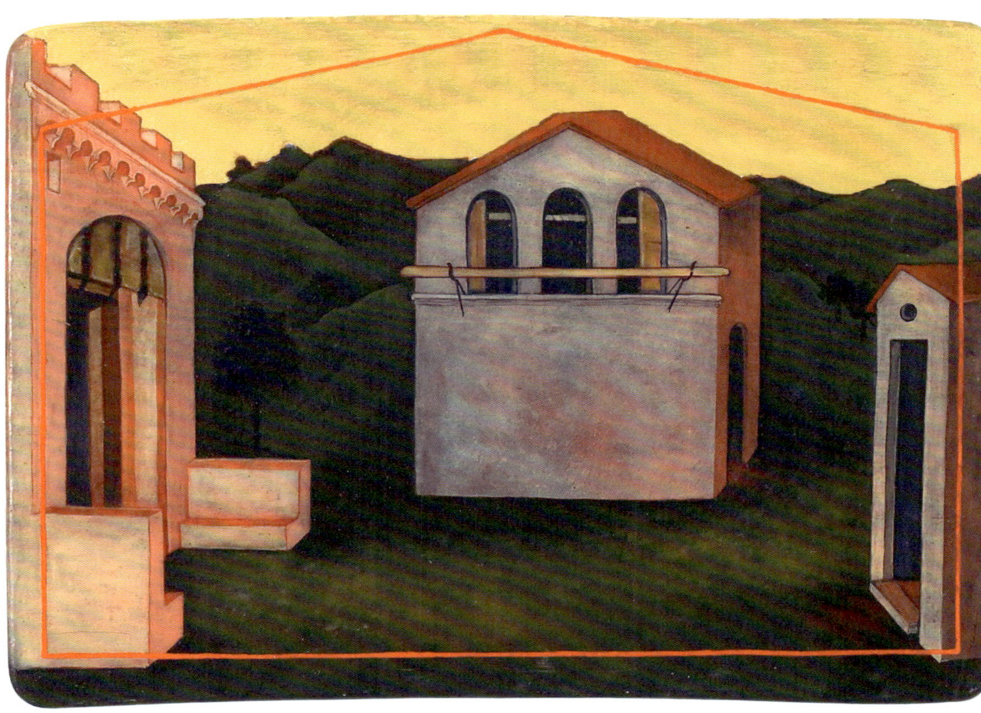

Patricia Miranda
Siena Landscape Series #4, 2005
Egg tempera and 22-carat gold leaf on panel, 15.2 x 22.9 cm
Patricia Miranda

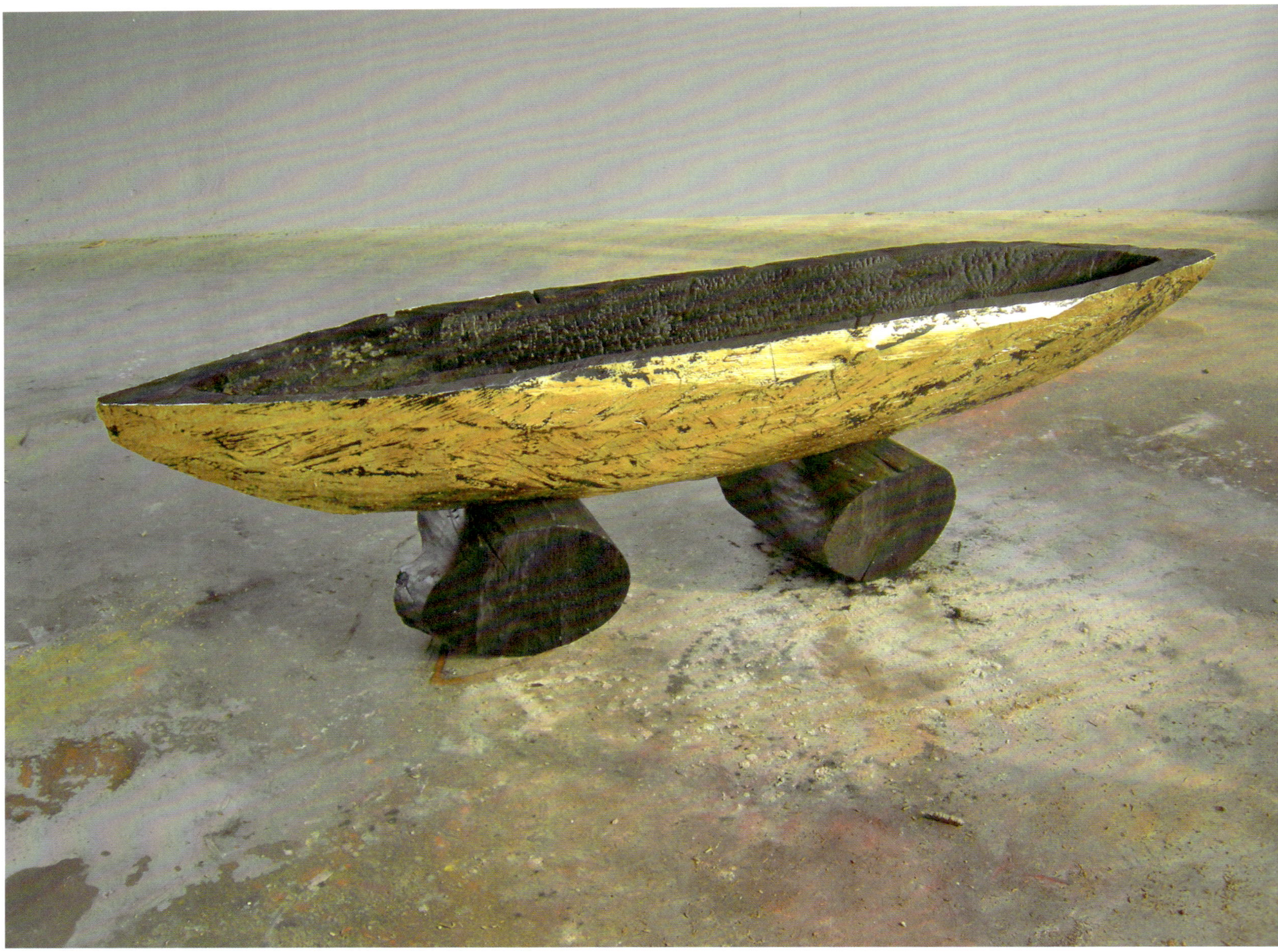

Karl Manfred Rennertz
Goldenes Boot, 2002
Tulip tree, gold leaf, blackened in fire, Length 2.70 m
Property of the artist

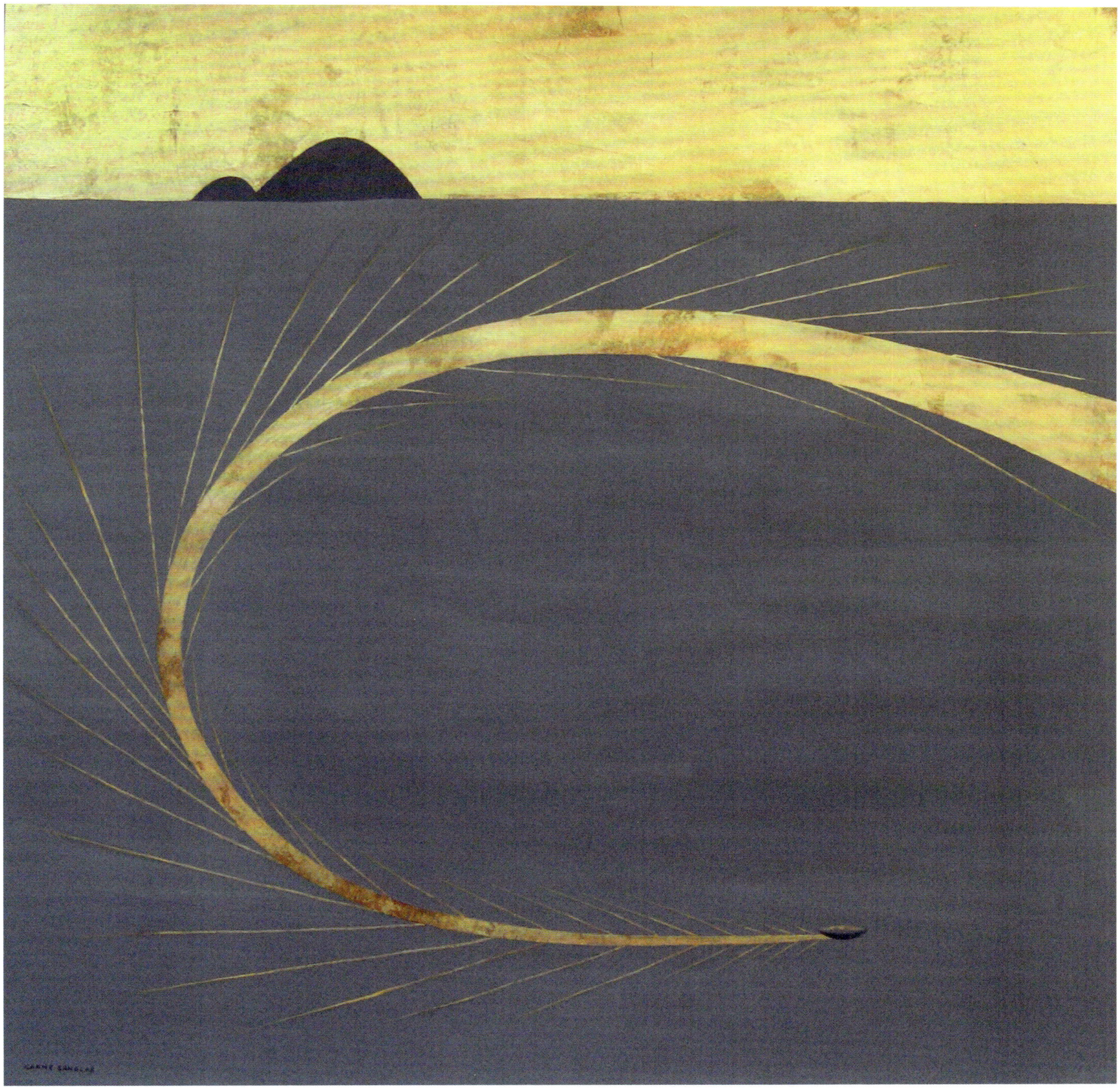

Carme Sanglas
Solstice, 2008
Mixed media and gold leaf on paper, 56 x 56 cm
Carme Sanglas

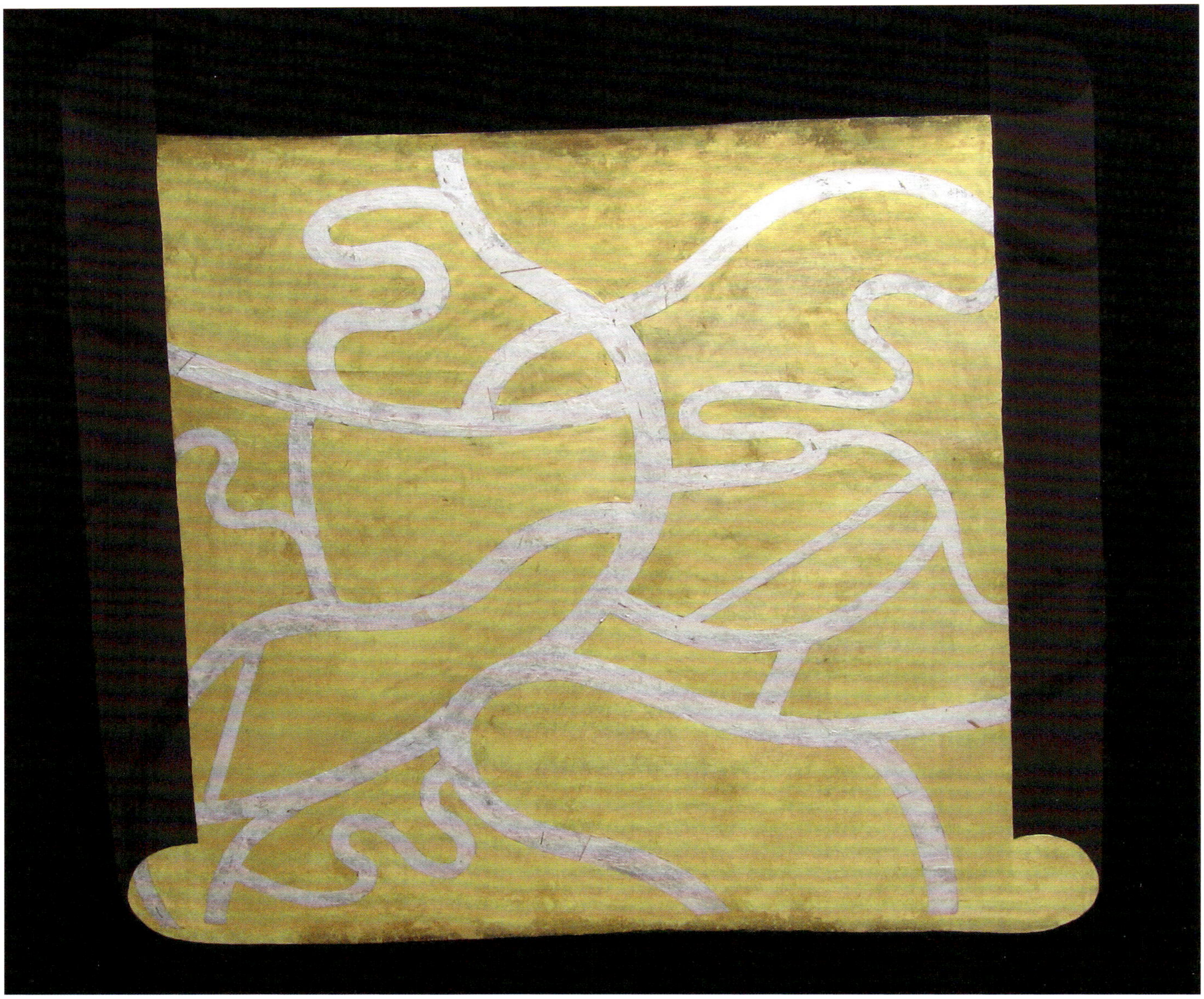

Carme Sanglas
Paths and Roads
(Camins i carreres), 2003
Mixed media, silver leaf and gold leaf on paper, 60 x 71 cm
Carme Sanglas

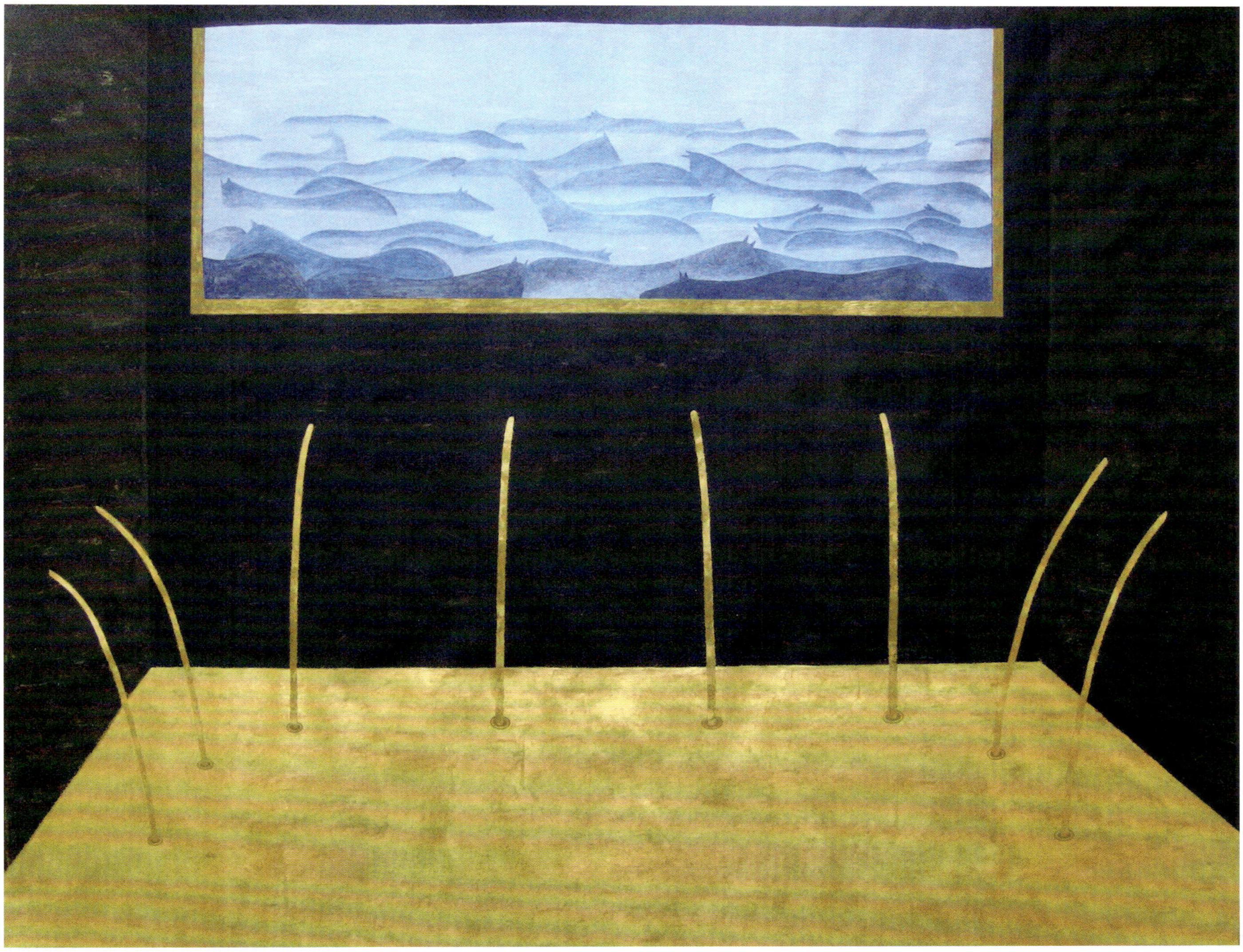

Carme Sanglas
Factory (Fàbrica), 2005
Mixed media and gold leaf on paper, 177 x 150 cm
Carme Sanglas

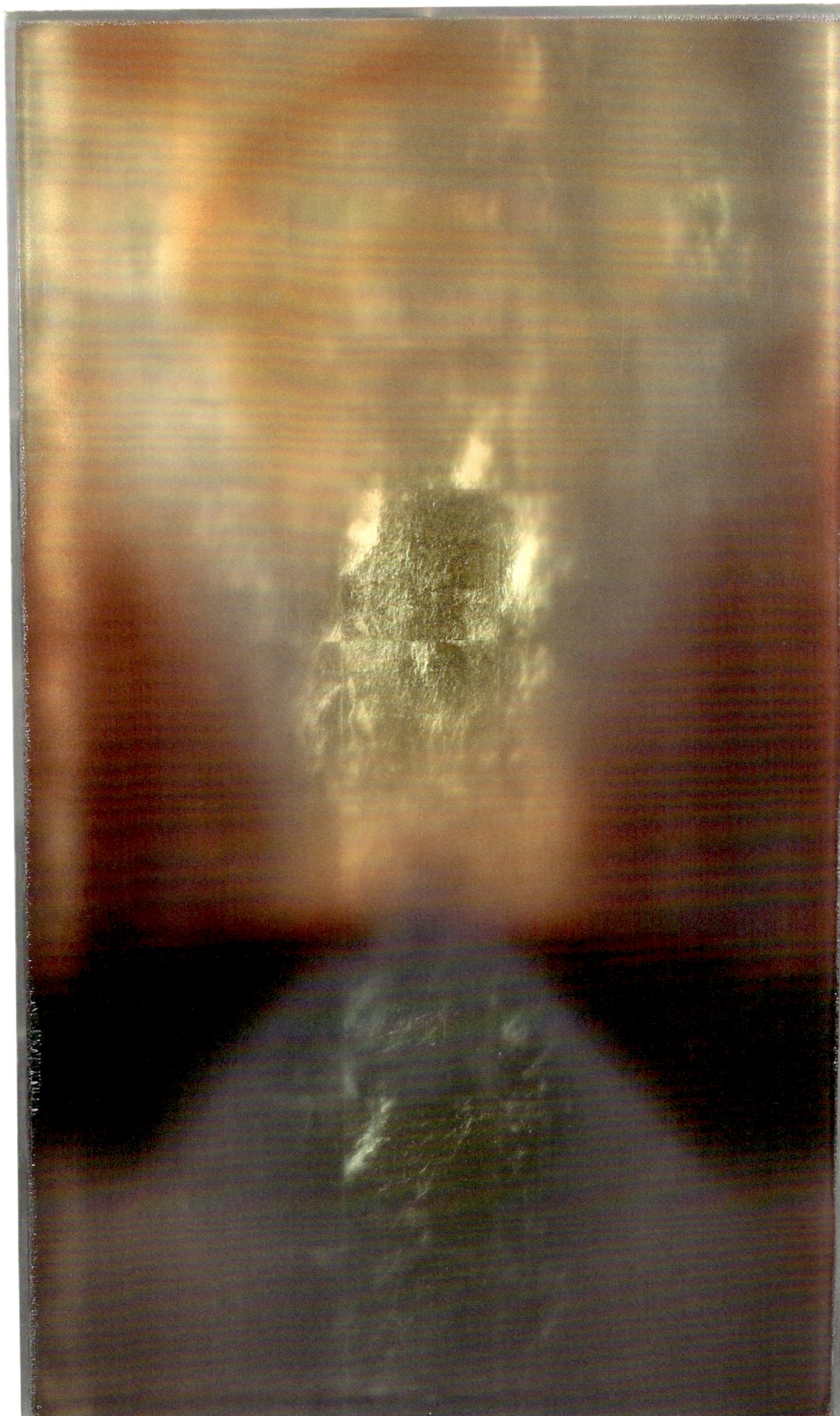

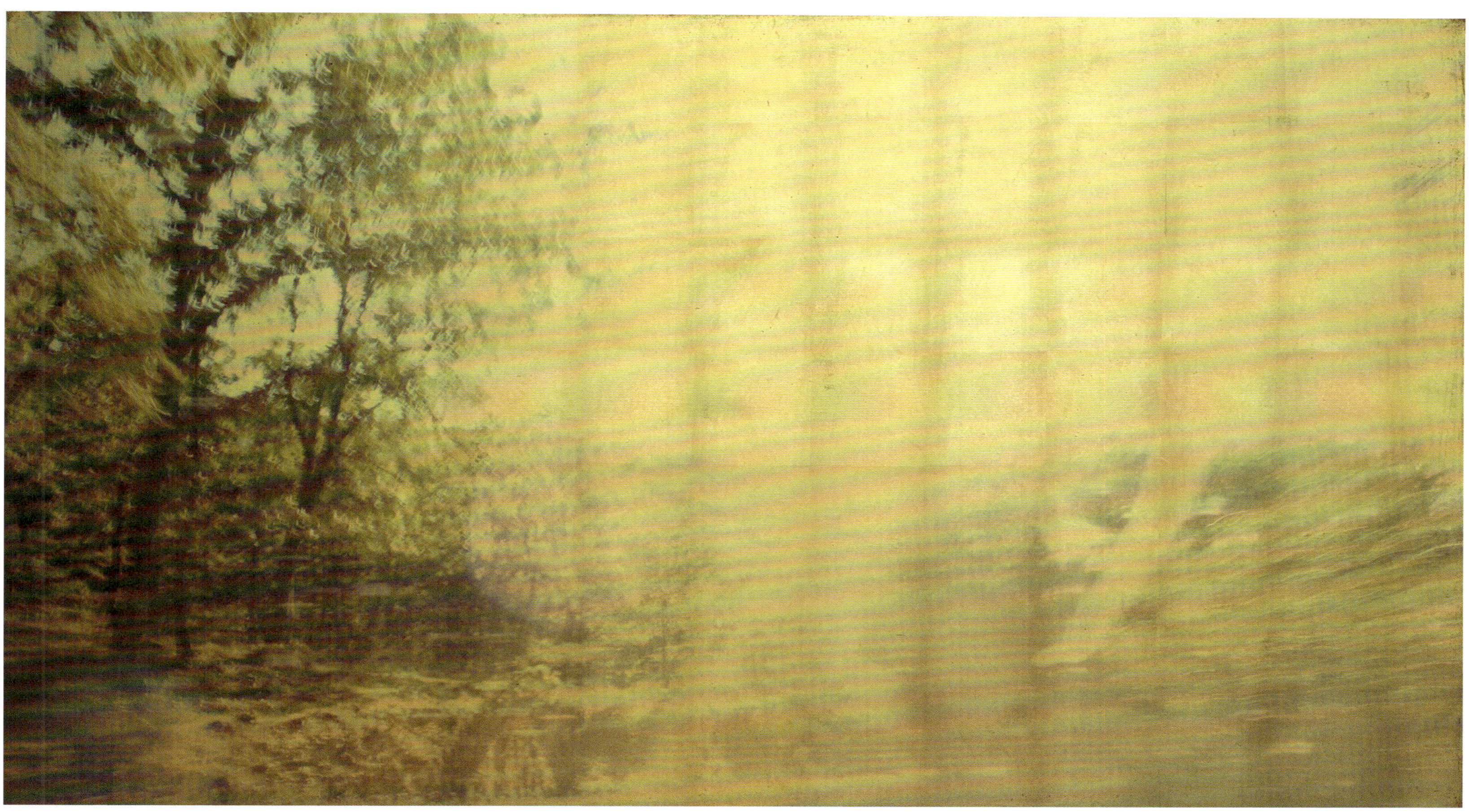

Hermann Staudinger
Gargano, 2010
Ed. 10, Digital print on 23-carat gold on wood, 56 x 100 cm
Courtesy the artist

Hermann Staudinger
Fossanova, 2011
Ed. 10, Digital print on acrylic glass, mounted on 23-carat gilded wood, 72 x 41 cm
Courtesy the artist

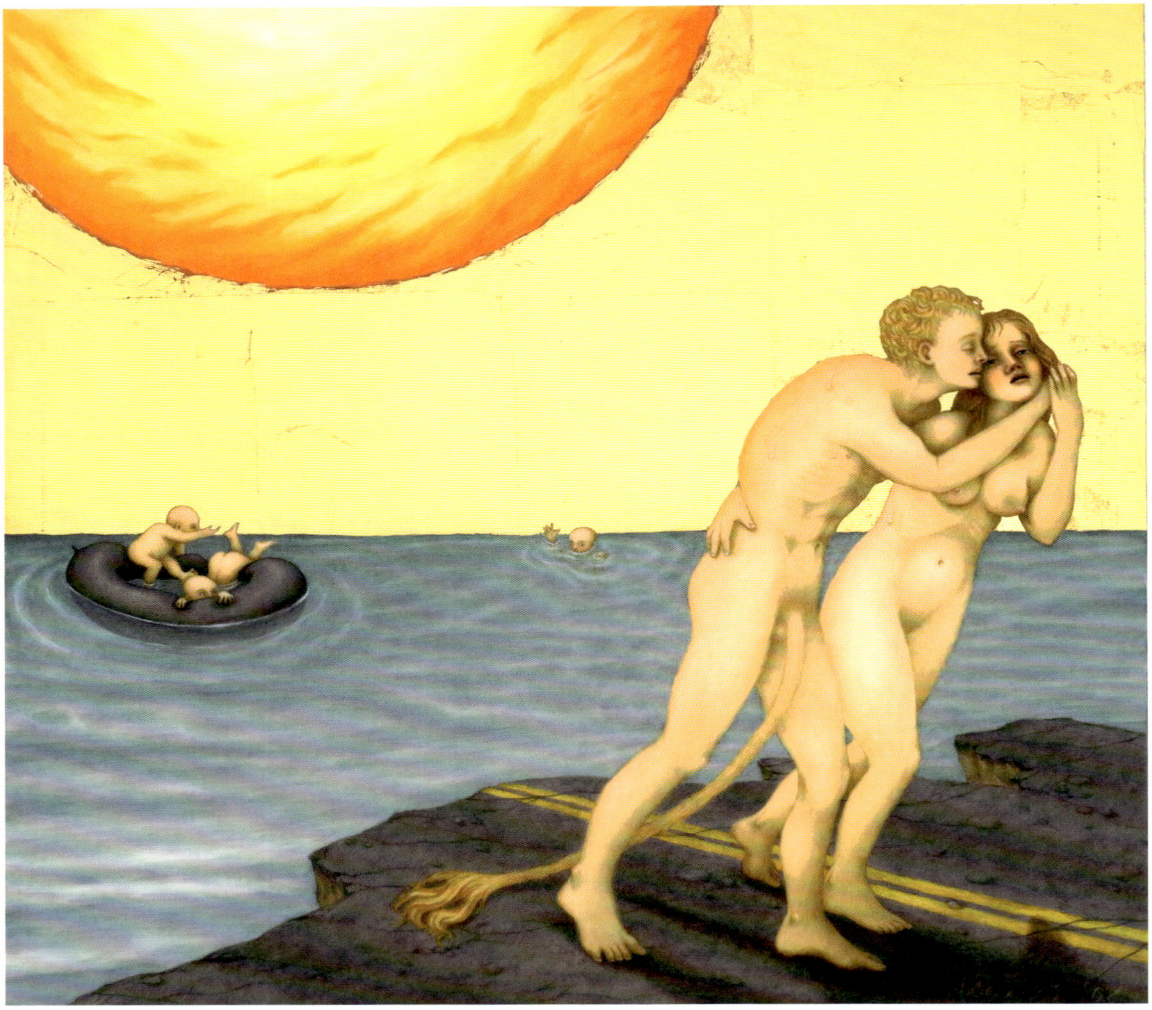

Tabitha Vevers
Eden (Expulsion), 2008
Oil and gold leaf on Ivorine,
35.6 x 38.1 cm
Private collection

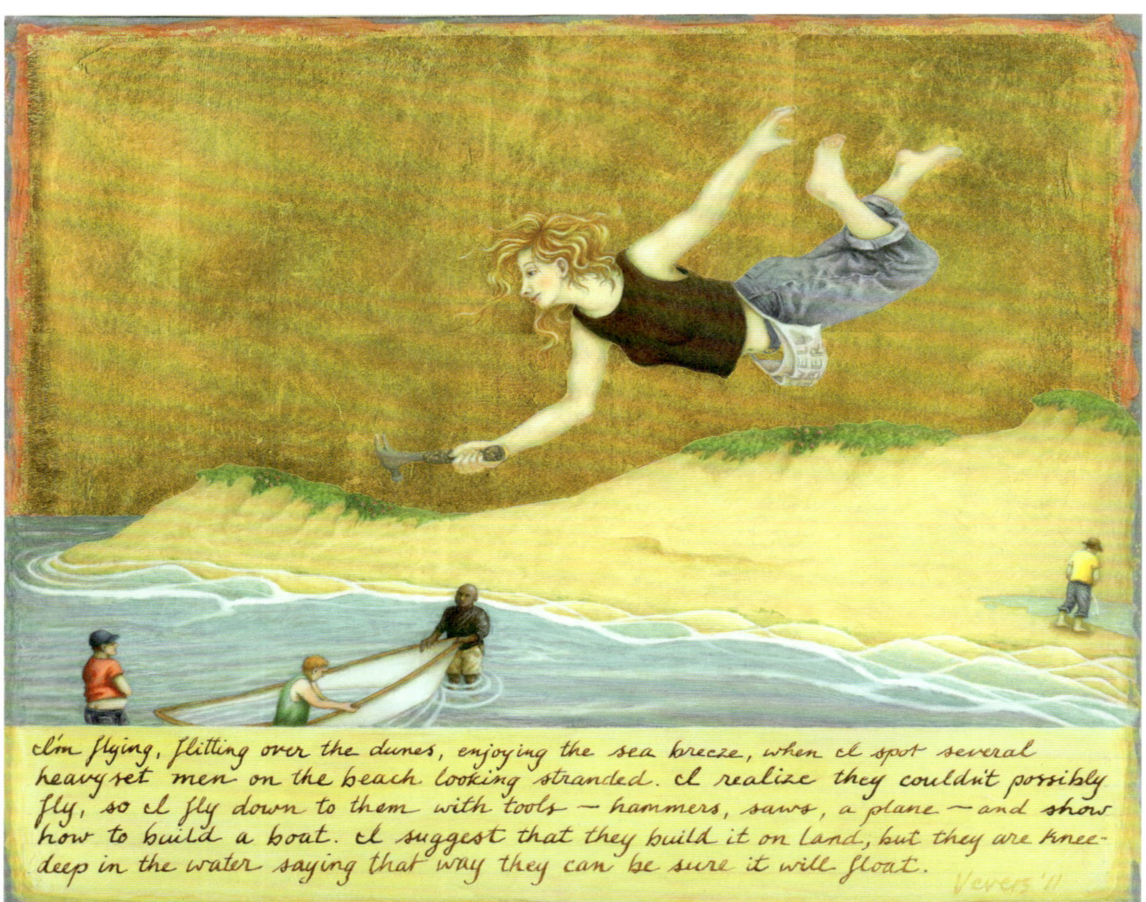

Tabitha Vevers
Flying Dream (In the Dunes II), 2011
Oil and gold leaf on galvanized
steel, 25.4 x 30.5 cm
Private collection

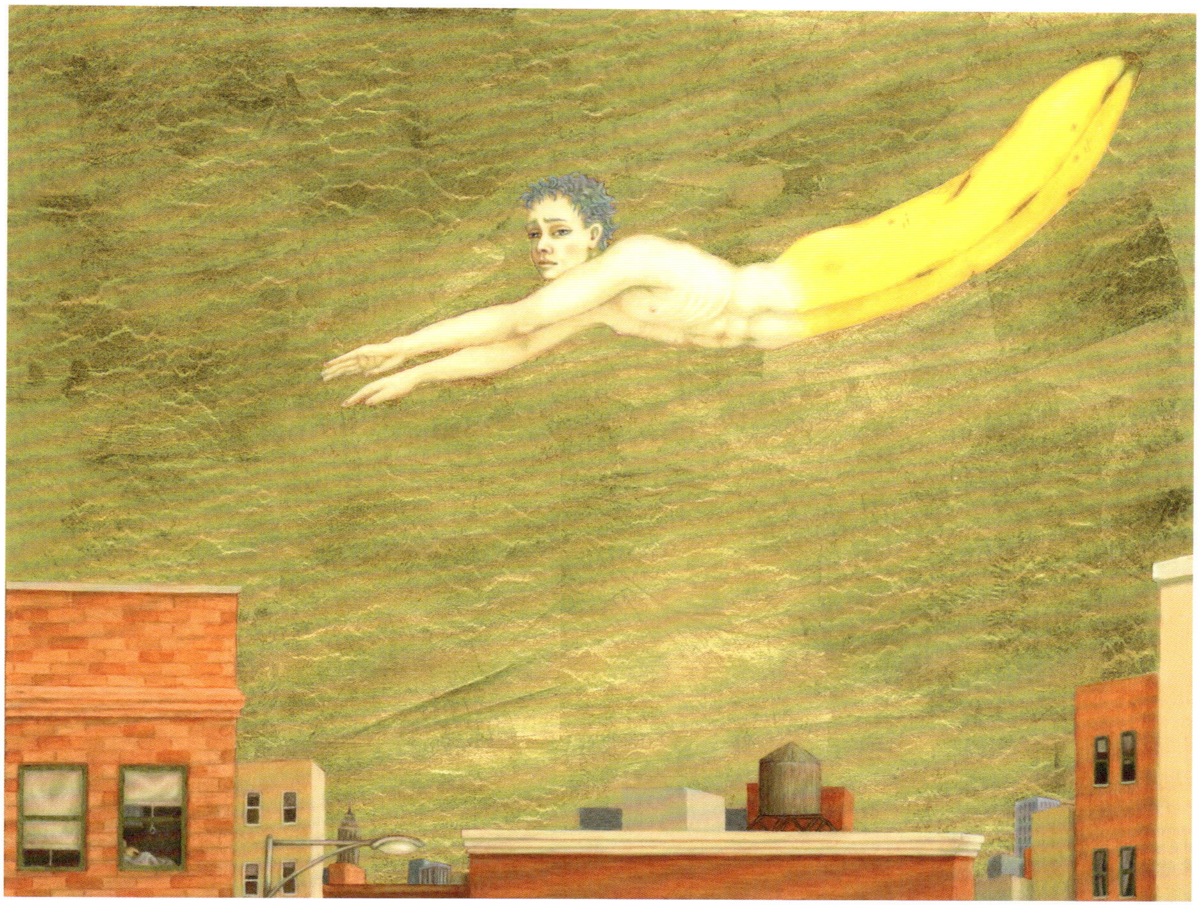

Tabitha Vevers
Bananaman, 2010
Oil and gold on Mylar foil,
27.9 x 36.2 cm
Tabitha Vevers

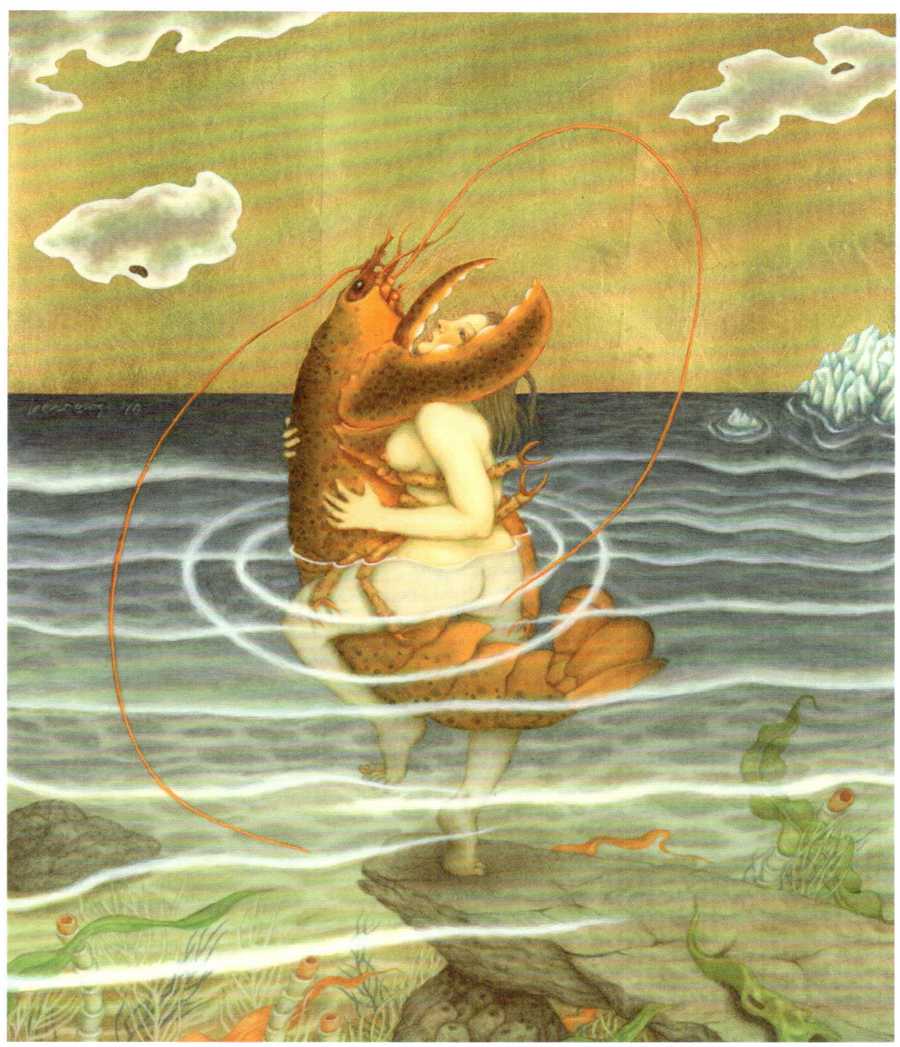

Tabitha Vevers
Shiva (While We Dance), 2010
Oil and gold on Mylar foil, 24.1 x 29.2 cm
Kim Manocherian

Seen in the city

If you raise your eyes when you walk around Vienna's streets and squares, you will see not only vast legions of sculptures, dancing allegorically on balustrades and cornices, but also a multitude of gilded emblems. Bodo Hell's essay sketches the crowns and eagles, coats of arms and swords, which form a network of references to kingship and domination.

On display in the corridor are examples from this association-rich field: the three designs for the parliament by Eduard Lebiedzki show the allegorical apparatus that represented the parliamentary ideals and functions, and also paid homage to 'Austria' both in the interior of the building along a 120-metre frieze (1907–1911) and on the exterior (1902) on a gold ground. Often categorized as revivalist, this work is more appropriately assigned to the very varied Viennese Jugendstil.

The fire-gilded business-sign *Zum Auge Gottes* (late eighteenth century) from the Alsergrund district is a reference to civic history and everyday culture.

The use of gold by present-day artists oscillates from *Ehrenzeichen* ('mark of distinction') by Cornelius Kolig (1980s), which wanders along all the byways of the imagination, via the abstract solar emblem as *Wandobjekt* ('wall object', 2009) by Gerwald Rockenschaub, to the well-balanced *Kreuz* ('cross', 1996) by Gerhard Richter.

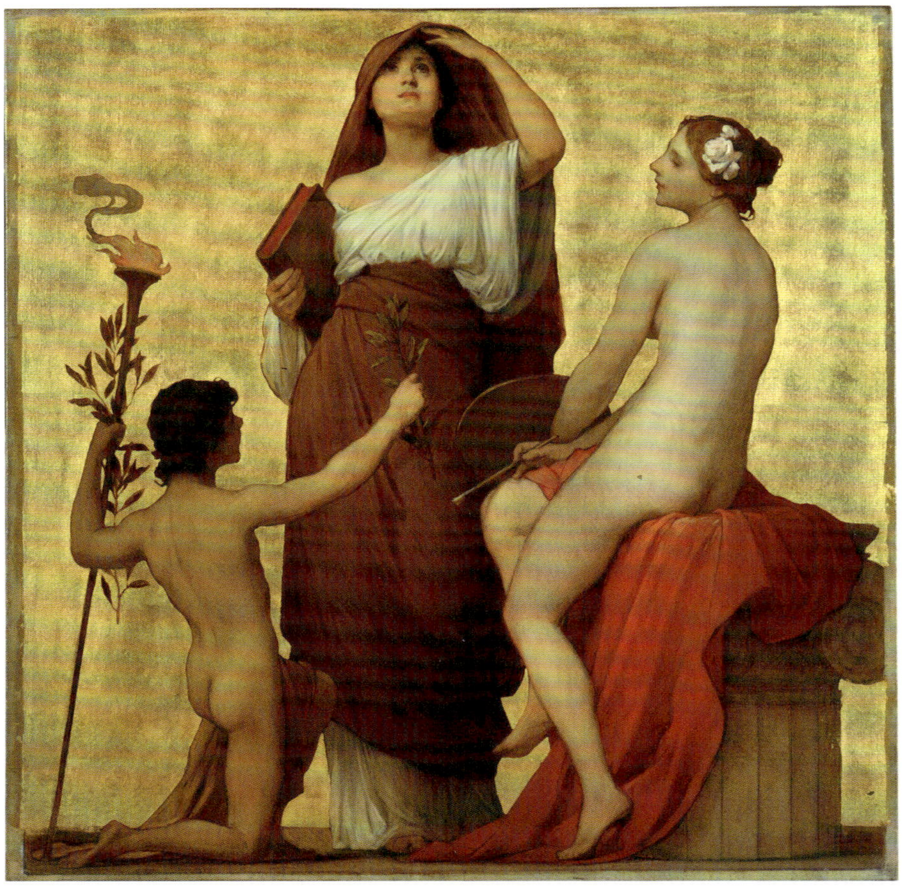

Eduart Lebiedzki
Design for the mosaic frieze in the portico of the Reichsrat building in Vienna, 1900
Oil on canvas, 79 x 77 cm
Wien Museum

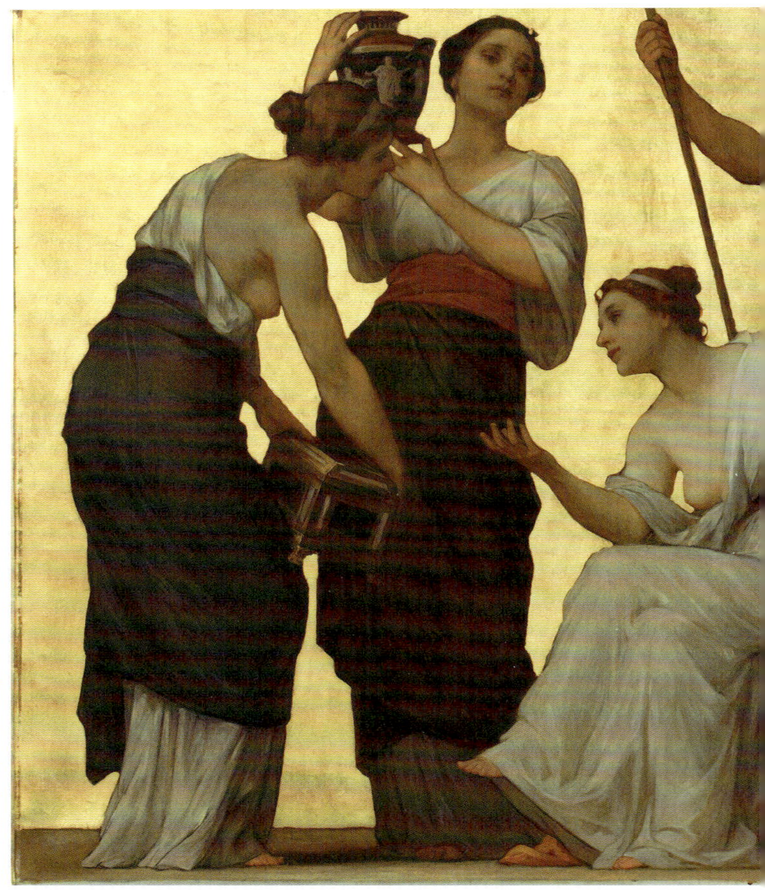

Eduart Lebiedzki
Design for the mosaic frieze in the portico of the Reichsrat building in Vienna, 1900
Oil on canvas, 73 x 136 cm
Wien Museum

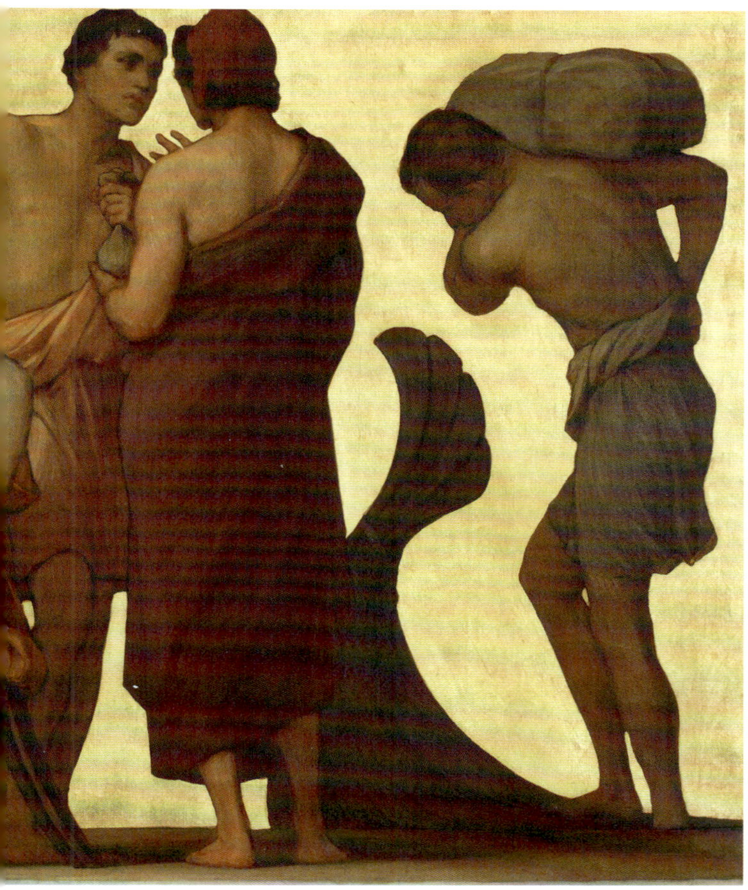

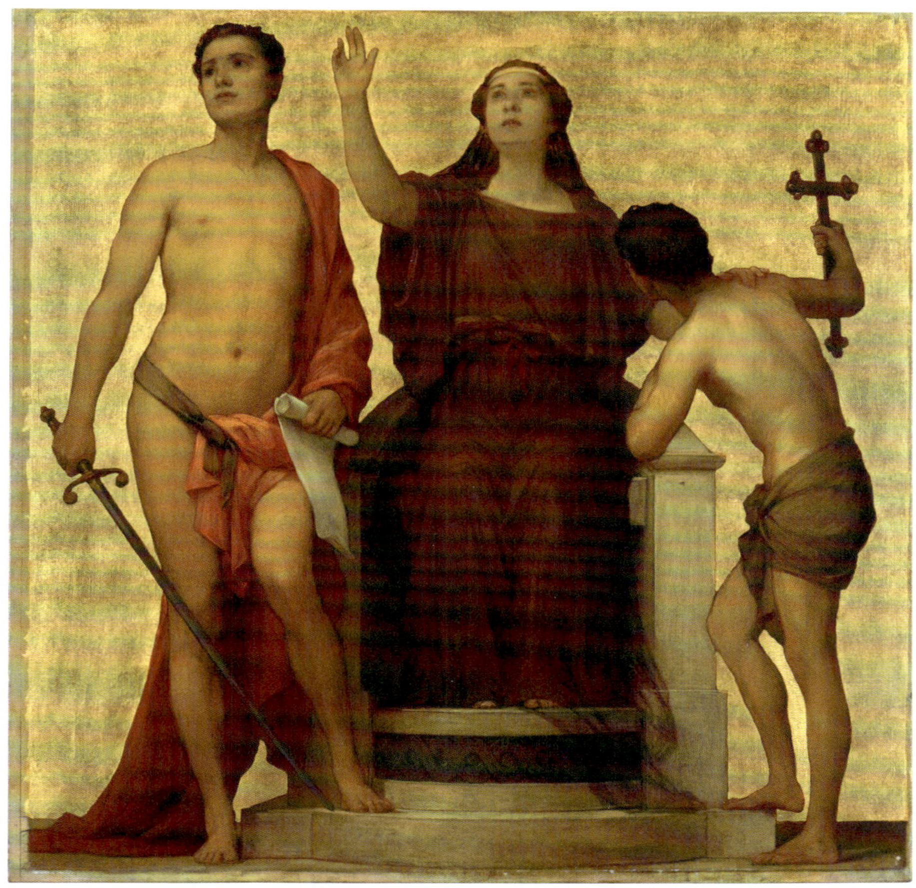

Eduart Lebiedzki
Design for the mosaic frieze in the portico of the Reichsrat building in Vienna, 1900
Oil on canvas, 79 x 77 cm
Wien Museum

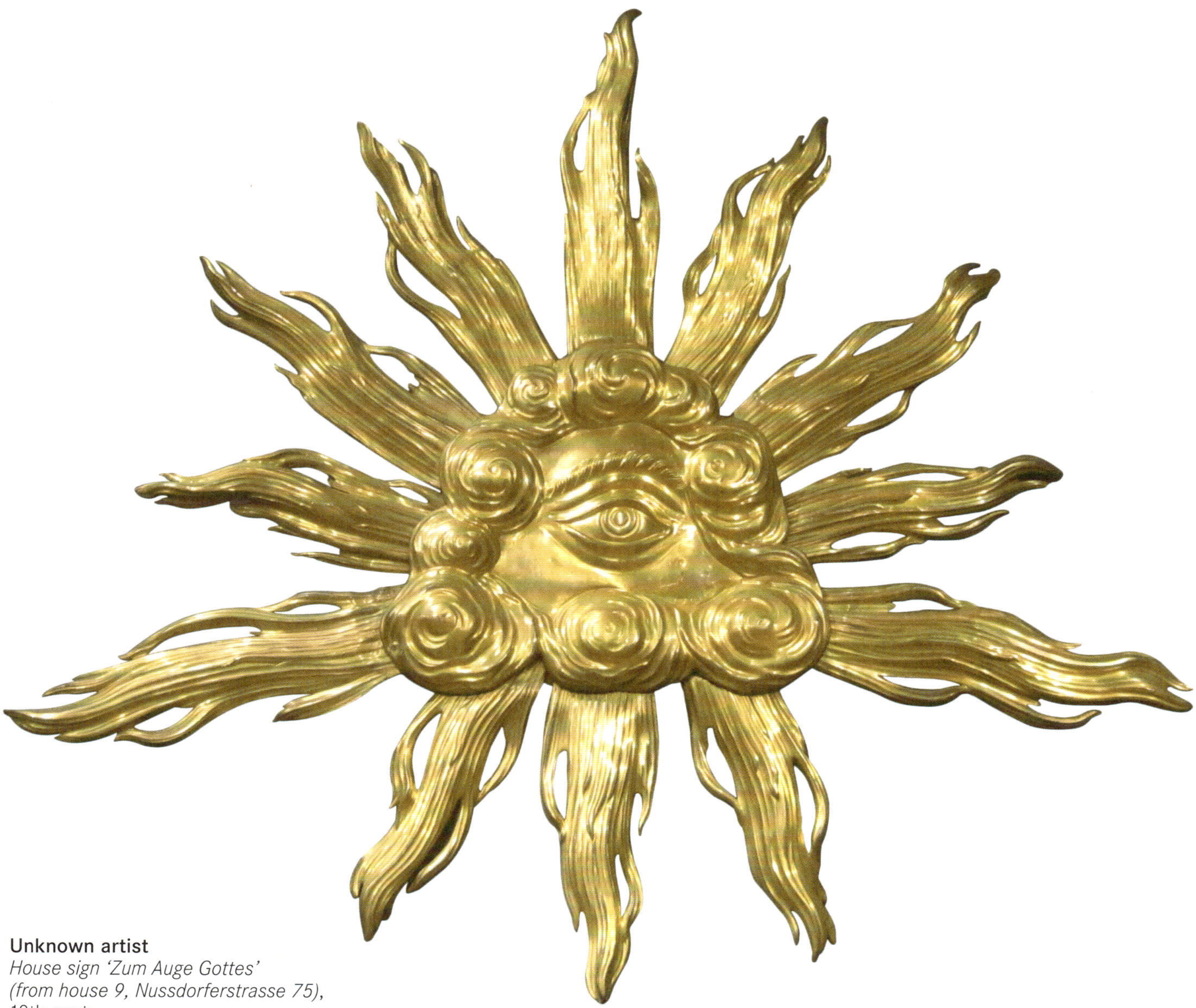

Unknown artist
House sign 'Zum Auge Gottes'
(from house 9, Nussdorferstrasse 75),
18th cent.
Fire-gilded copper
Wien Museum, Vienna

Gerhard Richter
Kreuz, 1996
Gold, 19.5 x 19.5 x 1.5 cm
Winterthur, Kunstmuseum Winterthur

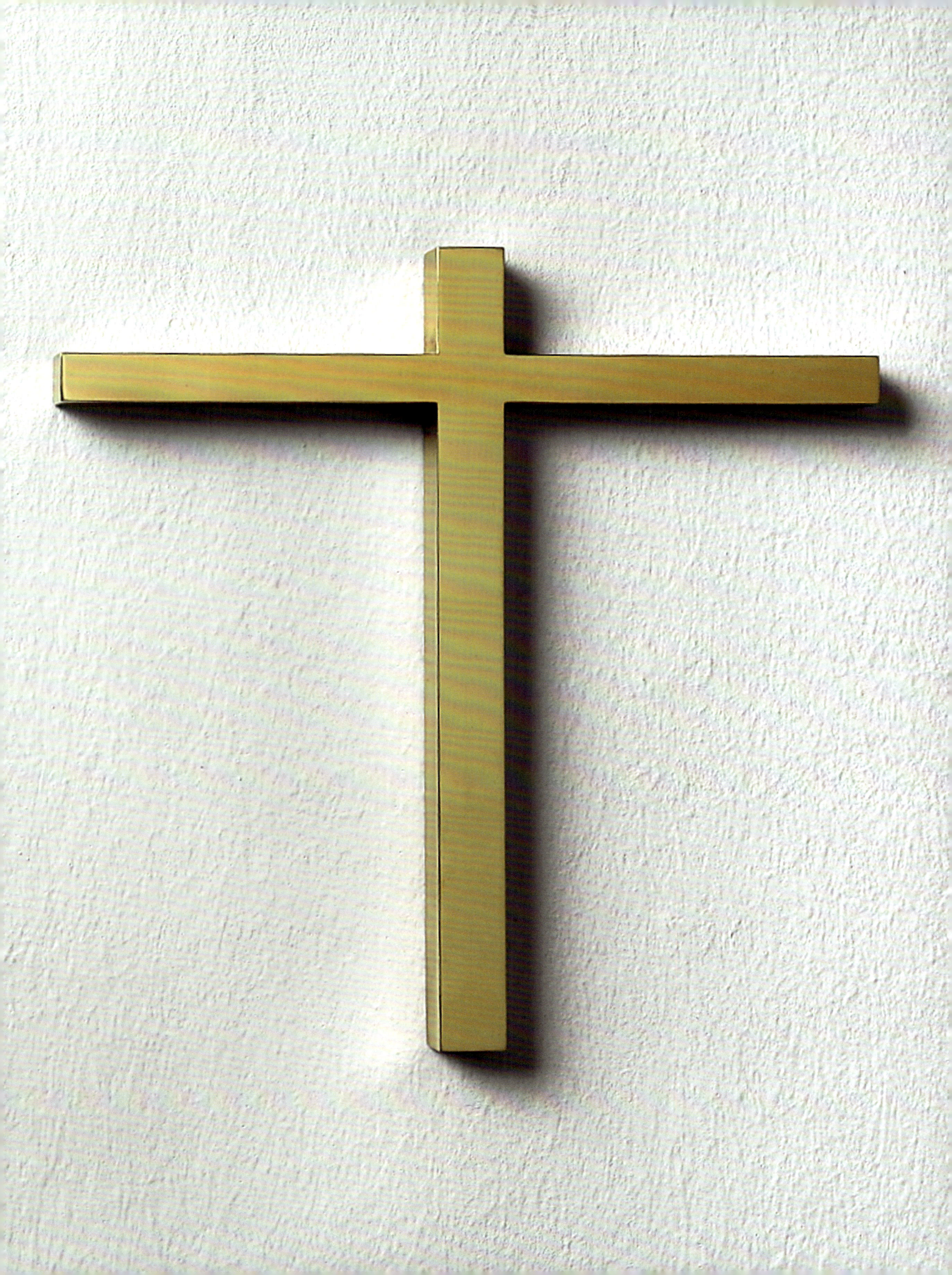

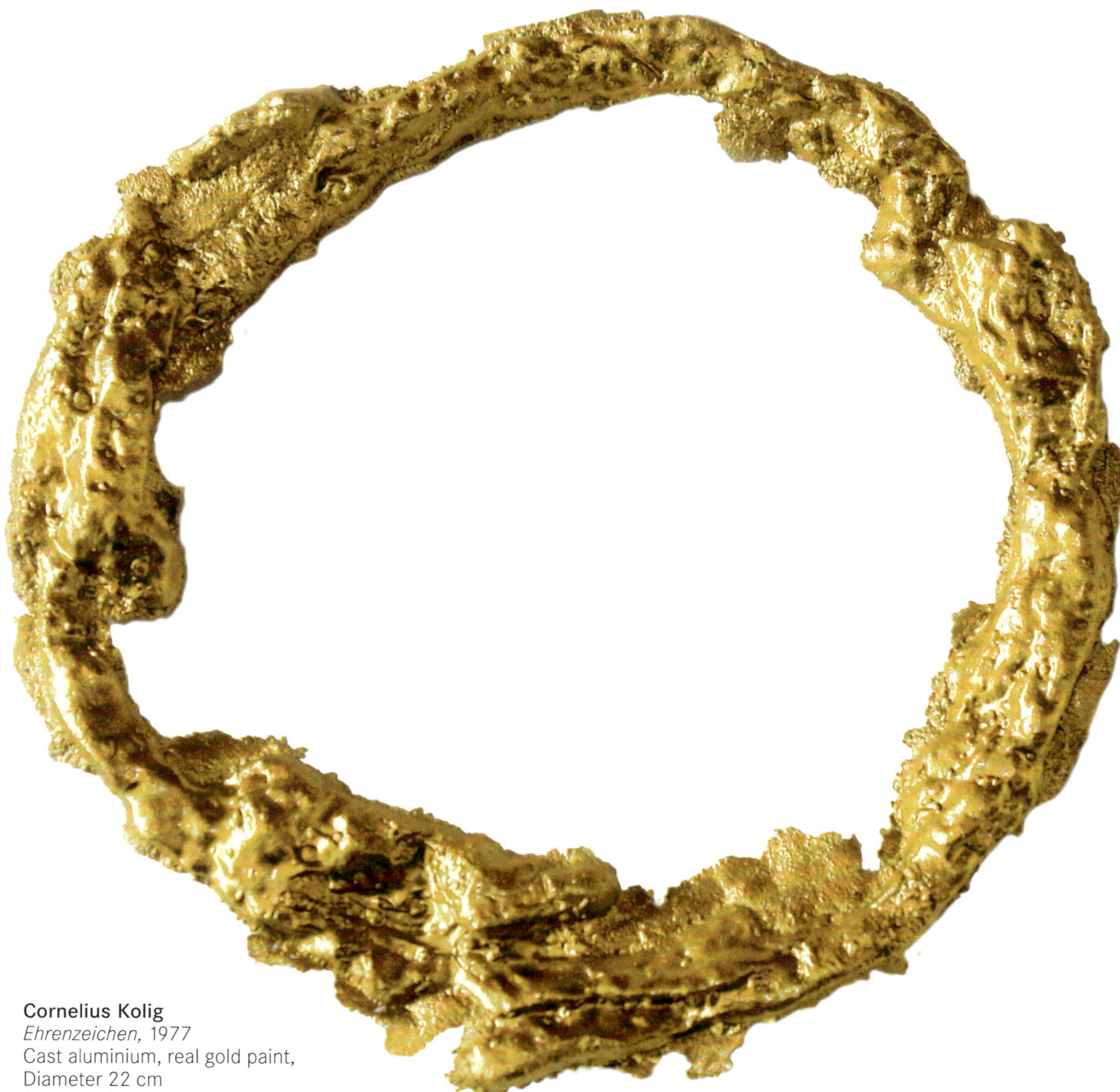

Cornelius Kolig
Ehrenzeichen, 1977
Cast aluminium, real gold paint,
Diameter 22 cm
Cornelius Kolig

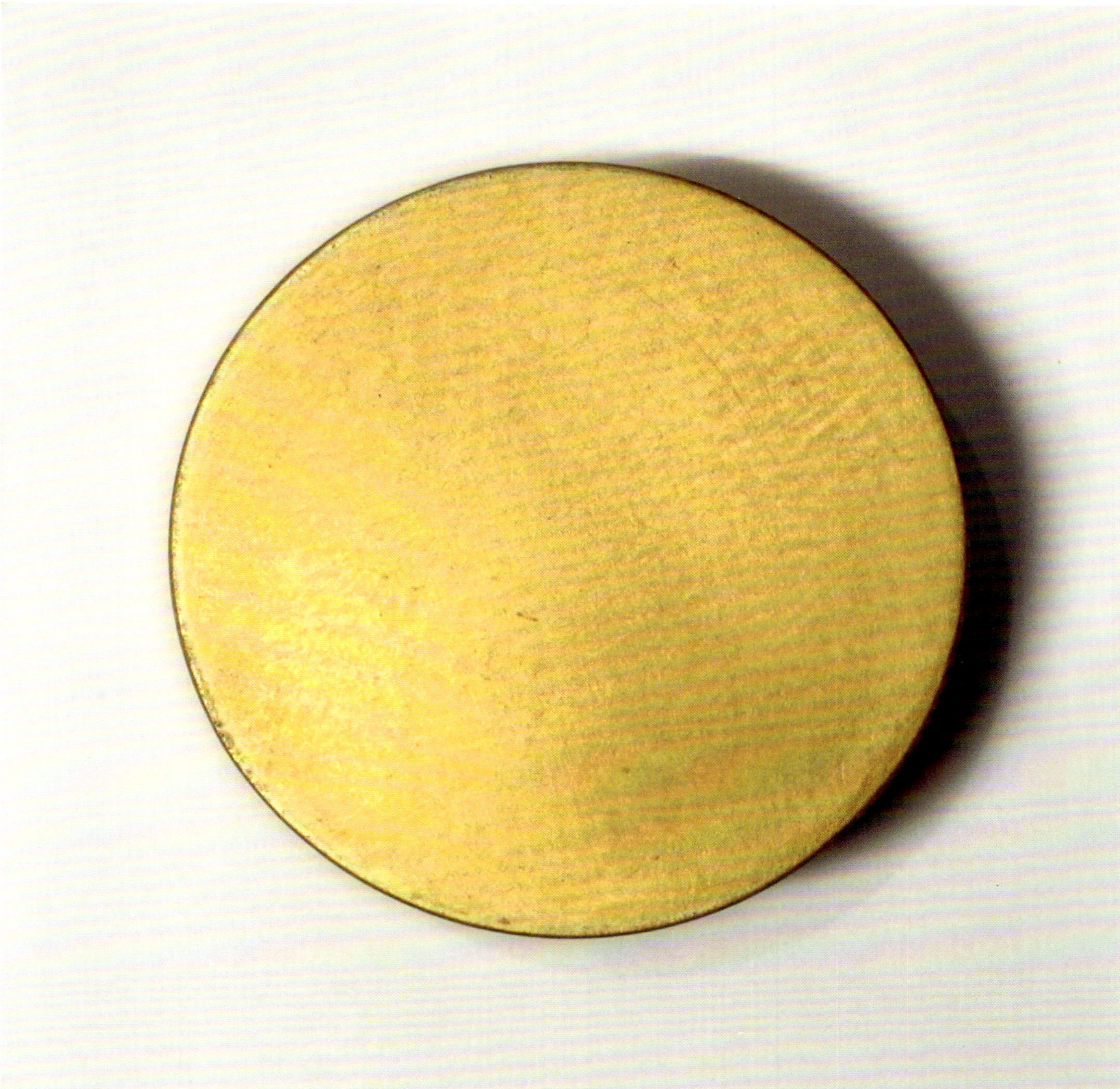

Gerwald Rockenschaub
Painting, 2009
MDF, gold leaf, 10 x 10 x 3 cm
Courtesy Georg Kargl Fine Arts, Vienna and Gallery Mehdi Chouakri, Berlin

Holiness: Triptychs and Haloes

Excursus: triptychs

The triptych is the best-known special case of the multi-part altarpiece, which can range from the two-part diptych to the many-winged polyptych. Art-historians like to call the triptych a 'pathos formula'. The term goes back to Aby Warburg, who used it for the formulaic representation of gestures and facial expressions, and it was applied to the triptych by Klaus Lankheit.[1] Since then the aura-tic quality of the format has been emphasized, and the fact "that the triptych assigns a very specific role to beholders. It leaves them few opportunities for men-tal confrontation, it is a decidedly authoritarian picture format, which, in its complexity, also exerts power, assigning to the beholder the subordinate position."[2]

Recently the attestation of the auratic effect has been supplemented by the formal analysis of the corporeality and the semanticization of the triptych.[3] It has rightly been subject to the criticism that in triptychs what was seen was not bodies, but mostly nothing but flat surfaces. Crucial for the format, though, is precisely the possibility of moving it, closing it, and opening it again. "A pictorial art form which tells a story, but which through its opening and shutting effects also gains new dimensions of action."[4] With the complexity of the different views, the interpretation can gain momentum. As the triptych often determined the form of medieval altars, artists have used it right down to the present day as an association-rich context for heavily meaningful paintings. As with the gold ground, a holy element is projected on to this multi-part picture in memory of Gothic altarpieces, mostly with an emphasized, somewhat larger central panel. Artists are also inspired by the possible gradation of the themes. Not only the middle section as the centre, but also the fronts and reverses of the movable wings sometimes allow a structured narrative.

The historical connexion between gold ground and triptych lies however not in any thematic analogy, but in the fact that they both disappeared in the post-Renaissance period. "So the triptych after a heyday from the fifteenth to the early seventeenth century vanished almost entirely as a picture type. Even in the early nineteenth century, in other words at a time when as a genuine genre of religious painting it would have fitted very well into the zeitgeist [...] there can be no talk of a revival."[5] Even the eventual renaissance of the triptych in the late nineteenth century has its parallel in the concurrent re-introduction of the gold ground and the corresponding 'art-religious fantasies'.

As a co-founder of the Secession, Rudolf Bacher was also its president (1904/05 and 1912-1914), in addition to being a professor at, and for many years rector of, the Akademie der bildenden Künste in Vienna. Today he is almost forgotten. There is no catalogue of his works, let alone a monograph. The large-format altar triptych attributed to him is dedicated to St Severin (c. 410-482), the apostle of Noricum, and to Klemens Maria Hofbauer (1751-1820), the patron saint of the city of Vienna. Their portraits are on the outsides of the wings. The work was begun after Hofbauer's canonization in 1909. Six panels accompany the central Crucifixion with the Virgin and St John. On the left are the Annunciation and the Adoration of the Magi and the Shepherds, and below is the Last Supper. A conspicuous feature is the black halo for Judas, iconographically very rare. On the right can be seen the Resurrection, Pentecost and the Coronation of the Virgin.

On the occasion of its purchase by the Belvedere, there appeared a report in the 1987 *Mitteilungen der Österreichischen Galerie*[7] listing the iconographic details. The triptych was not commissioned by the church. The great patron of the Secession, Karl Wittgenstein, bought the work. Afterwards, the report says, it was used as 'one of the rare examples of a domestic altar in private use'. This work has not been on public display for 100 years. The first and last occasion was in Vienna in 1912, at the *Exhibition of Church Art*, on whose 'honorary committee' Rudolf Bacher is named as rector of the imperial academy of visual arts. The catalogue lists those pupils of Bacher who executed the altarpiece: Karl Borschke, Andreas Harsch, Ferdinand Kitt, Karl May, Leopold Zlabinger. The gilding was carried out by Anton Schellerer.[8]

The conservative format and the gold ground should not deceive us into thinking that artists are unable to develop innovative depictions. Kurt Regschek, in his *Triptychon* (1960) leaves the subsidiary figures of Mary and John the Baptist on the gold grounds of his side panels. The Trinity is suggested by the motif of the Throne of Mercy. Christ with the Crown of Thorns appears however over and in front of the bearded God the Father, who, with the plumage, also embodies the Holy Spirit. The Viennese fantasist is here following no particular faith idea, but allows a variety of interpretations. However, he does "give the Holy Spirit priority insofar as he has God the Father rest on [the Spirit's] wings, while the Father in His turn brings His son Jesus Christ into the world by means of a mystical mind-game."[9]

In *Die Lösung des Welträtsels ist die Hinwendung zu unserer höchsten Möglichkeit* ('the solution of the world riddle is to turn to our highest possibility', 2002) Rudolf Leitner-Gründberg stresses the importance of the number three by the three crowns in the golden box. Even though the triple crown suggests the papal tiara and the golden chair in front of it the Holy See (*Sancta Sedes*: 'holy chair') (or of the empty throne waiting for the Second Coming of Jesus at the Last Judgement), the artist is concerned with the personal development of the individual. "The crown is his trademark [...] The ambivalence of the much-used symbol between real power and fairy-tale exaggeration plays a conscious role too."[10]

Not every triptych is an altarpiece. Three pictures juxtaposed can also be called a triptych, although they may have no common ground. In the parish church in Vienna's Hetzendorf district there are three large-format pictures by Ernst Fuchs on three panels consisting of sewn-together goat-skins. In accordance with the dedication of the church to 'Mary Queen of the Rosary', it depicts the *Mysteries of the Rosary* (1958–1961). In 1979 the pictures were badly damaged by an 'iconoclast', and subsequently expensively restored, before being rehung in 1990. For the first time since the Sao Paulo Biennale (1970) the *Glorious Rosary* is being shown once more at an exhibition. On the left stands the Angel of the Annunciation, depicted in blue. In the front to the right of the angel, Mary, on the crescent moon, presents the crowned child in a golden aureole, giving the sign of blessing, in the temple. To the right of her there rears up the dragon of hell with the seven heads and seven crowns, ten horns and a tail which sweeps a third of the stars from the sky. It is being stabbed in the back by an angel with an amethyst column.

Not every triptych is in the Christian tradition. Yoko Grandsagne refers on the one hand to European art, on the other she acknowledges the Japanese Rimpa tradition.[11] The gold ground she uses has its roots not in Gothic pictures, but in the Japanese technique, unbroken since the fifteenth century, of using gold leaf. One recognizes calligraphic influences in her motifs, and then the next minute sees informal elements, and then again is reminded by the lilies of van Gogh, who himself was influenced by Japanese models (Kōrin Ogata, 1658–1716), so the horizon broadens out into an East-West nexus. In accordance with Buddhism, the central panel of her *Zen Garden* (2010) proves to be the place of emptiness (*shunyata*).

The triptych appears from the Japanese point of view to be more a chance reduction of a multipart folding screen. These screens (*byōbu* = wind-wall) were sought-after collector's items.[12] They can look back on an unbroken tradition of gold-leaf grounds. The spatial view of post-medieval European painting had no more effect here than it did in icons. In the Japonisme movement[13] of the late nineteenth century original English sound was precisely the two-dimensional decorative aspect that had a major influence.

The triptych, like gold, is, quite apart from any religious connotation, not least a convincing aesthetic means. "What attracts me most in using gold leaf in my paintings is to do away with the sacred aura of a material traditionally associated with or considered a symbol of wealth, and instead to have it express a purely esthetical value." (Wojciech Mucha)[14]

Excursus: Nimbus

The halo is also called a nimbus, which in Latin means a cloud. In the Christian tradition it was in the nimbus that the gold ground survived longest. The word, contrary to what its original meaning might suggest, has nothing to do with any darkening, but refers to a luminous manifestation. The golden nimbus returned in the nineteenth century as a reference to tradition. Among the Pre-Raphaelites, for example, it also distinguished the gods of Antiquity. Jiří Dokoupil retains the gold from the early-medieval original of his *Madonna* (1987), but the halo is cracked.

By now other people have assumed the roles of the saints. Venerated are those stars who are present in the new media and who enthuse their fans to the point of ecstasy. In the portraits of these fêted personalities, the golden nimbus as the sign of 'the chosen one' has returned. When gold is revered as a material for its lasting value, then a rock star can be portrayed as gold bars (Jessica Diamond: *Elvis Alive*, 1990). The stage becomes an ersatz altar. Because it is left to each individual to deify and worship the new heroes, charisma is democratized. Any person can become the role model for others.

Animals are only distinguished by haloes when they are themselves symbols, such as the Lamb for Christ or the (two-headed) eagle for the Holy Roman Empire.

The painter of animals and icons Mary-Frances Dondelinger comes up with a surprise with her *Canary in the Coal Mine* (n. d.). She transposes the birds, which were used in coal-mines until well into the twentieth century, into a trecento mountain landscape against a gold ground, in order metaphorically to recall their function as life-savers. Canaries warned miners in good time of deadly danger: if they ceased to sing, they had fallen victim to poisonous gases such as methane or carbon monoxide.

With his haloes, Godfried Donkor transfigures boxers, goddesses, Graces and originals from the porn industry. He combines them with old prints of slave ships and stresses creolization as a creative force which owes its existence to the mingling of languages, cultures and social contrasts. Born in Ghana, he has lived in London since 1973. He is inspired by the milieu and the history of boxing, as he is also by the life of Lord Byron.

Barkley L. Hendricks has, with *Fela: Amen, Amen, Amen* (2002), erected an altar to the fêted musician, in front of which the high heels of his wives lie as sacrificial gifts. The Nigerian trumpeter, composer and singer Fela (1938–1997) was the founder of Afrobeat. His band had up to 40 members. As he refused to repeat already recorded numbers at his concerts, and tended to make virulent speeches, he had no success in the USA. In 1979 he was forced to leave Nigeria for Ghana. Back in his homeland, he was arrested. The magazine *Rolling Stone* declared him to be 'the world's most dangerous musician'. In a mass ceremony he married 27 of his dancers and singers. He was notorious for his misogynist views, he demonized homosexuality and died of AIDS. Hendricks, himself a trumpeter and saxophonist, places many of his portraits of African Americans on a gold ground.

Peter Murphy dedicates his portraits to musicians ranging from Beethoven to the superstars of our own age. *Kurt Cobain Nirvanamind* (2005) is, like the other icons, reproduced with wings as a supernatural being on a gold ground. In his right hand he holds a guitar, in his left the shotgun he shot himself with. In the pediment above is the portrait of his wife Courtney and his daughter Frances. In the other picture, *Jimi Hendrix (Are You Experienced?)* (2005), the title is a reference to the debut album of 1967. The other two members of the band, the *Jimi Hendrix Experience*, Mitch Mitchell (percussion) and Noel Redding (bass), are portrayed in the pediment. The inscription 'Bold as Love' in the halo refers to the second album dating from the same year.

In Pete Nawara's portraits, such as *St. Peter* (2004), the consecration comes about through the halo in a girl-next-door context. The English artist dispenses with hieratic compositions and chooses everyday poses for his portraits. Sometimes he also uses gold for the colour of blonde hair or as a background. Gold, like the nimbus, is for him an aesthetic sign designed to give pleasure to the beholder.

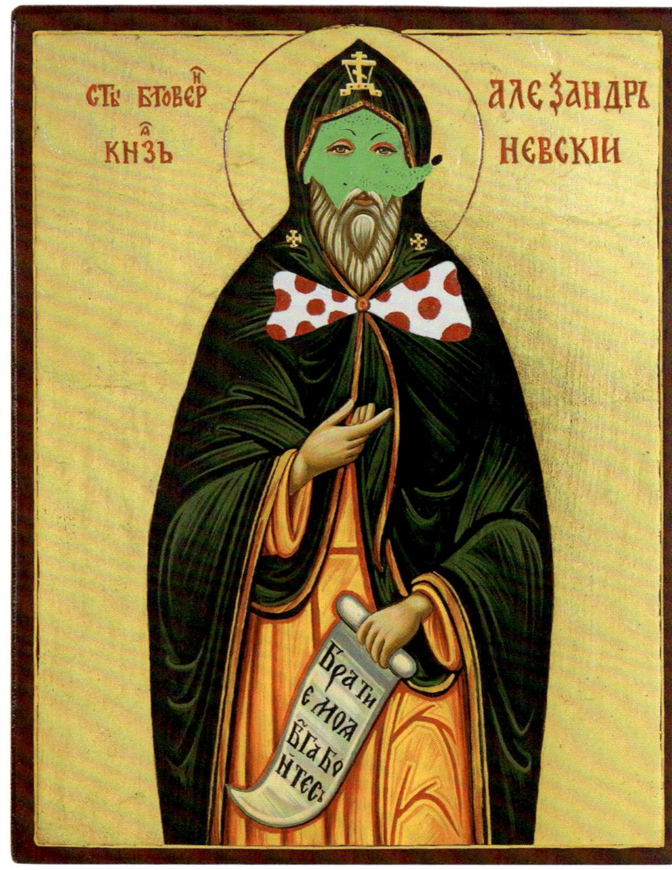

Fig. 1 Paul Insect, *Icon 8* (2008)
Gold leaf and mixed media on wood, 22x17 cm

Like the halo, wings too are a sign of a different reality. The Greek painter Edouard Sacaillan has studied medieval book illumination and Armenian art. He identifies reductionist archaism with idealism. In two designs for *The Angel IAN* (2007) the angel, looking upwards, is surrounded by the Armenian alphabet. On one occasion he triumphs over a demon, on another he places his foot on Mt Ararat. This explains his posture in the picture.

Contemporary artists either take the use of sacred symbolism seriously, or they caricature it, or turn it into something banal. Paul Insect, in his series of *Icons* (fig. 1, 2008) has literally trans-formed Byzantine originals of saints, the Holy Family and the Virgin Mary on gold grounds into blasphemous comic figures through the use of make-up, long noses and wigs. In the works of Insect and Pepo Pichler, it is the union of the incompatible that makes the pictures. The San Francisco-based Carinthian-born Pichler turns the circular form of the nimbus into a circular saw that cuts the putto on the console in half. In the meeting of Baroque splendour with technology, the beholder is confronted with a gruesome dramatization of historical contrasts. Art history has accustomed us to attributing particular meanings to particular forms. Even though John Armleder rejects iconographic interpretations, his gilt-silver crown of thorns *Illico* (2011) willy-nilly suggests the Passion.

The pathos formulae and narrative styles of the triptych, the nimbus and the altarpiece open up the tension field of the sacred. Even so, it is not easy to understand this. Art criticism cuts the Gordian knot by asserting: "The picture is always *holy,*"[15] and in the same breath stresses the difference between the holy and the religious. The "dazzling, excessive visibility," heightened by gold, "is there to conceal that there is an invisible sphere, and that there is no mediation between the visible and the invisible."[16]

We should not forget the transformation of light by the polarity of electric light source and gold leaf, as presented by Stephen Antonakos in *Arimathea* (2001).

1 Klaus Lankheit, *Das triptych als Pathosformel*, Heidelberg 1959.
2 Marion Ackermann in *art* in conversation with Hans-Joachim Müller about the 2009 exhibition *Drei. Das triptych in der Moderne* in the Kunstmuseum Stuttgart. URL: <http://www.art-magazin.de/kunst/14586/triptychon_in_der_moderne_kunstmuseum_stuttgart> [read on 12 Dec. 2011]
3 Marius Rimmele, *Das triptych als Metapher, Körper und Ort. Semantisierung eines Bildträgers*, Munich 2010.
4 Ibid., p. 315.
5 Wolfgang Ulrich, "Autoritäre Bilder. Die zweite Karriere des triptychs seit dem 19. Jahrhundert", in: Wolfgang Ulrich, *An die Kunst glauben*, Berlin 2011, p. 114.
6 Hanna Egger/Ruth Wenckheim (eds.), *Ikonen. Bilder in Gold. Sakrale Kunst aus Griechenland* (exhib. cat. Kunsthalle Krems, Minoritenkirche), Graz 1993, cat. no. 9.
7 "Wichtige Neuerwerbungen 1987", in: *Mitteilungen der Österreichischen Galerie*, vol. 31, no.75, Vienna 1987. For information about this triptych I am grateful to Katinka Gratzer-Baumgärtner of the Research Center Archive of the Belvedere.
8 Cf. *Ausstellung für Kirchliche Kunst* (exhib. cat. K. K. Österreichisches Museum für Kunst und Industrie, Vienna 1912), September-December 1912, cat. 42.
9 Peter Diem-Anton Wladar, *Kurt Regschek—Meister der Gegensätze. Leben und Werk des 'Wienerischsten' der phantastischen Realisten*, Vienna 2006, p. 120. Translator's note: 'mind-game' is in German 'Kopfgeburt', literally 'head-birth', but this pun is not reproducible in English.
10 Maria Rennhofer, "KRON-ZEUGE Rudolf Leitner-Gründberg", in: *Parnass*, vol. 19, no. 2, 1999, p. 118.
11 Toshinobu Yasumura (ed.), *Rimpa. Decorative Japanese Painting*, Tokyo 2011.
12 Rosina Buckland, *Golden fantasies: Japanese screens from New York collections*, New York 2004. - Janice Katz (ed.), *Beyond Golden Clouds - Japanese Screens from the Art Institute of Chicago and the Saint Louis Art Museum*, London 2009.
13 "The term 'Japonisme' was coined in Paris c. 1870 by Philippe Burty (1830-1890) to refer to works whose principal themes were Japanese woodblock prints (ukiyo-e)" Johannes Wieninger, *Gustav Klimt und die Kunst Ostasiens*, 2009, quoted from: URL: <http://wieninger.com/publikationen/2010/> [read on 12 Dec. 2011]
14 Wojciech Mucha, *Carnet d'atelier*, Amsterdam 2008, p. 32.
15 Jean-Luc Nancy, "Das Bild—Das Distinkte", in Doreet Le Vitte-Harten (ed.):*Heaven* (exhib. cat. Kunsthalle Düsseldorf/Tate Gallery Liverpool), Ostfildern-Ruit 1999, p. 44.
16 Thierry de Duve, "Auf, Ihr Menschen, noch ein Versuch, post-christlich zu werden!", in: Le Vitte-Harten 1999 (as note 15), p. 90.

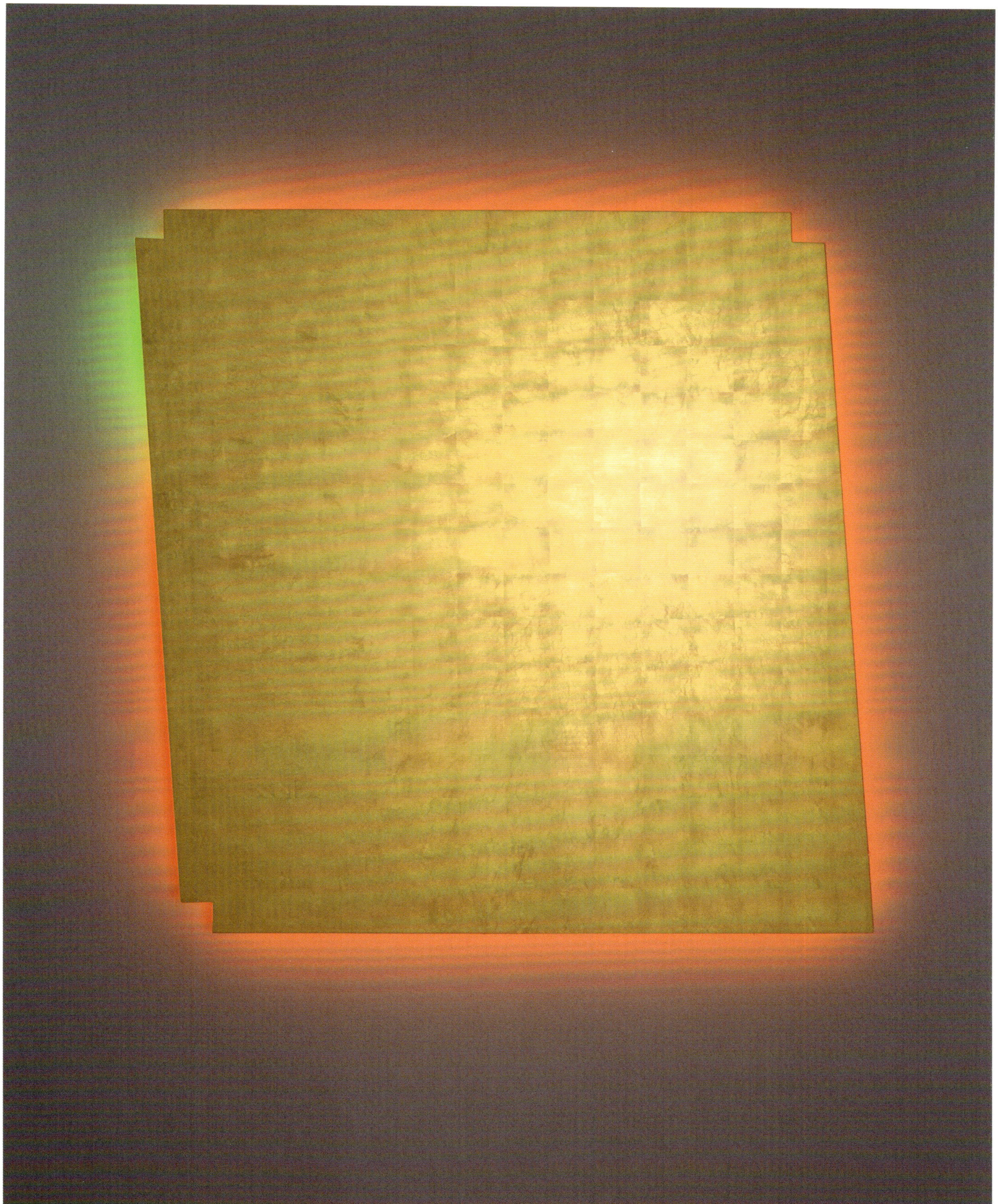

Stephen Antonakos
Arimathea, 2001
Gold leaf on wood and neon light, 122 x 122 x 15 cm
Courtesy Kalfayan Galleries, Athens—Thessaloniki

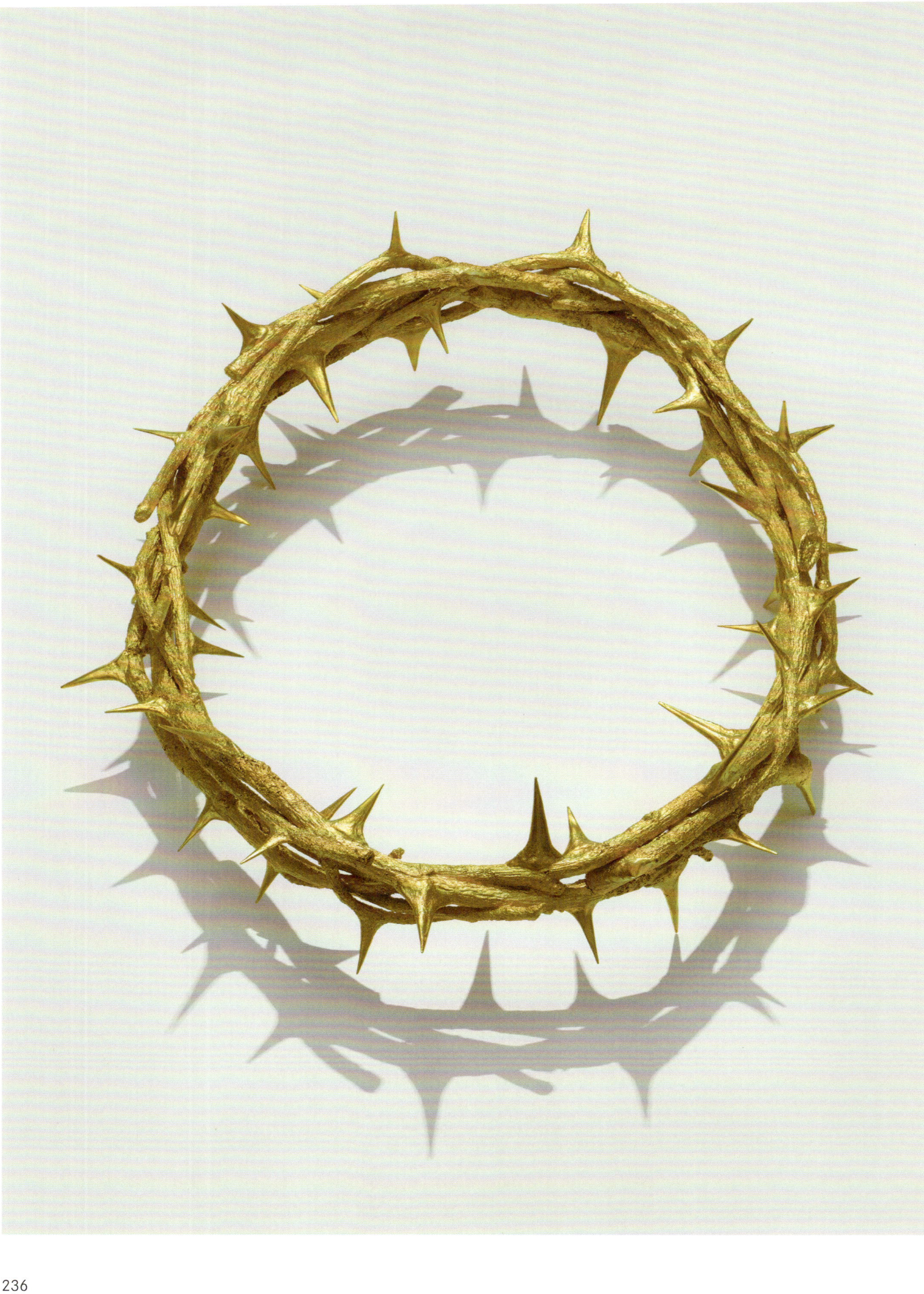

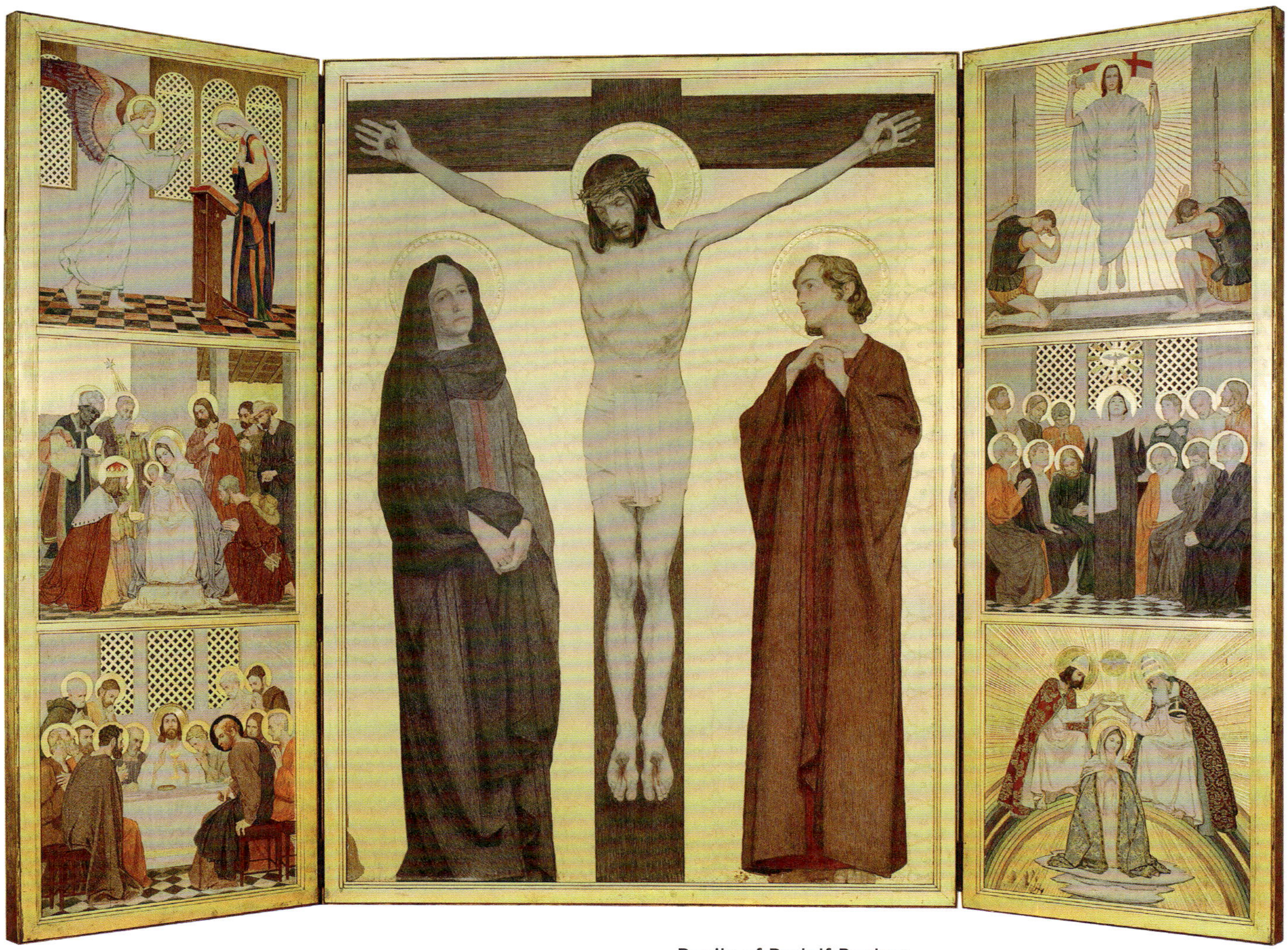

Pupils of Rudolf Bacher:
Karl Borschke, Karl Maria May, Leopold Zlabinger, Andreas Harsch, Ferdinand Kitt, Anton Schellerer (gilder)
Altar-Triptychon, pre-1912
Oil on tempera on wood, partly gilded, central section: 211 x 154 cm; wings: each 211 x 68 cm
Belvedere, Vienna

John M. Armleder
Illico, 2011
Silver-gilt, 7 x 23 x 27 cm
Galerie Andrea Caratsch, Zurich

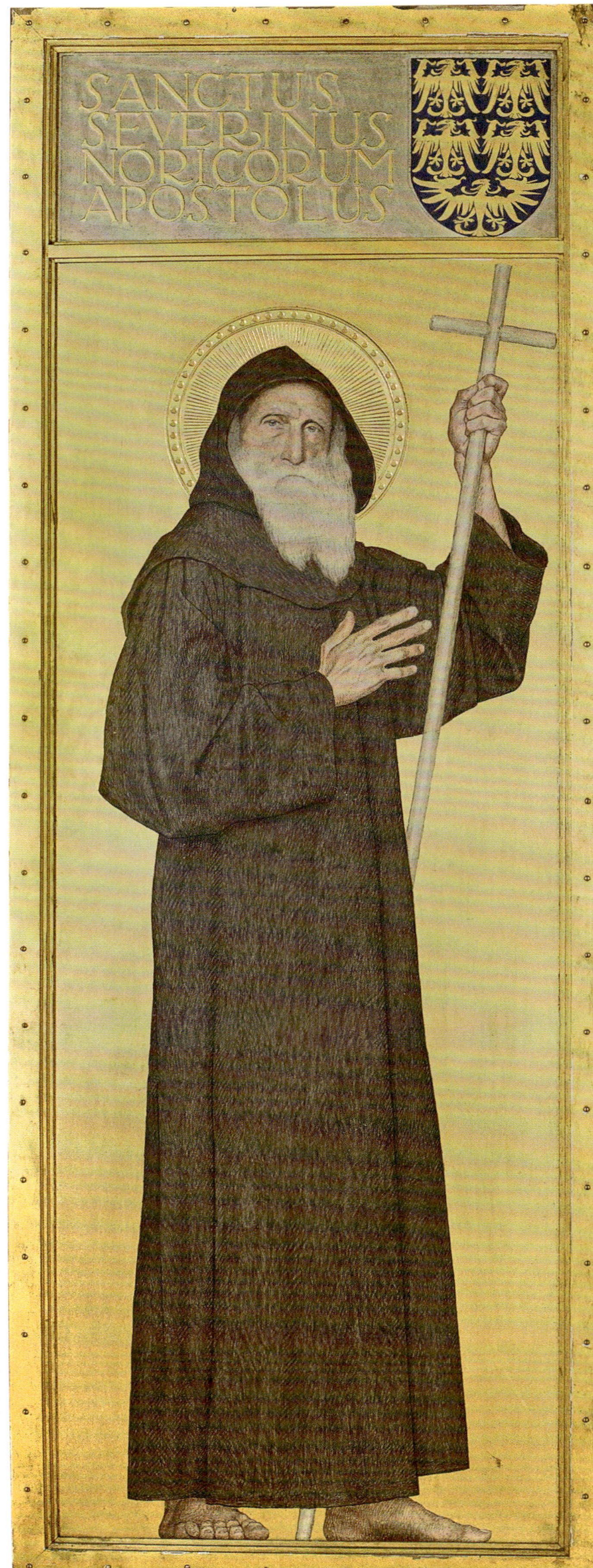
SANCTUS
SEVERINUS
NORICORUM
APOSTOLUS

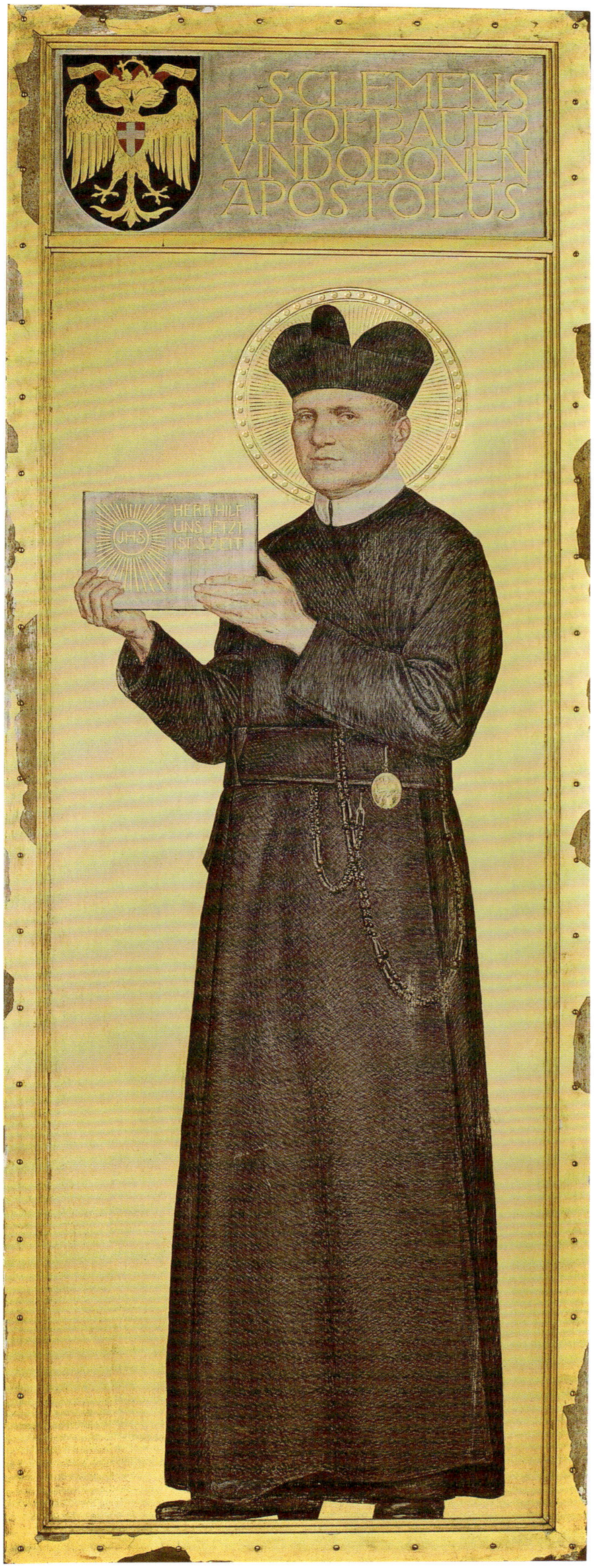

Pupils of Rudolf Bacher:
Karl Borschke, Karl Maria May, Leopold Zlabinger, Andreas Harsch, Ferdinand Kitt, Anton Schellerer (gilder)
Altar-Triptychon, pre-1912
Oil on tempera on wood, partly gilded, Wings: each 211 x 68 cm
Belvedere, Vienna

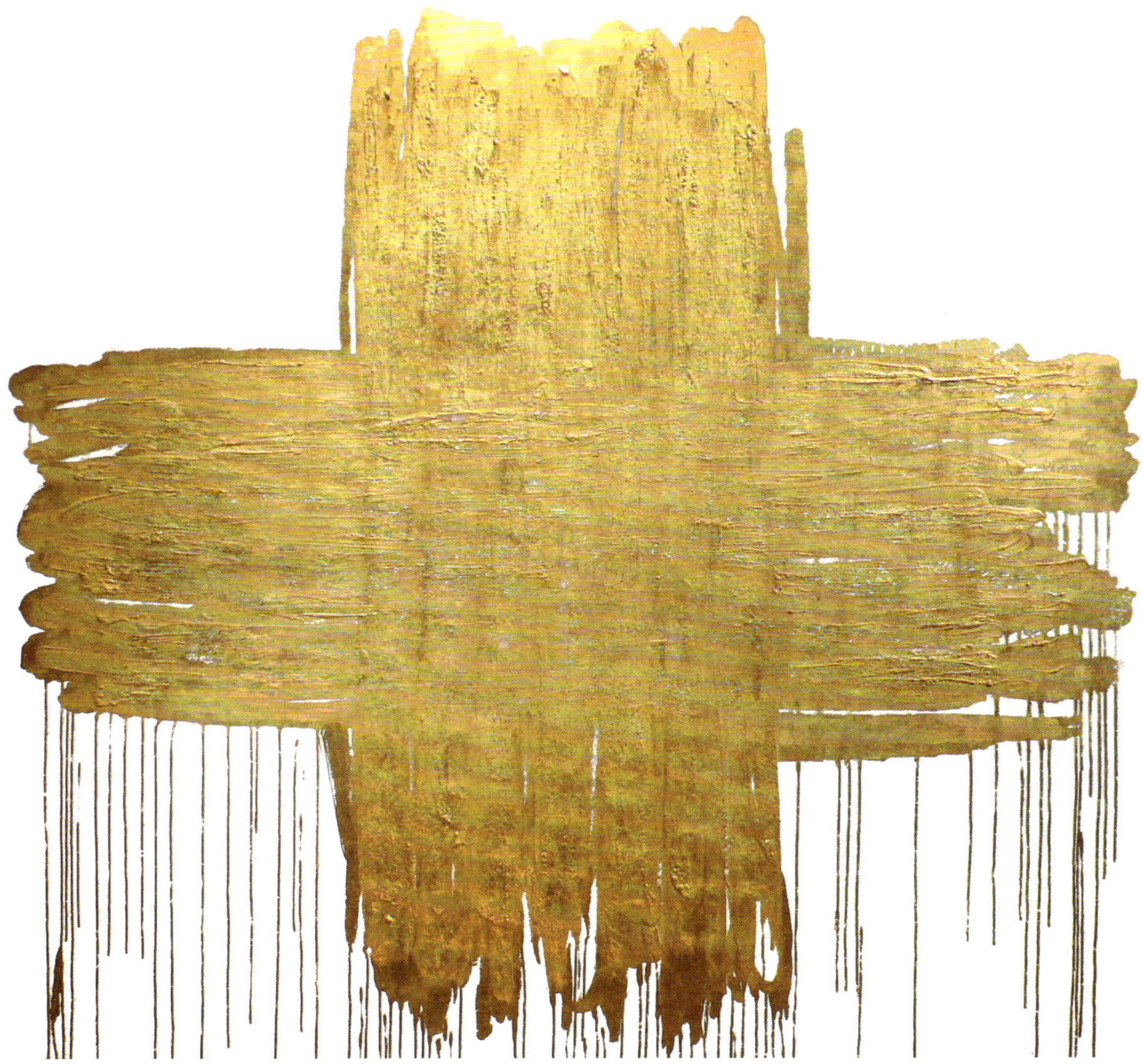

Gor Chahal (Гор Чахал)
Fat Cross, 2008
Gold on canvas, 170 x 180 cm
Gor Chahal

Jiři Georg Dokoupil
Madonna, 1986
Acrylic and gold foil on linen, 180.5 x 140.5 cm
Galerie Andrea Caratsch, Zurich

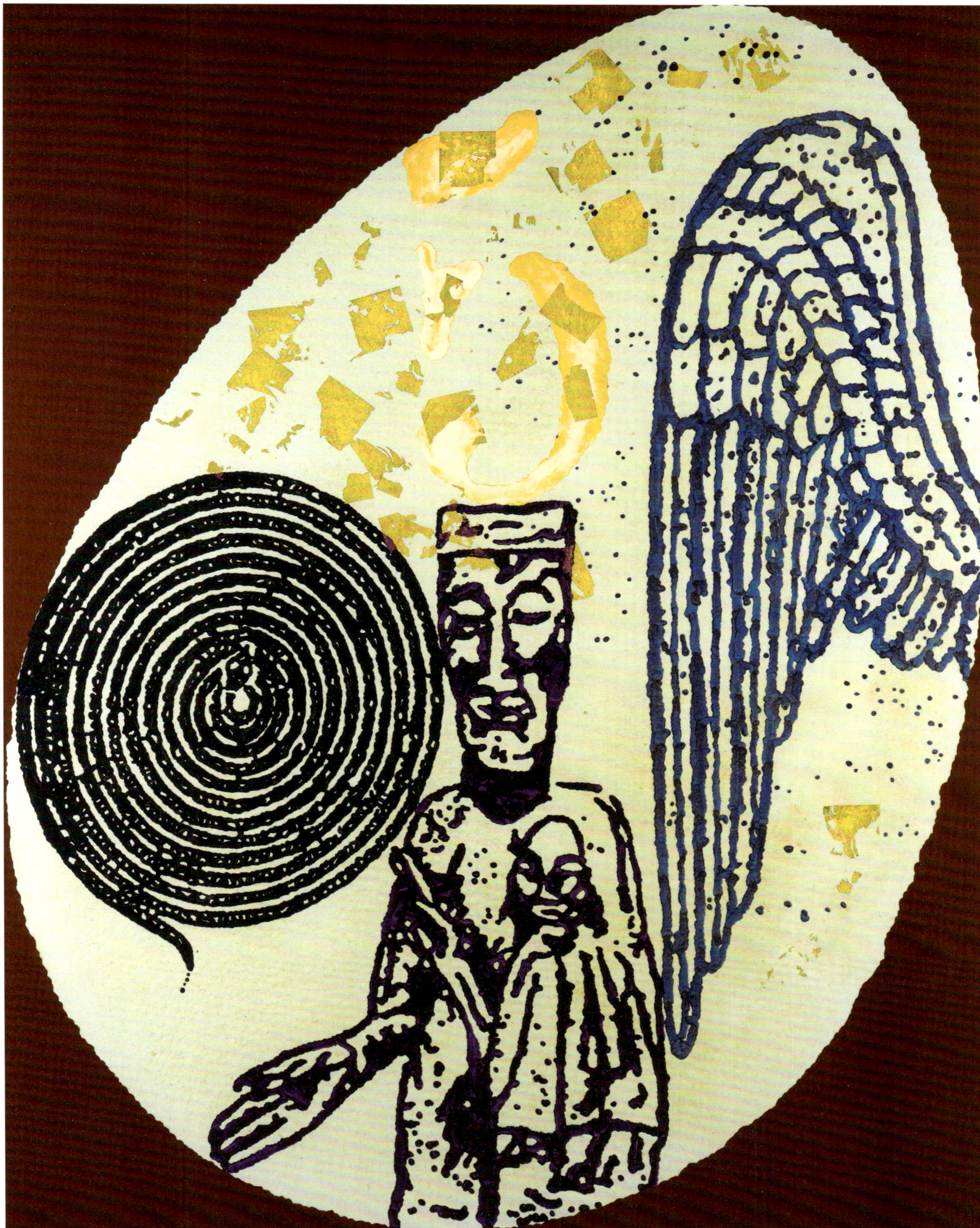

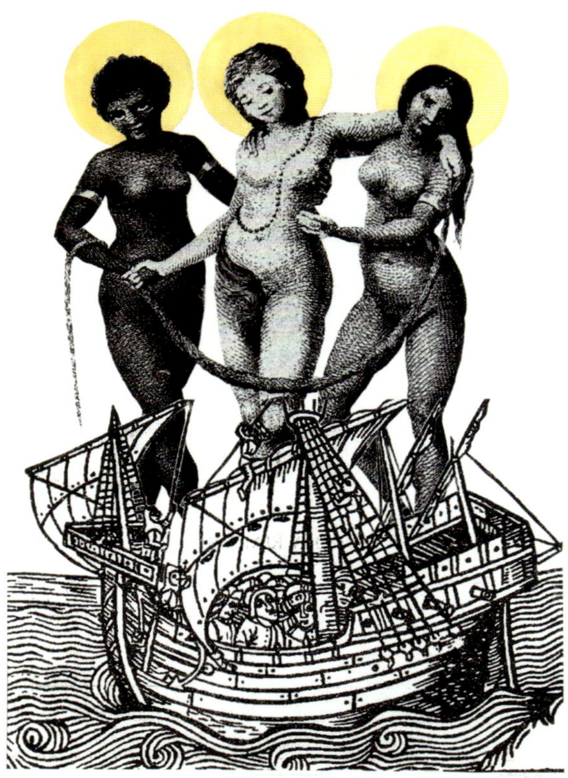

Godfried Donkor
Three Madonnas, 2009
Collage/giclée-print on heavy paper,
hand-finished with real gold leaf, 100 x 70 cm
Collection Kempen, Aachen

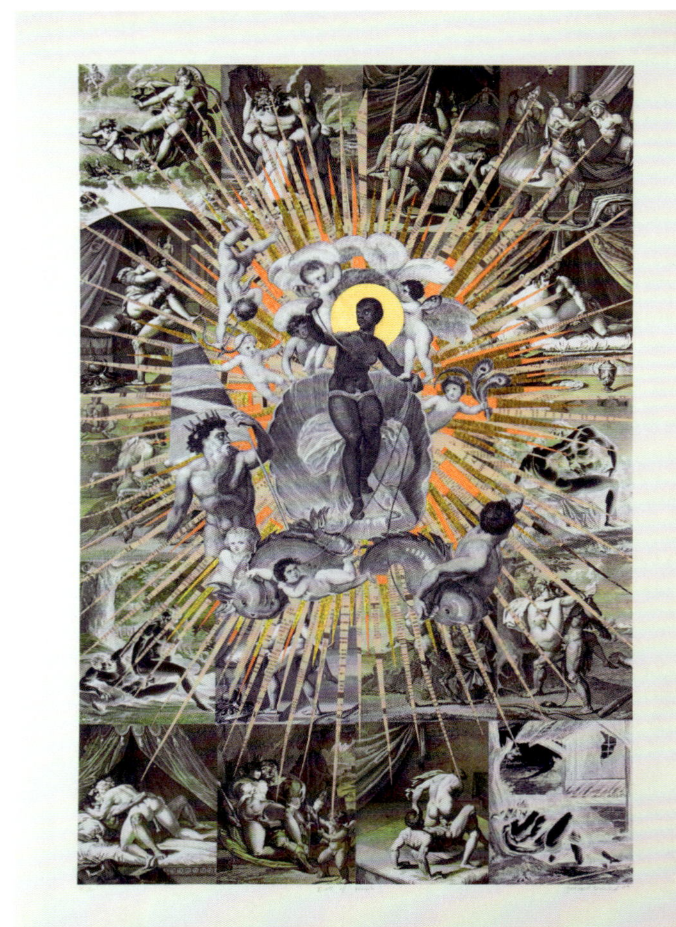

Godfried Donkor
Birth of Venus, 2009
Collage/giclée-print on heavy paper,
hand-finished with real gold leaf, 100 x 70 cm
Collection Kempen, Aachen

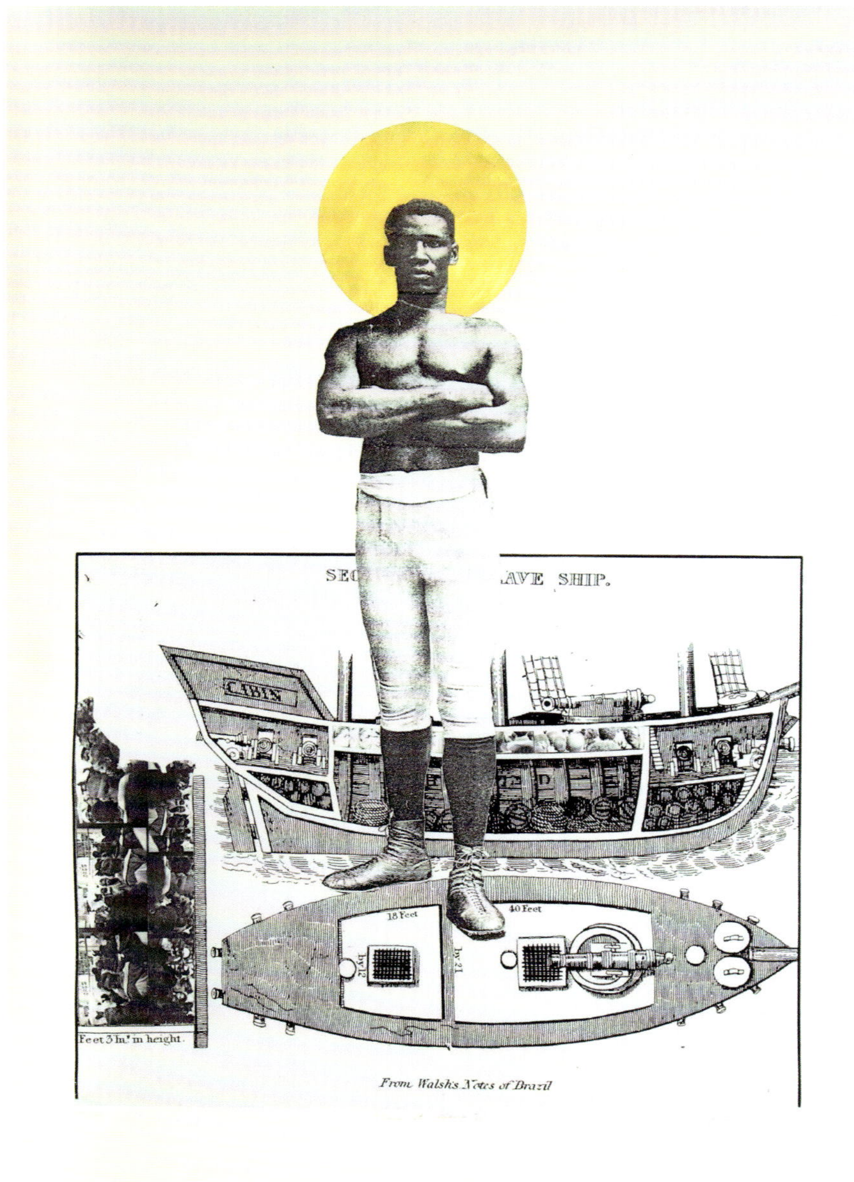

Godfried Donkor
Slave To Champ, 2009
Collage/giclée-print on heavy paper,
hand-finished with real gold leaf, 100 x 70 cm
Collection Kempen, Aachen

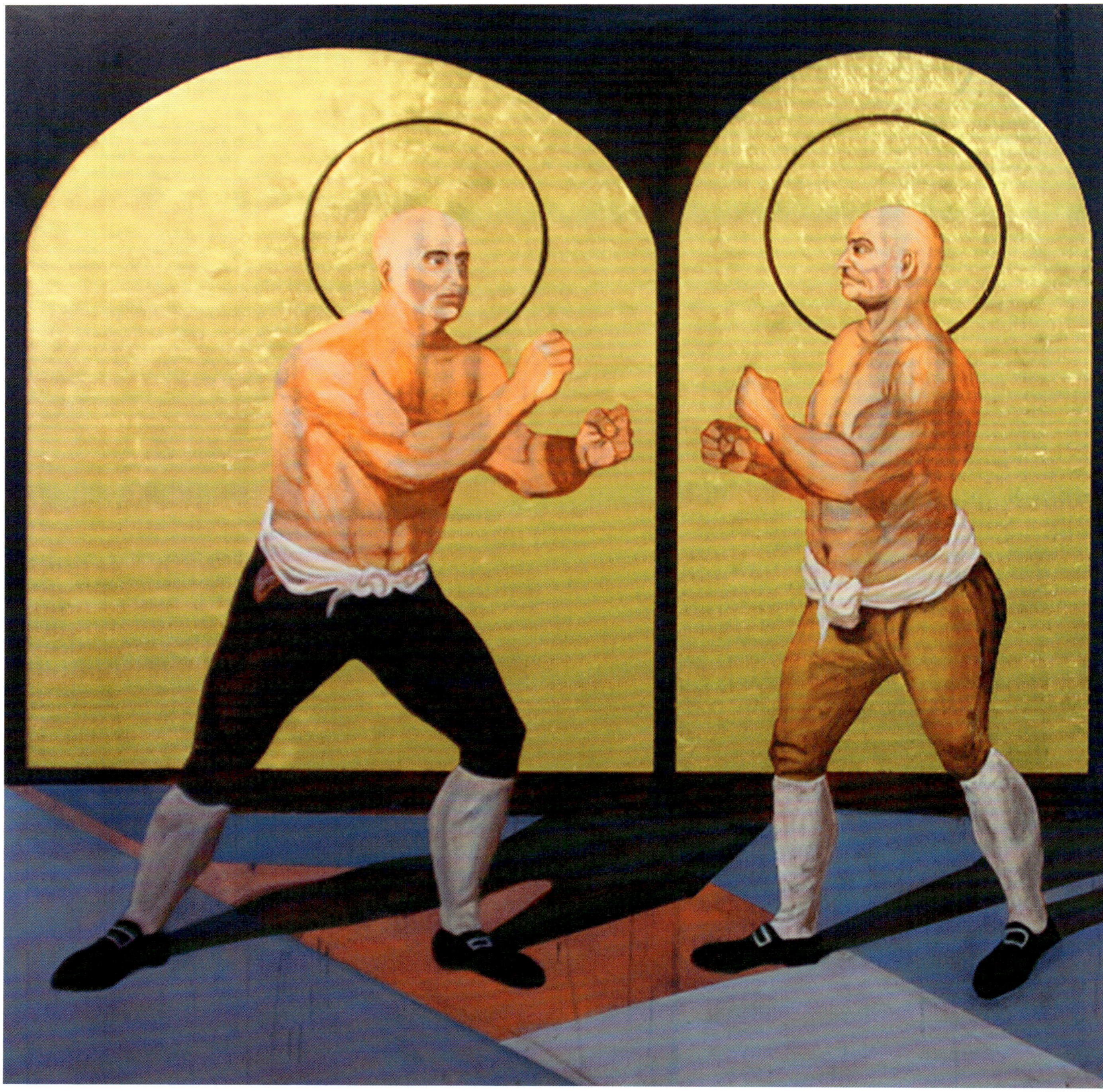

Godfried Donkor
St Georg Stevenson vs. St Jack Broughton, 2010
Oil and gold leaf on canvas, 200 x 200 cm
ARTCO Galerie Herzogenrath Germany

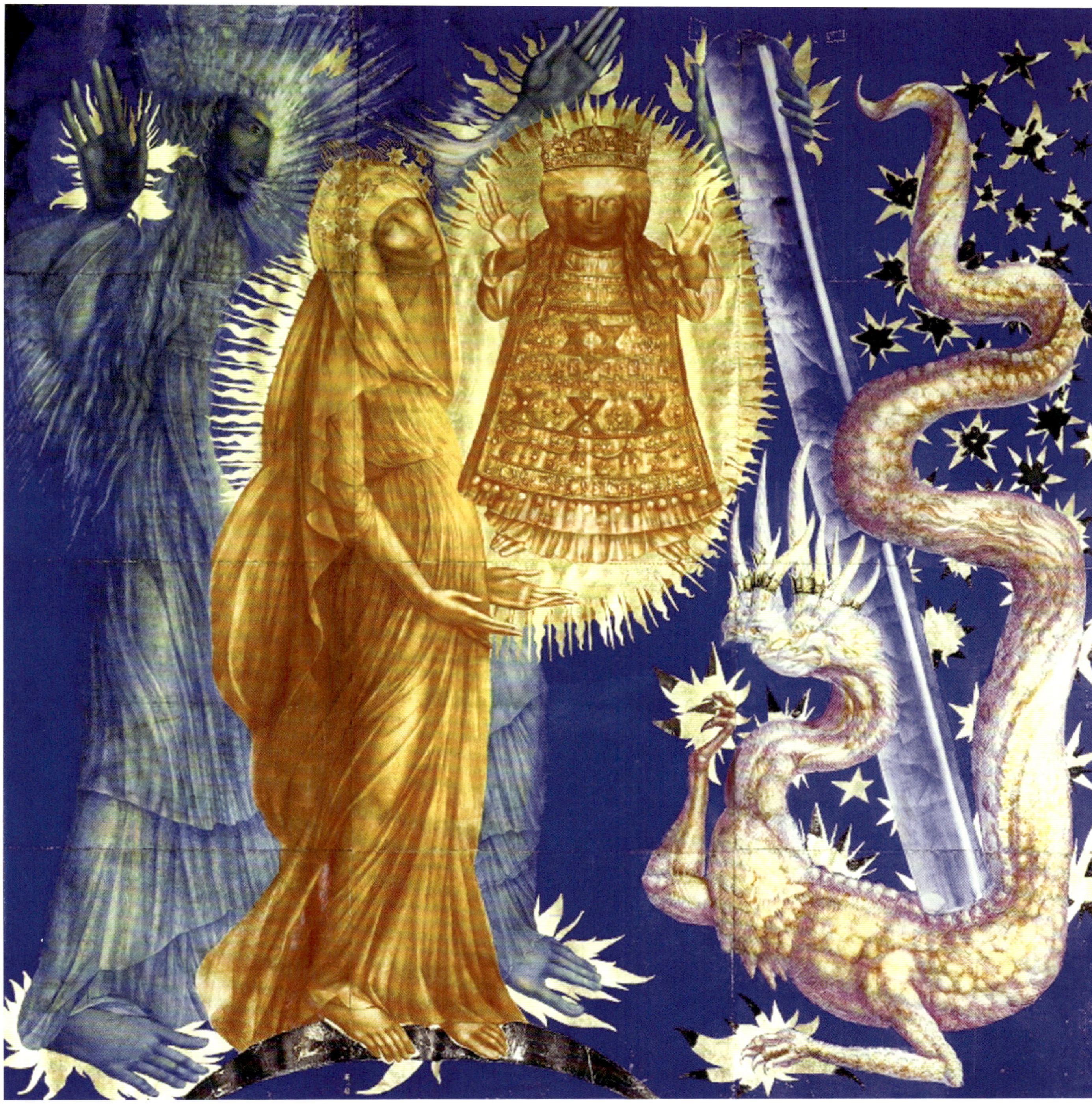

Ernst Fuchs
Die Geheimnisse des hochheiligen Rosenkranzes, linkes Bild des Triptychon, 1958–1961
Mixed media, gold leaf on goat-skins, 300 x 300 cm
Erzdiözese Wien, Hetzendorf parish church

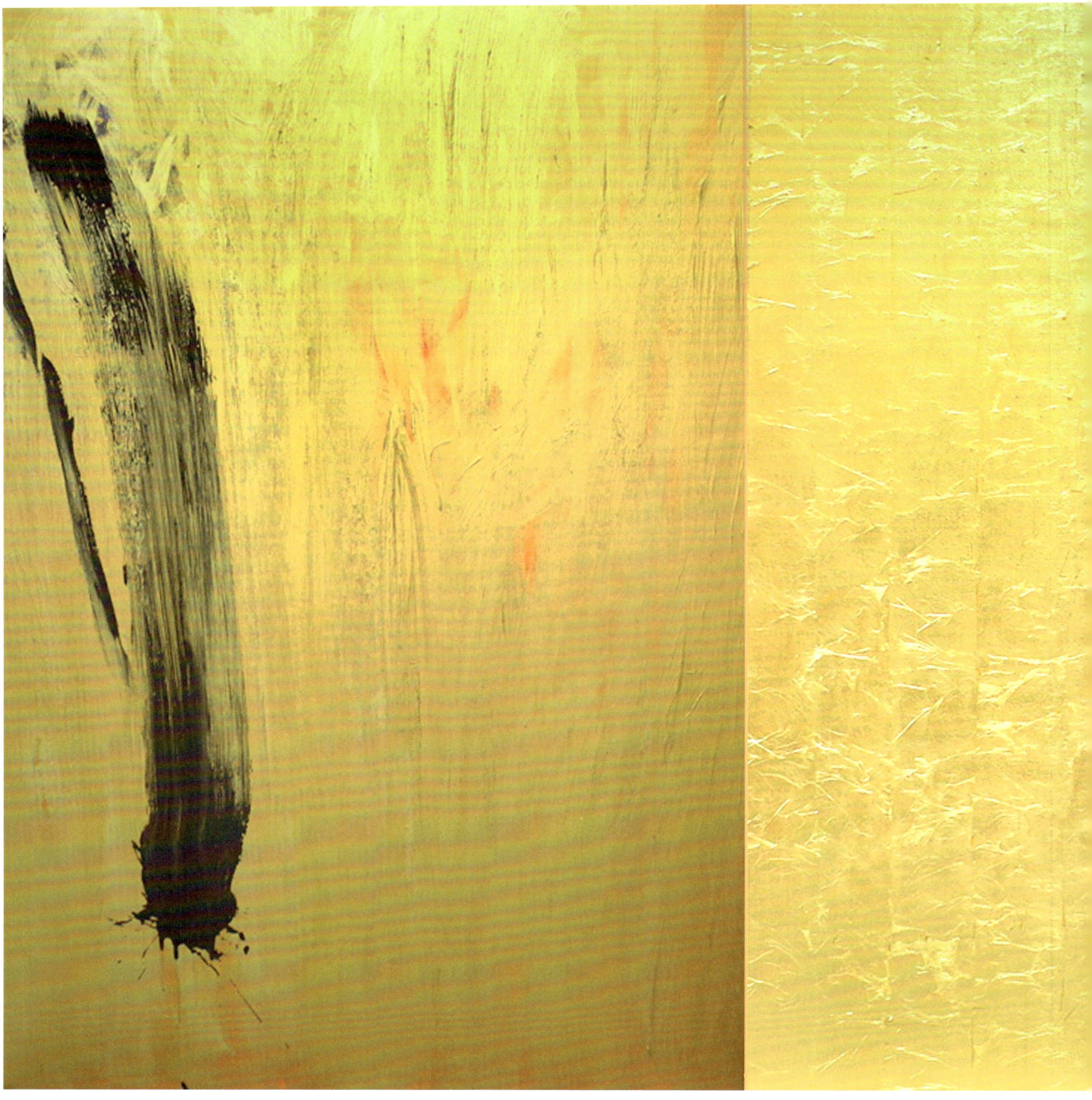

Yoko Grandsagne
Zen Garden, 2010
Mixed media, gold leaf on canvas, 195 x 390 cm
Yoko Grandsagne

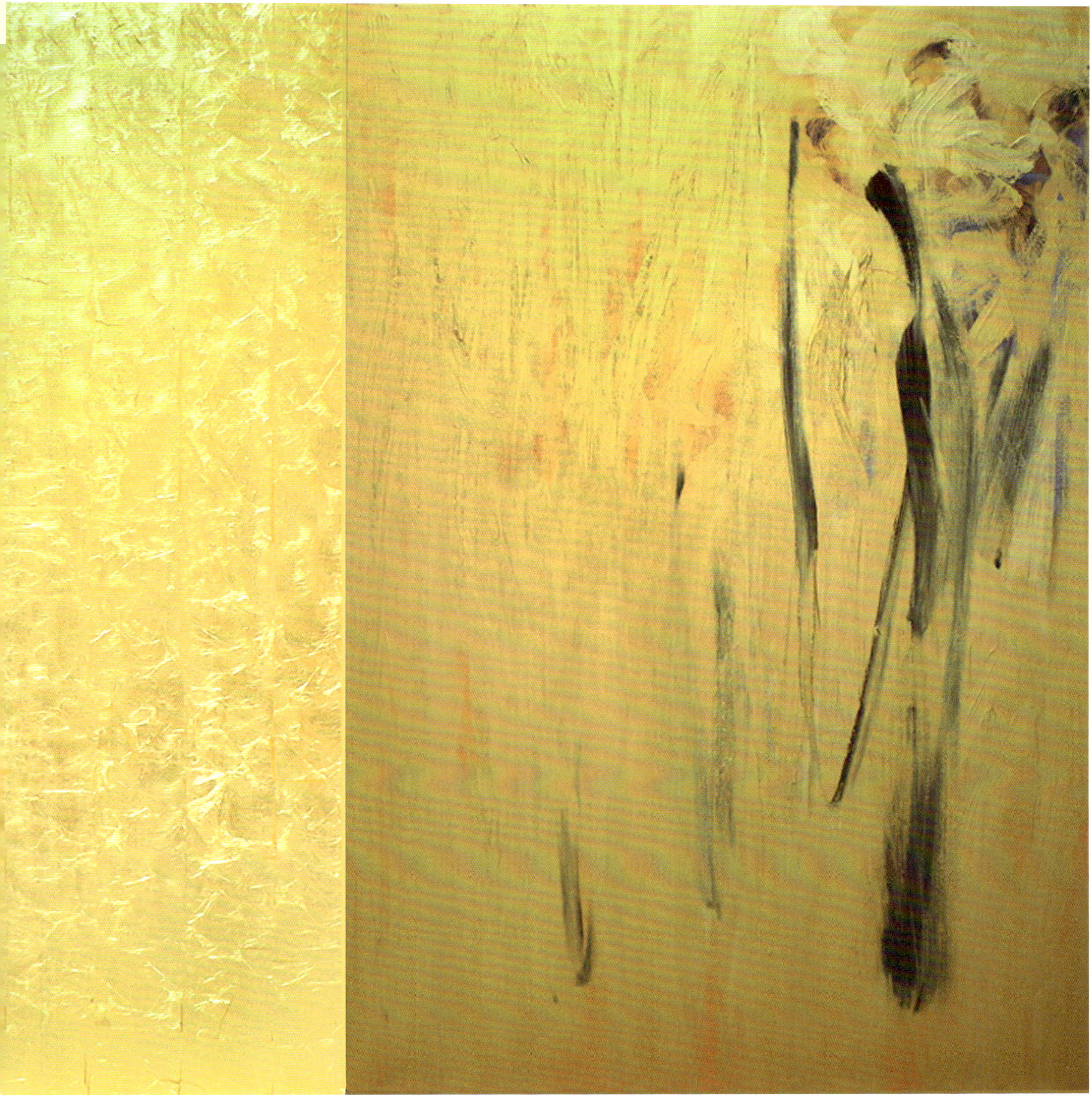

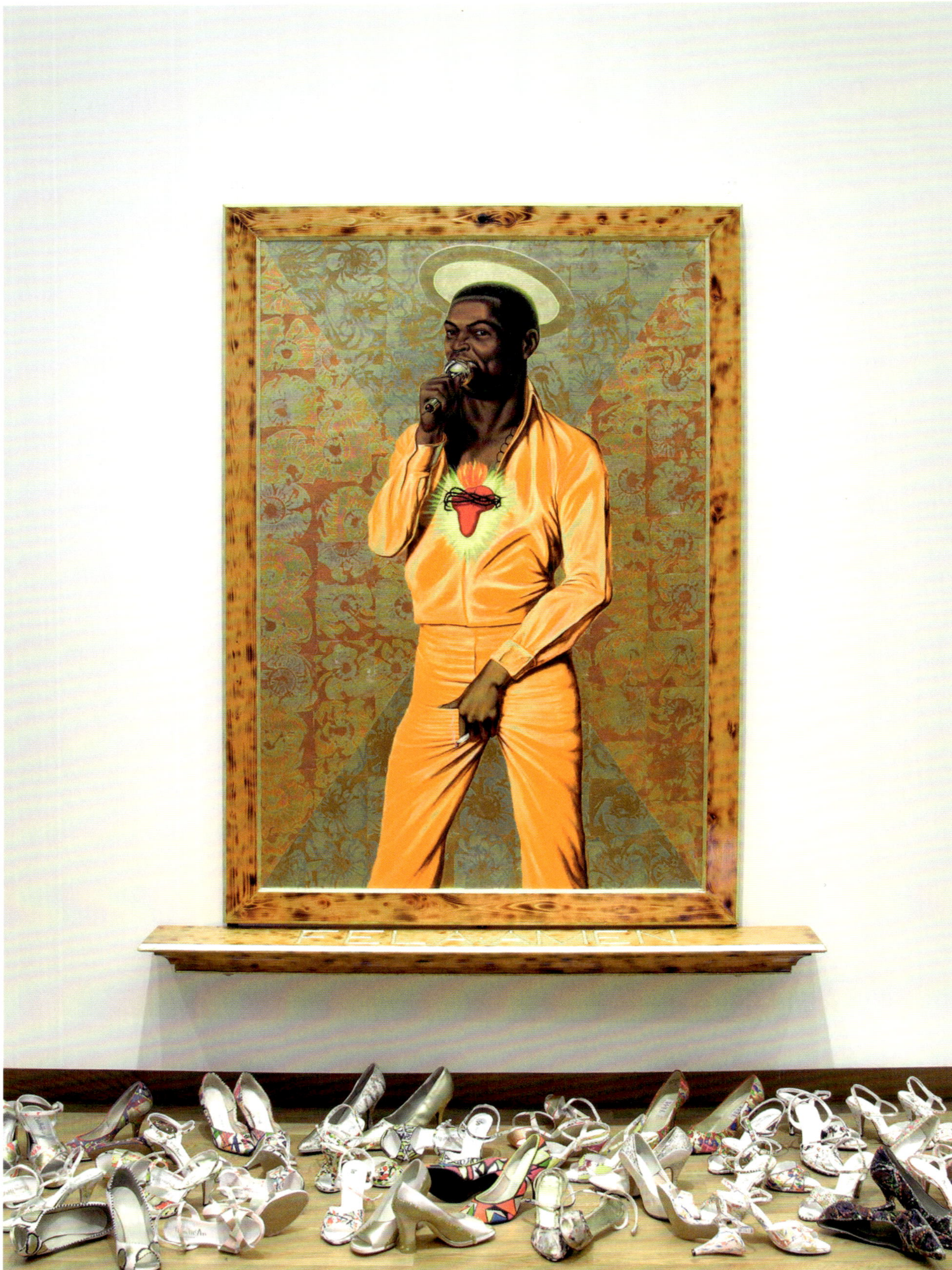

Barkley Hendricks
Fela: Amen, Amen, Amen, Amen, 2002
Oil and variegated metal leaf on canvas, wooden frame, armature,
27 pairs of high heels, 169.6 x 118.7 cm
Private Collection

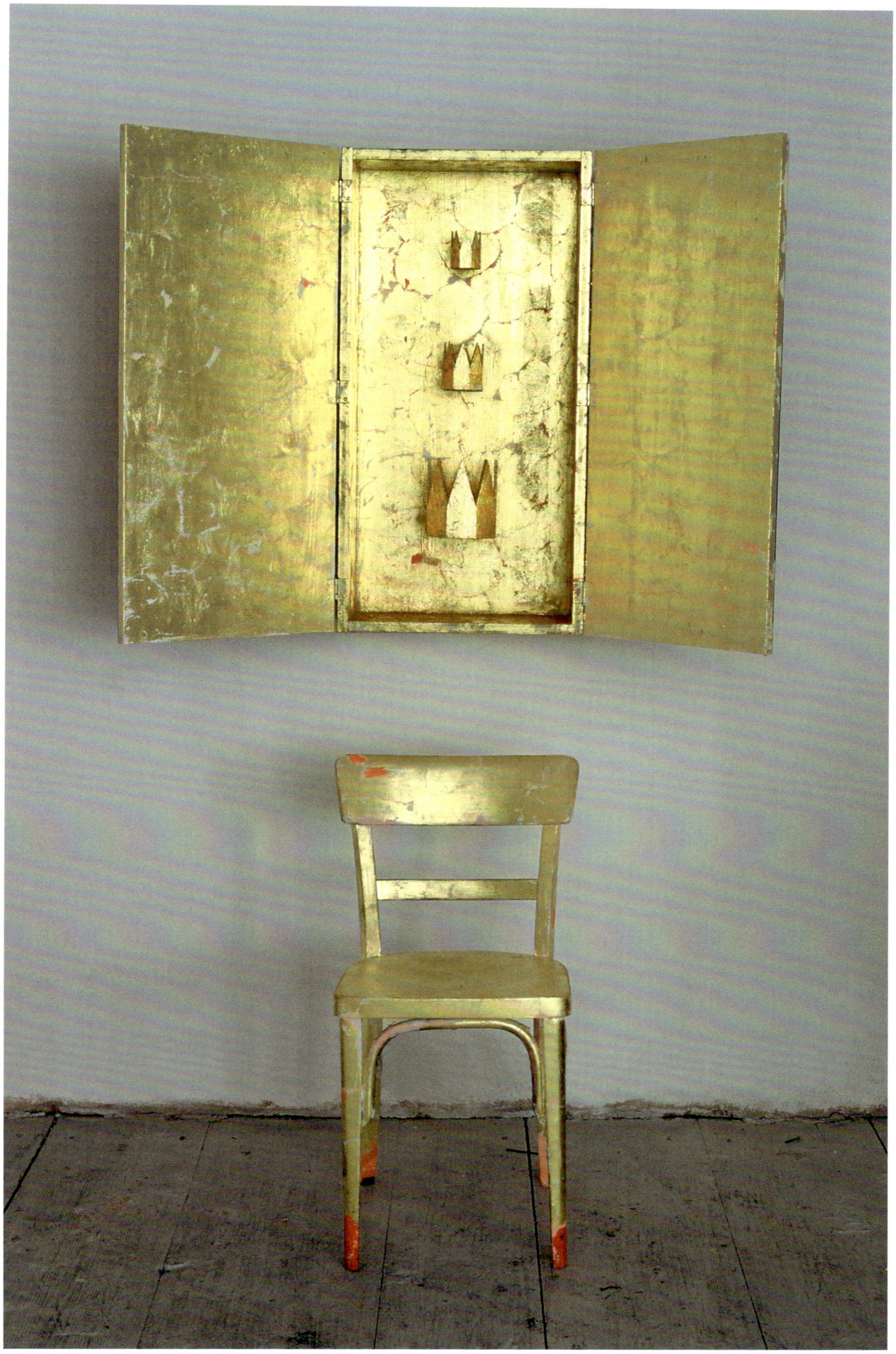

Rudolf Leitner-Gründberg
Signaturschriftbild, 2010
Gold leaf, oil, packing paper, painted gilt solid-wood frame
(part of the artwork), behind glass, 114 x 149 cm
Rudolf Leitner-Gründberg

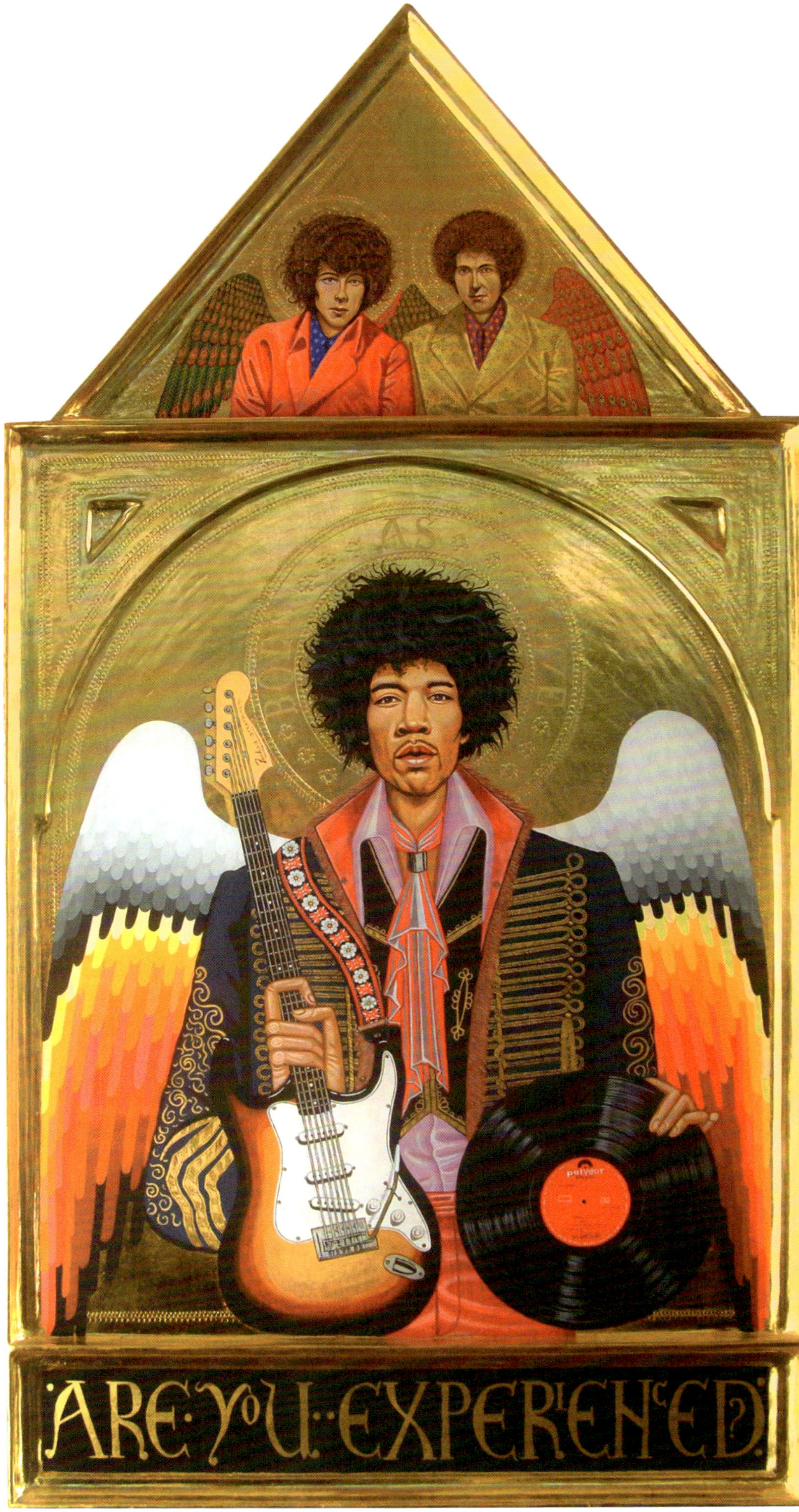

Peter Murphy
Jimi Hendrix, Are you experienced?, 2008
Egg tempera and gold leaf on wood, 92 x 48 cm
Peter Murphy,
Courtesy the artist

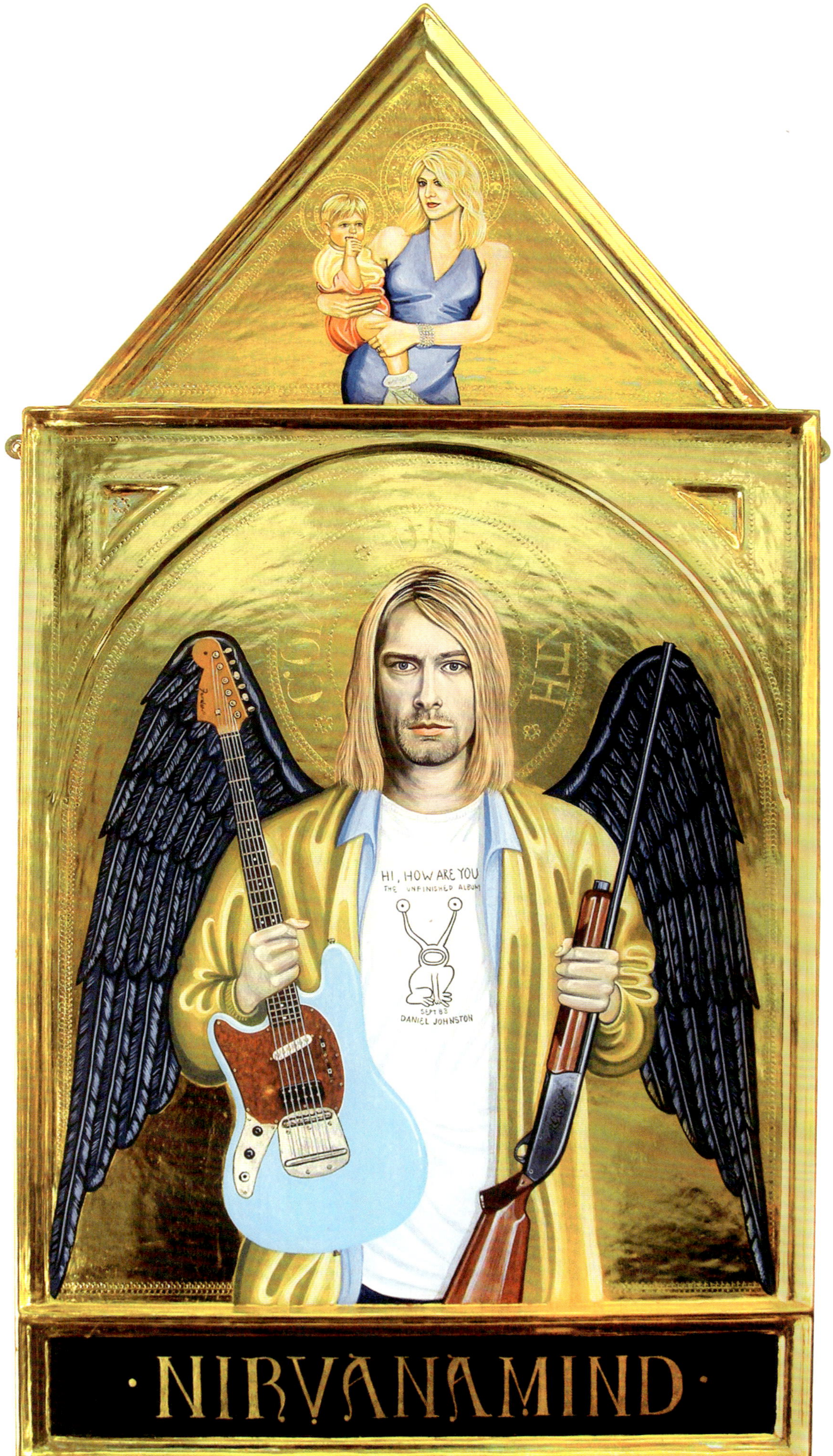

Peter Murphy
Kurt Cobain Nirvanamind,
2009
Egg tempera and gold leaf on wood, 92 x 48 cm
Peter Murphy,
Courtesy the artist

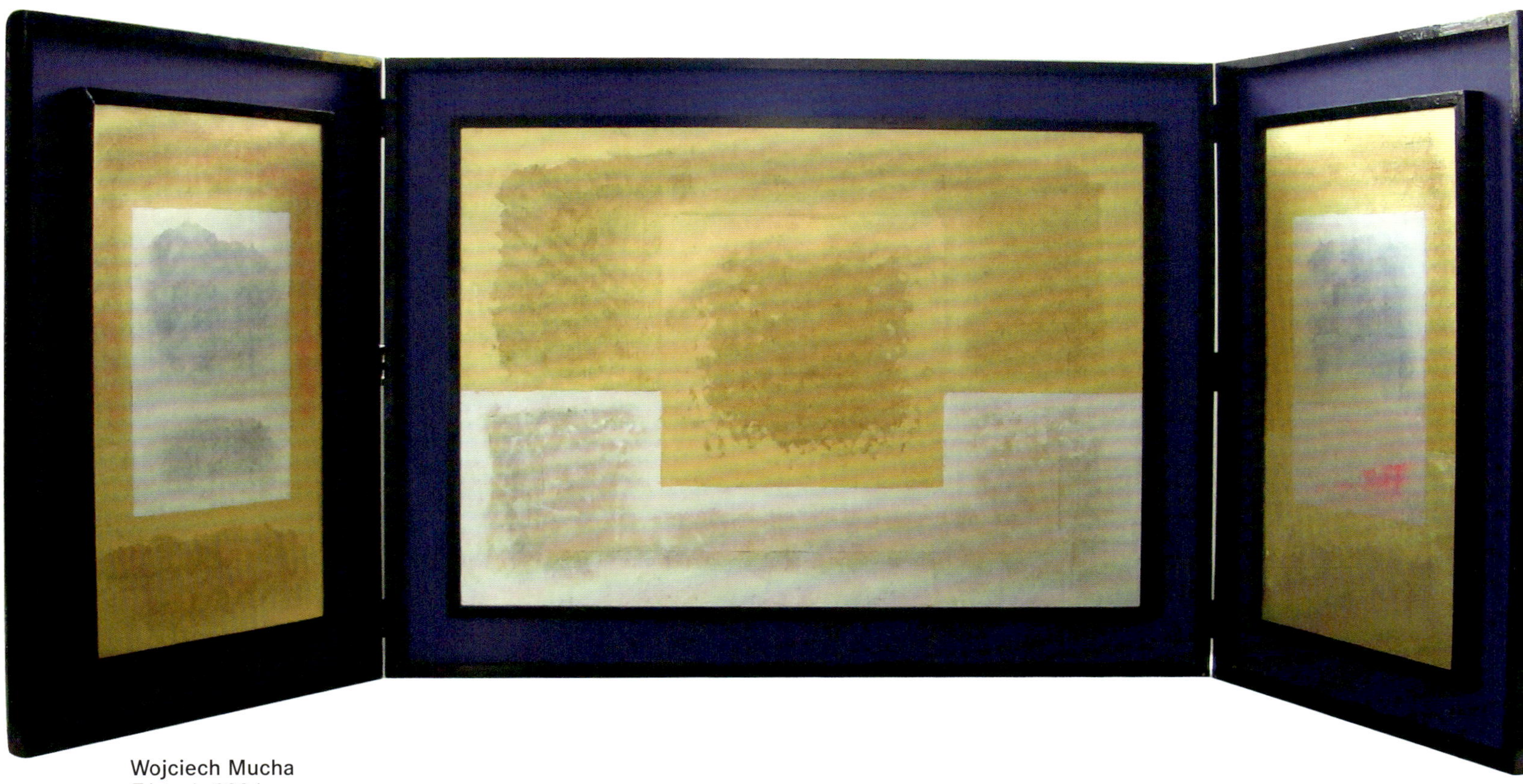

Wojciech Mucha
Triptych, 2006
Oil and gold leaf on wood, 78 x 200 cm (open, with frame), 84.5 x 105 x 12 cm (closed)
Private collection/estate of W. Mucha, Geneva

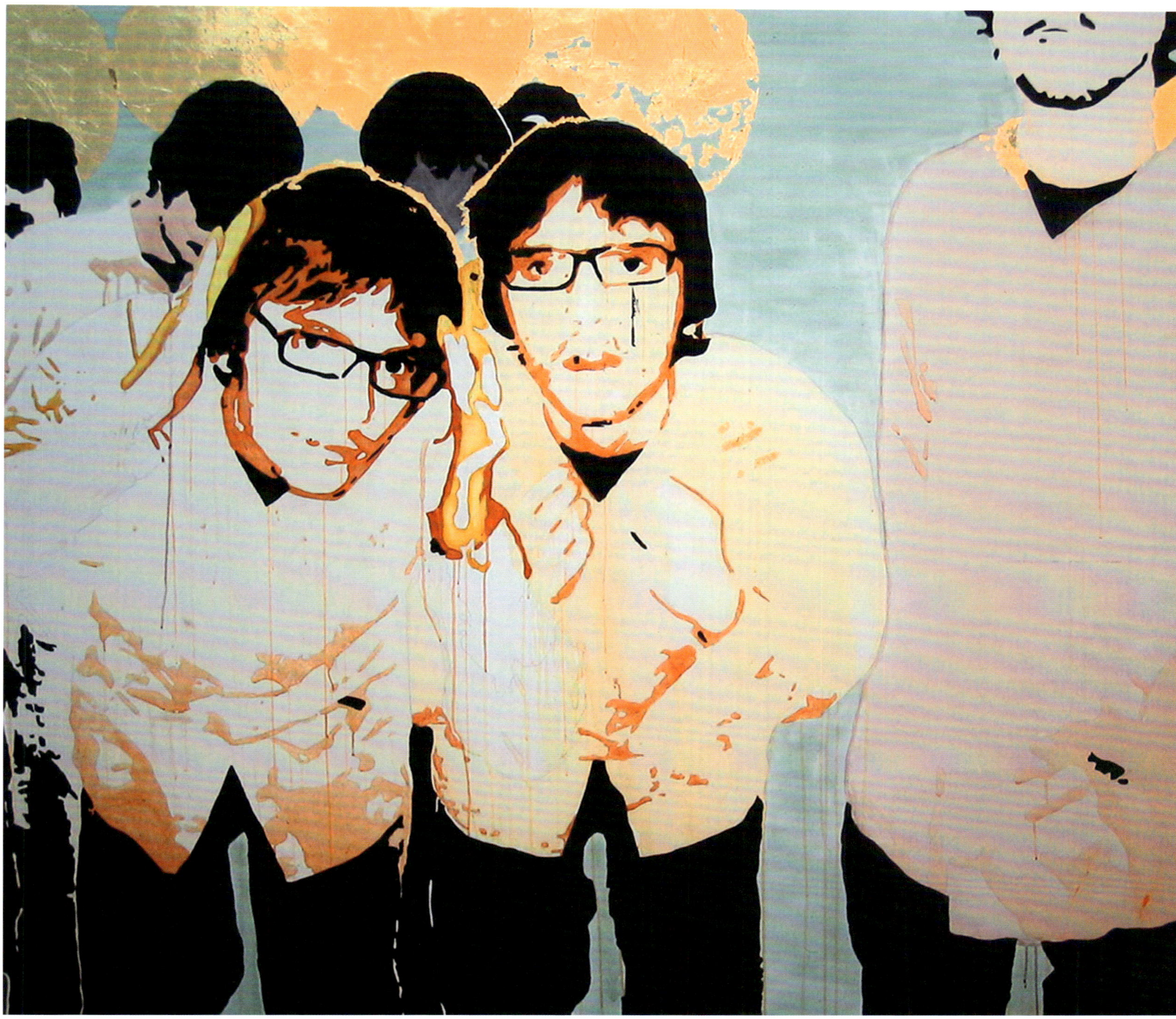

Pete Nawara
St. Peter, 2004
Mixed media on canvas, 180 x 203 cm
Pete Nawara

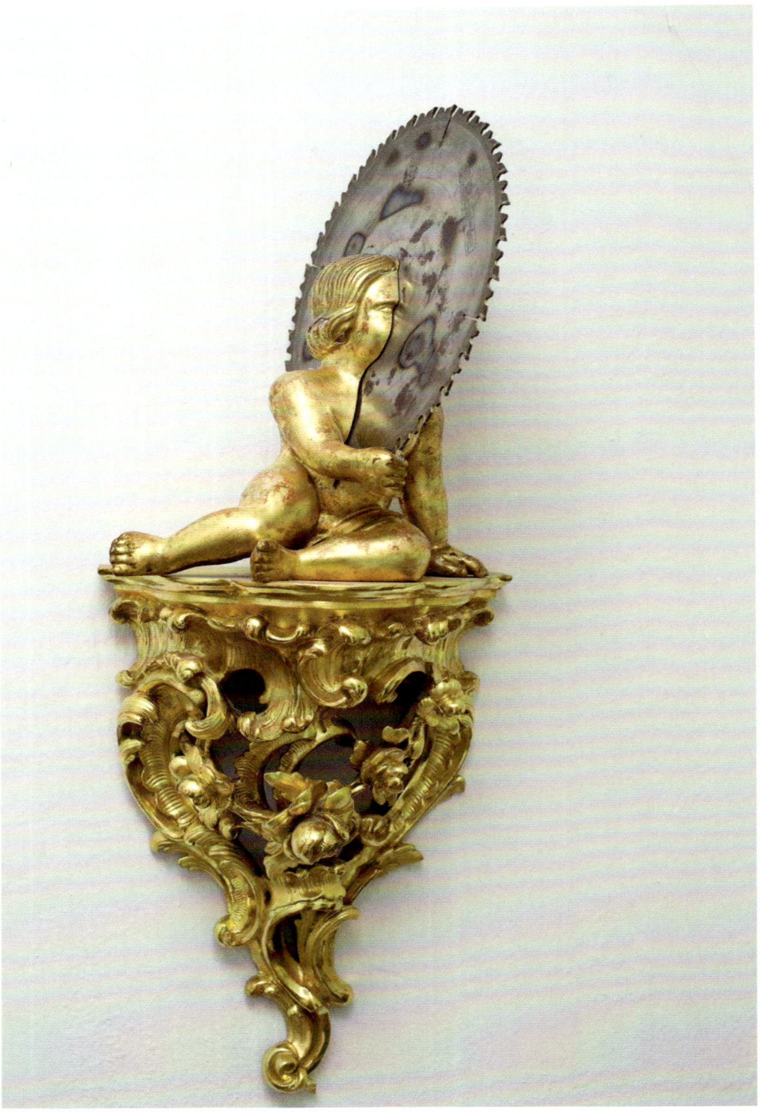

Pepo Pichler
Boy with Halo, 2002
Wood, gold leaf, steel, 98 x 40 x 34 cm
Private collection

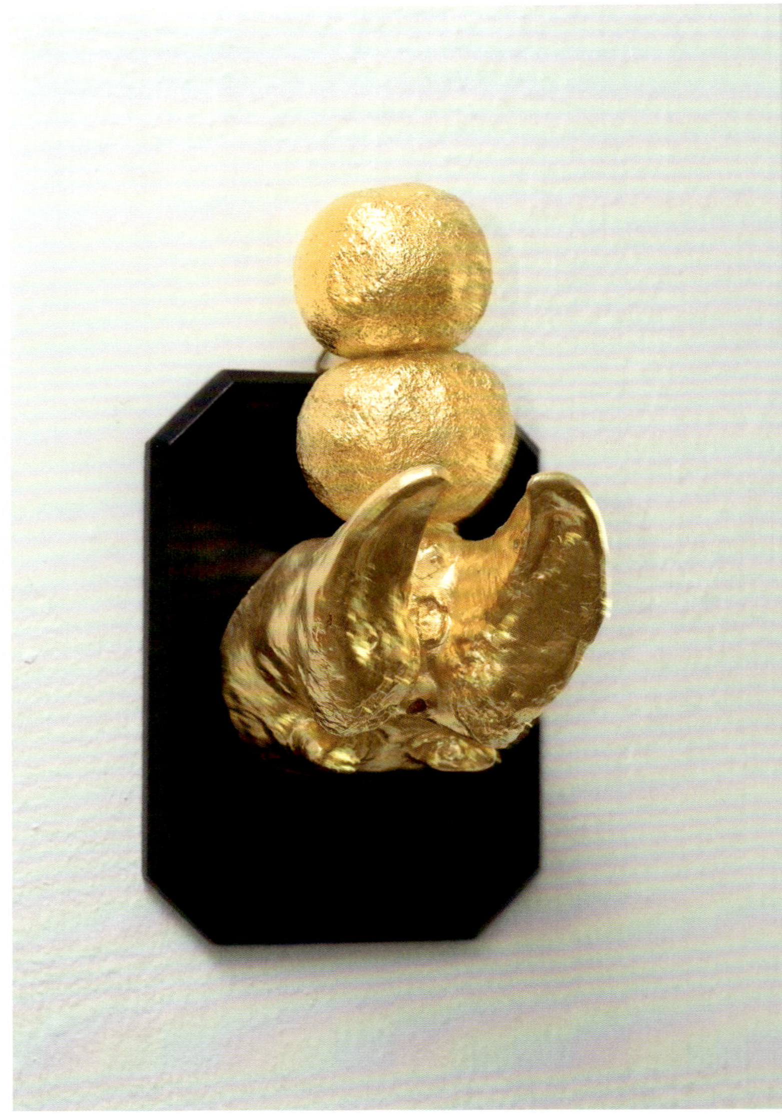

Pepo Pichler
Vache sabot jongleur, 2010
Wood, plaster, gold leaf, concrete, 35 x 18.5 x 18.5 cm
Private collection

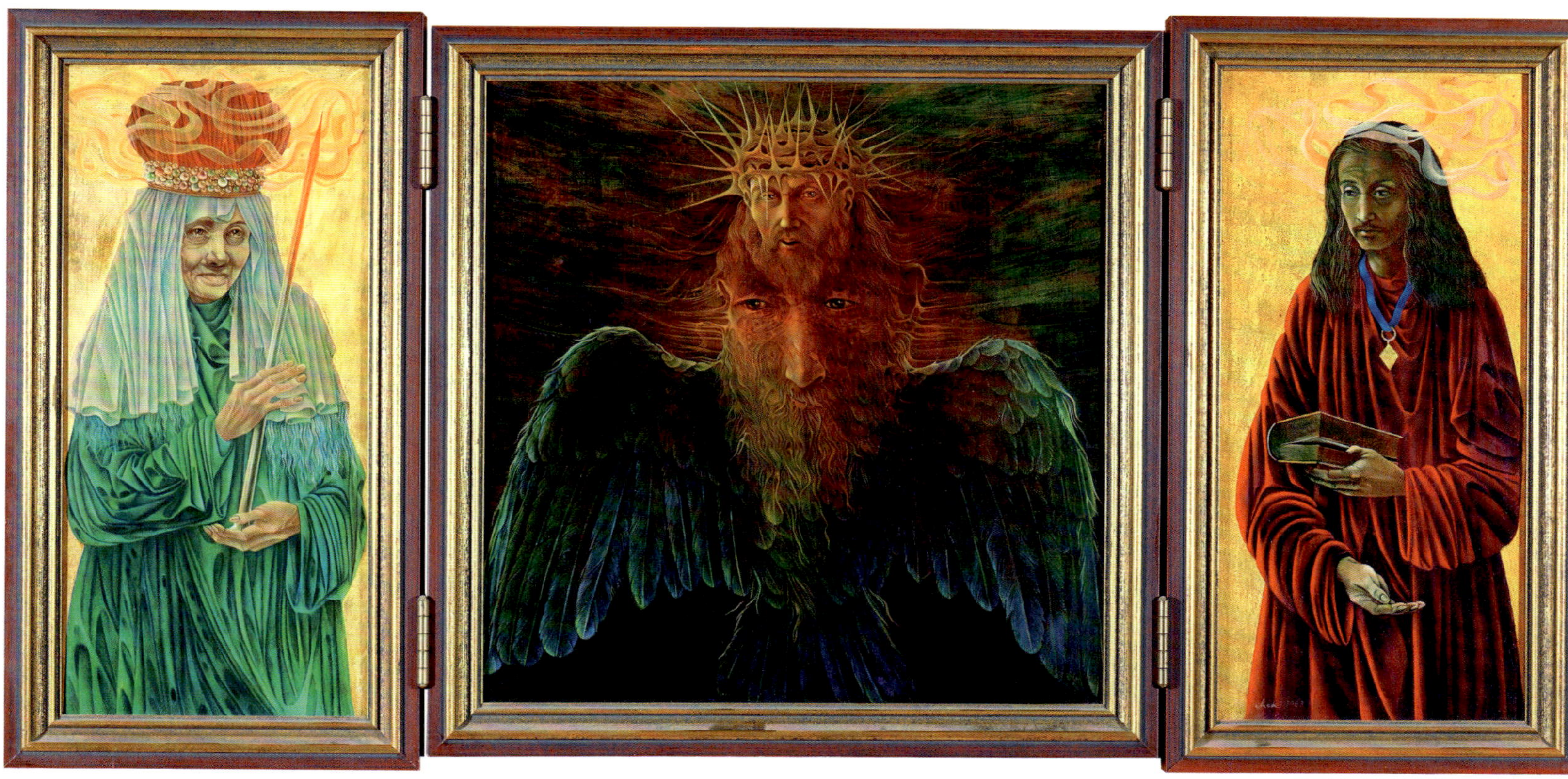

Kurt Regschek
Triptych, 1960
Mixed media on gold leaf and wood, 25 x 32 cm
Franziska Muliar, Gross-Enzersdorf

Ground Gold and other Haloes

Prince Eugen's former bedroom in the Lower Belvedere has, since the mid 1760s, been the site of the 'Golden Cabinet' transferred here from the City Palace. All golden cabinets owe their existence to the exemplar in Ver-sailles, they were paraphrases of the great Hall of Mirrors.

The valuable Chinese porcelain on the pedestals was sold at the time and found its way to Turin. Prior to 1708, it was not known in Europe how to make porcelain, so that the 'white gold' had to be imported from China. The European discovery of porcelain was an accident resulting from the alchemists' search for gold. After a porcelain factory was established in Vienna, too, in 1718 (the second in Europe, after Meissen), doubtless the ancient pieces from China lost their allure.

The iconography of the golden cabinet with the painted personifications of the four known continents, the four seasons and the four elements beneath the stucco relief of Luna and the sleeping hunter Endymion (appropriate to the prince's bedroom) reflects a time-space cosmology which provides the ideal frame for the arrangement. The differences in awareness and physical state evoked here lend the gold an extravagant complexity.

Access to this golden cabinet, filled with gold and its iconographic ramifications, is blocked by Clemens Wolf's *Opened Space (Golden Fence)* (2010). The paradoxical title relates to the gap having been opened by force, using the forceps lying on the floor. With his barriers, Wolf defines the limits of public spaces.

Behind stands *Electra* (2009) by Elisabeth von Samsonow. We know Electra from Greek mythology: she has inspired so many writers from Sophocles and Euripides to Hugo von Hofmannsthal, Ezra Pound and Gerhart Hauptmann. The creator of this gilded limewood sculpture has accompanied and explained the work with mises-en-scène, performances and texts. Of central importance to the mummy-like totem pole are two elements. The name Electra is derived from ἤλεκτρον (ē*lektron*), which means amber. Amber was the first material on which the phenomenon of electrostatic charge was noted. For a long time electricity remained a mysterious force. In the first half of the eighteenth century, at the time when the Golden Cabinet was moved to the Lower Belvedere, it was discovered that there are two forms of electric charge, to which the terms positive and negative were applied.

The artist links the thus-charged *Electra* with the Egyptian belief in the afterlife. The most important act in the Egyptian cult of the dead consisted, following mummification, in the ritual of the opening of the mouth. This completed the actual re-animation of the dead. This rite of the animating mouth-opening was performed on the mummy while it was standing and facing south; among the utensils was a 'finger of gold'. The statue was produced in a workshop called 'Goldhouse'. Its temporary installation in the Golden Cabinet thus recalls its creation. Before it was gilded, a Horus eye was still visible on the statue, so that the iconographic range was more obvious. The high crown is reminiscent, if anything, of a basket, or reproduces the first impression of *Electra* in the play by Euripides, where she declares: "O sable night, nurse of the golden stars! beneath thy pall I go to fetch water from the brook with my pitcher poised upon my head..." [1]

The artist broadens the range of meanings by a re-reading of Friedrich Nietzsche's *Birth of Tragedy from the Spirit of Music or: Hellenism and Music* by transforming the 1900 text through 'sculptural' interventions. Brigitte Borchhardt-Birbaumer calls the text, changed by the addition of diaphanous white highlights, and re-titled *Die Geburt des Mädchens aus dem Geiste der Plastik* ('Birth of the Girl from the Spirit of Sculpture') as the 'primal story of the twenty-first century'. On page 168 of Nietzsche's classic are directions for use: "Here it will be needful to thrust ourselves with a bold run-up into a metaphysics of art in that I repeat the early statement that existence and the world only seem justified as an aesthetic phenomenon."[2]

The author whites out the train of thought that follows, namely that the ugly and discordant also constitute an artistic game; these categories do not concern her, for after all, 'Harmonia' is, according to Pausanias, a daughter of Electra. The multiply ambiguous *Electra* (2002) is surrounded by 16 gilt-bronze heads by George Condo. The eclectic variety of the visible influences of pre-Columbian, classical and neo-classical models creates a confusing sum. "Perhaps George Condo is the alter ego of art history." (Thomas Kellein)[3]

Ground Gold

Since 2009 the Lower Belvedere has been connected with the Orangery by a steel-and-glass corridor along the Kammergarten. The visitor's tour of the exhibition is obstructed here, at first in three places, by gold-leaf on the floor. The thresholds will remain for different lengths of time: the first will only be available to walk over on the opening day, the next for a few days, and the third for the whole duration of the exhibition. The three armoured-glass panels coated in gold will, afterwards, be put together by Johannes Angerbauer to form a triptych. It will document a process of gold being trodden underfoot by visitors. The individual reactions will be very different. Doubtless the prevailing emotion will be that a taboo is being infringed.

Excursus: The taboo of gold on the floor

There are several reasons for the reluctance to tread on gold. For one, the noble sheen is damaged, for the skin of the thin gold leaf is very sensitive. For another, there is a reluctance to walk into, or on to, a work of art. Thirdly, there is a culture-historical link between gold and shadow. The medieval gold ground and the post-medieval shadow or shading are incompatible. Three-dimensional thinking put an end to the gold ground. The word 'ground' in this sense always referred to the background and not to the ground beneath one's feet. The link between gold and shadow, however, became apparent once more with the rediscovery of the Middle Ages, albeit not in painting but in literature. In *Peter Schlemihls wundersame Geschichte* (1814) by Adelbert von Chamisso, the eponymous hero exchanges his shadow for an inexhaustible bag of gold. In a key scene, he strews more and more gold on the floor, "trampled upon it, feasted on its very sound and brilliancy [...] rolling and revelling on the gorgeous bed, until I sank exhausted. [...] night found me still lying on the gold, where, at last, sleep overpowered me."[4] Here for the first time we have a horizontal gold floor, which, by the swallowing action of the shadow, turns the human being into a 'nobody'. A person without a shadow has no identity, because in the world of the bright-dark, no space forms around him.[5]

A fourth reason for the reluctance to tread on the golden floor lies in the fact that the idea of durable value will be undermined. "Gold as the abstract basis of our market economy, gold as the symbol of power and thus the main aim of our society is imagined as an earthy material that has neither symbolically nor factually an eternal value, but like any other element is bound into changing movements."[6] By leaving our traces on the gold, we become accomplices in the social process of an intention aimed against the durability of gold.

Angerbauer's "gold objects want to be trodden underfoot, scuffed, scratched, rejected and deconsecrated, in order thus to bring 'truth to light.'"[7]

On display in the Orangery is also a diptych by Angerbauer, who, after a break, has been using the pseudonym 'Johannes Goldhoff' once more since 2008.[8] This makes his ideas, which can also be read in the internet, clear.[9] In this *Kristalltag, 10. Nov. 1998,* which, like all of his works, had its origin in an 'action', the traces of expellees are left on the gold between its armoured-glass panes. The second picture shows, behind the broken glass, the yellow picture of the German-Polish Jew Herschel Feibel Grynszpan, whose shooting of a German diplomat in Paris in 1938 unleashed the pogrom known as the 'Kristallnacht'[10]. The similarity of Grynszpan's name with that of Alan Greenspan is welcome. He was accused in the USA in 2001 of manipulating the gold price, but for the legendary banker, no ill consequences came of it. Any profits from an 'action' by Angerbauer go back into socially useful projects (AIDS, tsunami victims etc.); the 'actions' on principle are always directed against the speculation in gold.

Other Haloes

The 'Prunkstall' in the Lower Belvedere is a museum within a museum, an exhibition area for medieval art. In this treasure house there are numerous examples for the discussion of the gold ground. The important collection was on display in the Orangery from 1953. After the latter was renovated and enlarged, there was a restructuring. Part of the collection was presented in a museum context in the Upper Belvedere, and since 2007 in the Prunkstall. The medieval art of the altarpieces and sculptures concentrates on depictions of the Passion and the Saints. During the Gold exhibition, they are confronted with new pictures on a gold ground and two sculptures by Jan Fabre that span the eras. The *Triptych with Philoxenia in the central section* (Crete, sixteenth century) from the Kunsthistorisches Museum in Vienna is set in a positively Mannerist superstructure with a crowning pine-cone motif. Placed in the centre between Abraham and Sara is Philoxenia, the personification of hospitality. The meanings of the scene range from the Trinity to the depiction of Pentecost. On the left-hand wing the Virgin Mary appears in the Burning Bush with the handing of the Tablets of the Law to Moses. On the right sits Christ, with the woman of Samaria approaching.

The tradition of icon painting continues unbroken to this day. The iconographic motifs repeat themselves, which corresponds to the timeless of the themes. Imaginative effects are often limited to the external form of the framing. Even today as with Mary-Frances Dondelinger's Nativity (2008), the artistic freedom is manifested in exuberant shapes of frame. Religions have long since been supplemented, if not replaced, by esoteric traditions. Jan Fabre's *Holy Dung-beetle* (2011) with a cross referring to the Passion was, as the scarab, the symbol of the Egyptian sun god Re. The *Golden Human Brain with Angels' Wings* (2011) recalls the postulated end of the soul in present-day neuro-science.

In the other pictorial works, which owe more to the Japanese tradition, the cast of manga heroes has taken command. The fighting machines and heroes from computer games are replacing the Western drama of the fight between the powers of darkness and Christian redemption. In *Saints* (2008), by Gajin Fujita, the gold leaf is used just like the spray-paint of graffiti artists. Fujita is a member of the graffiti teams K2S (Kill 2 Succeed) and KGB (Kids Gone Bad) in Los Angeles. The signs in the background proclaim the secret message of a fearsome masked samurai, which is inspired by a portrait by Utagawa Toyokuni (1807) of the actor Danjuro VII in the role of Shibaraku.[11] Fujita also reproduces in red drawing the Kumadori make-up, which represents virtue and strength. Shibaraku means, literally, 'stop for a moment' and is called out in the Kabuki repertoire before the appearance of the actor. His flowing black garment, in Fujita's picture, is decorated with lilies. Since 2000, the lily has been the logo of the New Orleans Saints football team. Fujita painted the picture to support the Saints, who had lost their home after hurricane Katrina.

Hisashi Tenmyouya calls his painterly work 'Neo Nihonga' ('new Japanese painting'). His motif-repertoire of fighting machines and 'spirits' has been made famous by its publication in an opulent book of tattoo art patterns that appeared in 2003 and was quickly sold out.[12] The combination of old ideologies and Japanese pride has made him popular. In the monumental pictures, the skin is replaced by gold leaf, which sets them moving in a new lustre. The exotic ideas, nurtured by manga and anime[13] open up the metaphorical world of spiritual struggles. The sentai (warriors) form a contemporary counterpoint to the Christian world of the saints.

1 Euripides, *Electra*, translated by E.P. Coleridge, available under http://classics.mit.edu/Euripides/electra_eur.html
2 Elisabeth von Samsonow on Friedrich Nietzsche, *Die Geburt des Mädchens oder: Griechenthum und Futurismus,* Vienna 1900/2011. The title means: 'The birth of the girl or: Hellenism and Futurism'.
3 Thomas Kellein, *George Condo—Sculpture*, Zürich 2003, p. 4.
4 From the translation *Peter Schlemihl The Shadowless Man*, translator unknown, print-on-demand from Kessinger Publishing, 2010, p. 11
5 Victor I. Stoichita, *Eine kurze Geschichte des Schattens*, Munich 1999, pp. 165 ff. The fact that we are dealing here not with a gold ground, but with a floor strewn with gold ducats, is irrelevant in our context.
6 Peter Assmann, in: *Dokumentation der Aktion "T.A.09605—Tsunami"*, URL: <http://www.ta-art.com/D/texte/assmann.htm> [read on 12 Dec. 2011]
7 Raimund Locicnik, "Das Denken verschüttet die Wahrheit", in: *Passagen*, 5/98.
8 URL: <http://www.socialgold.com/D/biografie.htm> [read on 24 Nov. 2011]
9 URL: <http://www.ta-art.com/kristalltag.htm> [read on 24 Nov. 2011]
10 Translator's note: 'Kristallnacht' (literally 'crystal night') is usually translated 'night of broken glass'. It refers to 9 November 1938. The title of the work 'Kristalltag' can be translated analogously, substituting 'day' for 'night', and of course the date is relevant.
11 Illustration in: Klaus Albrecht Schröder (ed.), *Japan—Die fließende Welt. Aus der Sammlung Drennig (*exhib. cat. Kunstforum Wien), Vienna 1991, p. 13.
12 *Japanese Spirit Tenmyouya Hisashi OOP tattoo art book,* Tokyo 2003.
13 Mangas are Japanese comics, anime is derived from *animēshon*, i.e. animation.

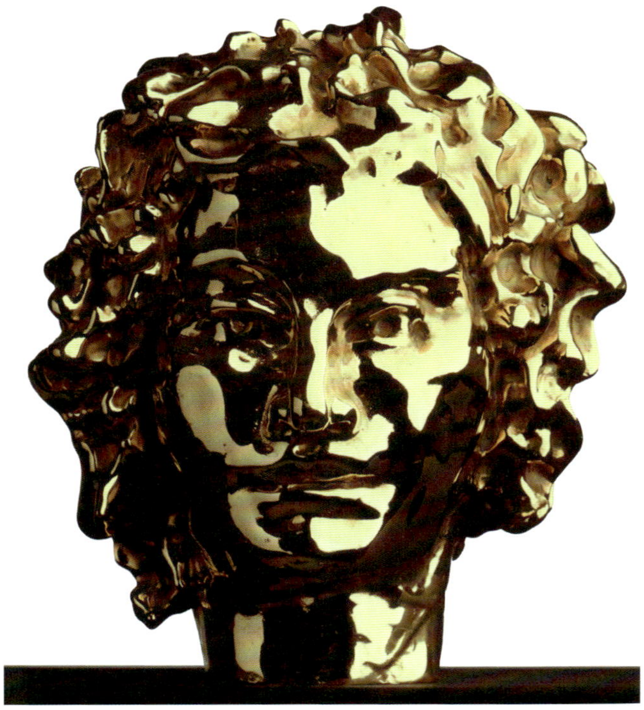

George Condo
Dionysus, 2002
Bronze-gilt, 23.5 x 20.5 x 20 cm
Private collection, Courtesy Galerie Bruno Bischofberger, Zurich

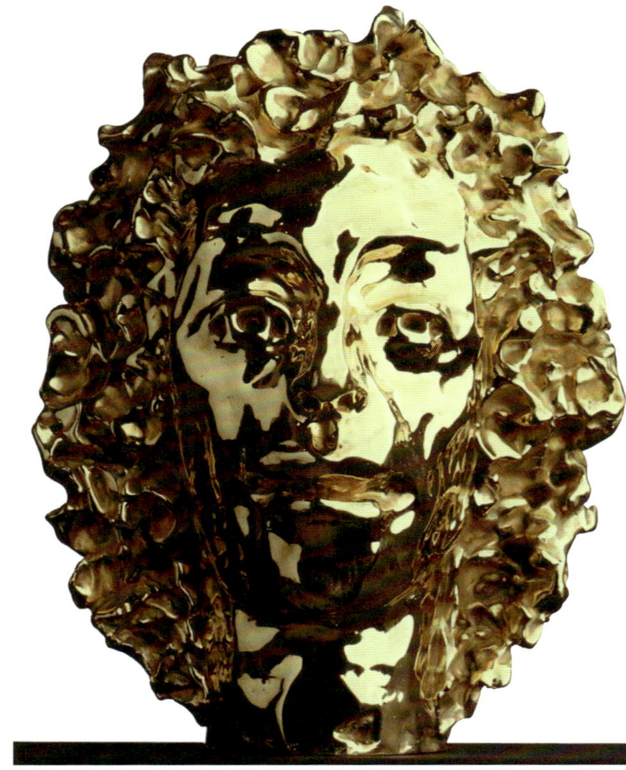

George Condo
The Old Water Nymph, 2002
Bronze-gilt, 31 x 23 x 22.5 cm
Private collection, Courtesy Galerie Bruno Bischofberger, Zurich

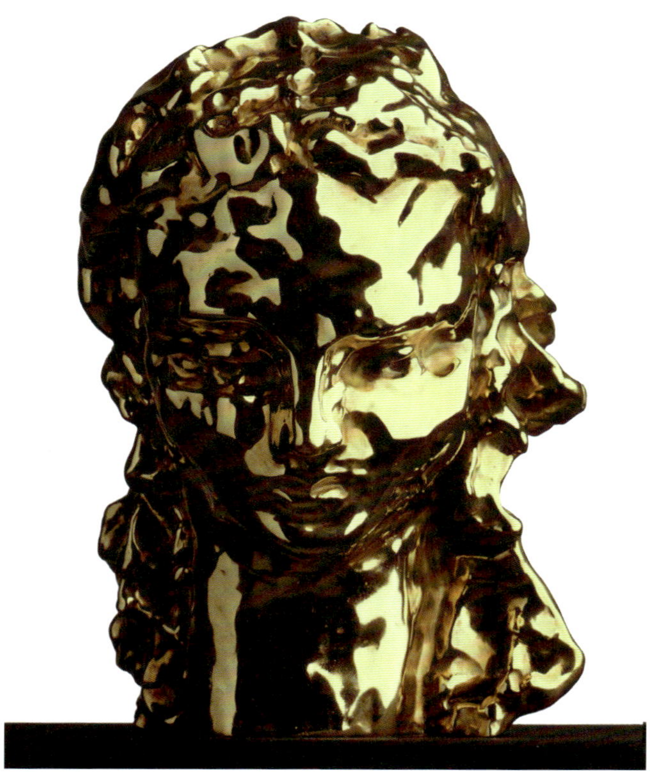

George Condo
Lamentation, 2002
Bronze-gilt, 29 x 20 x 17.5 cm
Private collection, Courtesy Galerie Bruno Bischofberger, Zurich

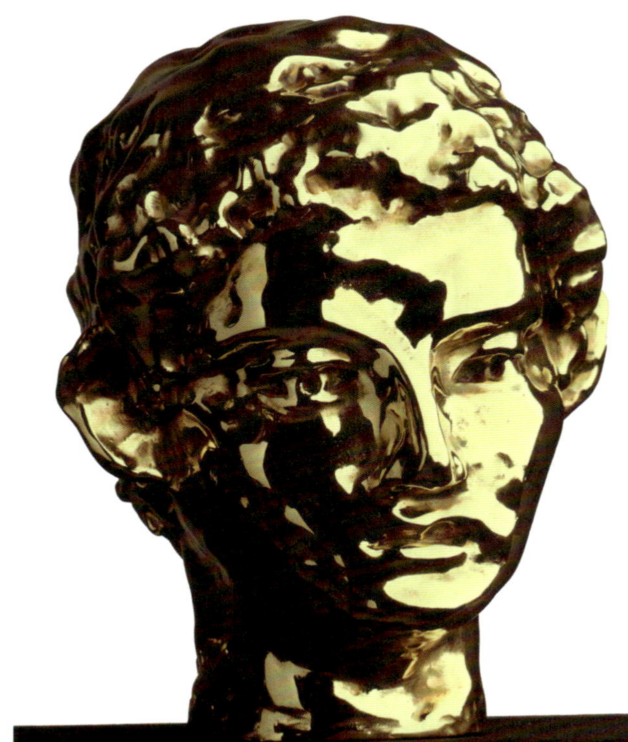

George Condo
The Linden Boy, 2002
Bronze-gilt, 21 x 18 x 19 cm
Private collection, Courtesy Galerie Bruno Bischofberger, Zurich

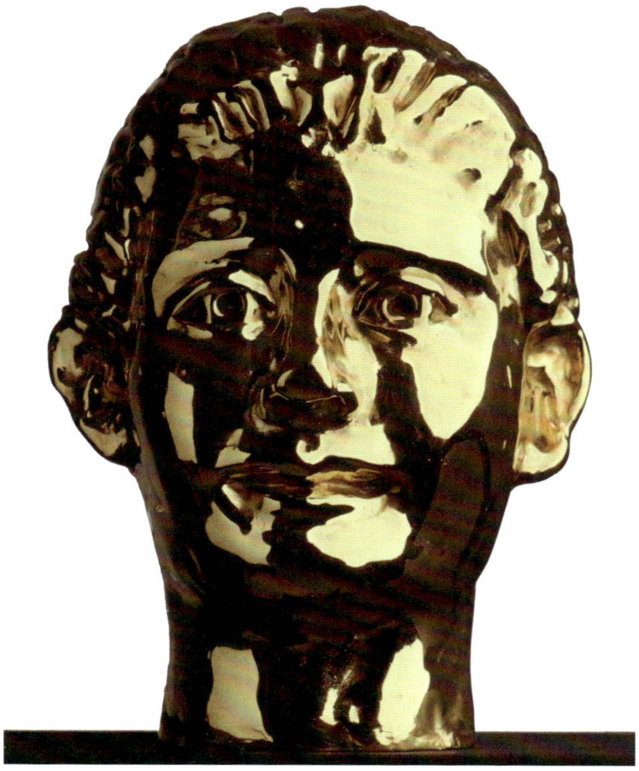

George Condo
Young Warrior, 2002
Bronze-gilt, 22.5 x 17.5 x 20.5 cm
Private collection, Courtesy Galerie Bruno Bischofberger, Zurich

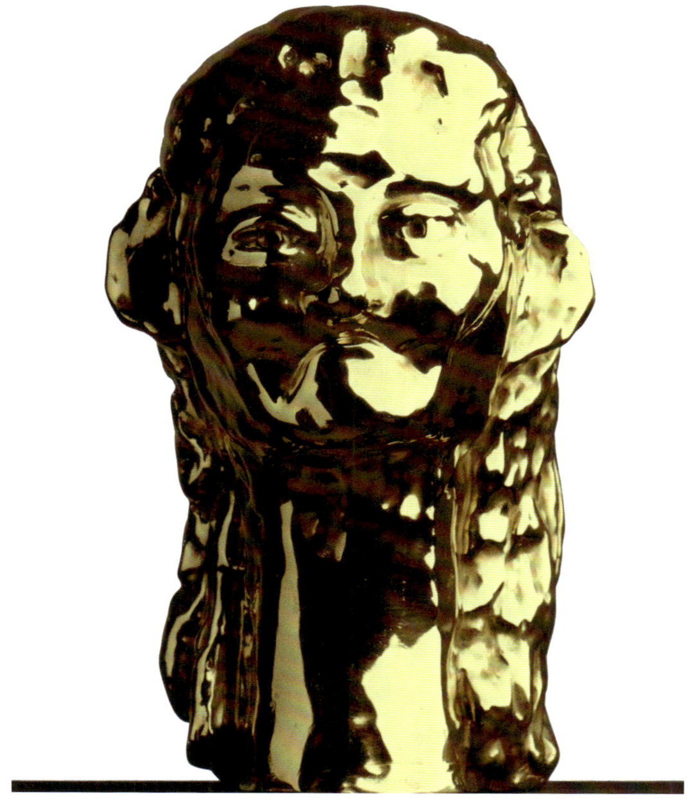

George Condo
The Old Sea Hag, 2002
Bronze-gilt, 27 x 16 x 20.5 cm
Private collection, Courtesy Galerie Bruno Bischofberger, Zurich

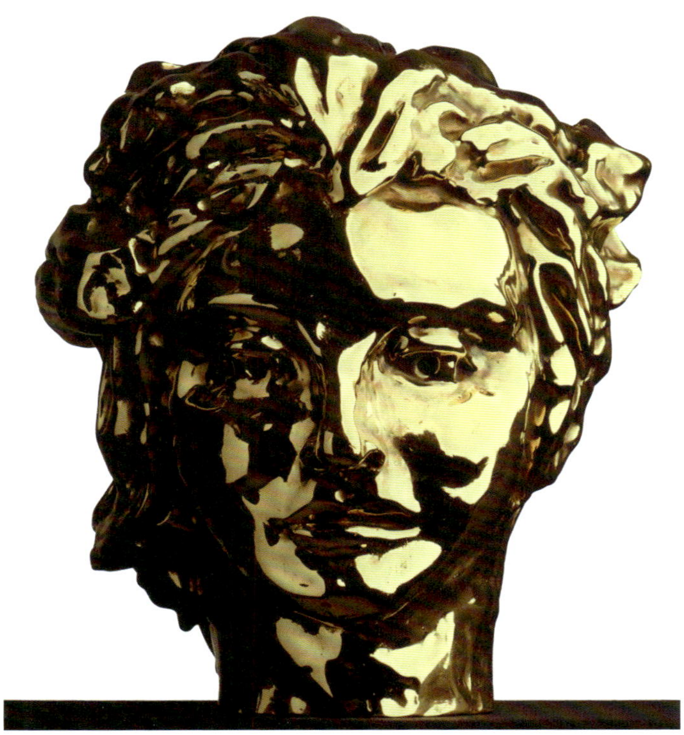

George Condo
Perseus, 2002
Bronze-gilt, 23 x 20.5 x 19 cm
Private collection, Courtesy Galerie Bruno Bischofberger, Zurich

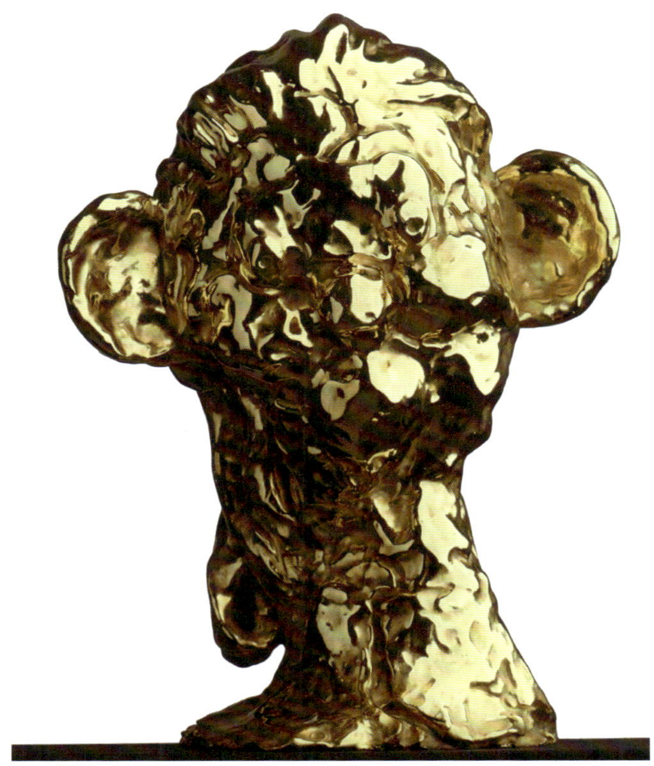

George Condo
The Alcoholic, 2002
Bronze-gilt, 36 x 27 x 21.5 cm
Private collection, Courtesy Galerie Bruno Bischofberger, Zurich

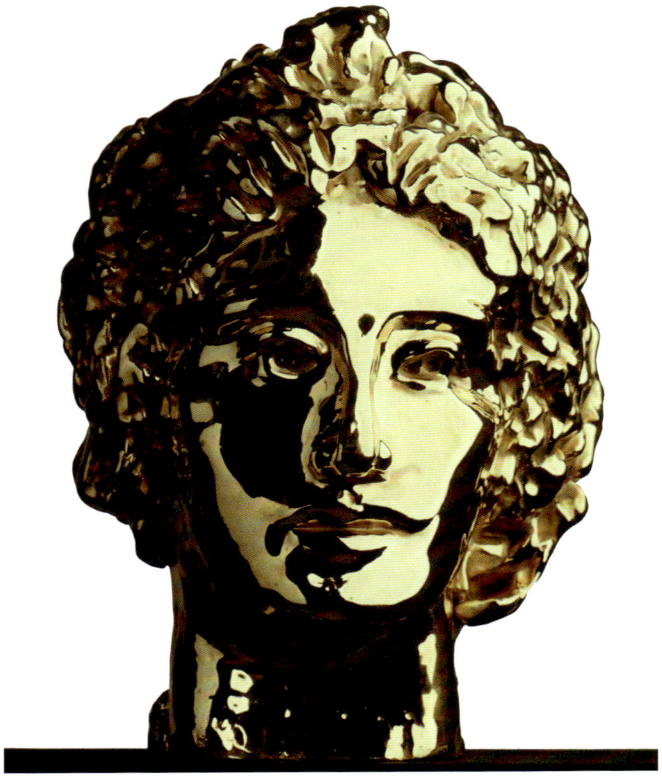

George Condo
The Alexandrian Women, 2002
Bronze-gilt, 32 x 22 x 26 cm
Private collection, Courtesy Galerie Bruno Bischofberger, Zurich

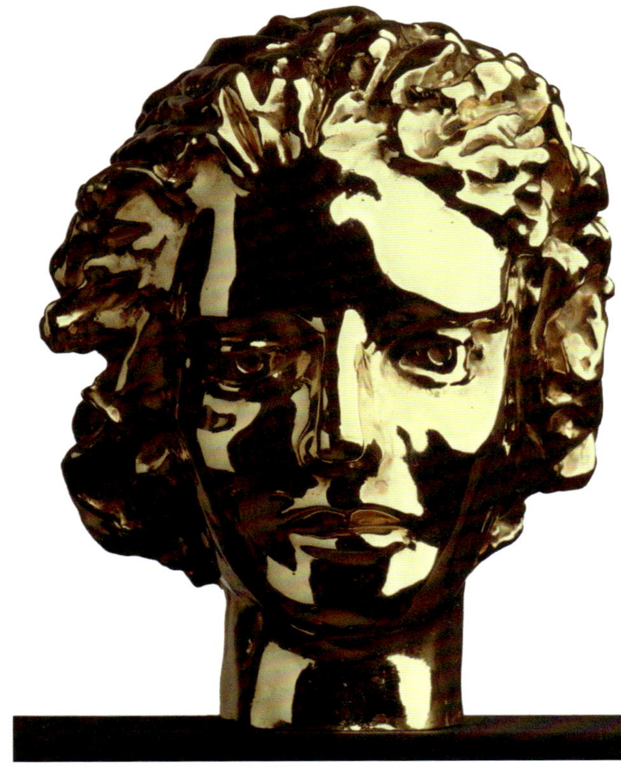

George Condo
The Crying Girl, 2002
Bronze-gilt, 26 x 19.5 x 23 cm
Private collection, Courtesy Galerie Bruno Bischofberger, Zurich

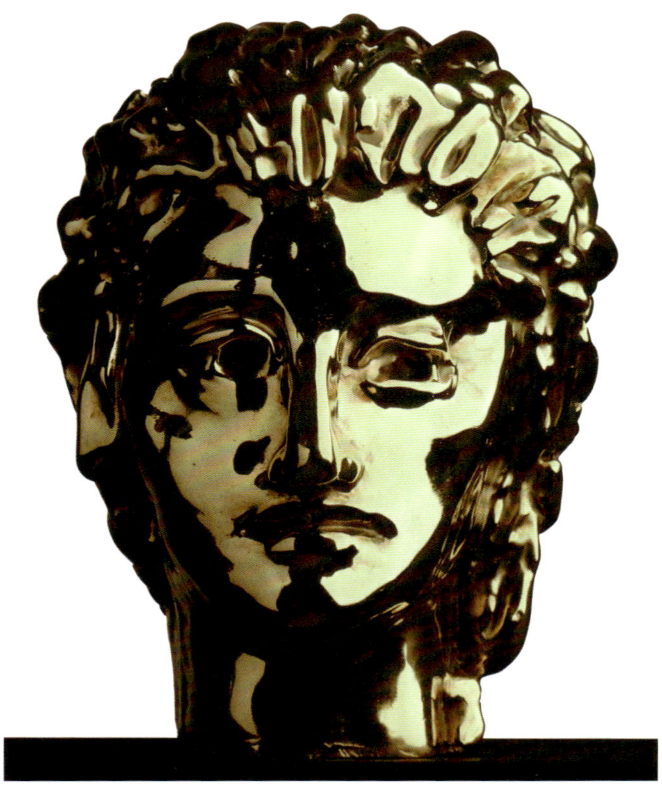

George Condo
The Frenchman, 2002
Bronze-gilt, 23 x 17 x 20 cm
Private collection, Courtesy Galerie Bruno Bischofberger, Zurich

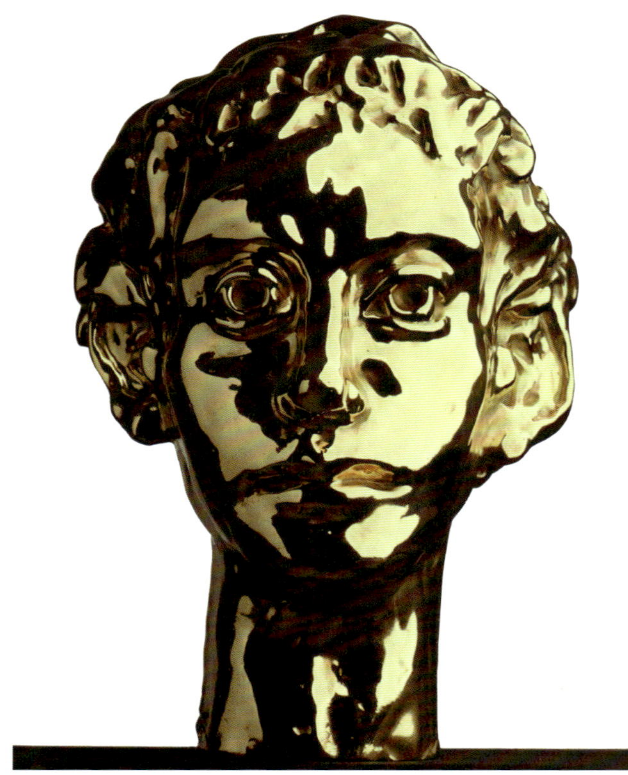

George Condo
The Senate Council, 2002
Bronze-gilt, 30 x 20 x 22.5 cm
Private collection, Courtesy Galerie Bruno Bischofberger, Zurich

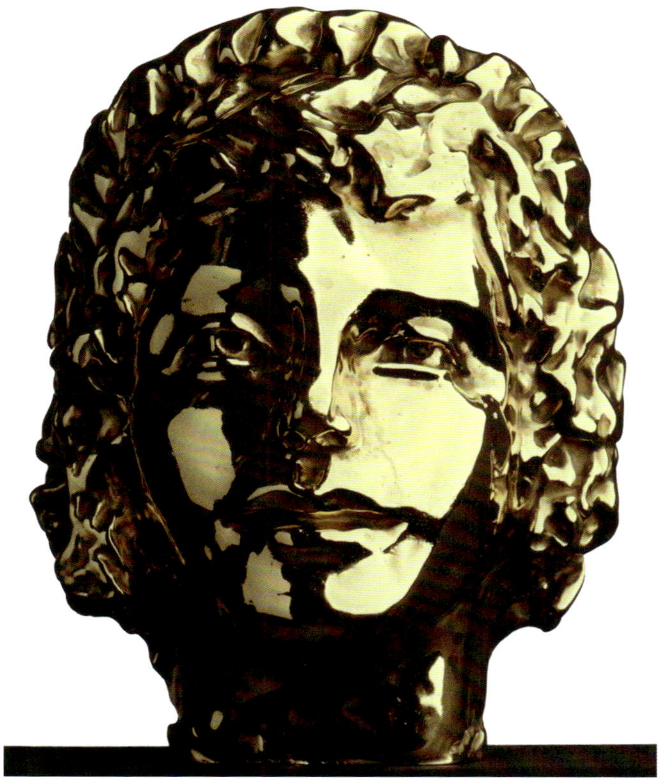

George Condo
Pan, 2002
Bronze-gilt, 22 x 17 x 19.5 cm
Private collection, Courtesy Galerie Bruno Bischofberger, Zurich

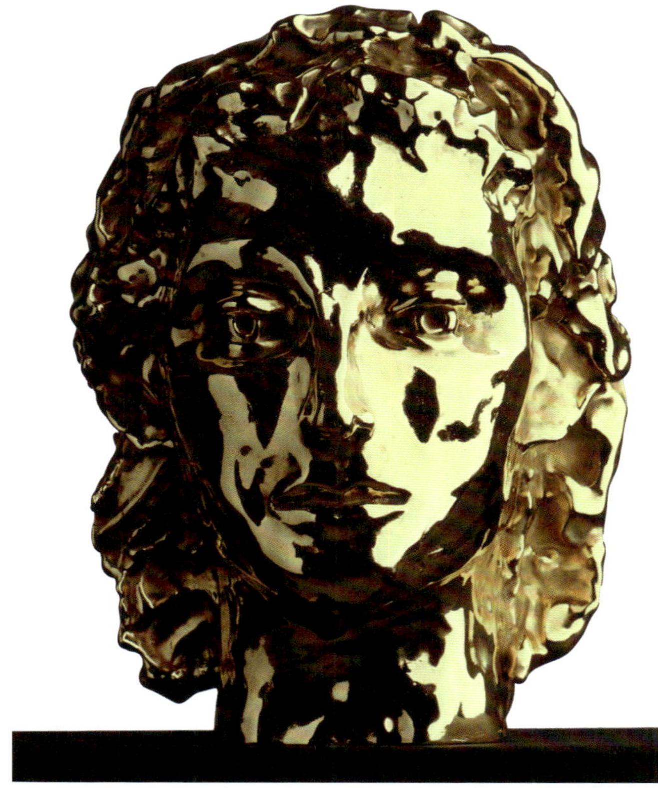

George Condo
The Virgin, 2002
Bronze-gilt, 26.5 x 20 x 21 cm
Private collection, Courtesy Galerie Bruno Bischofberger, Zurich

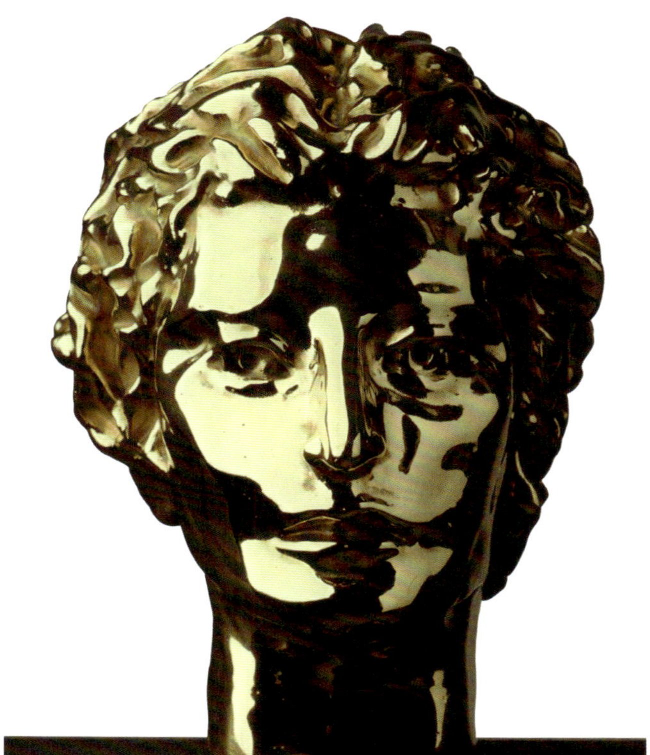

George Condo
Tristessa, 2002
Bronze-gilt, 21 x 15.5 x 18 cm
Private collection, Courtesy Galerie Bruno Bischofberger, Zurich

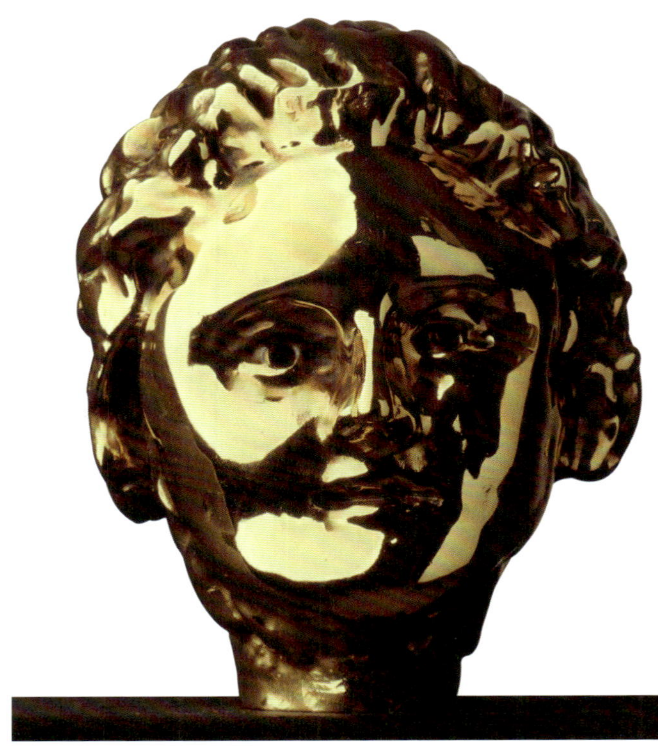

George Condo
Atlanta, 2002
Bronze-gilt, 15 x 11 x 13 cm
Private collection, Courtesy Galerie Bruno Bischofberger, Zurich

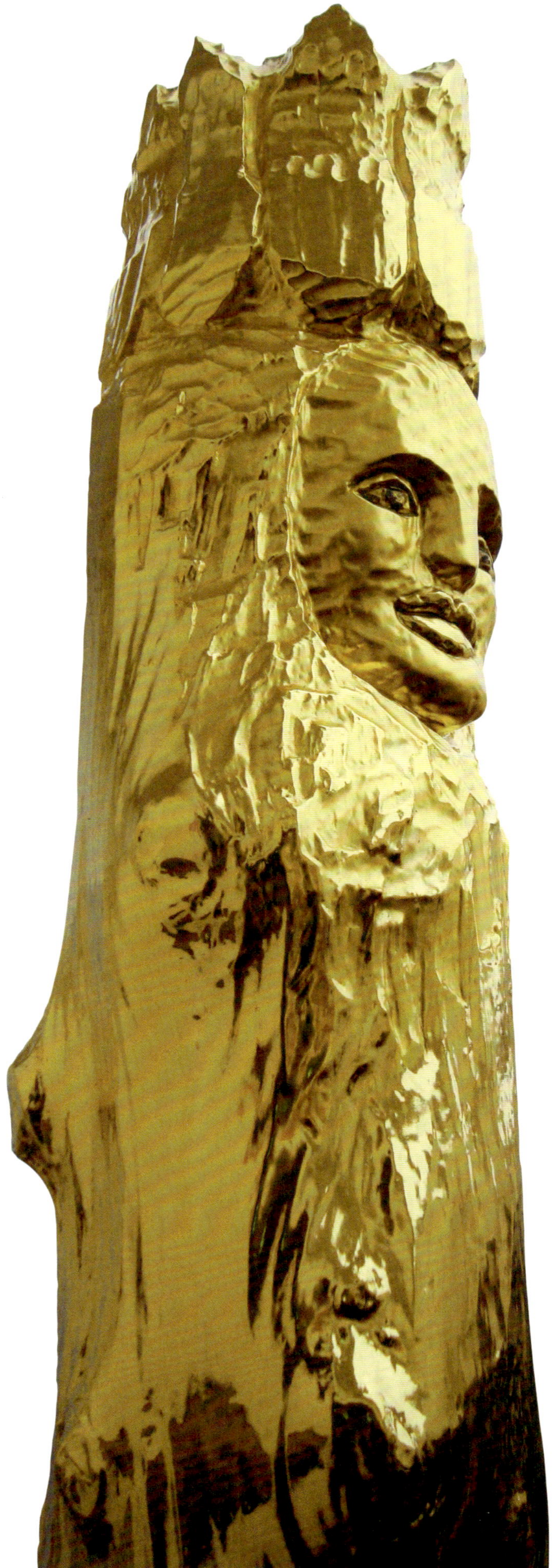

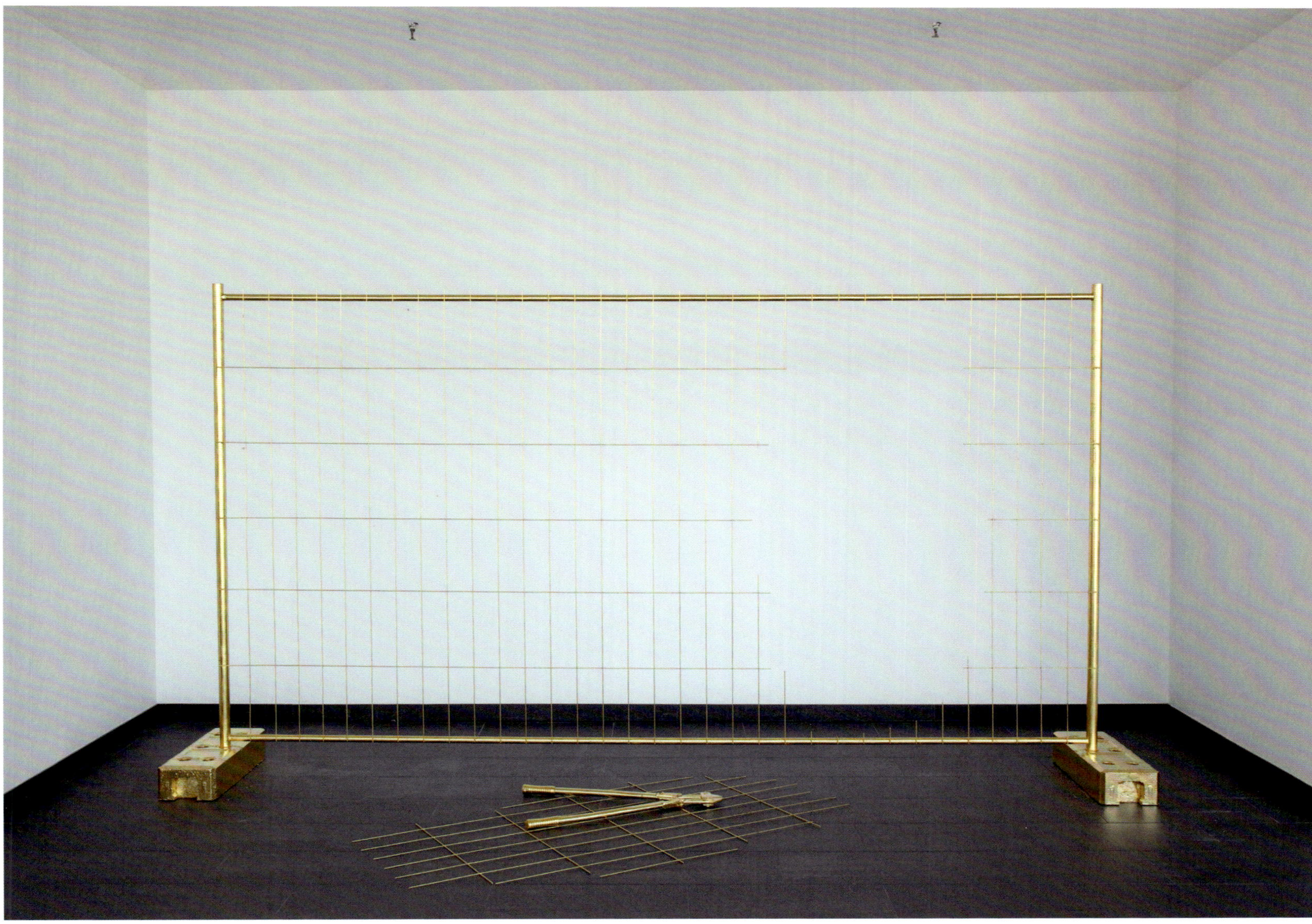

Clemens Wolf
Opened Space (Golden Fence), 2010
24-carat gold leaf, metal fence, concrete blocks, bolt cutter, 200 x 380 x 150 cm
Galerie Steinek

Elisabeth von Samsonow
Elektra, 2009
Lime wood, gold leaf, oil glaze, height 300 cm
In the artist's possession, Courtesy the artist

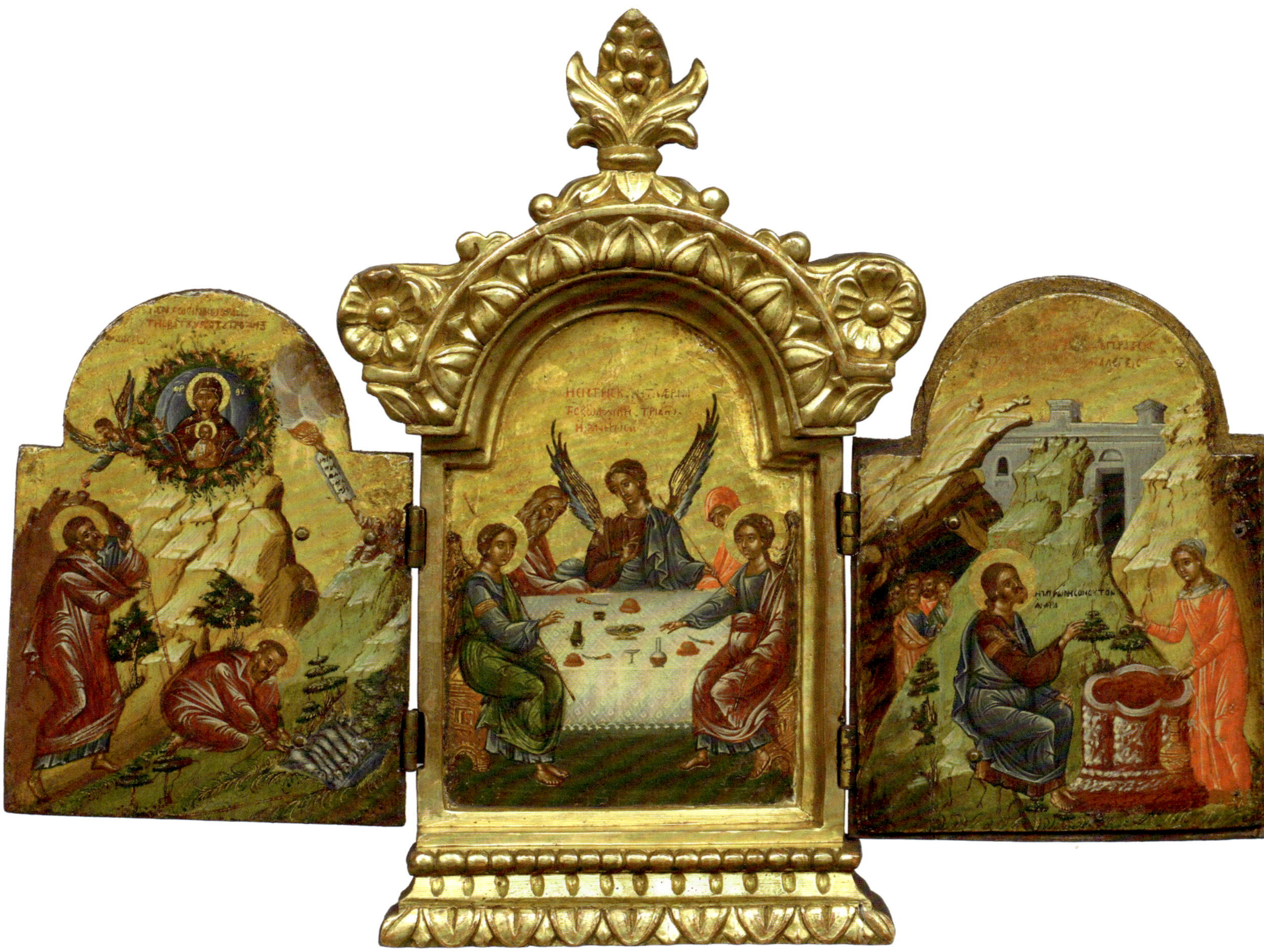

Greek (Cretan painter)
Hinged altarpiece (triptych) with
Philoxenia in the central section, 16th cent.
Wood, painted, 25 x 34 cm (open)
Kunsthistorisches Museum, Vienna, Kunstkammer

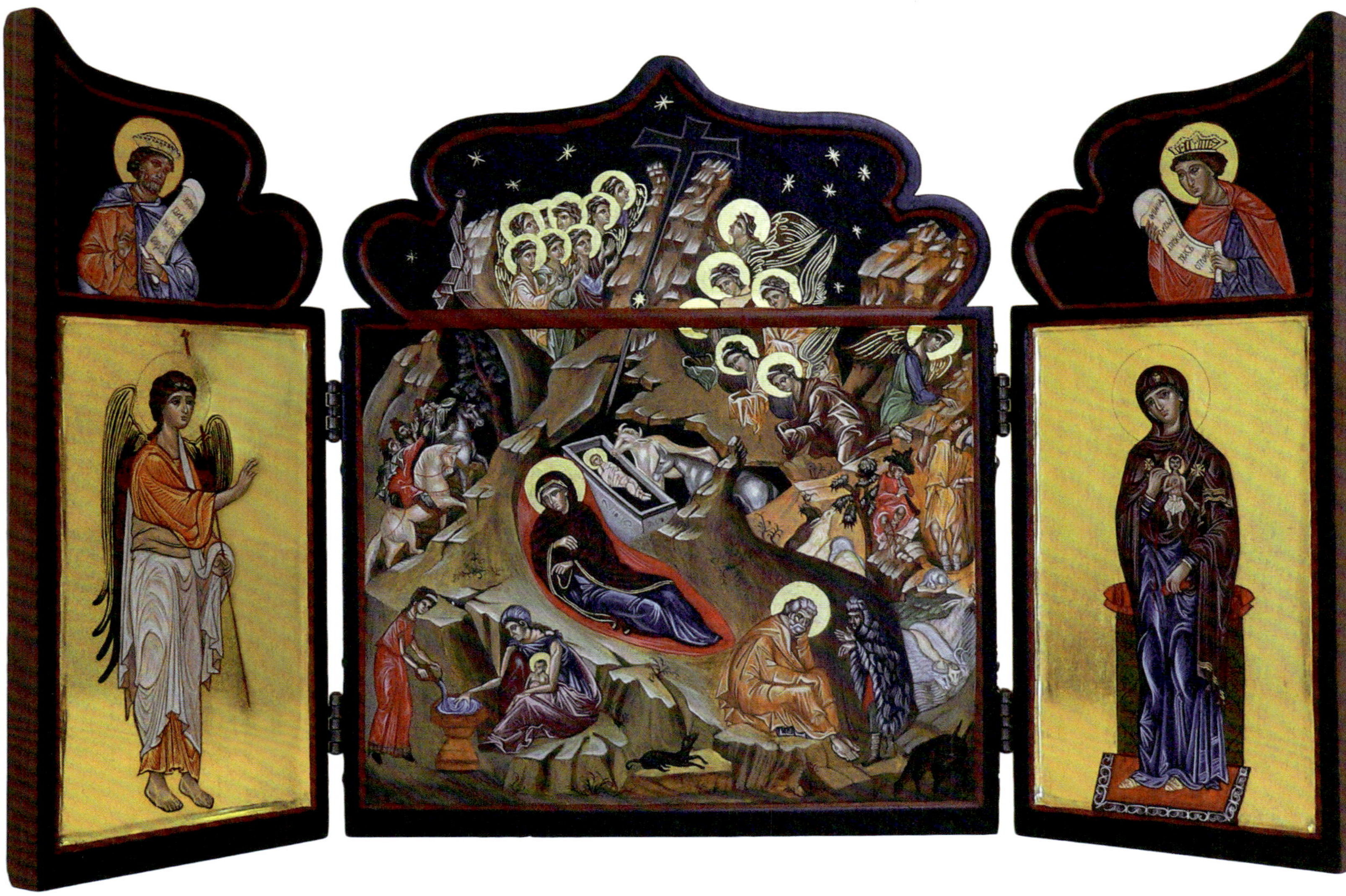

Mary Frances Dondelinger
Nativity, 2008
Egg tempera, 23 and 24-carat gold on
wood panels, 43.2 x 72.4 cm
Courtesy the artist

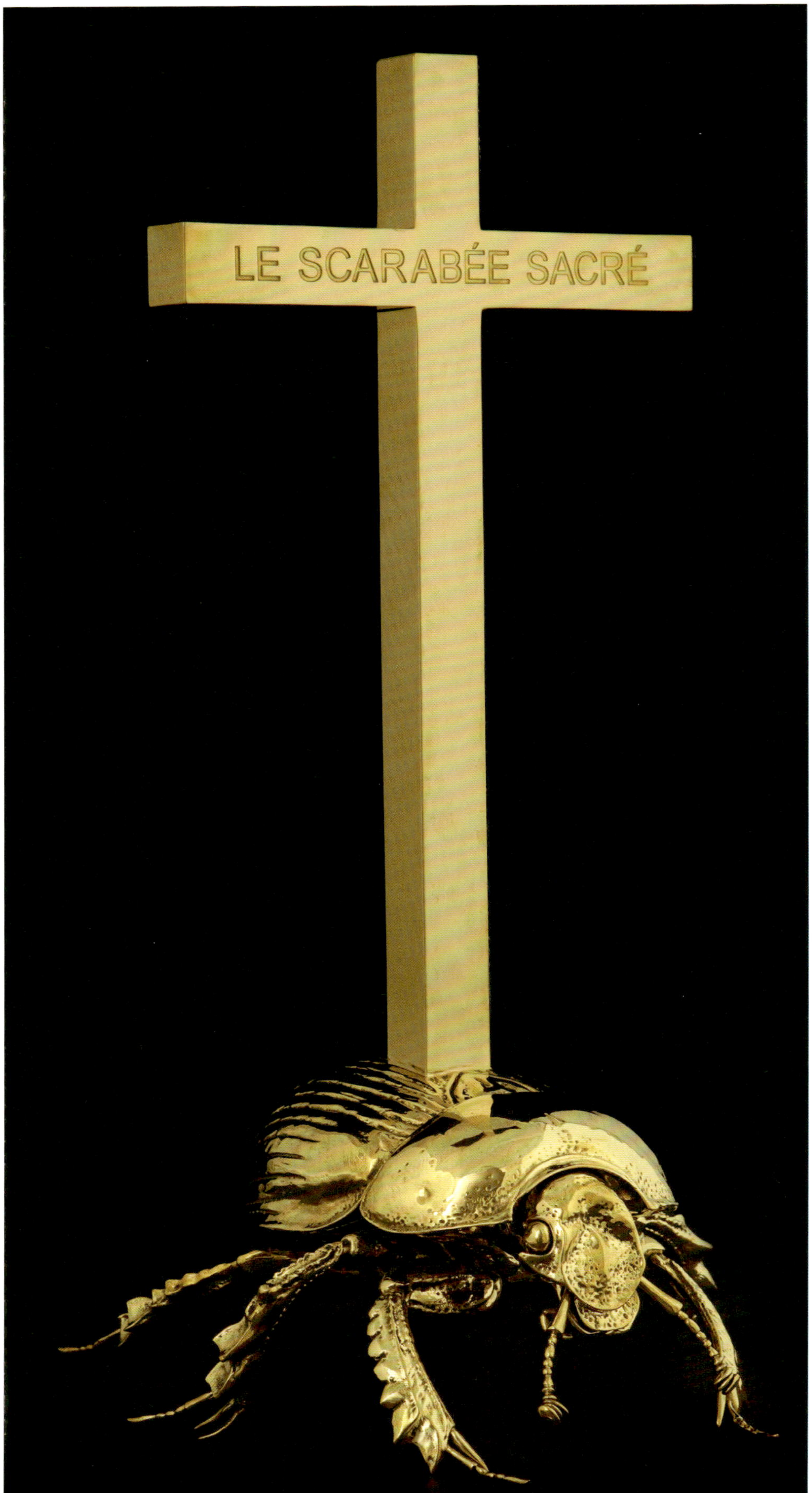

Jan Fabre
Golden holy dung beetle, 2011
Silicone bronze, 24-carat gold,
90 x 45 x 45 cm
Courtesy Mario Mauroner
Contemporary Art Gallery, Vienna

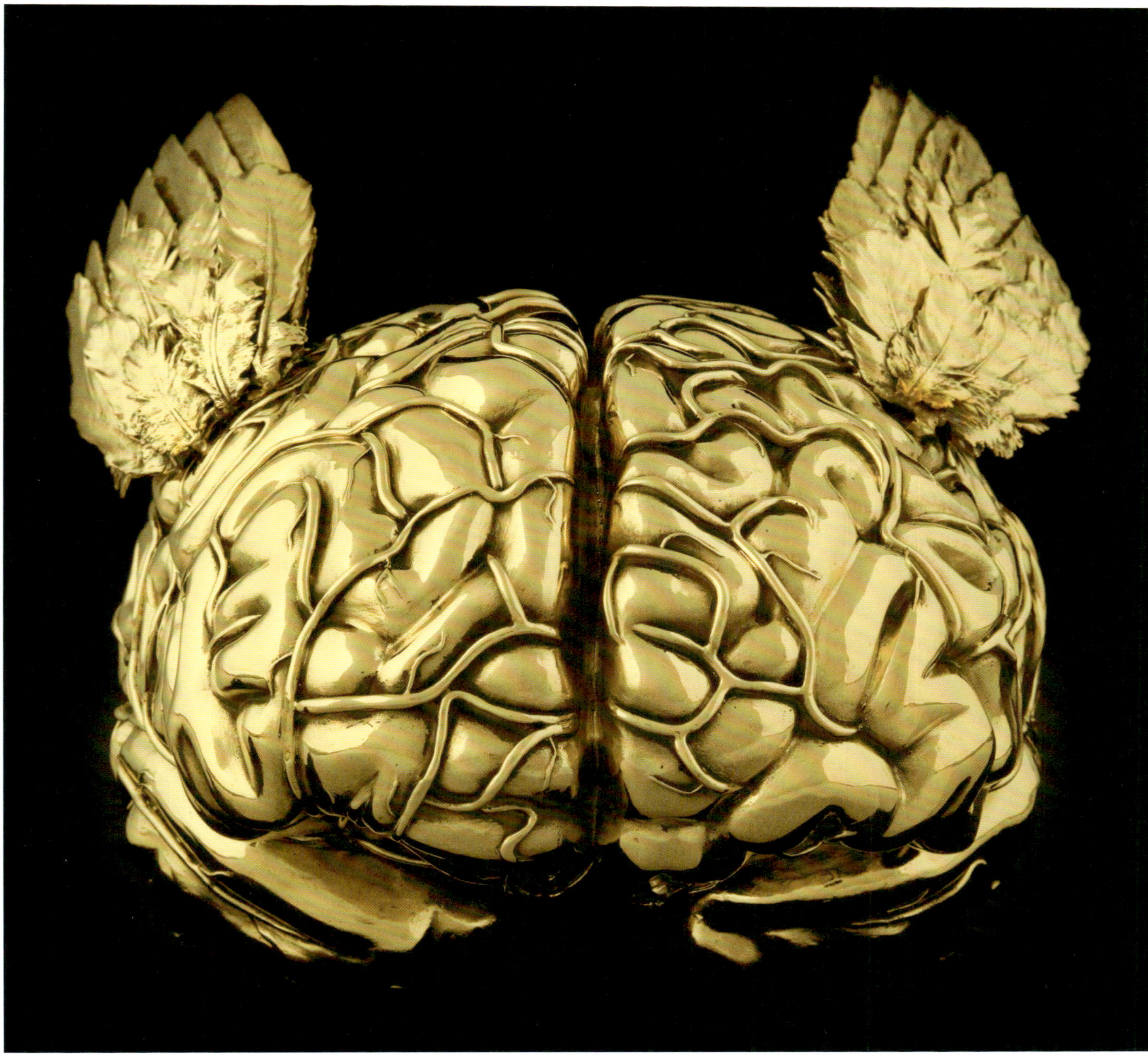

Jan Fabre
Golden human brain with angel wings, 2011
Silicone bronze, 24-carat gold, 22 x 30.5 x 26 cm
Courtesy Mario Mauroner Contemporary Art Gallery, Vienna

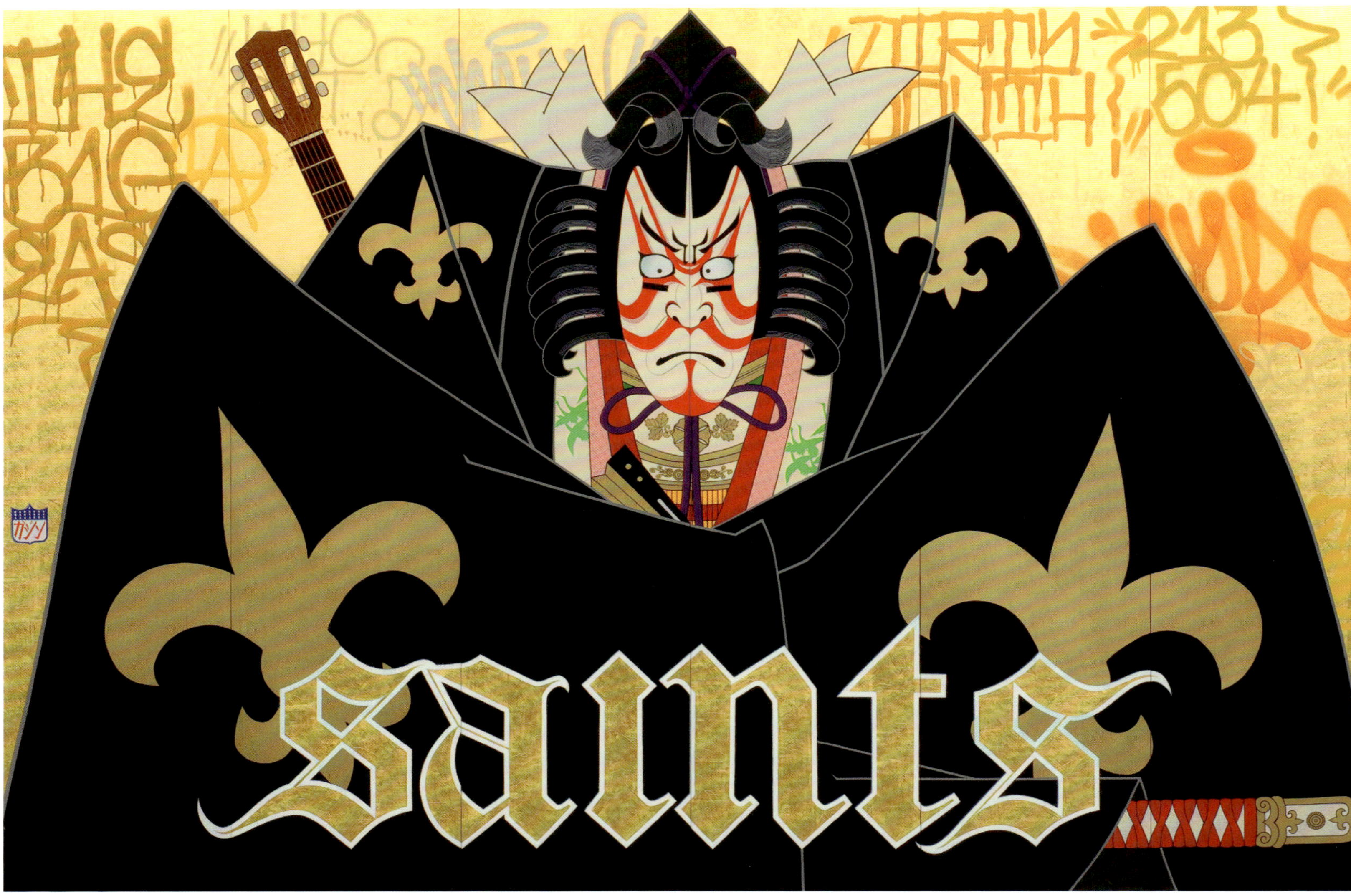

Gajin Fujita
The Saints, 2008
Gold leaf, acrylic, spray-paint and paint-markers on wood, 182.9 x 274.3 cm
U.S. Biennial Inc., New York, Prospect New Orleans collection

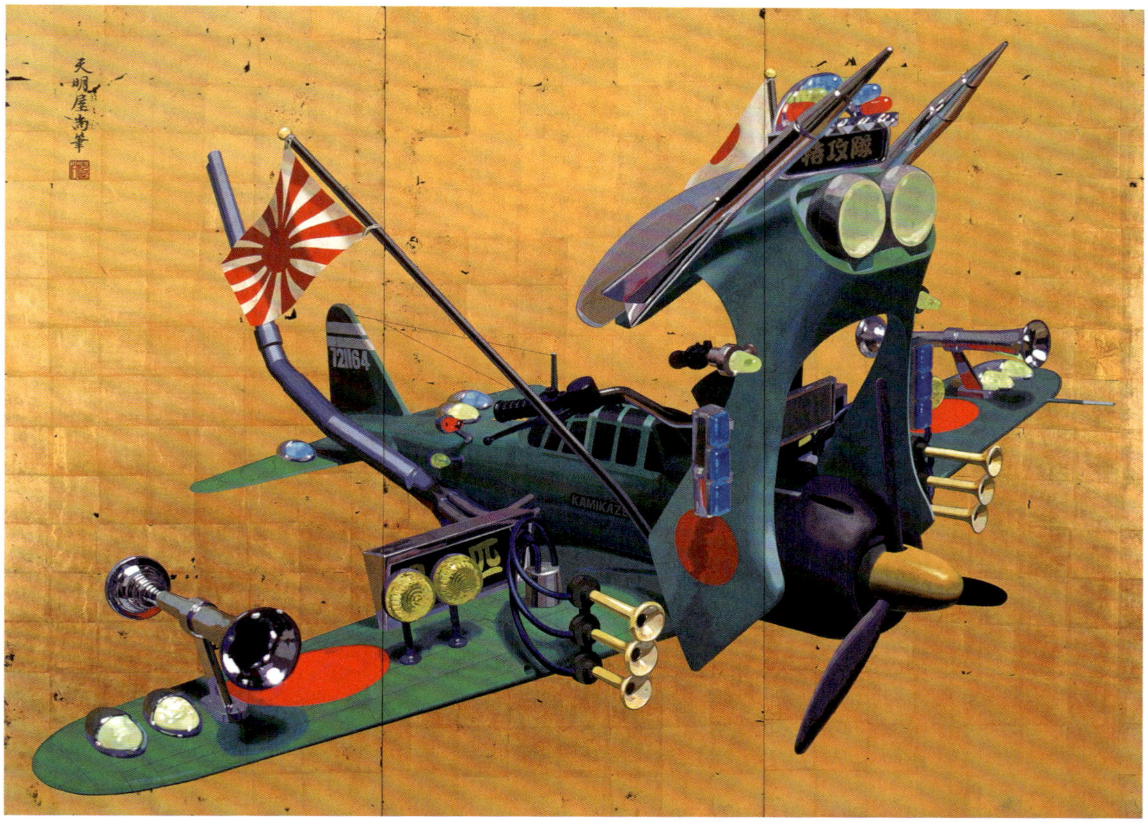

Hisashi Tenmyouya
Kamikaze, 2003
Acrylic and gold leaf on wood
Courtesy of Mizuma Art Gallery, Tokyo

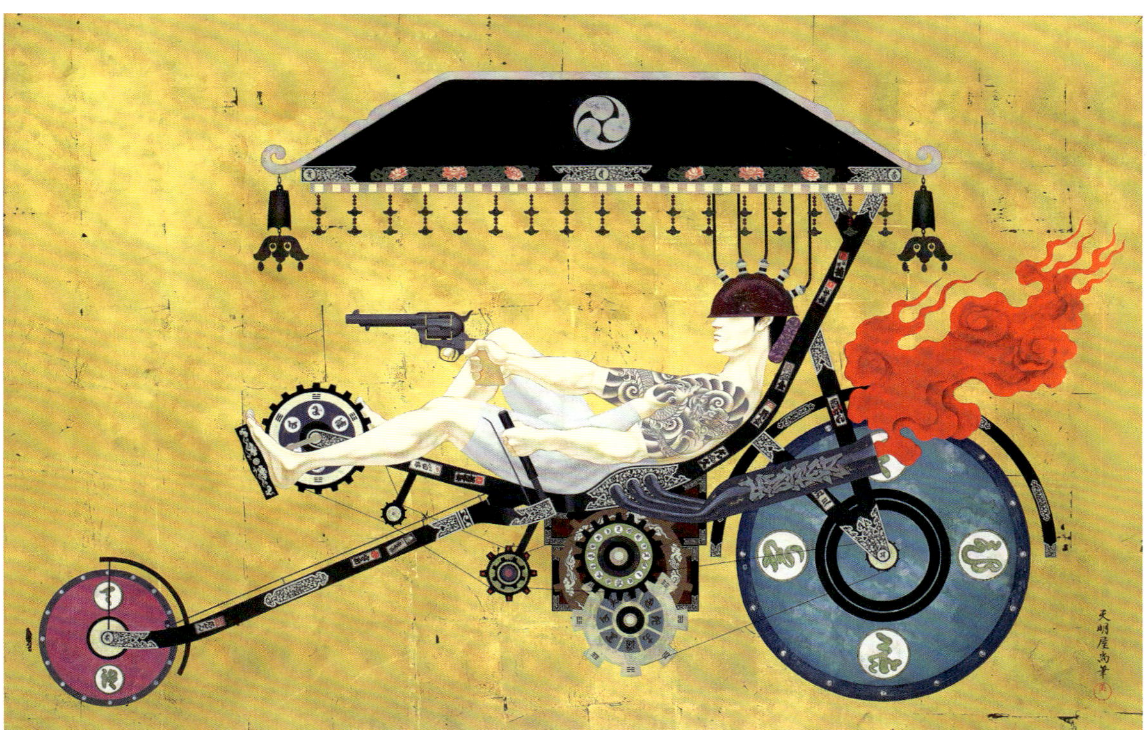

Hisashi Tenmyouya
Japanese Spirit #1, 1997
Acrylic and gold leaf on wood
Courtesy of Mizuma Art Gallery, Tokyo

Figuration/Abstraction

"[...], but unfortunately everything is always given a meaning."[1]

(Marcel Duchamp)

In the rooms of the Lower Belvedere the gold phenomenon is arranged by topic. In the Orangery by contrast, the rich spectrum of Modernist and contemporary gold art corresponds to a treasure house. The iconographical filter (frame, still-life, landscape, pathos formulae) is replaced by generalizing orientation fields. They represent important areas of Modernism: 1) writing, 2) object art 3) figuration, 4) abstraction. This structure is permeable and remains open. It does not represent a systematic pattern, but makes it clear that gold, the material, is not ignored in any field. This is the actual surprise. The resulting stories are seen in different ways, and reinvented, by different visitors. "I honestly believe that the picture is made no less by the beholder than by the artist."[2] (Marcel Duchamp, 1960)

Every exhibition reaches its limits where only a fraction of the totality of relevant works can be included, and in the fact that associations are a construct. The items not on show can always be considered along with those that are.[3] "Laws may be involved somehow—like the interplay of intellectual and economic developments or the perspectival factor in the visual arts, for instance—but there is always something else present too that is just this way just this one time."[4] (Robert Musil)

Excursus: 'The Meaning of Gold'

No other material seems to have better suited itself to be a projection surface for the creation of 'meaning' than gold. In the process, the narrative mostly remains hidden behind the iconic. In other words, the aura of the noble metal has often done its work of transfiguration without a story. This would make the function, content and context particularly important. To speak 'of the meaning of gold'[5] in medieval art presupposes the knowledge that the meaning was not attributed to the material, but that the material formed, anagogically, the basis of a sacred iconography. As St Boniface said as early as the eighth century: "Wooden chalices in the hands of 'golden priests' are more blessed than golden chalices in the hands of wooden priests."[6]

The seductive power of gold can be seen precisely in those places where for no reason it occupies an associative space, asserts its rank, and gets unconditionally into the headlines. In September 2011, during a period of gold speculation that had been going on for many years, the international media were telling us where gold came from. Science reports took on the tone of religious exegesis: 'Gold probably comes from space',[7] 'Gold rain from space'[8]. According to the geological explanation, in the early geological history of the earth, metals must have sunk into the core. Geochemists from Bristol and Oxford now explained that a swarm of meteorites brought metals to the earth's surface almost four billion years ago. Note, incidentally, that not just gold was involved, but all precious metals. The story was evidently only interesting because in this way gold could be assigned a non-earthly origin.

As chance would have it, not only the theory of the origin of gold was published, but at the same time also a proposal to explain how gold was made in the first place. The origin of none of the metals heavier than iron (atomic weight 56) can be explained by slow stellar nuclear fusion. Scientists from Brussels and Garching near Munich presume that the framework conditions can only be found in collisions between neutron stars. The origin of about 80 percent of the elements was, in the headlines, reduced once more to a gold story: 'Extreme star collisions—the true source of gold'.[9] The lustre of gold can lead to wondrous conclusions. In order to measure the aesthetic dimensions of the precious metal, we must shake off these frothy fantasies.

With collage, Cubism opened up pictures to all materials and to writing. Writing, as a fringe area of painting, was exempt from the taboo on gold. Since Antiquity, out of respect for the text, 'gold was poured over the letters'. Alongside the aspects of decoration and language integration came the inversion of the figure/ground relationship: today writing too (like landscapes, still-lifes, portraits) is placed on a gold ground. The aestheticization of writing is enriched by language games (Ian Anüll, Marcello Jori), alphabetization (Jill Biskin, Edouard Sacaillan), conceptualization and the abandonment of the 'linearity dogma'. "It is a matter of 'ideography' [...] a visualization of thematic aspects that have no equivalent on the level of sounds."[10] The spectrum ranged from the integration of individual letters and figures, newspaper cuttings and adverts to the gesturally individual shaping of the Informel of the 1950s and the monumentalization of the colourful 1960s. Even the subculture of the graffiti has long since been integrated.

Fonts

AIDS by the artist group General Idea consists of four letters reproduced in gold. Some of the background is black, the rest white. One letter, the D, appears not to be upright. The diseases, newly discovered in the 1980s, which at the time claimed the lives of many creative individuals, appears here in a solemn presentation. The picture has to be seen against the background of a work in which expression is given to a colourful joie de vivre and the 1960s' feeling, which has been varied countless times and has also been translated, and also consists of four letters: *LOVE* by Robert Indiana (1966 ff.). The work also exists as a monochrome sculpture, red, or, in the Sanskrit version (*PREM*) gold. All the life in this work, which unambiguously derives from colourful love, has disappeared from the work by General Idea. The fact that in one variant the lively colours have returned to the AIDS sphere (1987) shows that the debate the debate on the love-AIDS context cannot be reduced to simple explanatory patterns.

Gajin Fujita's *East versus West* (2010) must not be understood as a symbolic *Clash of Civilisations* (Samuel Philips Huntington, 1993). In this clash, on the left we have the Eastern representative of New York on the chestnut horse, who is storming towards the Western combatant from Los Angeles on the black steed. They are separated by the letters 'vs' for 'versus'. The Western horseman claims the 'originality' of an artistic confrontation. The merciless battle depicts the best-known hip-hop 'beef' or feud. The two recording companies Bad Boy Entertainment from New York and Death Row Records from L.A. played the lead roles. In the course of the gang warfare in the 1990s, well-known rappers such as Tupac Shakur and Notorious B.I.G. were shot dead.

Object Art

The emancipation of sculpture has been pushed forward by the integration of objects. Pioneers of this still effective integration of the everyday were Pablo Picasso and Marcel Duchamp. Picasso was the master of the formal transformation and integration of objets trouvés into important masterpieces. Thus toy cars can be put together to form a monkey's head (Picasso: *La guenon et son petit*, 1951). In the process, the original forms are lost, or they are left to unfold their poetic charm. Oswald Oberhuber's golden combs float with the stars in the firmament of the wooden panel. Marcel Duchamp, unlike Picasso, tried in vain to banish taste and iconography from his work. Sherrie Levine, in her 1991 work *Fountain (after Marcel Duchamp)*, references his famous urinal of 1917 and gilds it. This refinement results in its iconization. Using her affirmative appropriation she replaces the 'aspect of exaggerated meaning-attribution' by radiant lustre. The revolutionary potential of the readymade had become a pseudo-archaic sculpture reminiscent of Constantin Brancusi.

Gold leads less to a metamorphosis than to a negation of all materials. In his *Passtück* (1983–1985), Franz West coats the plaster and papier mâché with gold leaf. He associates gold with the memory of Vienna c. 1900. Thus *Freude* (1985) with gold foil points to Sigmund Freud, and the ornamental Ab*gabe* (1987) with gold leaf to Gustav Klimt.

The confrontation of other materials with gold has its basis not just in aesthetics. As a street artist in South Africa, Robin Rhode has used charcoal drawings in a socially critical manner to draw attention to conditions on the fringes of society. The black pigment transfers the *Spade* (2007) into a contrast to the precariousness of life, not least to the country's mining industry, in which gold and diamonds are produced.

Gilding extends not just to the grounds of paintings or sculptures, but, alongside object art, also to performances and happenings with an emphasis on the body, and even to minimalist Concept Art (e.g. Walter de Maria, *The Equal Area Series*, 1976–1990, Dia: Beacon Museum). The integration of everyday materials can simply leave these as they are, as in the fetishization of objects (Sylvie Fleury), or exaggerate them, as in Robert Weber's Ethiopian shoe-cleaner's box, which is dedicated to the dignity of the shoe-shine children.

Historical references characterize the mask *Tamerlano* (1969) by Luciano Fabro. He references the Mongol ruler variously known as Timur Lenk, Tamerlane or Tamburlaine, who achieved fame as the general commanding the 'Golden Horde' in the fourteenth century. He was the subject of a play by Christopher Marlowe (1587) and an opera by Handel (1719). The curious effect derives from the straws projecting out of the nose. The gilt mask was taken from a live model, and Fabro left the necessary breathing tubes in place.

Some cross-references arise from biographies. One cannot see from the *Soothsayer* (2009) by Małgorzata Chodakowska that the Polish-born sculptress was a pupil of Bruno Gironcoli, who only once, in his *Blattähre* (2003) used gilding as a technique.

Figuration

There is a conflict underlying the concept of realism. However, it is otiose to try and decide whether a readymade is 'real' if anything as a mimetic sculpture. Gilding tends to detemporalize and to hieratic exaggeration. Stephan Balkenhol's *Man on Golden Globe* (2004) is a preliminary work for *Sphaera* (2007), created for Salzburg's Kapitelplatz. Apart from the opulent hair, Edward Lipski's *King* (2007) is gilded all over. "My work exists between the cultural and the visual level," says the artist. "I'm interested in the space between something you understand and something that is also abstract. This confusion creates a particular intensity."[11]

In the first half of the twentieth century, gold grounds were rare. Max Weiler gave a precise description of the old-masterly technique of applying a gold ground in his *Tag- und Nachtheften.*[12] As a member of the Neuland movement, he was seized by a deep piety and love of nature. He was shocked by the reaction to his religious pictures of the 1930s: "The usual church circles did not know what to say to these works, they were already using words like un-German, obsessed with originality, experiments deriving from the intellect, and that the common sense of the man in the street would reject such stuff."[13]

In figural painting, whether realist (Hung Liu, Frédéric Léglise, Fred Wessel), expressive (Georg Baselitz) or fantastic (Arik Brauer, Jason Clay Lewis, Tabitha Vevers) as well as the hybrid forms, gold is used both as a gold ground (William Cordova, Josef Kern) and to accentuate the pictorial elements (Tom Sachs). The motives for the use of gold leaf vary. Mostly the lustre is seen as costly and solemn.

For the realization of the *Safety Curtain of the State Opera* (1955) by Rudolf Hermann Eisenmenger, the public were asked to make a donation for the gold ground. "This is to be seen in connexion with the extremely difficult period of the reconstruction of Vienna—at the time more than 40% of the houses were badly damaged, and 15% totally destroyed, the reconstruction was difficult and the money for luxurious decoration in very short supply. Thus the population helped—for St. Stephen's cathedral, for example, the roof tiles were paid for by public donations, and for the State Opera, the people donated 'gold'. The interior of the rebuilt opera house was to be splendid. A symbolic building of the post-war period as a temple to the 'power of music'. [...] It was the artist's intention [...] to lend the space a particular note, and to harmoniously add the largest picture ever painted in Austria at that time, measuring 170 sq. metres [...]. For the execution, [...] half a kilogram of gold leaf of the highest quality was needed. [...] The composition of *Orpheus and Eurydice* was placed by the artist on the left-hand side as seen from the auditorium, and the top right-hand section is gilded [...]. The gold here is symbolic on the one hand of the purity of the couple, and on the other of the almost attained goal, while the warm gleam of the gold is also a symbol of the sun (the underworld will shortly be left behind—sun and warmth await the couple."[14]

Abstraction

The superficially most appropriate field for lustrous gold is abstraction. Yves Klein shows with *The Value of Gold* (1960) his esoteric glorification of the cosmological void at the start of the decade of 'New Realism'. Klein had made contact with the ZERO group in 1959, and had also met Heinz Mack, whose central theme was to be light. His gilt sculptures bear witness to this. Victor Vasarely, with his square *Taller* (1965, Spanish, literally: 'workshop') translates the field of perception into a kinetic resonance between interior and exterior. Christian Eckart supplements gold with an IKB (International Klein Blue) frame. Each of the works opens up a broad spectrum of portentous references to individual mythologems, art-historical relationships and, not least, technical variety in the production processes.

Whether gold is used as a collage-like partial element in a whole variety of formats (Sarah Cain, Hundertwasser, Lucy Skaer), as a surface application (Günther Förg), as a ground beneath a colour grid (Helmut Federle) or as blood-red outpouring (Hermann Nitsch), or extends over the whole surface of the picture (Johannes Domenig, Winfried Muthesius), it never fails to exert its hypnotic attraction. Even when it is sealed behind a transparent film or as a reverse glass painting (Michael Burges), gold leaf unfolds its glowing aspects with its own shadings. Gold leaf tends, through its radiance, to pervade or outshine the compositional elements.

The finale in the Orangery is the *Shield of Apollo* (2011) by Walter Schnabl. The monumental labyrinthine disc has, in spite of all appearances, no relief; the apparent relief is only an optical illusion created by the illuminated, smooth-structured surface. Gold has, since time immemorial, evoked a variety of reactions. The role of art is increasingly replacing traditional mythological and sacred foundations of meaning. Thus *Schwarzbuch Gold—Gewinner und Verlierer im neuen Goldrausch*[15] does not begin with the gold bubble of our own day, but attests 'the new glitter of gold' by reference to an artist's burial of fourteen ingots of the metal.[16]

1 Pierre Cabanne, *Gespräche mit Marcel Duchamp* (1967), Cologne 1972, p. 76.

2 Marcel Duchamp, *Interviews und Statements.* Gesammelt, übersetzt und annotiert von Serge Stauffer, Graphische Sammlung Staatsgalerie Stuttgart 1992, p. 111.

3 Currently the archive on which the exhibition is based comprises more than 730 artists. The list of works which the exhibition was unable to use is long. A vast number of artists worked with gold in the twentieth century. One need only think (in alphabetical order) of: Marina and Ulay Abramovic, Horst Antes, Arman, John Armleder, Hans Arp, Matthew Barney, Thomas Bayrle, Rudolf Belling, Anna-Eva Bergman, Joseph Beuys, Guillaume Bijl, Max Bill, Julius Bissier, Peter Blake, Barbara Bloom, Louise Bourgeois, Marcel Broodthaers, Daniel Buren, Alberto Burri, Michael Buthe, Ernst Caramelle, César, Francesco Clemente, Salvador Dalí, Charles Demuth, André Derain, Braco Dimitrijevic, Jiri Georg Dokupil, Gino De Dominicis, Ugo Dossi, Elmgreen & Dragset, Ian Hamilton Finlay, Lucio Fontana, Gilbert and George, Mathias Goeritz, Subodh Gupta, Keith Haring, Georg Herdold, Damien Hirst, Hans Hollein, Rebecca Horn, Roni Horn, Robert Indiana, Johannes Itten, Magdalena Jetelova, Anish Kapoor, Anselm Kiefer, Iannis Kounellis, Agnes Martin, Georges Mathieu, Gordon Matta-Clark, Marisa Merz, Yasumasa Morimura, Takashi Murakami, Christian Ofili, Luigi Ontani, Orlan, Eduardo Paolozzi, Mimmo Paladino, Claudio Paramiggiani, Francis Picabia, Otto Piene, Sigmar Polke, Michelangelo Pistoletto, Robert Rauschenberg, David Salle, Julian Schnabel, HA Schult, Kiki Smith, Florine Stettheimer, Rudolf Stingel, Philip Taaffe, Tim Ulrichs, Ben Vautier – to name but the best-known.

4 Robert Musil, "Helpless Europe", in Burton Pike and David S. Luft (eds., trans.), *Robert Musil: Precision and Soul* Chicago 1978, p. 118f.

5 Victor H. Elbern, *Goldschmiedekunst im frühen Mittelalter*, Darmstadt 1988, ch. 6. (I am grateful to Dieter Blume, Jena, for pointing this out to me.)

6 Ibid., p. 118.

7 *Stern*, 8 Sept. 2011, URL: <http://www.stern.de/wissen/kosmos/meteoriten-gold-stammt-vermutlich-aus-dem-all-1725405.html> [read on 10 Jan. 2012].

8 *Focus*, 7 Sept. 2011, URL: <http://www.focus.de/wissen/wissenschaft/astronomie/edelmetalle-goldregen-aus-dem-all_aid_663101.html> [read on 10 Jan. 2012]; also *Die Zeit*, 10 Sept. 2011, URL: <http://www.zeit.de/wissen/umwelt/2011-09/meteoriten-metalle-erde> [read on 10 Jan. 2012] and: *Neue Zürcher Zeitung*, 14 Sept. 2011, URL: <http://www.nzz.ch/nachrichten/hintergrund/wissenschaft/goldregen_aus_dem_all_1.12483007.html> [read on 10 Jan. 2012]. -

9 *Frankfurter Allgemeine Zeitung*. Natur und Wissenschaft, 14 Sept. 2011, p. N1.

10 Sybille Krämer, "'Schriftbildlichkeit' oder: Über eine (fast) vergessene Dimension der Schrift", in S. Krämer/H. Bredekamp (eds.) *Bild—Schrift—Zahl*, Munich 2003, p. 160.

11 Edward Lipski, zit. nach: URL:< http://www.undo.net/it/mostra/127877> [read on 10 Jan. 2012].

12 Max Weiler, *Tag- und Nachthefte* (1960–1991), quoted from: Robert Meyer/Max Weiler, *Meyer liest Weiler*, Audio-CD, 2002..

13 Max Weiler, *Tag- und Nachthefte*. A critical edition is in preparation (I am grateful to Yvonne Weiler for this information).

14 I should like to thank Maria Missbach, the artist's biographer, for this retrospect.

15 Brigitte Reisenberger/Thomas Seifert, *Schwarzbuch Gold. Gewinner und Verlierer im neuen Goldrausch*, Vienna 2011. (The title means: The 'Black Book' of Gold: Winners and Losers in the New Gold Rush.

16 Michael Sailstorfer triggered a gold rush thus in Pulheim near Cologne in 2009.

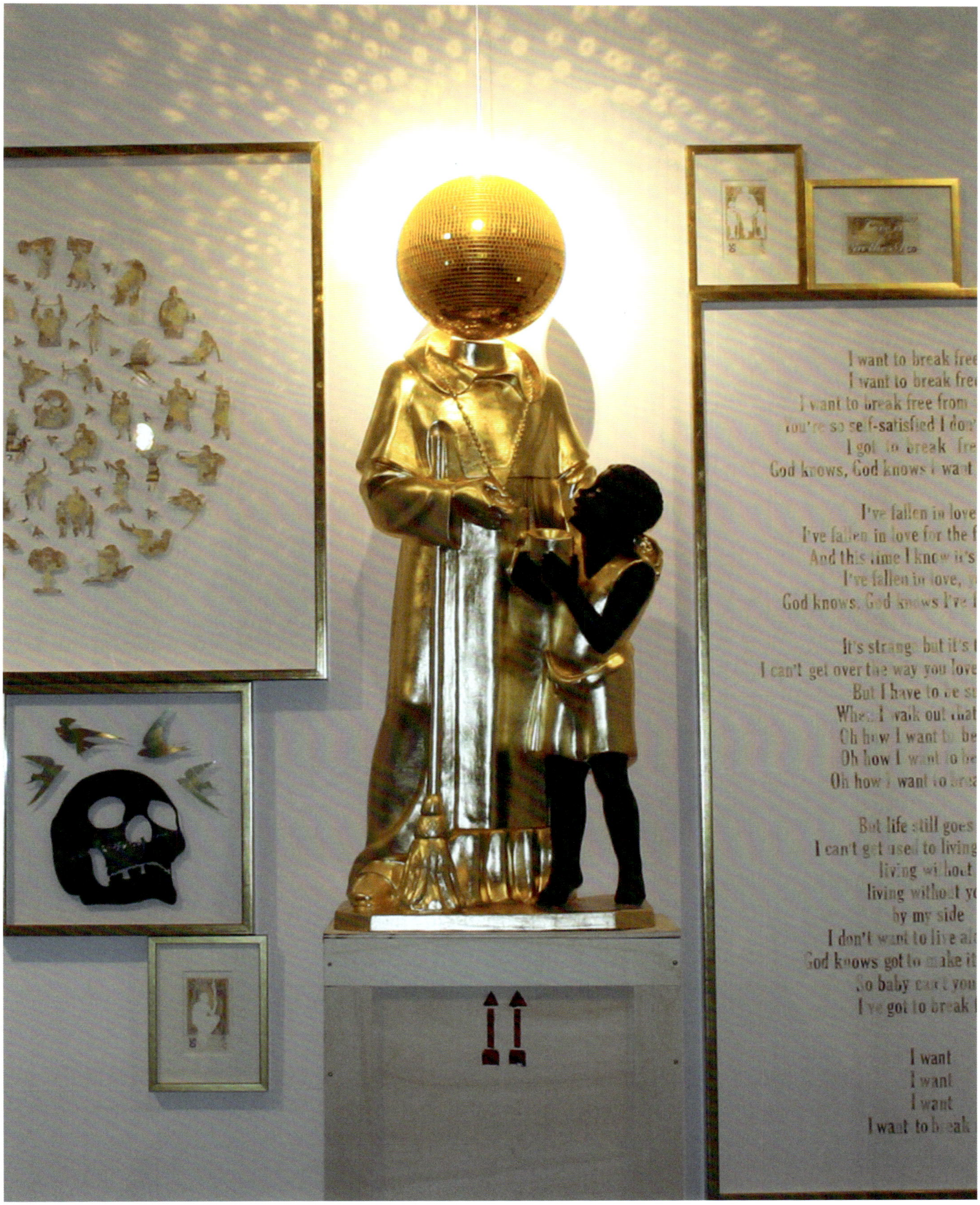

Carlos Aires
Black Requiem, 2010
Polychrome and gold leaf plaster and fibreglass sculpture, golden disco globe, wooden base, 110 x 40 x 40 cm
Courtesy Mario Mauroner Contemporary Art Vienna

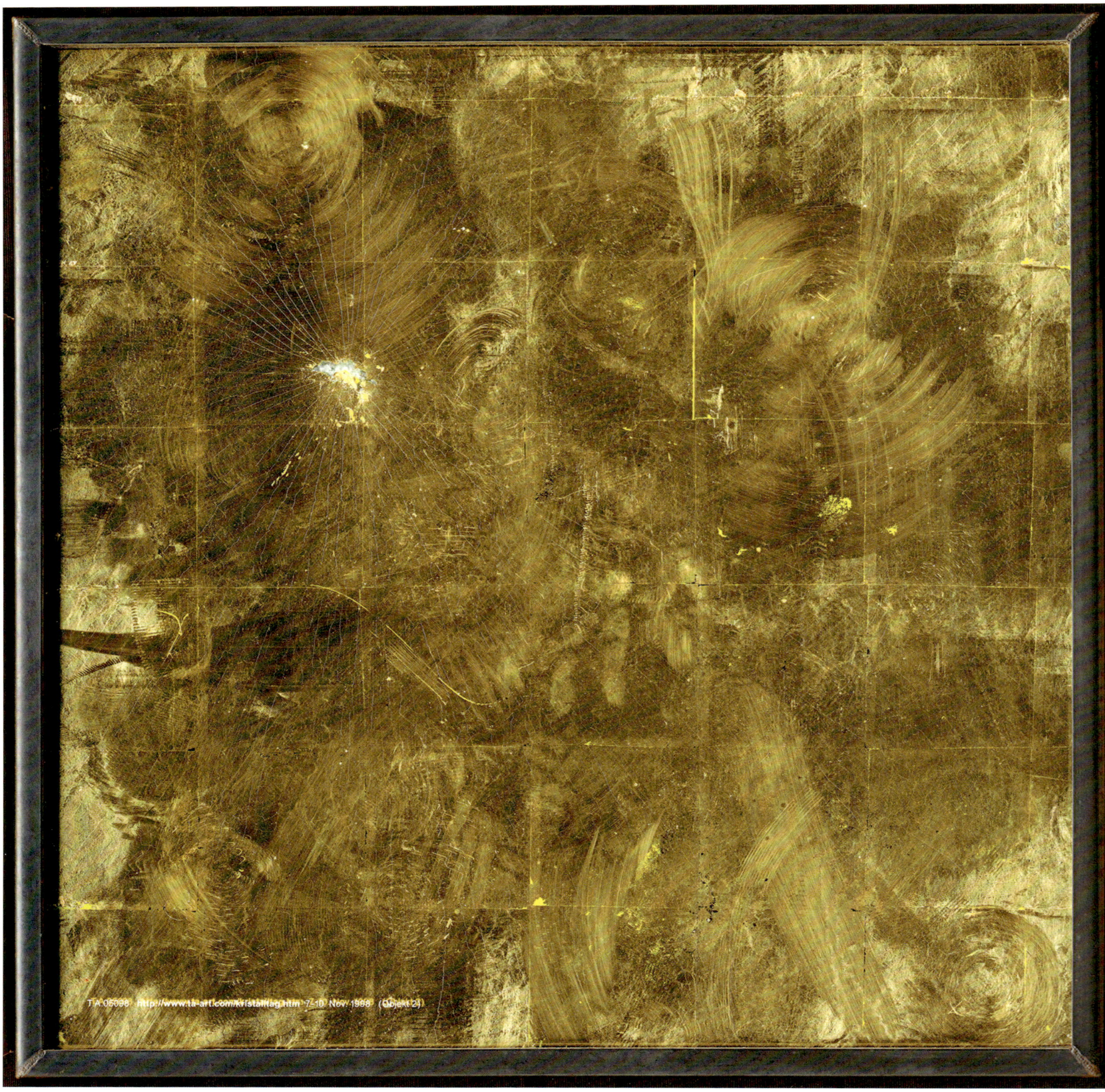

Johannes Angerbauer
Kristalltag, 10. November 1998
Silkscreen print, security glass, gold leaf, steelframe
50 x 50 cm
Courtesy the artist

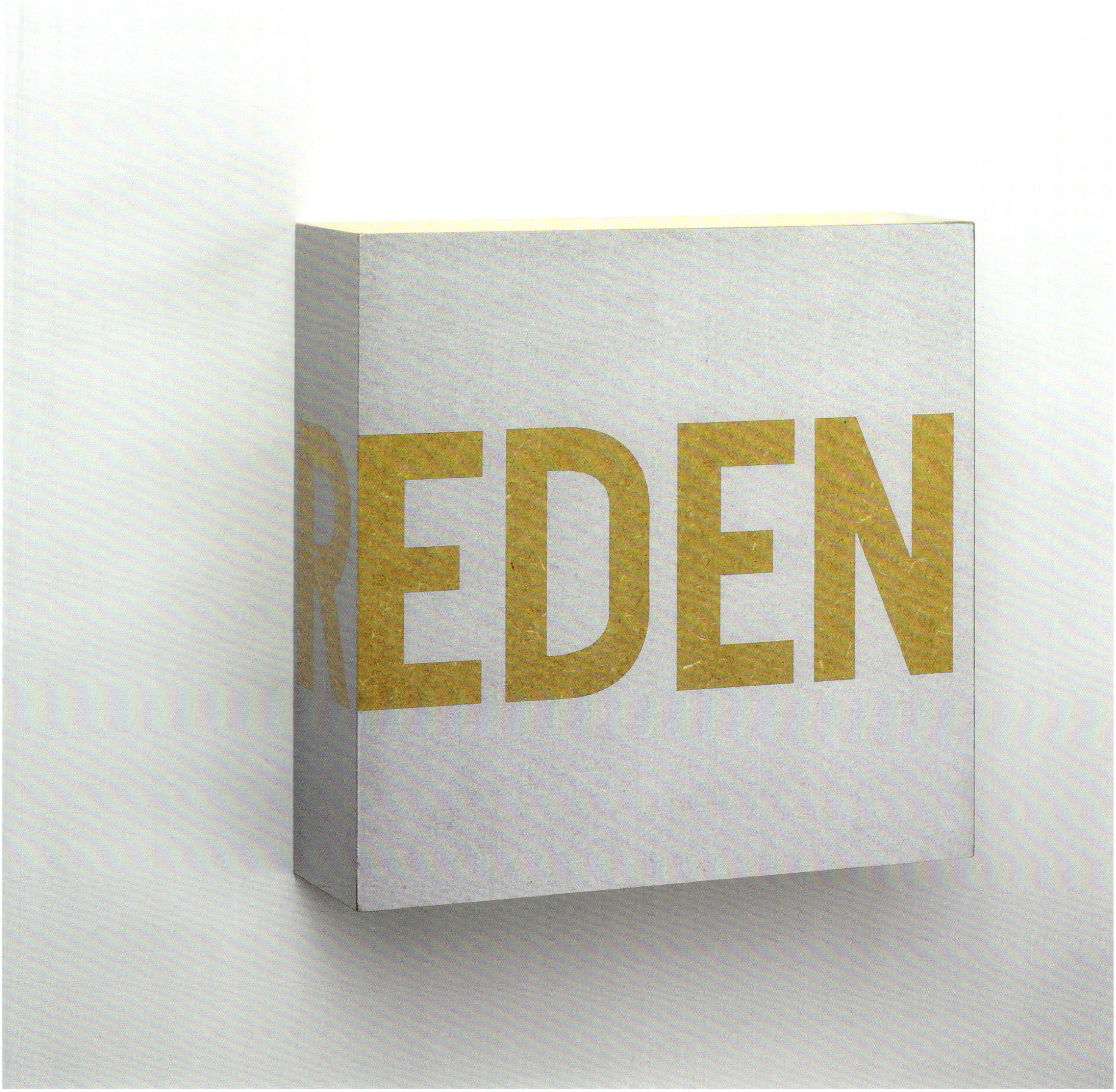

Ian Anüll
Reden ist Silber, Schweigen ist Gold, 2004
Screenprint on MDF, upper surface gilded,
25 x 25 x 7 cm
Courtesy Edition 5, Erstfeld, Switzerland

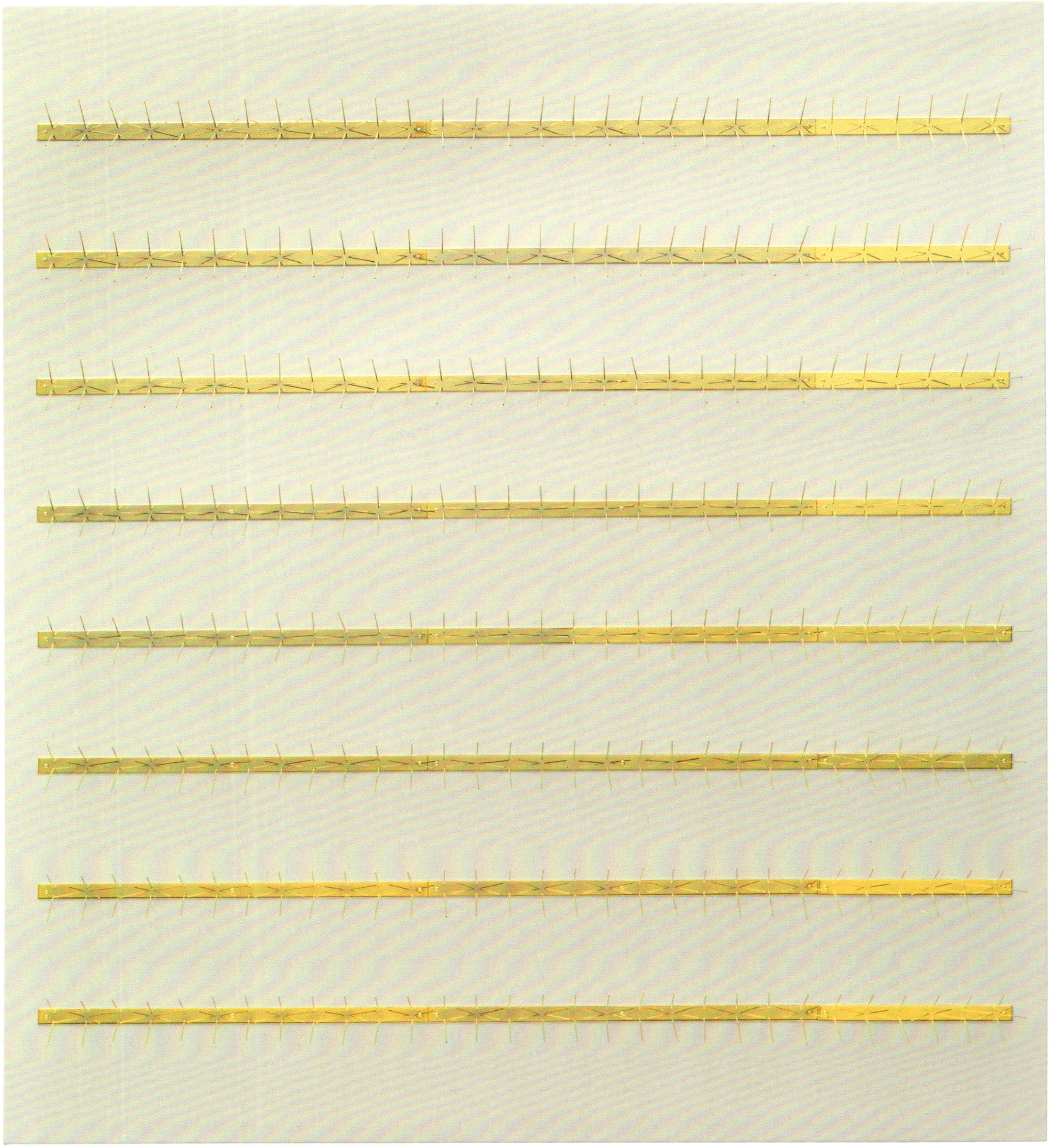

John M. Armleder
Siro, 2011
Acrylic on canvas with gilded anti-pigeon barbed wire, 180 x 160 cm
Galerie Andrea Caratsch, Zurich

Stephan Balkenhol
Mann auf goldener Kugel, 2004
Poplar wood, painted, gilded with gold leaf, 190 x 65 x 45 cm
Private collection, Germany

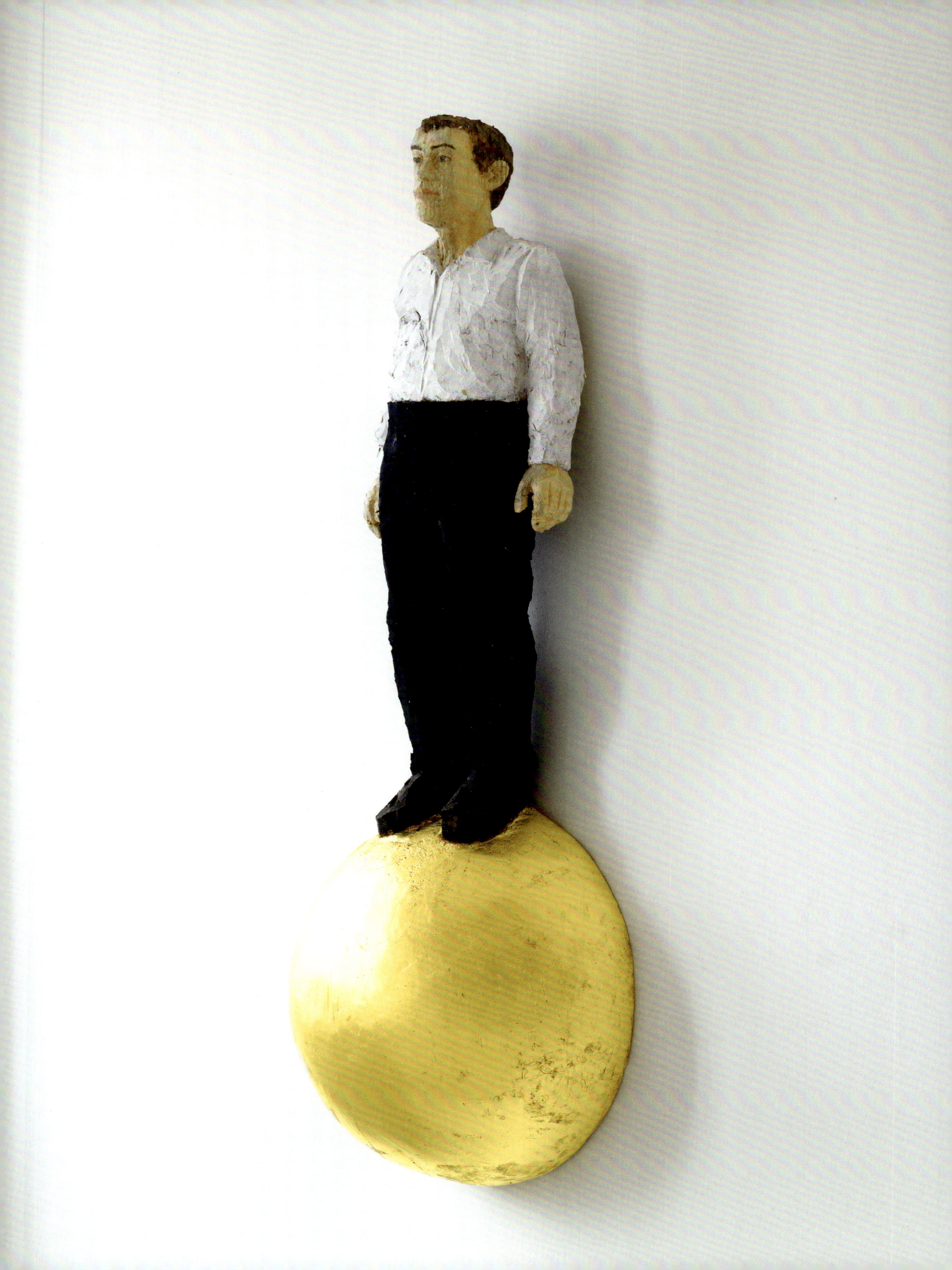

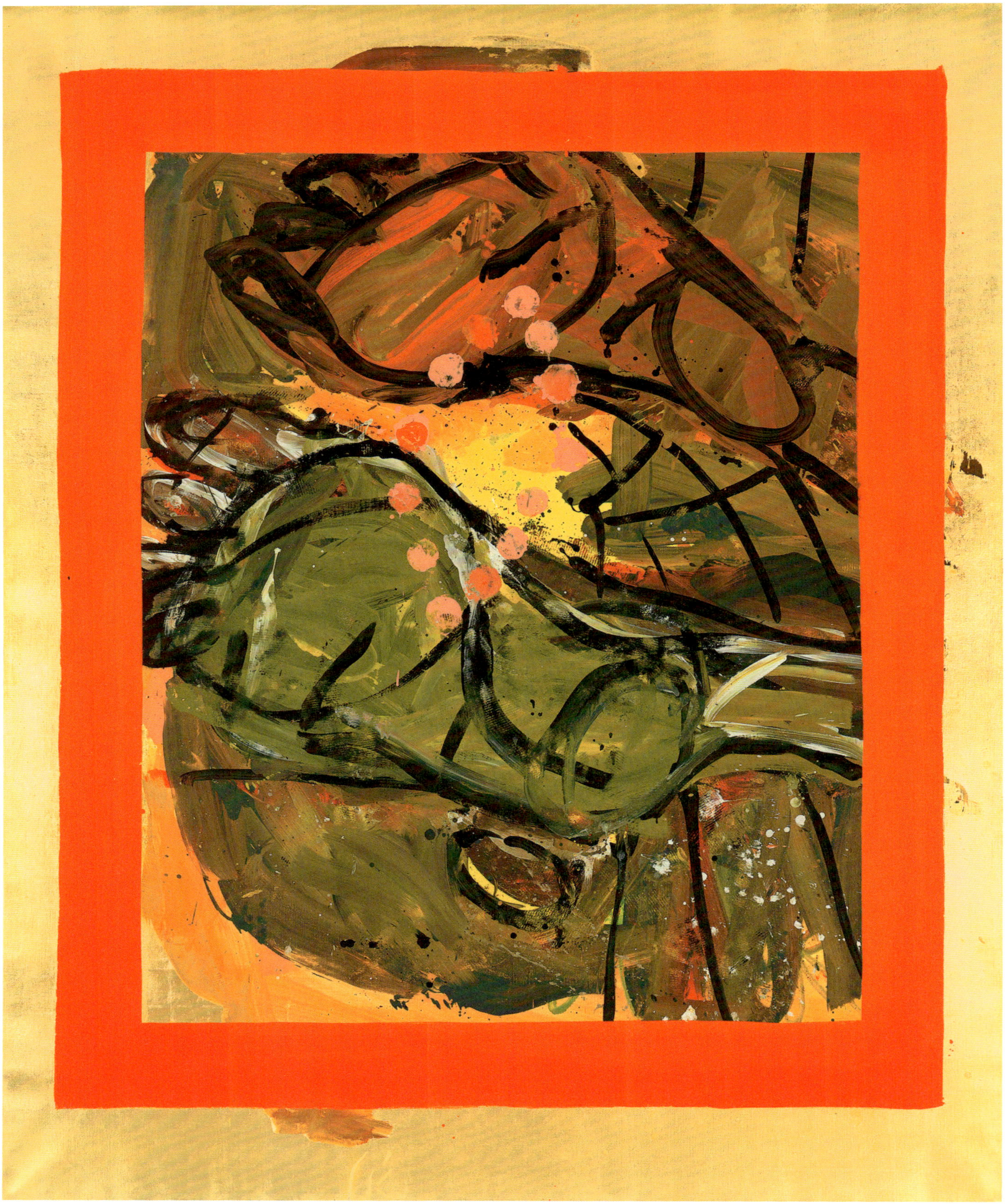

Georg Baselitz
De Chiricos Selbstbildnis, 1995
Gold leaf and oil on canvas, 200 x 162 cm
Private collection, Hamburg

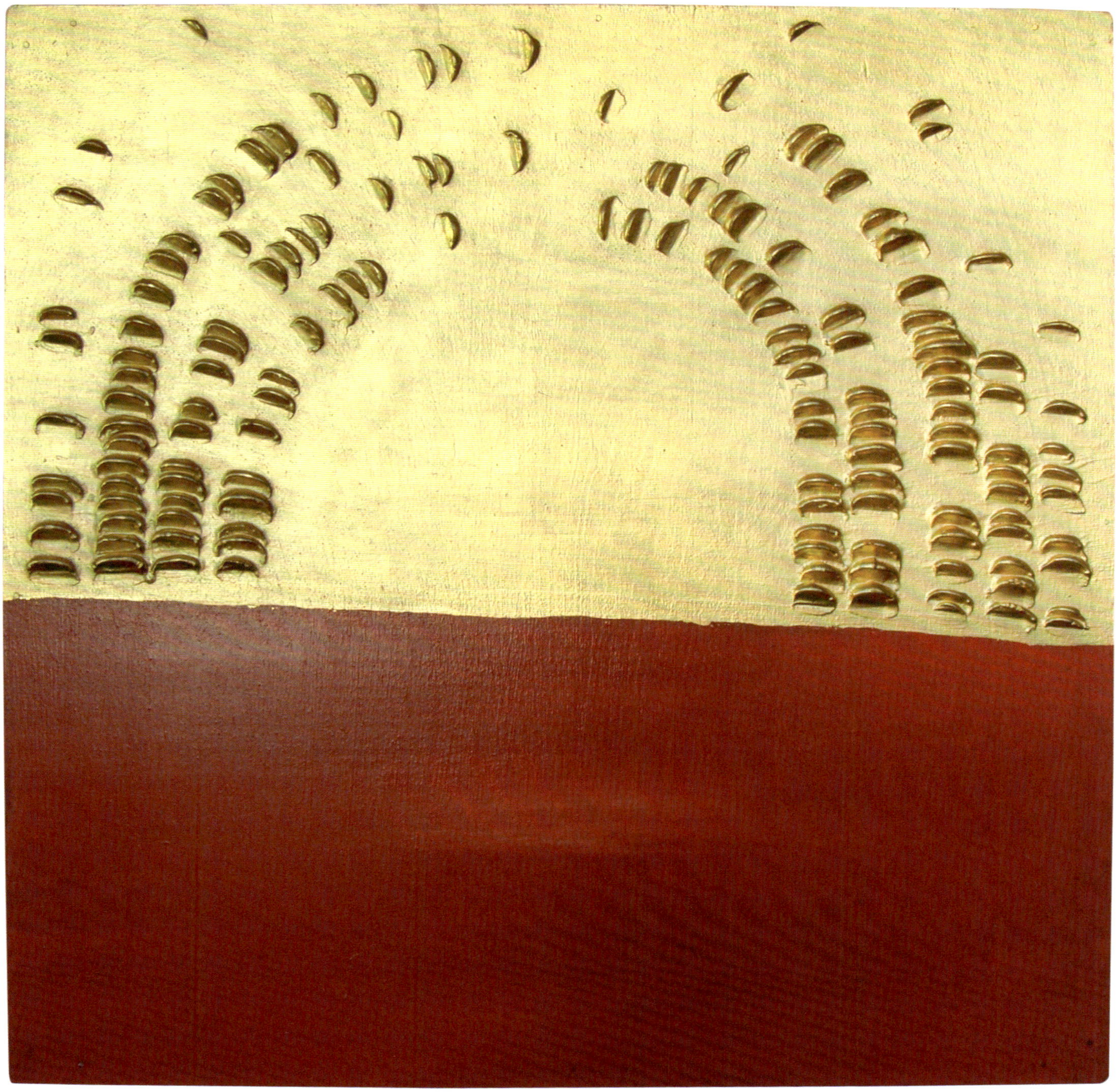

Hans Bischoffshausen
Invasion, 1973
Cell cement, gold, red lacquer, 60.5 x 60.5 cm
Galerie Walker Schloss Ebenau Rosental

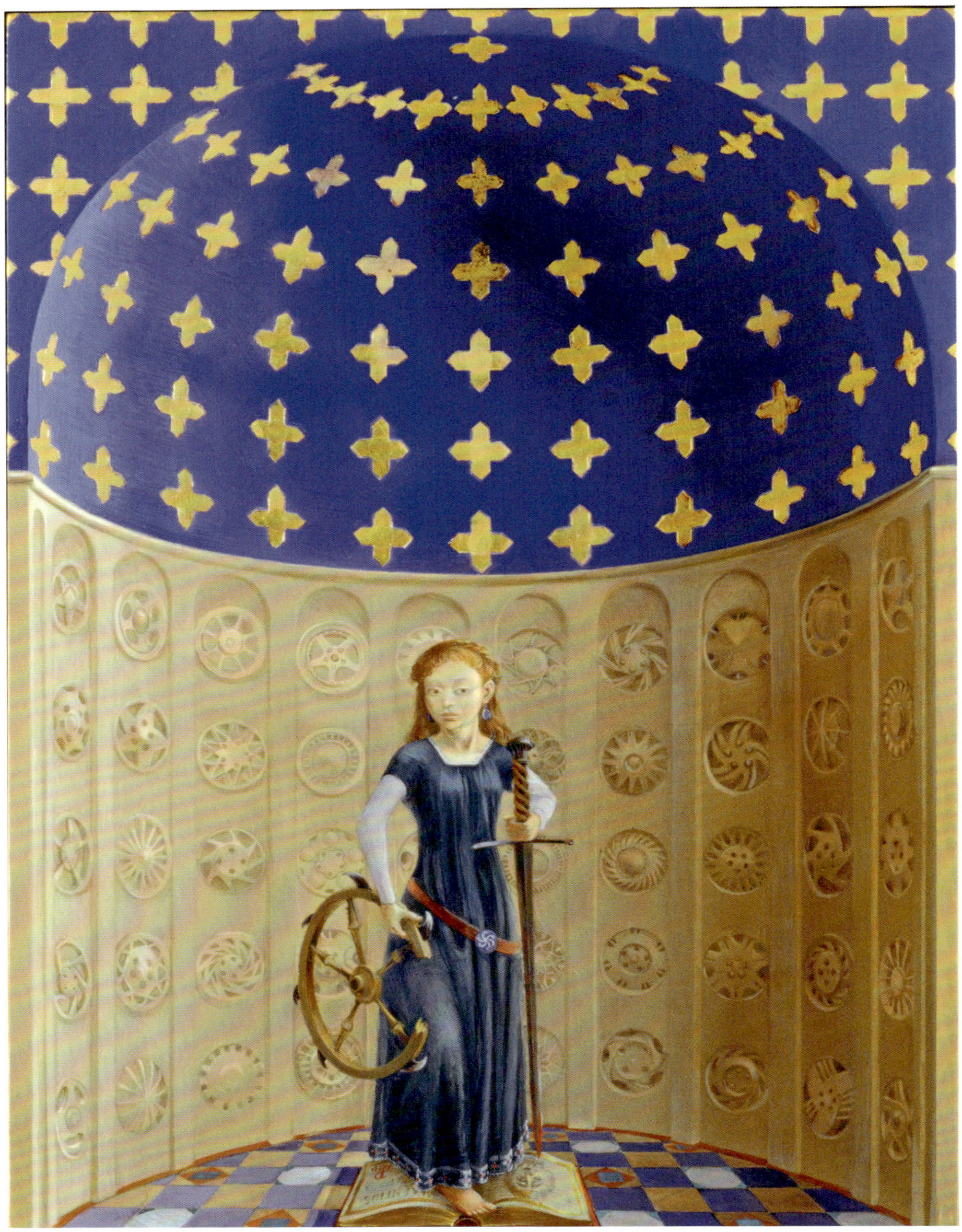

Jill Biskin
Saint Catherine, 1992
Egg tempera and gold leaf on wood, 16 x 12 cm
Courtesy of King & Spalding

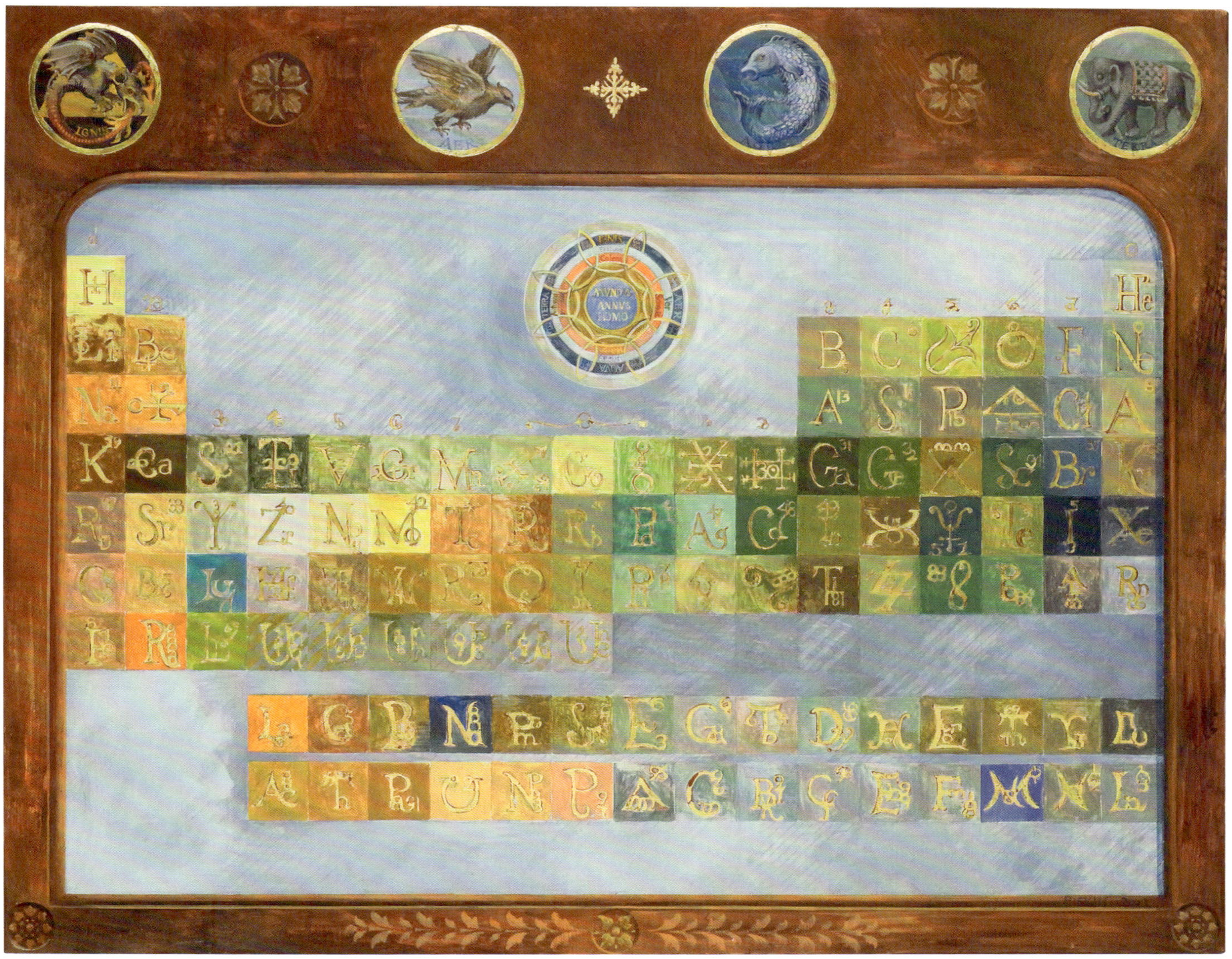

Jill Biskin
The Elements, 2003
Egg tempera and gold leaf on wood, 16 x 20 cm
Jill Biskin

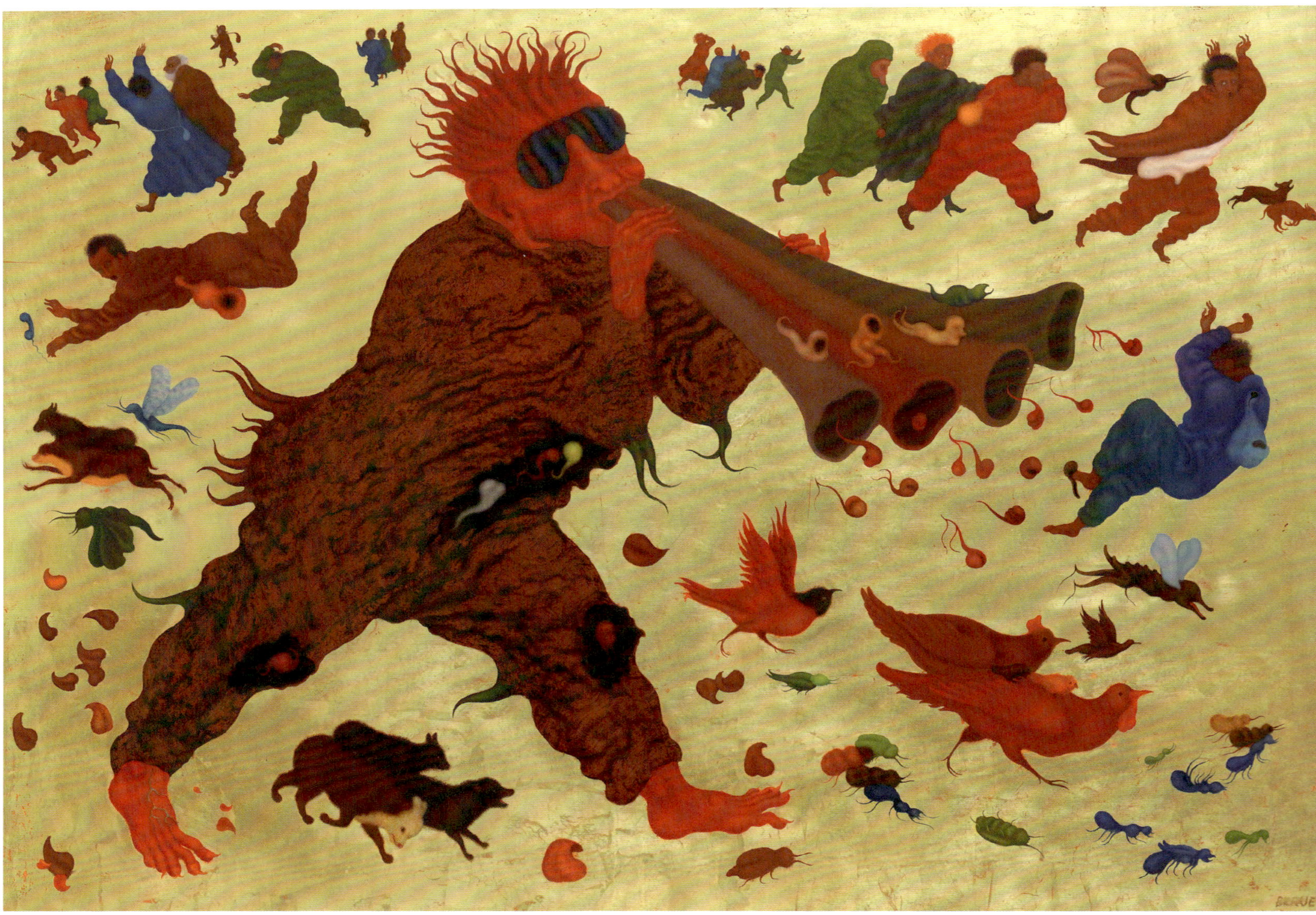

Arik Brauer
Pan, 2009
Oil, gold leaf on board, 105 x 160 cm
Private collection, Arik Brauer

Michael Burges
Reverse Glass Painting No. 10, 2011
Acrylic, Perspex, 23-carat gold leaf, aluminium, 180 x 150 cm
LS LandskronSchneidzik Galerie und Kunstagentur

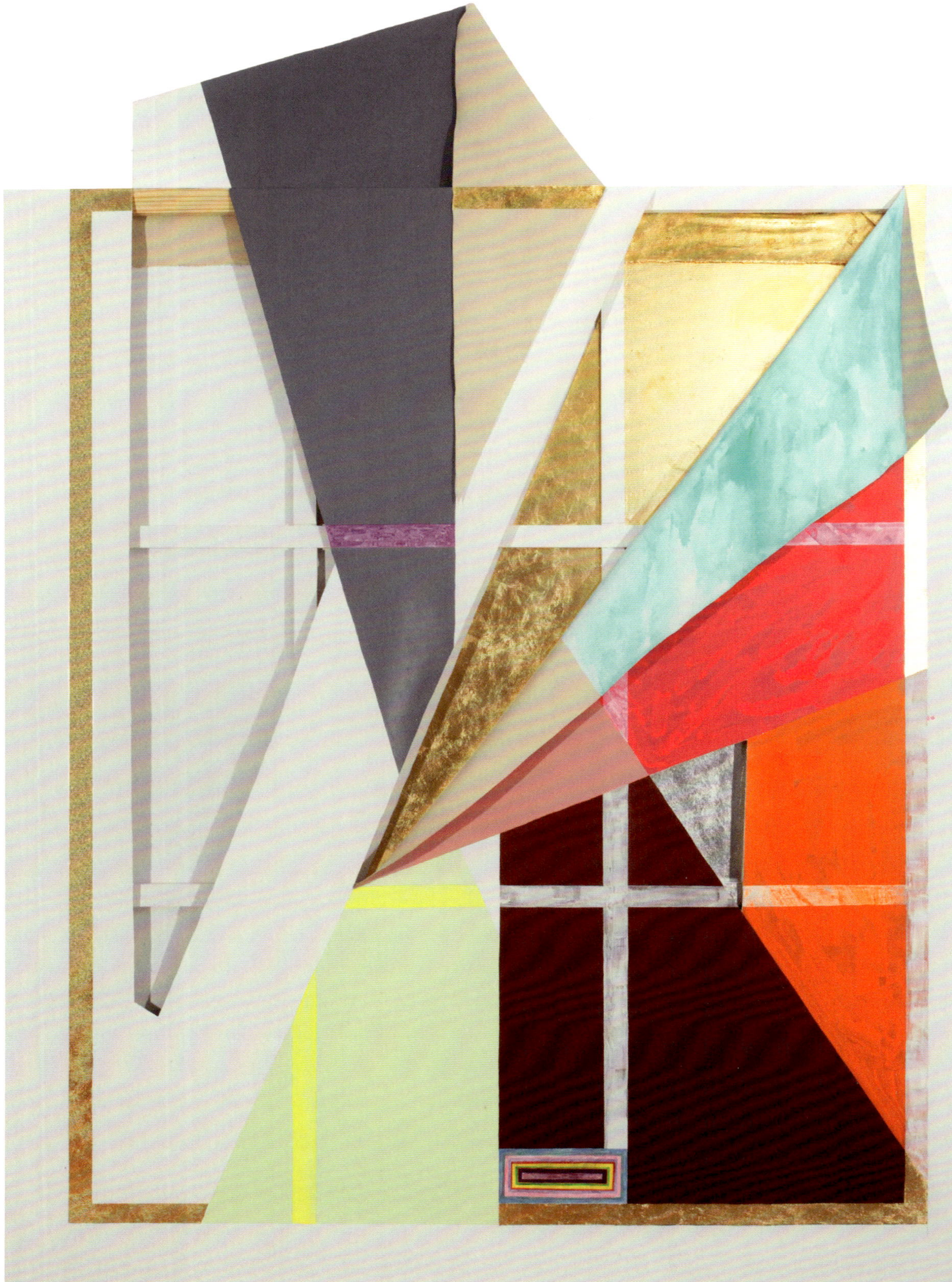

Sarah Cain
Breaking Light, 2010
Acrylic, gold leaf and silver leaf on canvas and Masonite plate,
259.1 x 205.1 cm
Courtesy the artist and Anthony Meier Fine Arts,
San Francisco

Małgorzata Chodakowska
Seherin, 2009
Bronze-gilt, Height 187 cm
LS LandskronSchneidzik Galerie und Kunstagentur

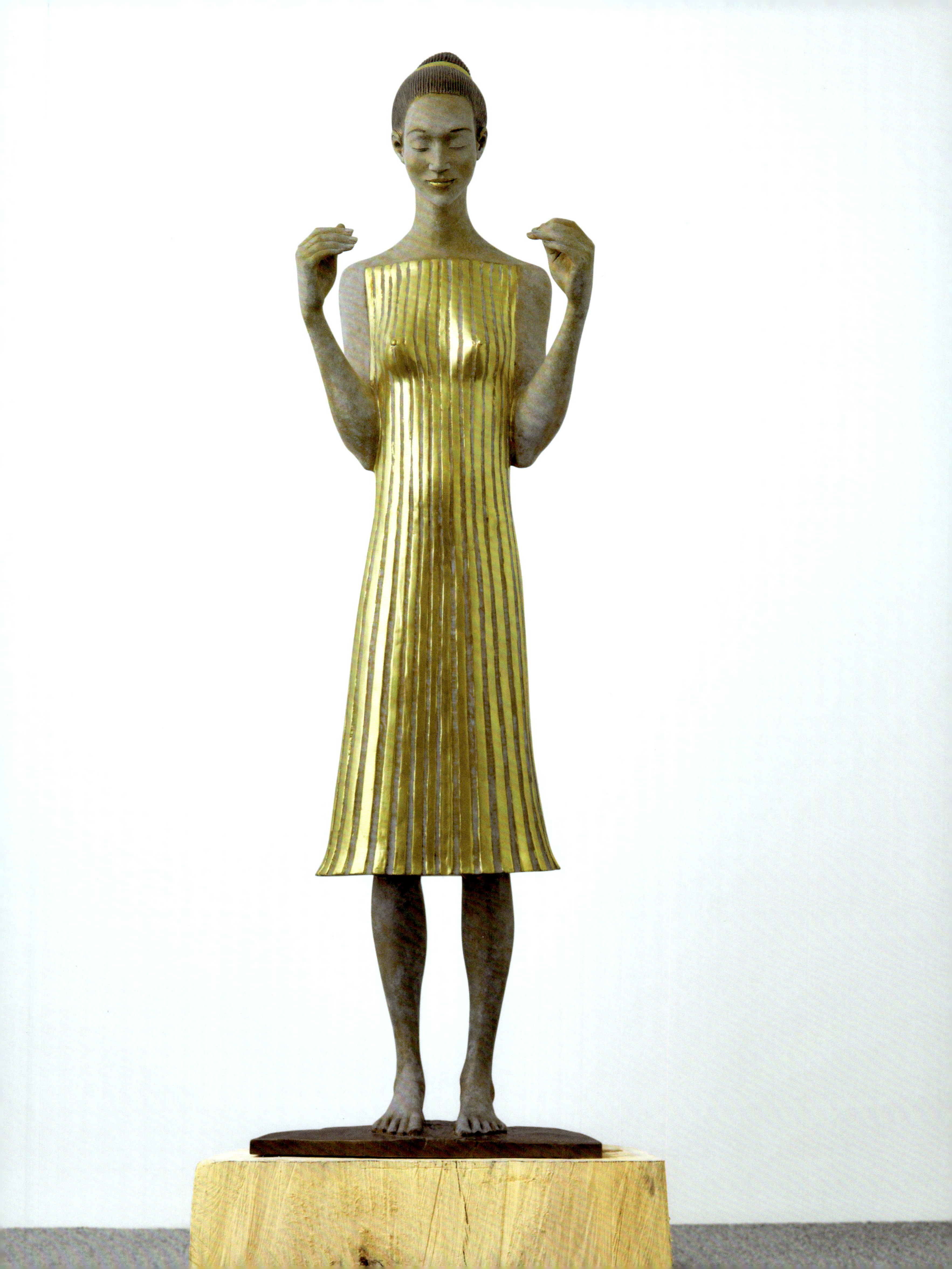

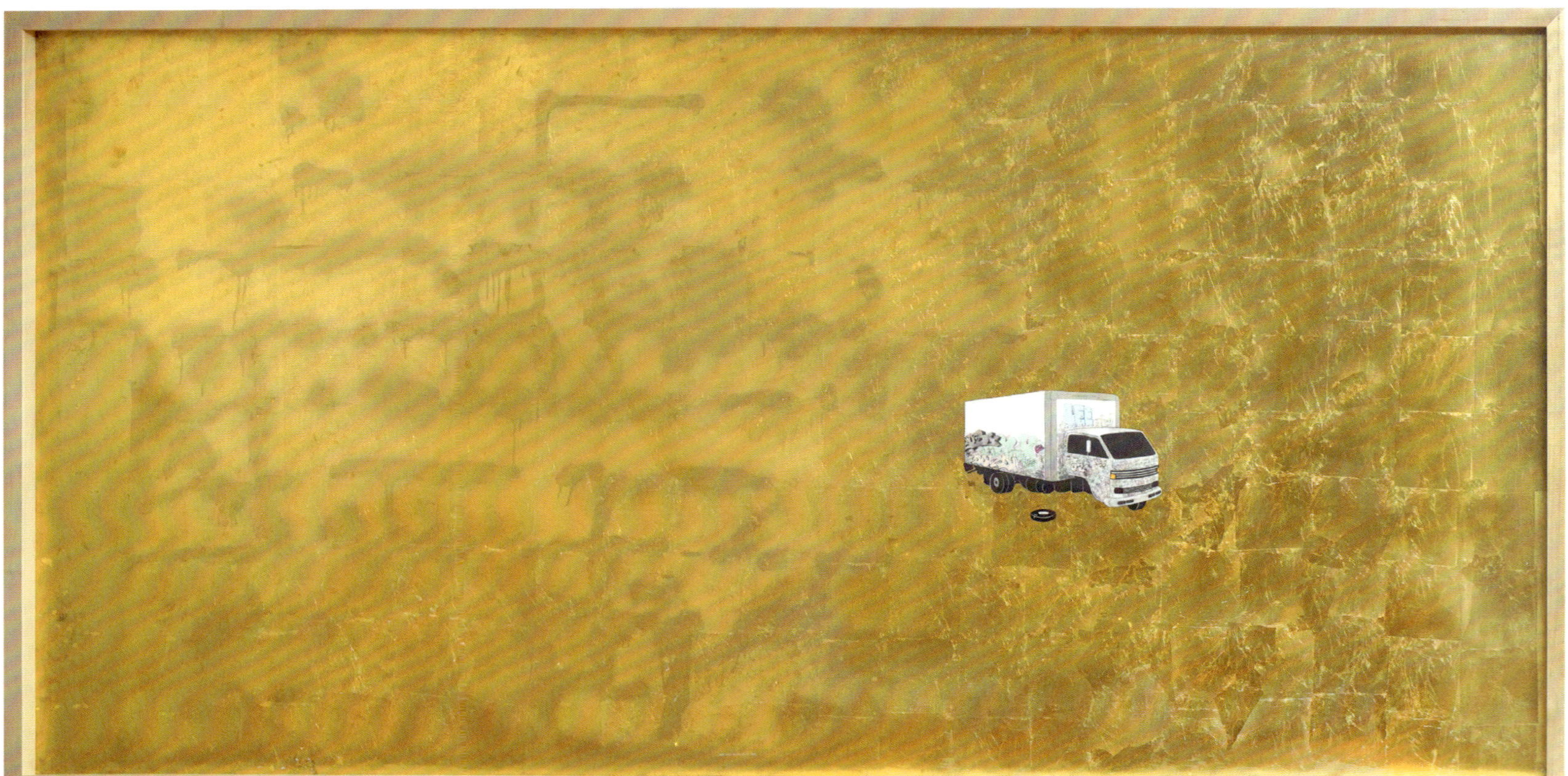

William Cordova
Wholesellers, retailers & bullshitters, 2004
Gold leaf, graphite pencil and ink on paper, 137.2 x 274.3 cm
Collection Peters-Messer, Viersen

Johannes Domenig
Untitled, 2008
Bole gilding on wood, rose thorns, 150 x 120 x 5 cm
In the artist's possession

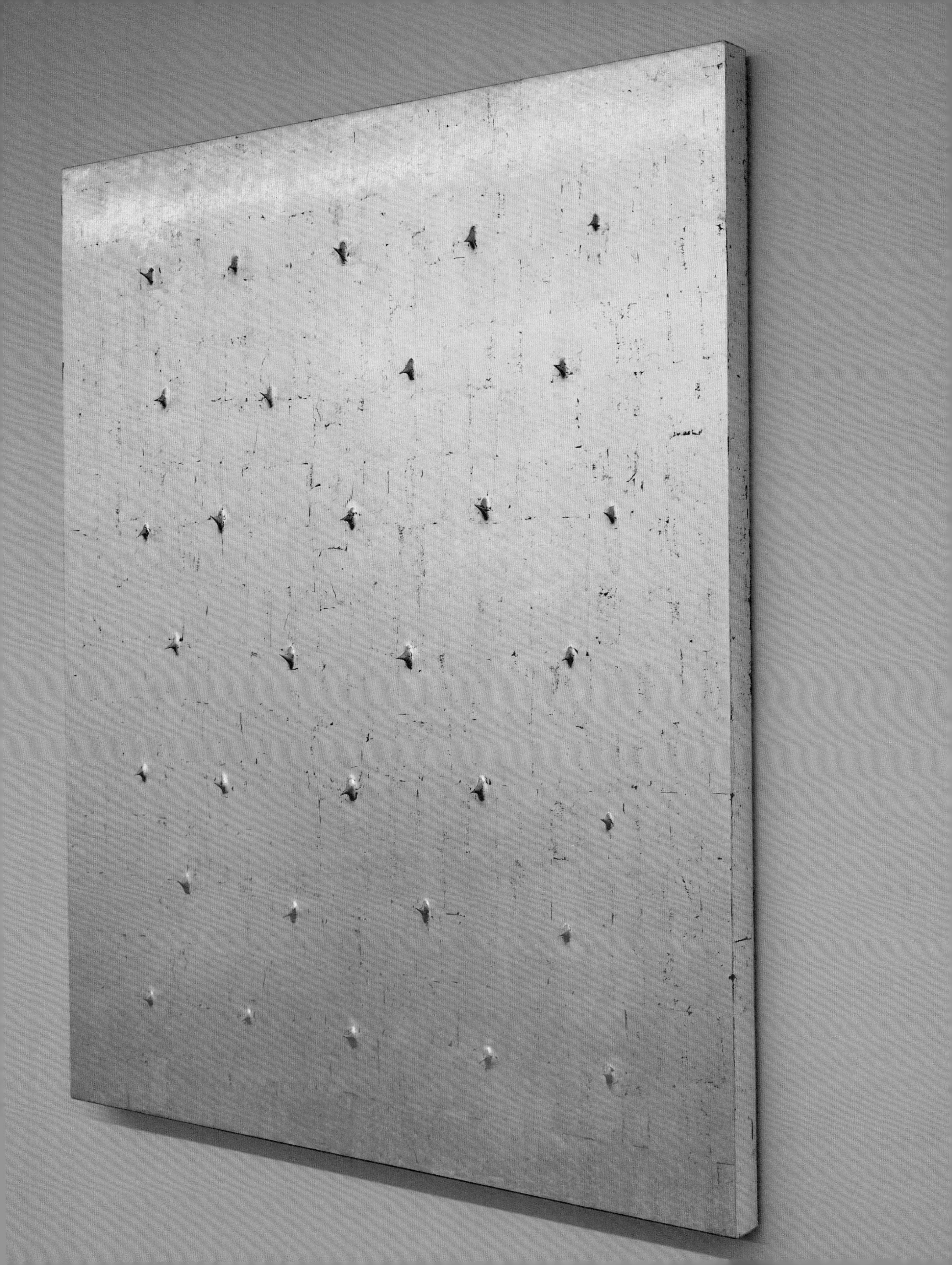

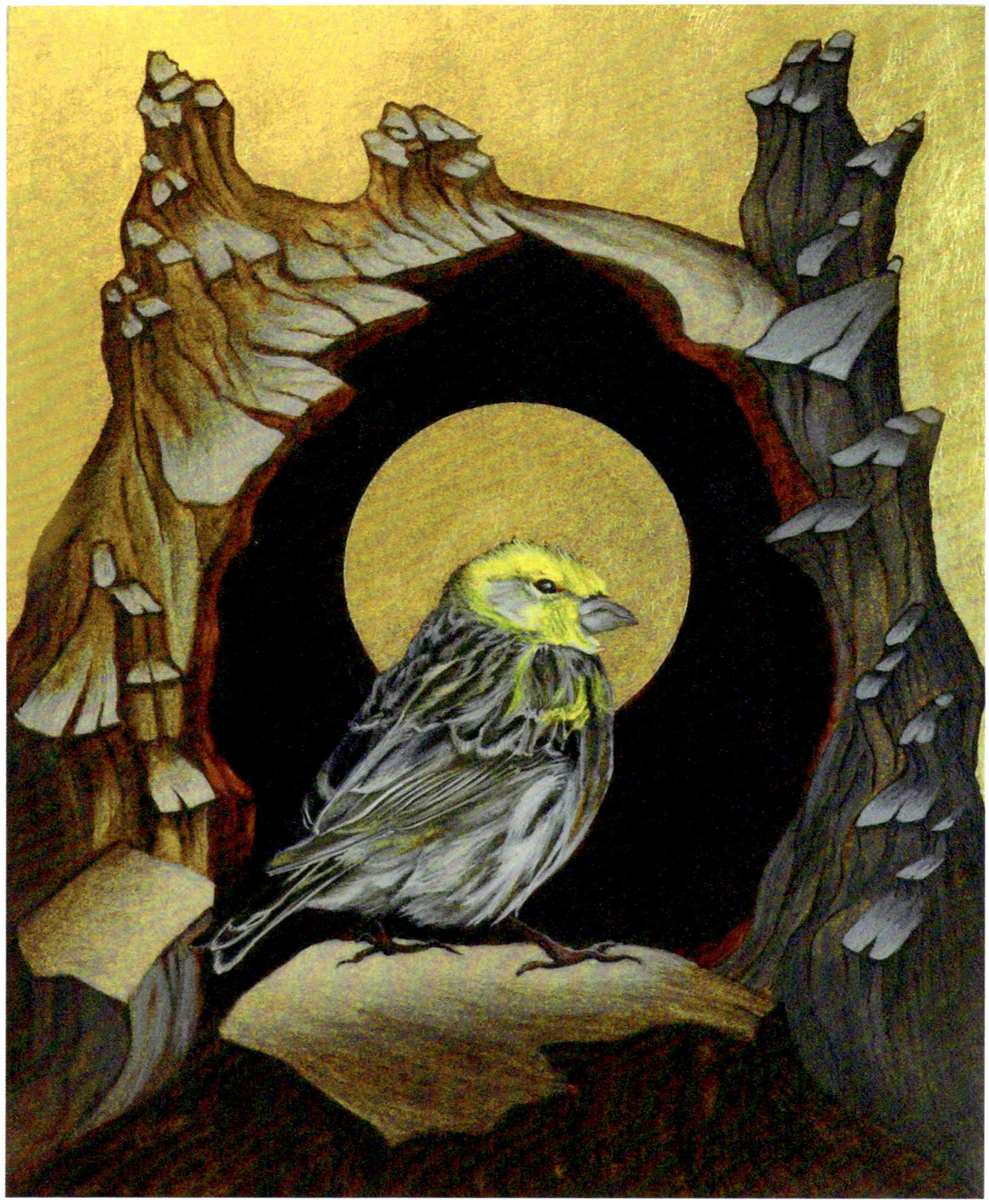

Mary Frances Dondelinger
Canary in the Coal Mine, 2010
Egg tempera and 23-carat gold on panel, 20.3 x 25.4 cm
In the private collection of G. Todd and R. Craig

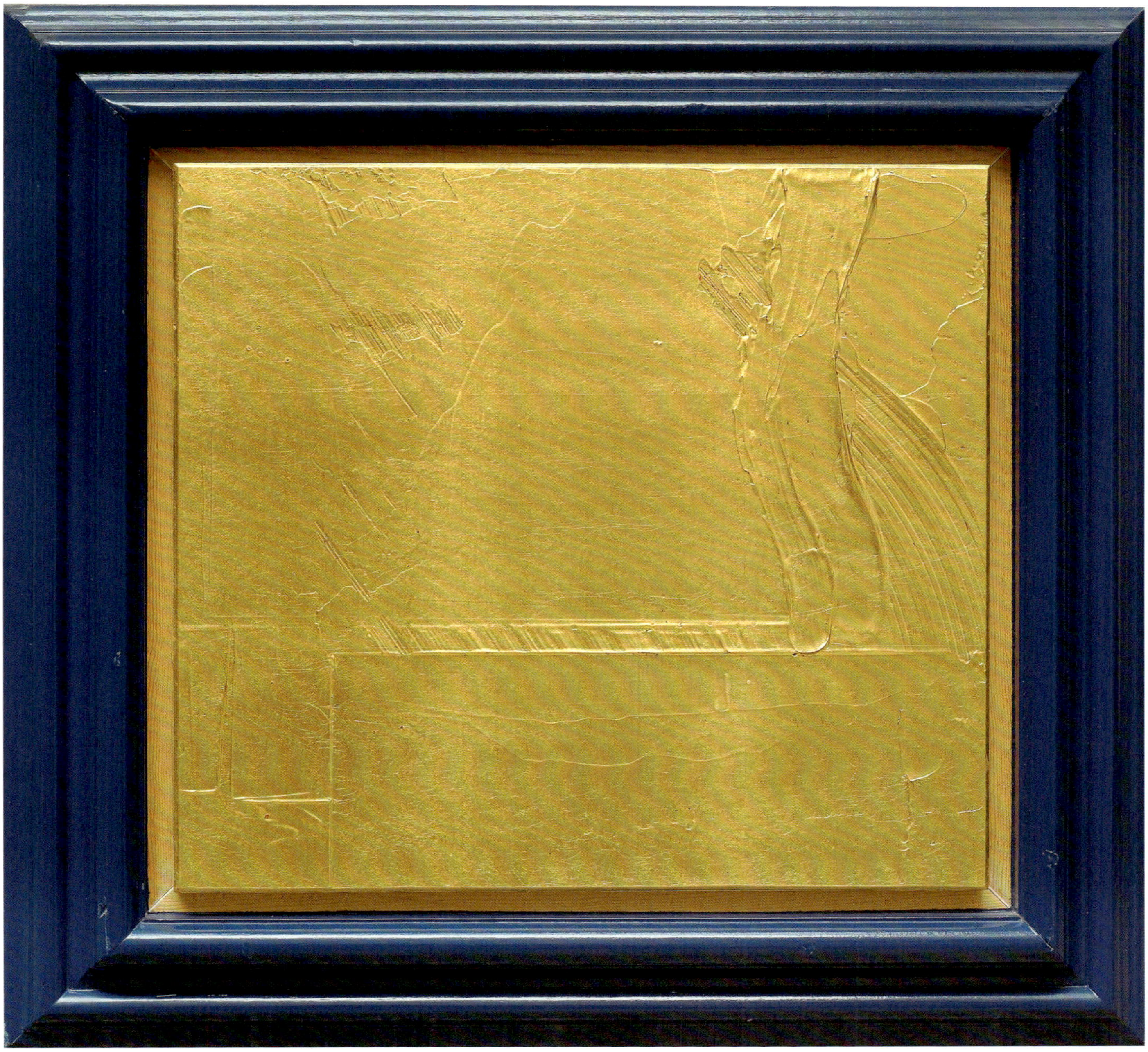

Christian Eckart
Detail Painting, No. 9, 1986
Acrylic and gold leaf on plywood, 39.5 x 35.5 cm
Collection Clo and Marcel Fleiss, Paris

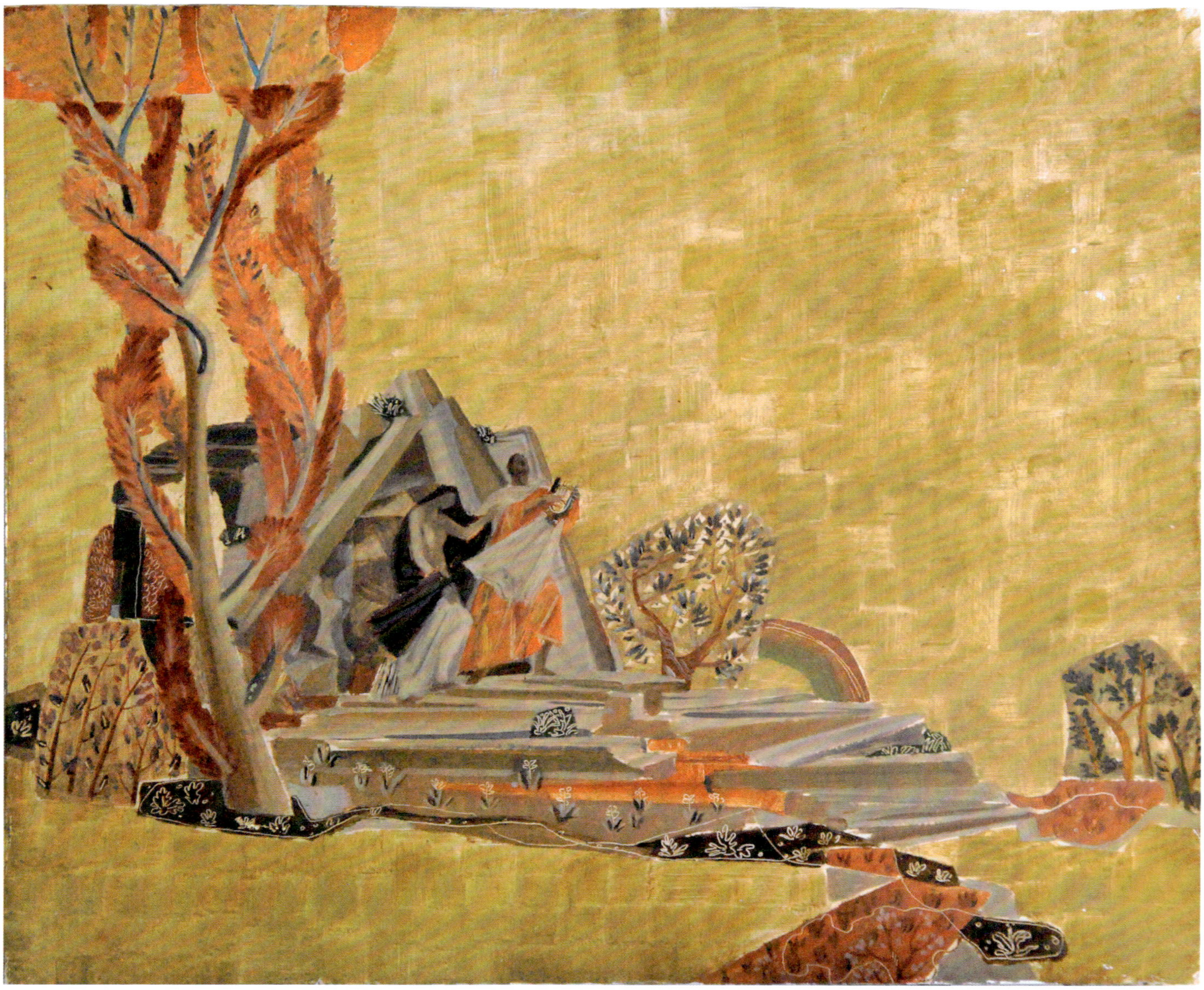

Rudolf H. Eisenmenger
Design for the safety curtain of the Staatsoper, 1954
Mixed media, gold leaf, 49 x 57 cm
Private collection

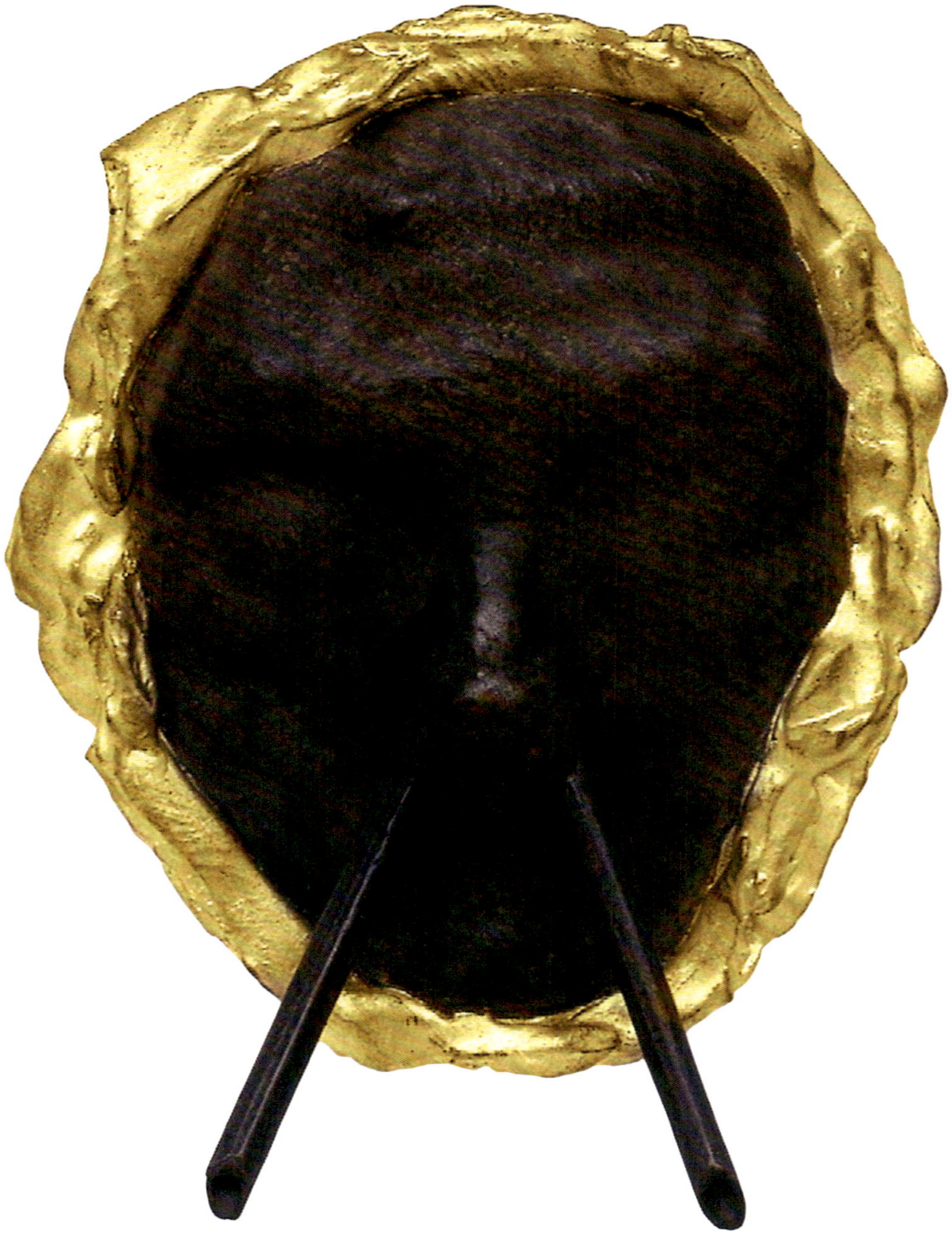

Luciano Fabro
Tamerlano, 1969
Bronze-gilt, 26 x 19 x 5.3 cm
Courtesy Collection Goetz

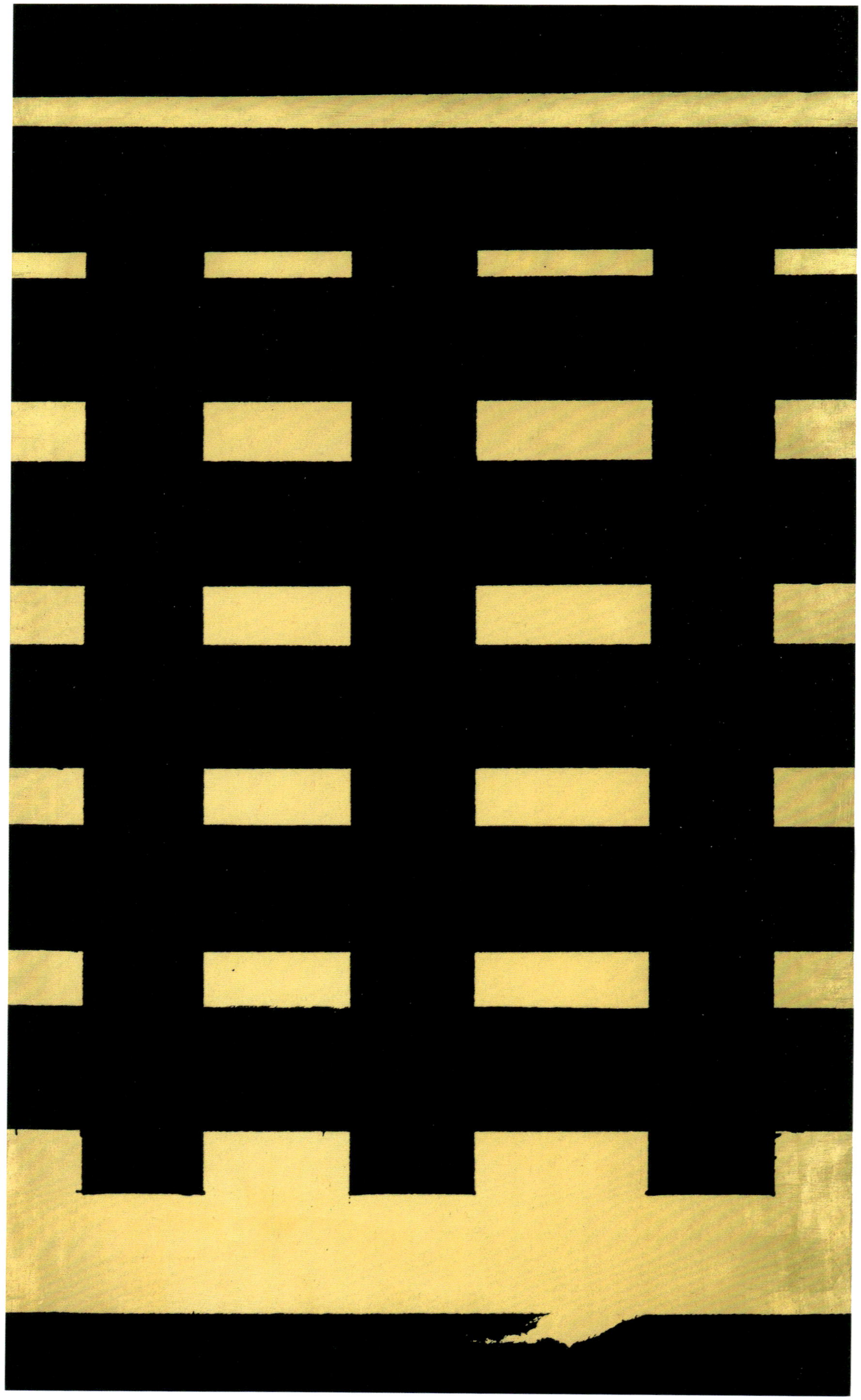

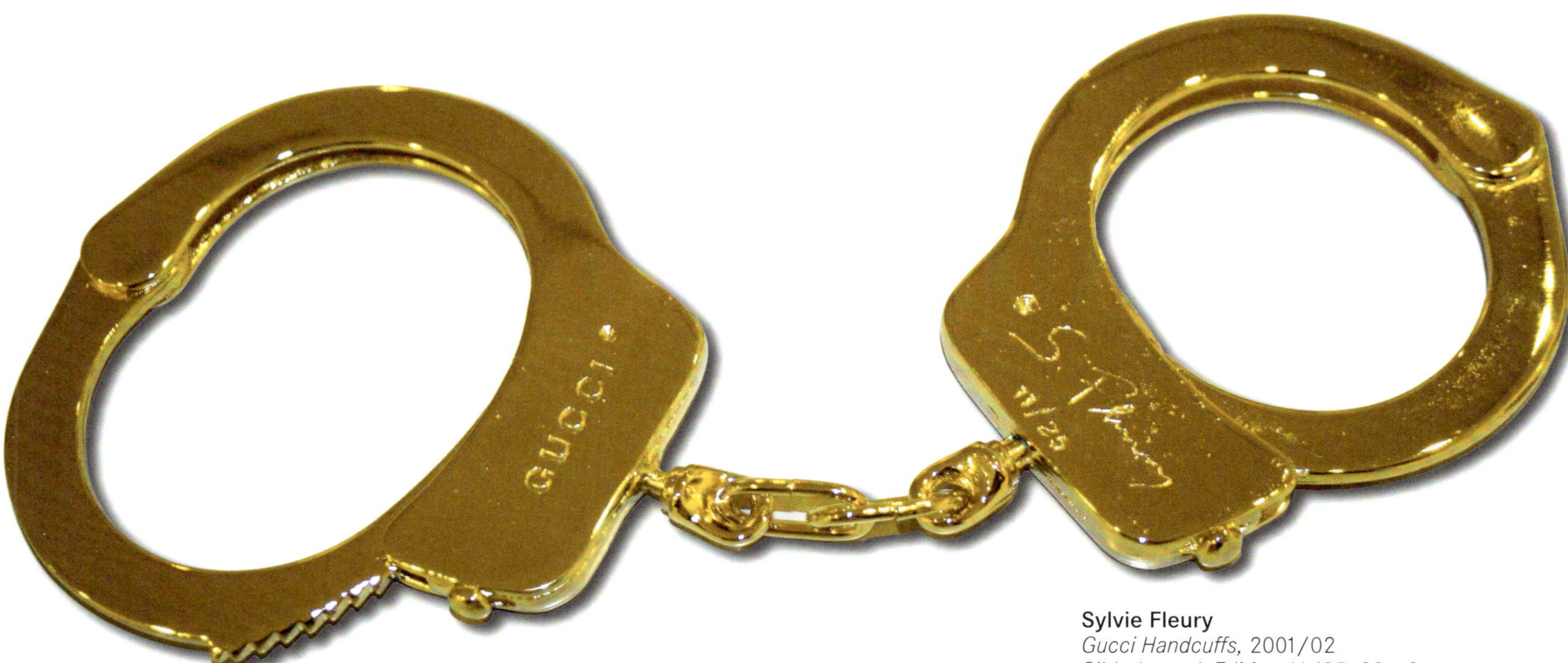

Sylvie Fleury
Gucci Handcuffs, 2001/02
Gilded metal, Edition 11/25, 23 x 9 cm
Galerie Thaddaeus Ropac, Salzburg · Paris

Helmut Federle
Untitled, 2002
Screenprint, gold leaf on aluminium, 193 x 116 cm
Collection Sigrid and Franz Wojda, Vienna

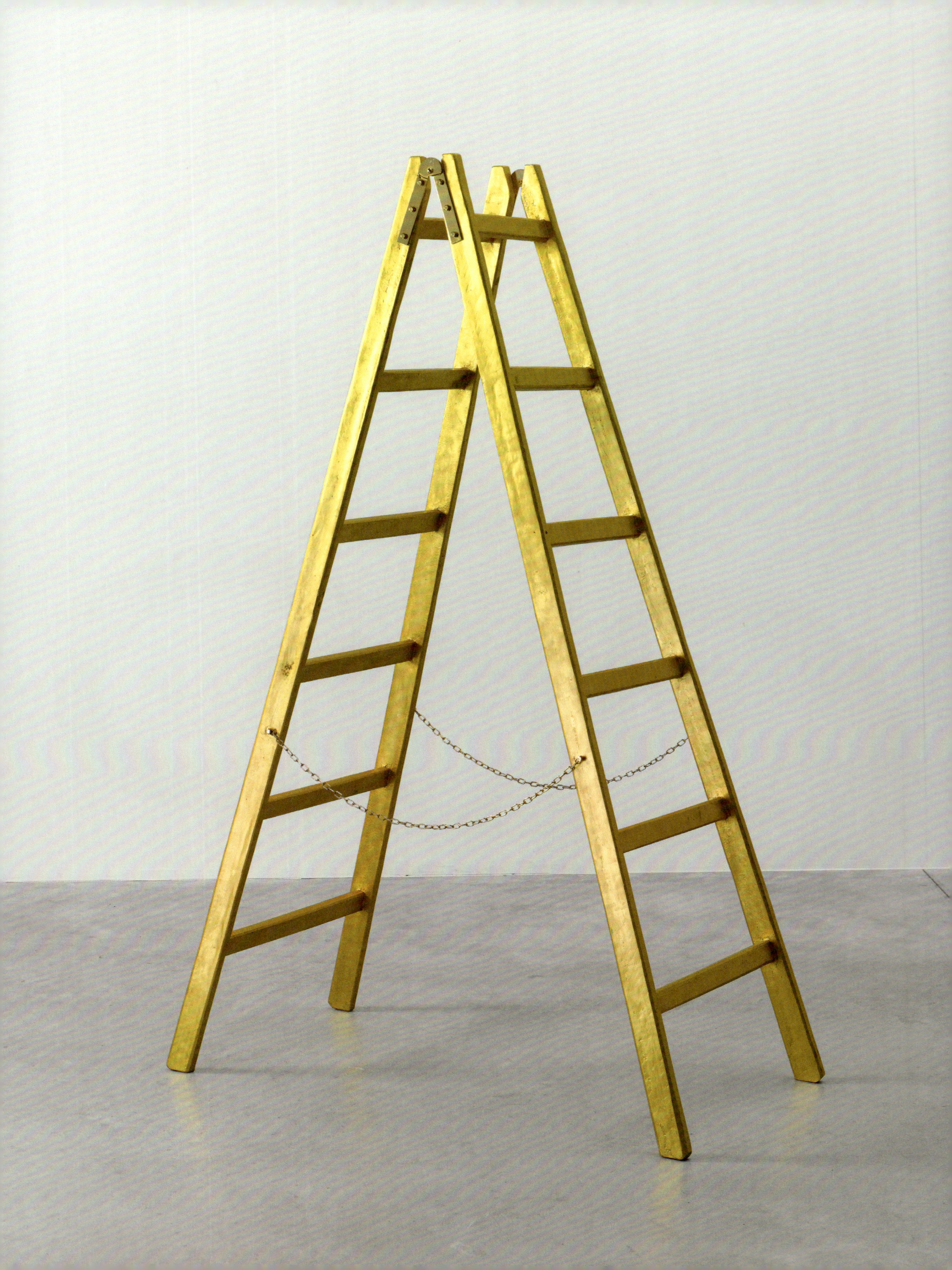

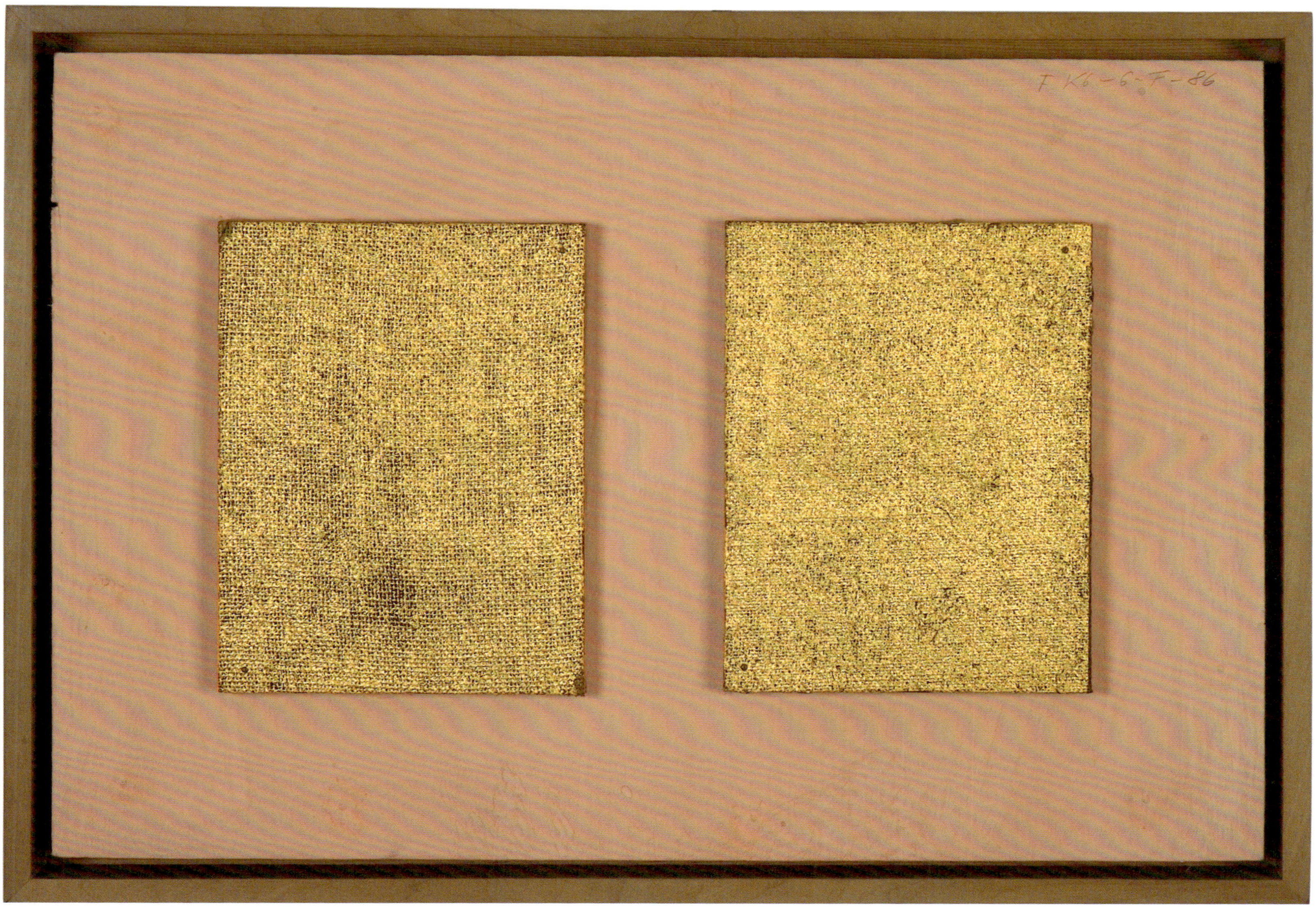

Günther Förg
Untitled, 1986
Acrylic, gold leaf on wood, 41 x 60 cm
Grässlin Collection, St. Georgen

Sylvie Fleury
Ladder, 2007
23.5-carat gold, Edition 2/8, height 196 cm
Galerie Thaddaeus Ropac, Salzburg · Paris

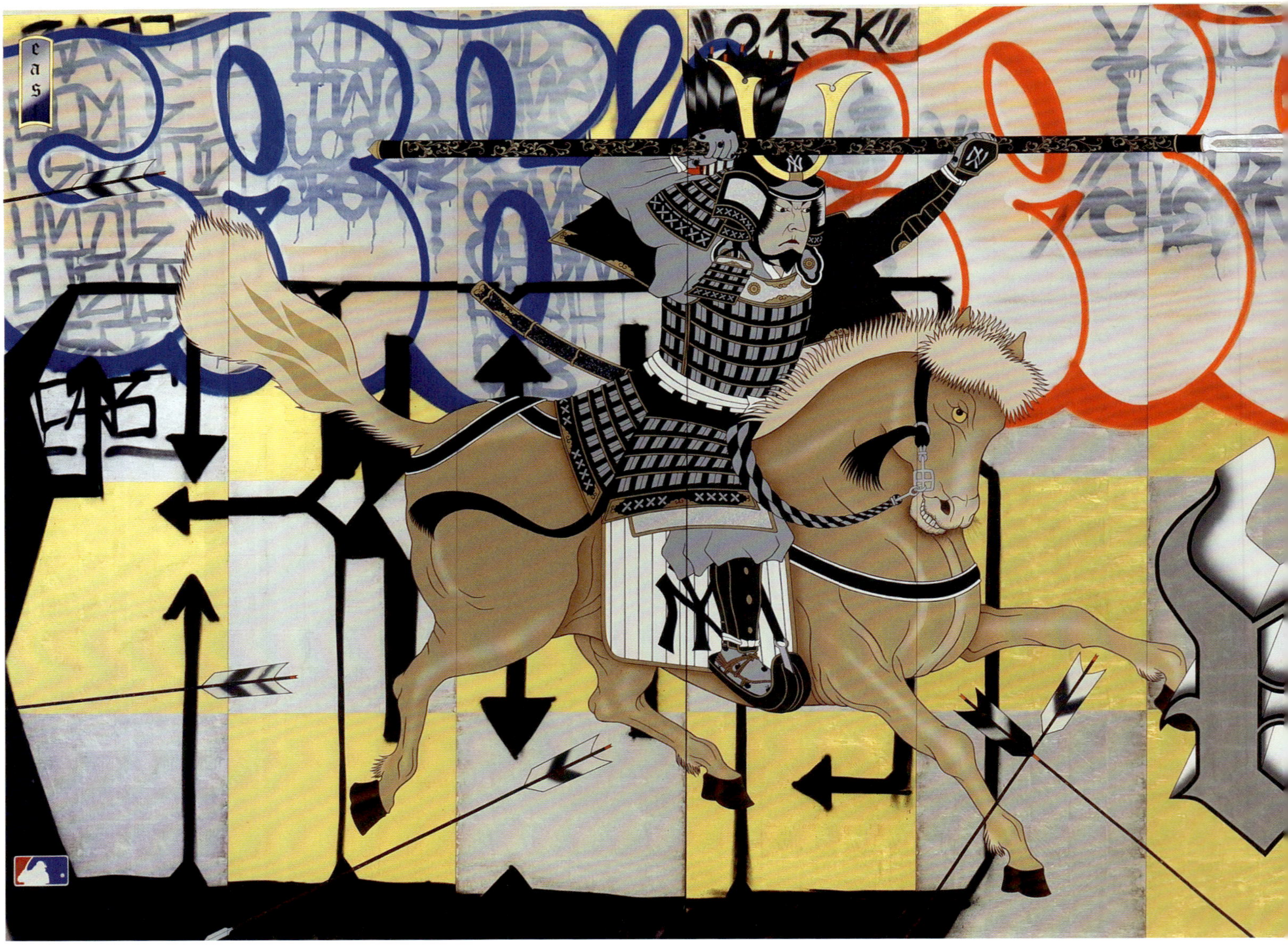

Gajin Fujita
East vs. West, 2010
Platinum foil, gold leaf, acrylic, paint spray and paint marker on wood, 12 panels, each 213.4 x 55.9 cm (in total 213.4 x 670.6 cm)
Private collection, Courtesy of L. A. Louver, Venice, California

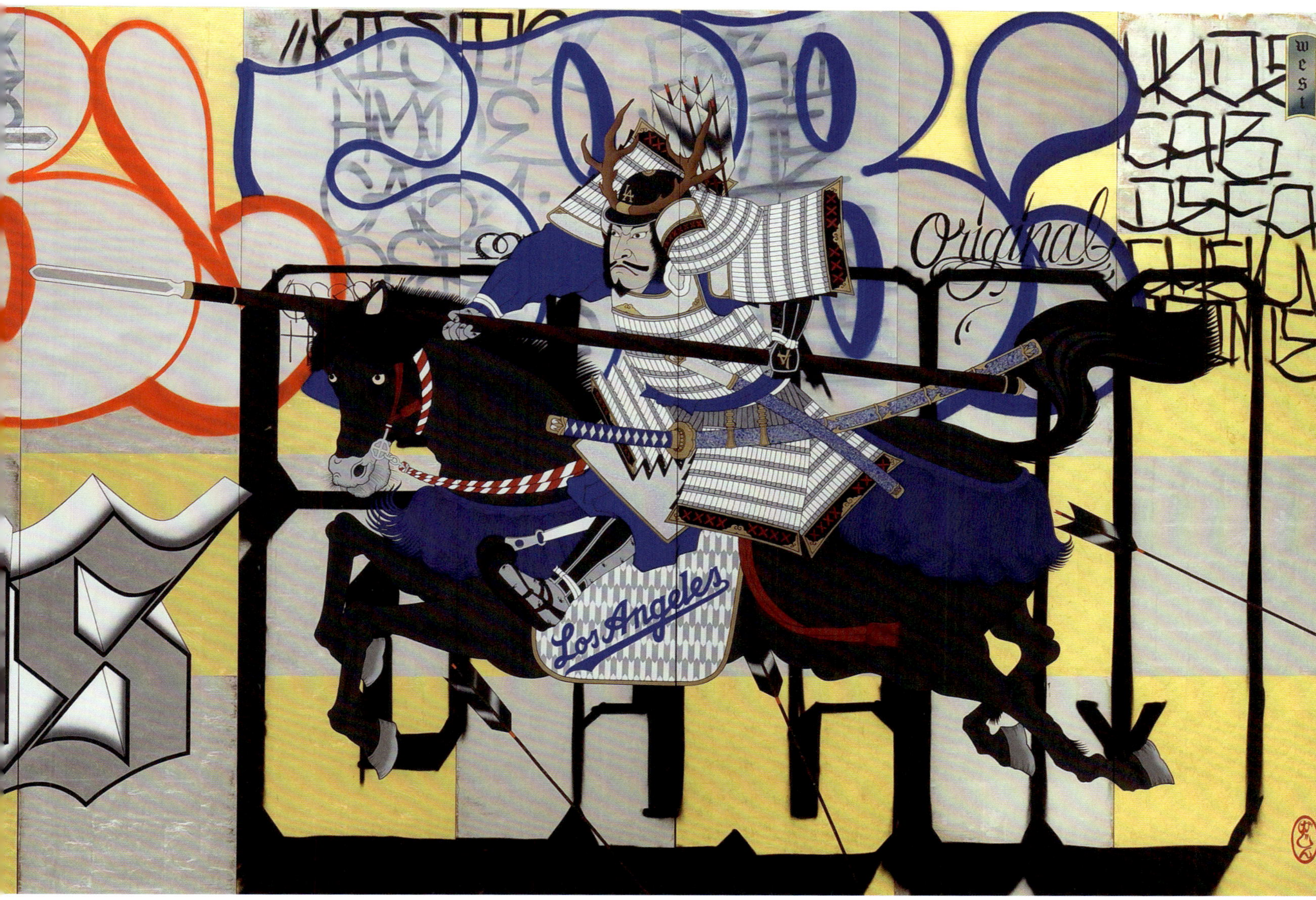
west
Original
Los Angeles

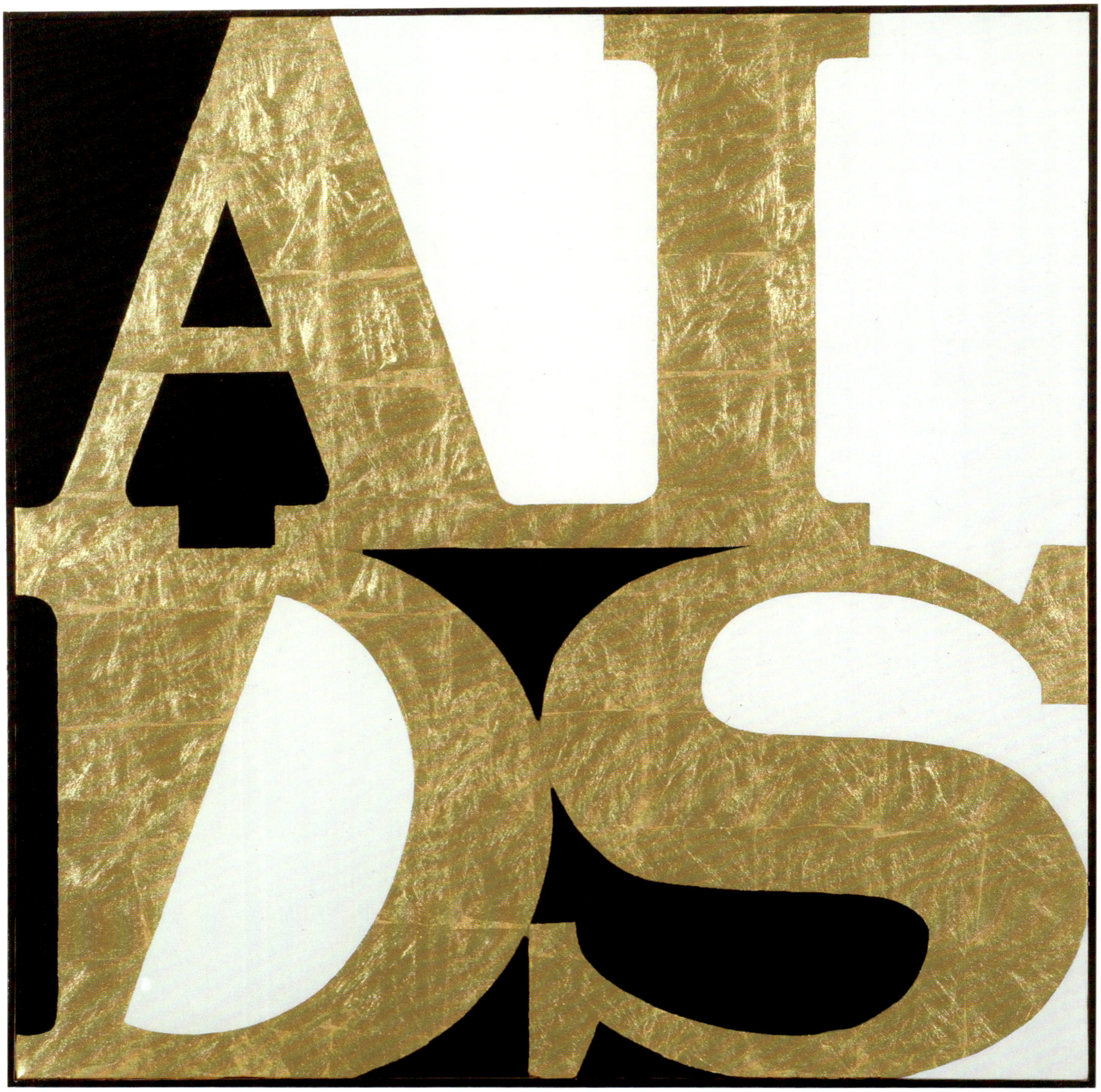

General Idea
AIDS, 1987
Acrylic, gold leaf on canvas, 76 x 76 cm
Collection Katharina Faerber, Geneva

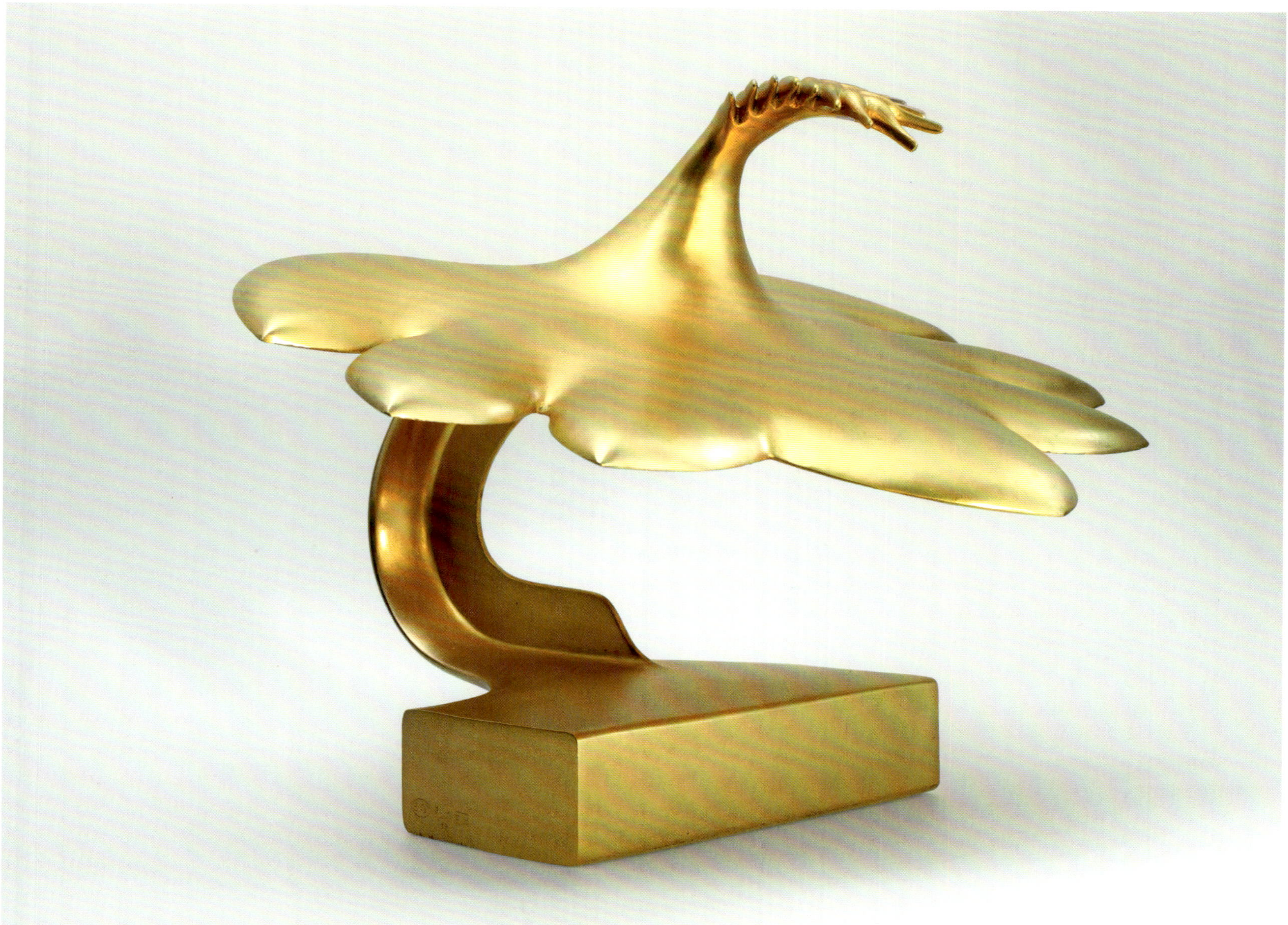

Bruno Gironcoli
Untitled (Blattähre), 2003
Gilded cast aluminium, 55 x 49 x 42 cm
Courtesy Galerie Elisabeth and Klaus Thoman, Innsbruck

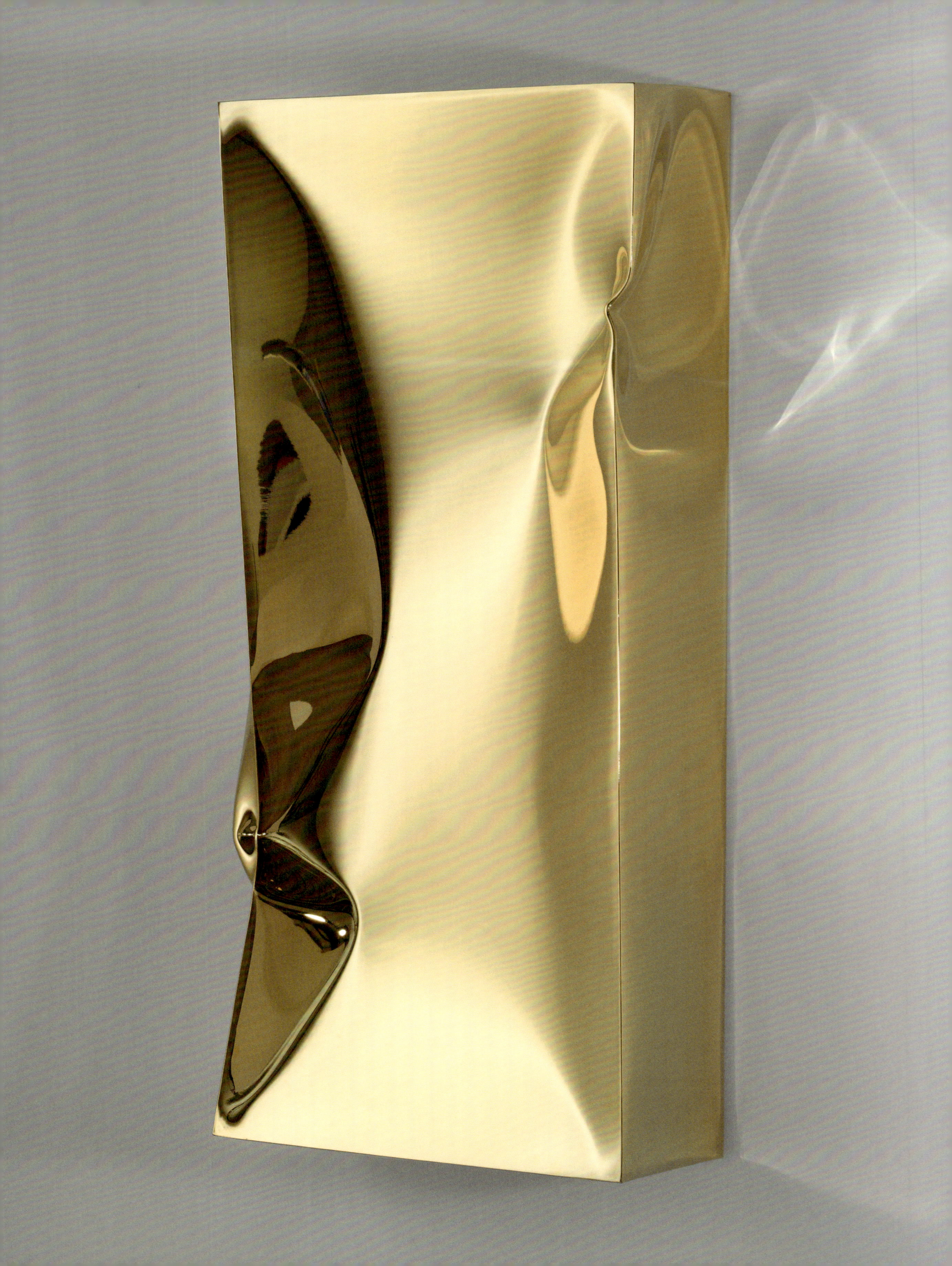

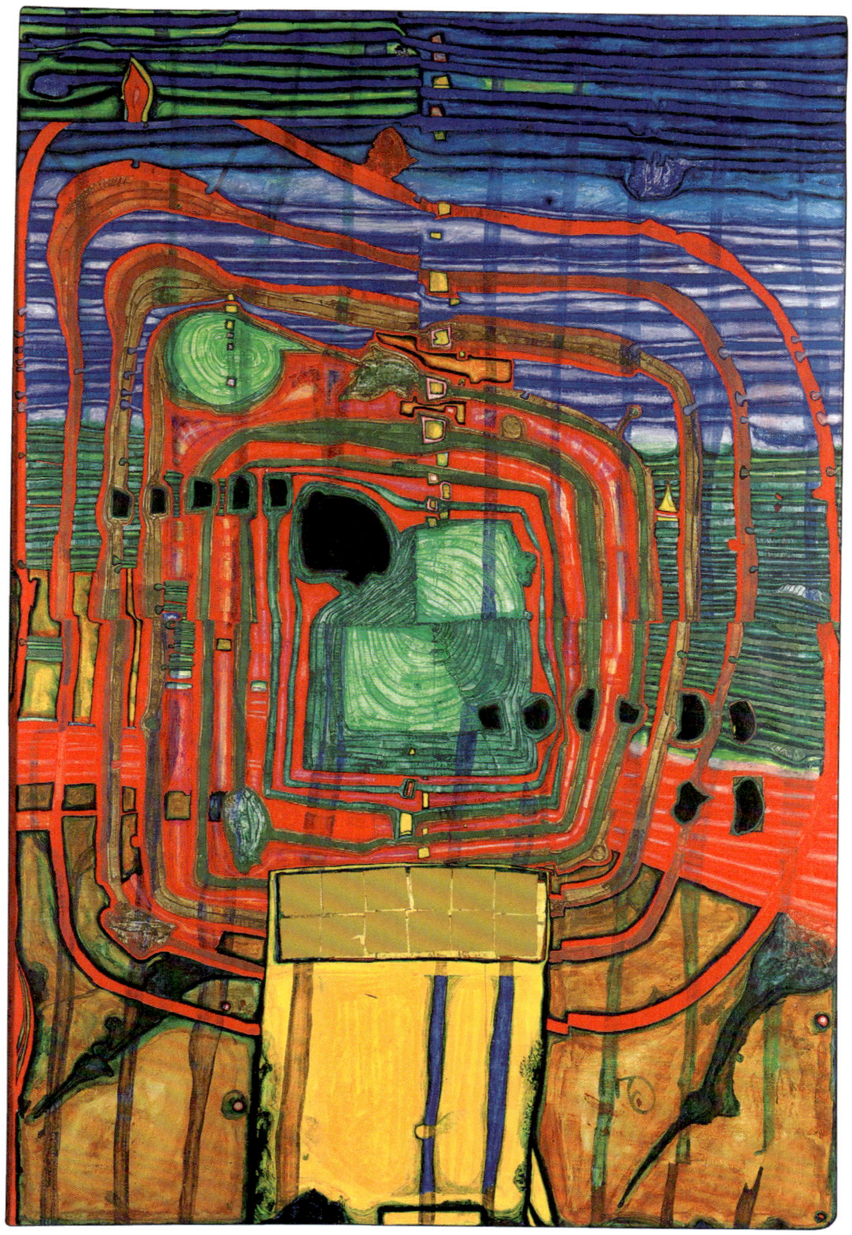

Friedensreich Hundertwasser
460 Hommage au tachisme, 1961
Mixed media: egg tempera, oil and gold leaf on packing paper primed with chalk and polyvinyl, mounted on jute with PV, 195 x 130 cm
Die Hundertwasser Gemeinnützige Privatstiftung, Vienna

Lori Hersberger
Intergalactic Totem, 2011
Steel, gold, 120 x 50 x 25 cm
Galerie Thaddaeus Ropac, Salzburg · Paris

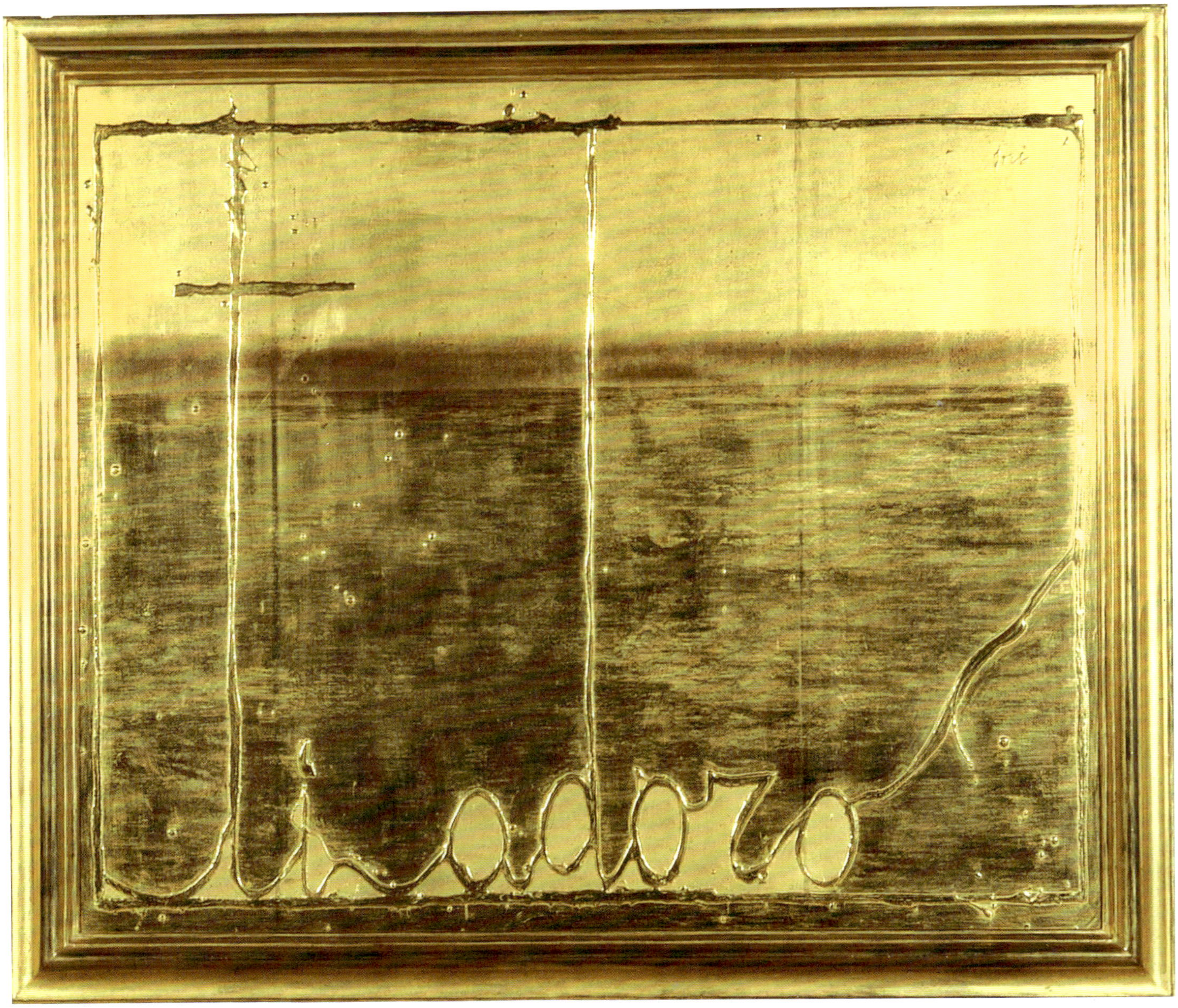

Marcello Jori
ti adoro, 2005
Gold leaf on wood, 104 x 120 cm
Private collection, Prato

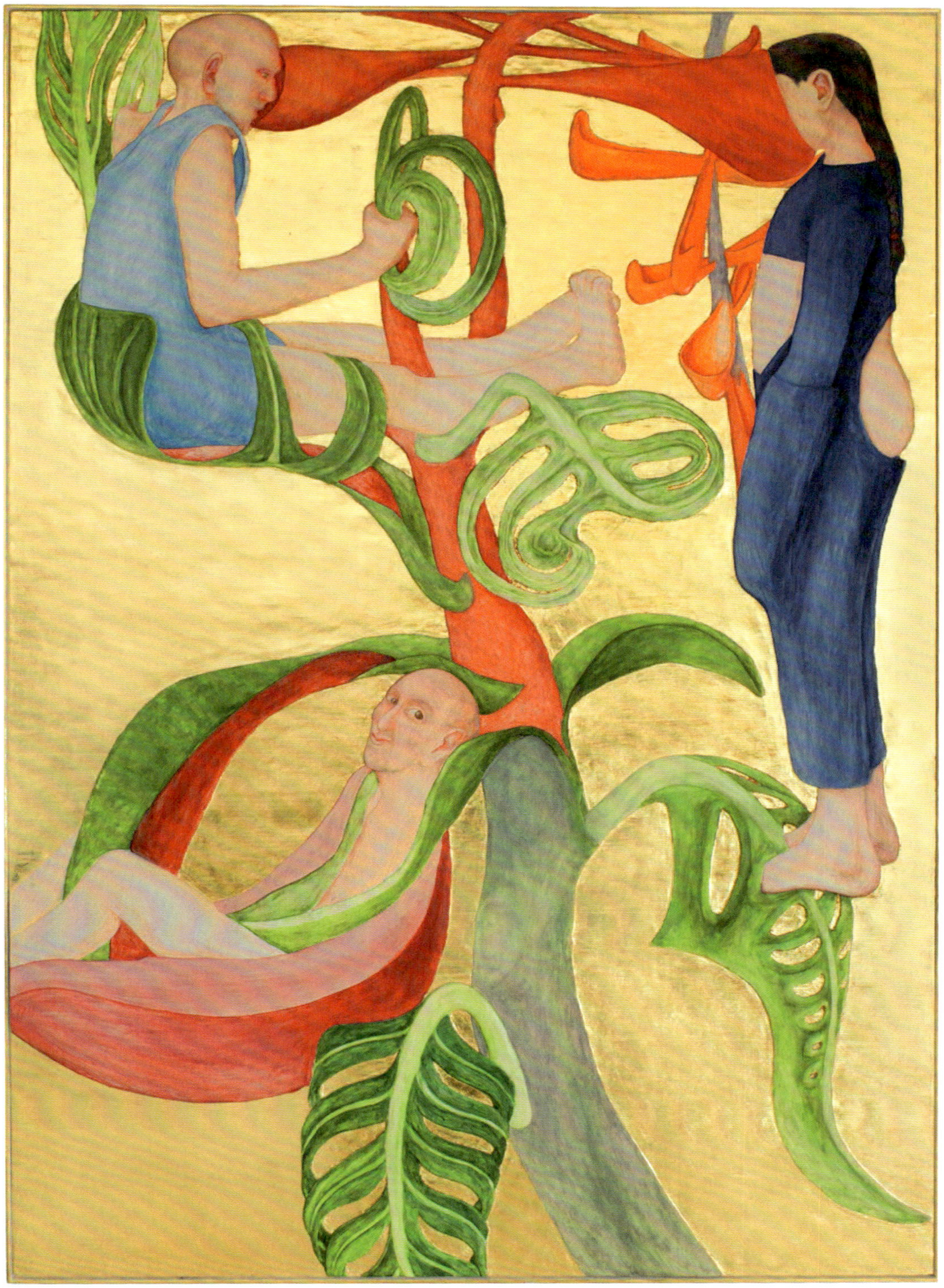

Josef Kern
Ein Sommer in Bäumen, 2006
Wood, gold leaf, 202 x 142 cm
Private collection

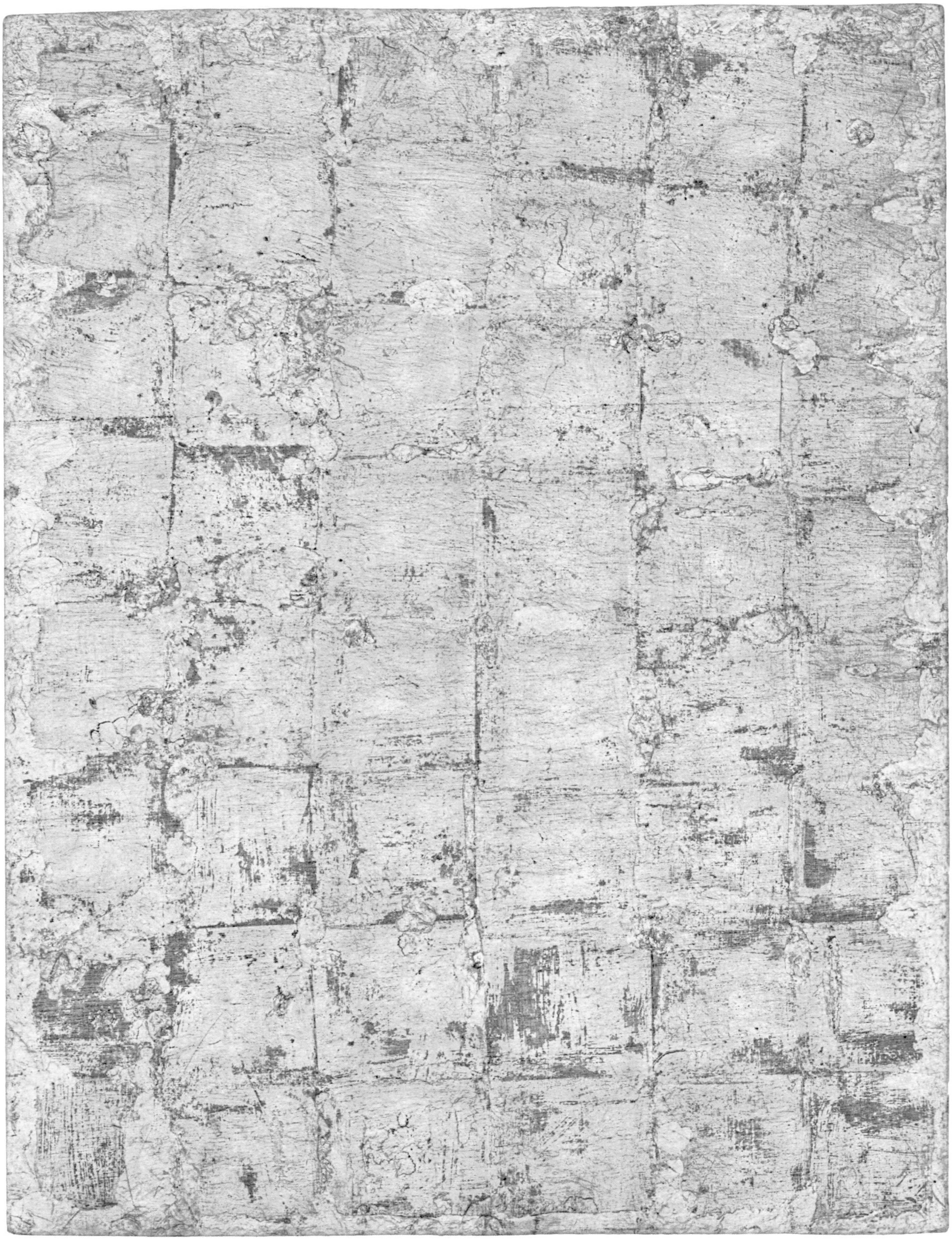

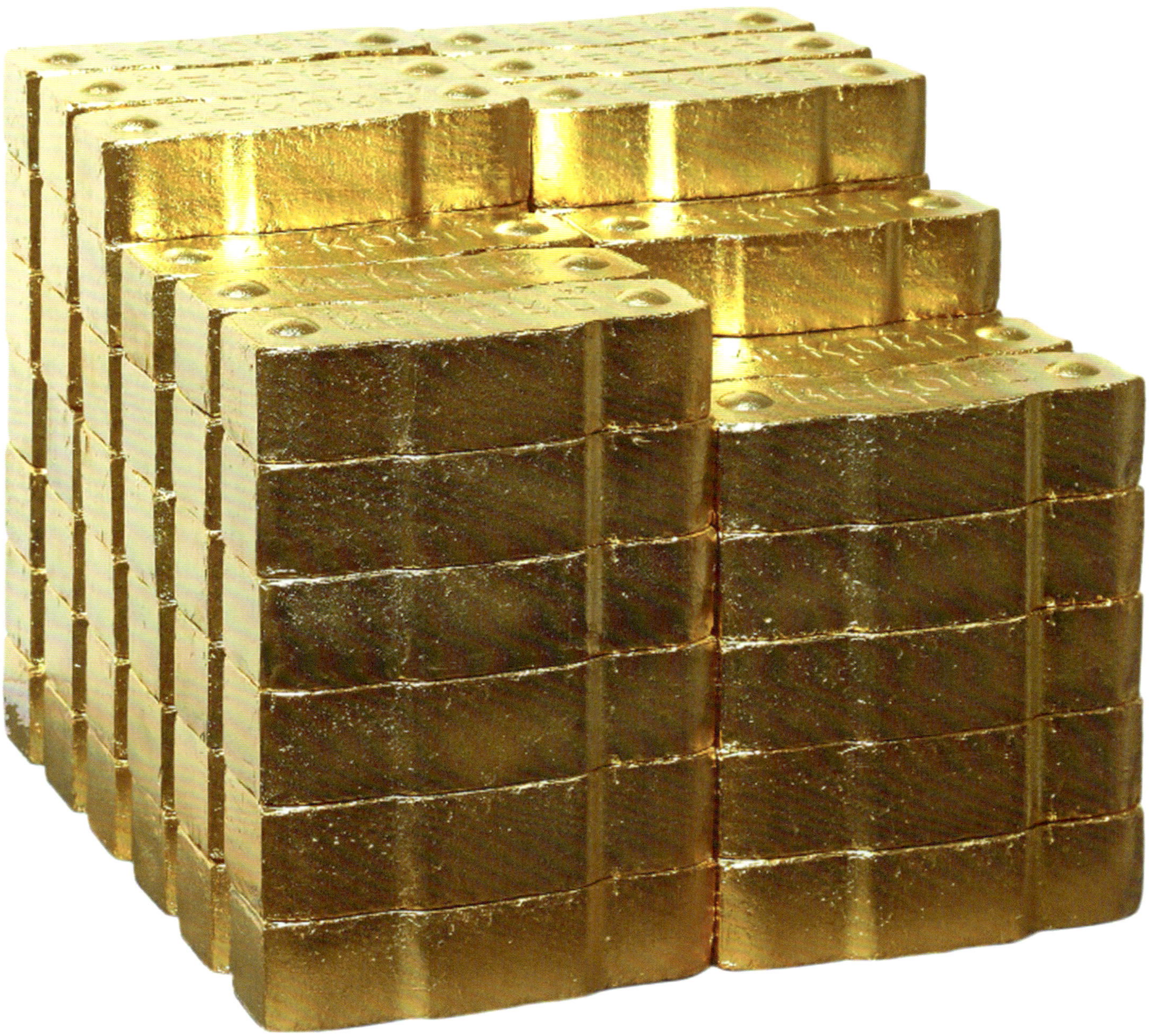

Alicja Kwade
Kohle, 2006
Cast bronze of coal briquette, gilded, 24-carat real gold, 77 parts, 33 x 34 x 35.8 cm
Private collection, Stuttgart

Yves Klein
Valeur or (Monochrom gold), 1960
Gold leaf on plywood, 59 x 44.5 cm
Kunstmuseen Krefeld

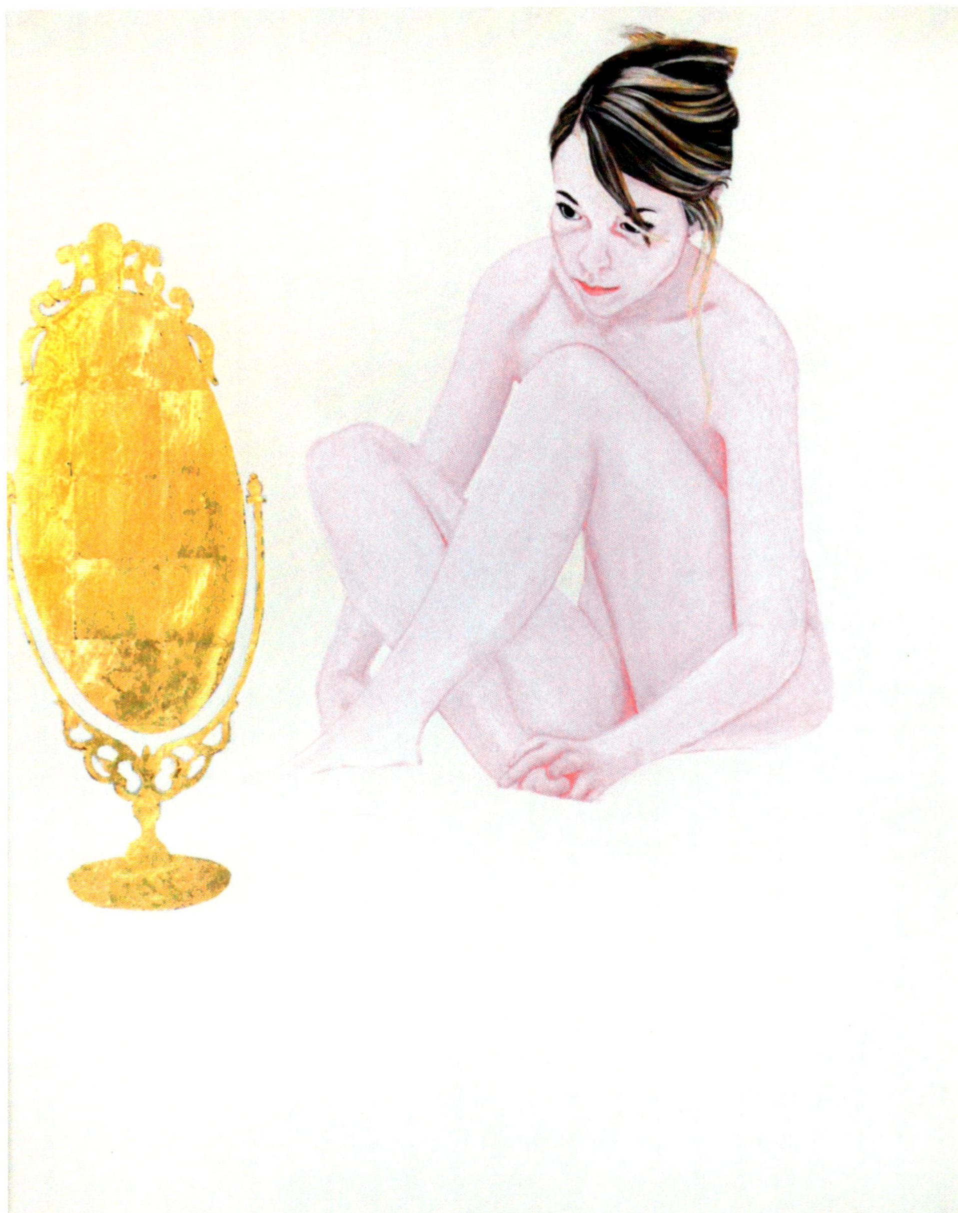

Frédéric Léglise
Charline, 2007
Oil, gold leaf and gloss paint on canvas, 115 x 88.5 cm
Collection David Marcel Fleiss, Galerie 1900–2000, Paris

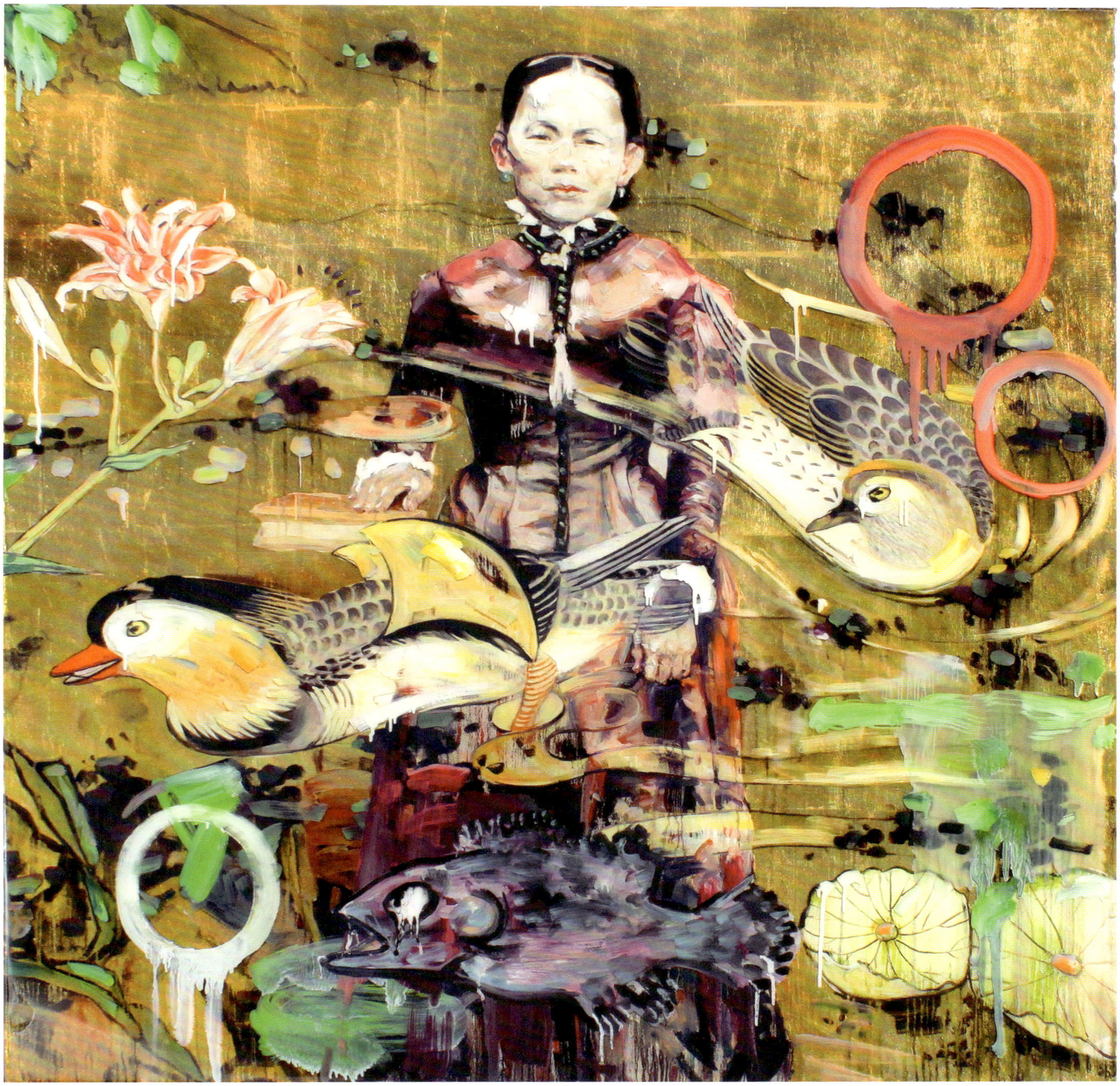

Hung Liu
Polly with Mandarin Ducks, 2008
Mixed media on wood, 104.1 x 104.1 cm
Turner Carroll Gallery, Santa Fe/Dallas

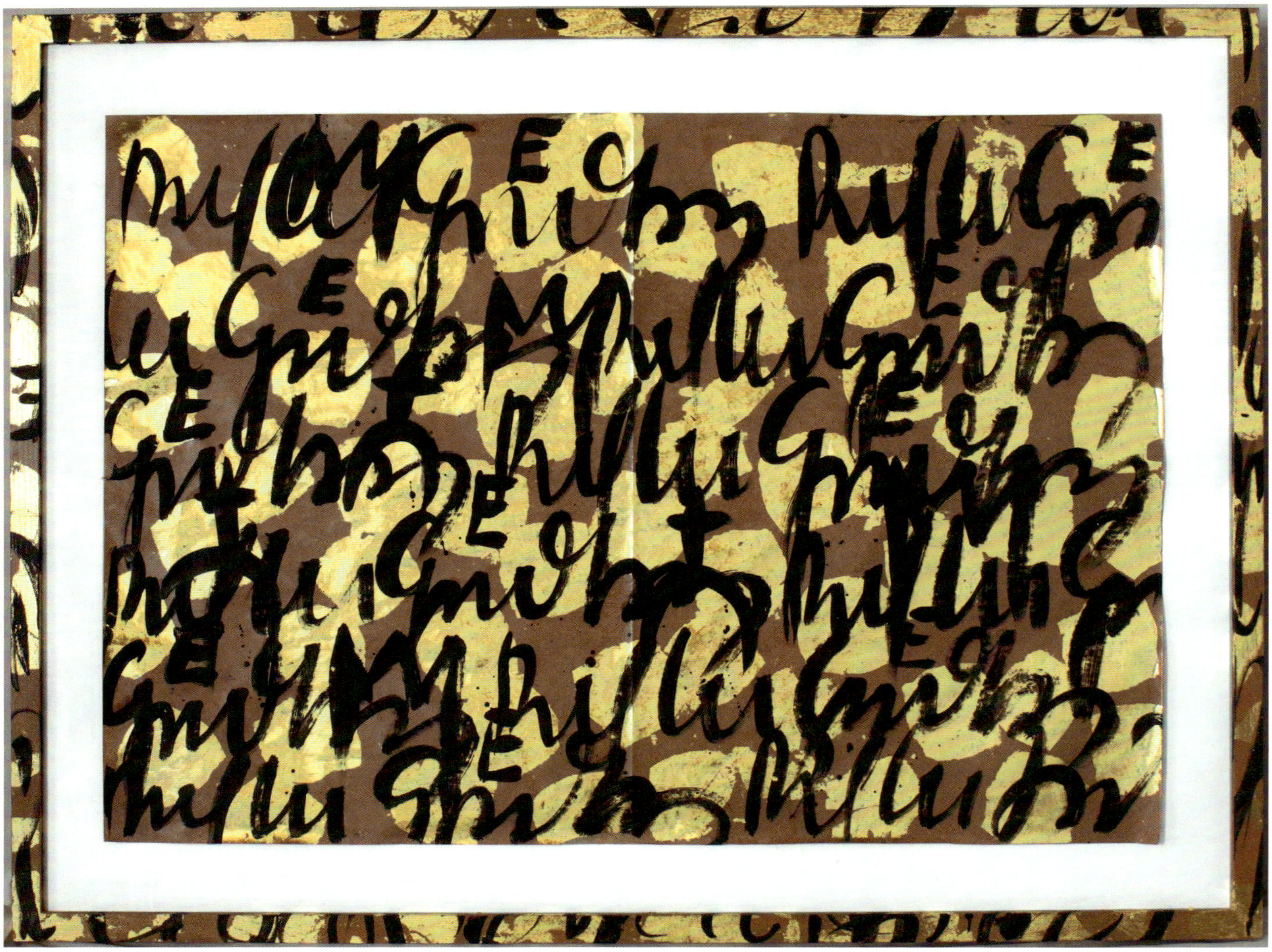

Rudolf Leitner-Gründberg
Die Lösung des Welträtsels ist die Hinwendung zu unserer höchsten Möglichkeit, 2002
Triptych—gold leaf, oil, solid wood case, 131 x 90 x 23.5 cm
Galerie Heike Curtze Vienna–Berlin

Sherrie Levine
Fountain (after Marcel Duchamp: A.P.), 1991
Cast bronze, 36.8 x 36.2 x 63.5 cm
Collection Walker Art Center, Minneapolis, T. B. Walker Acquisition Fund, 1992

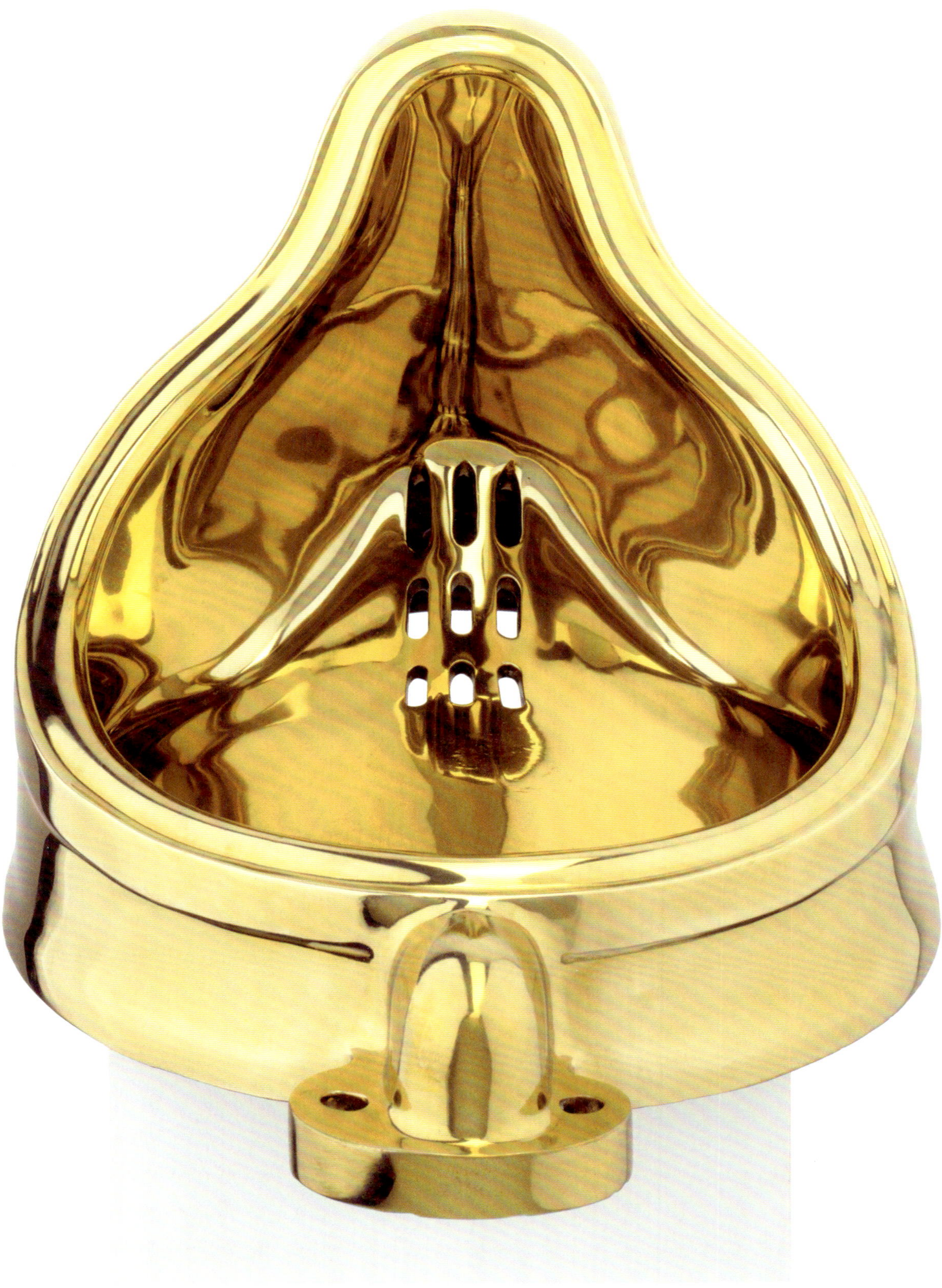

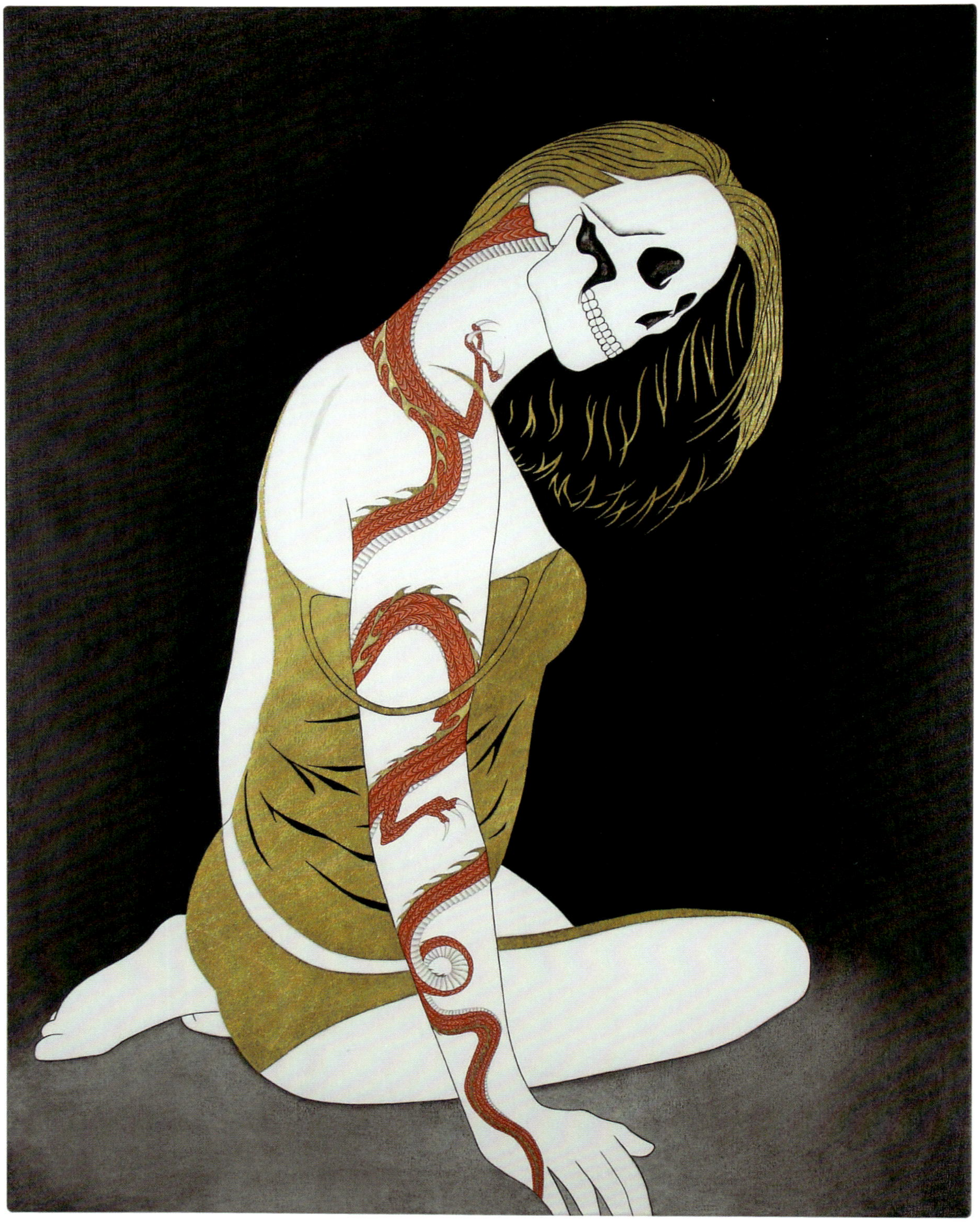

Jason Clay Lewis
Girl with dragon, 2007/08
Acrylic and gold leaf on canvas, 121.9 x 91.4 cm
Jason Clay Lewis

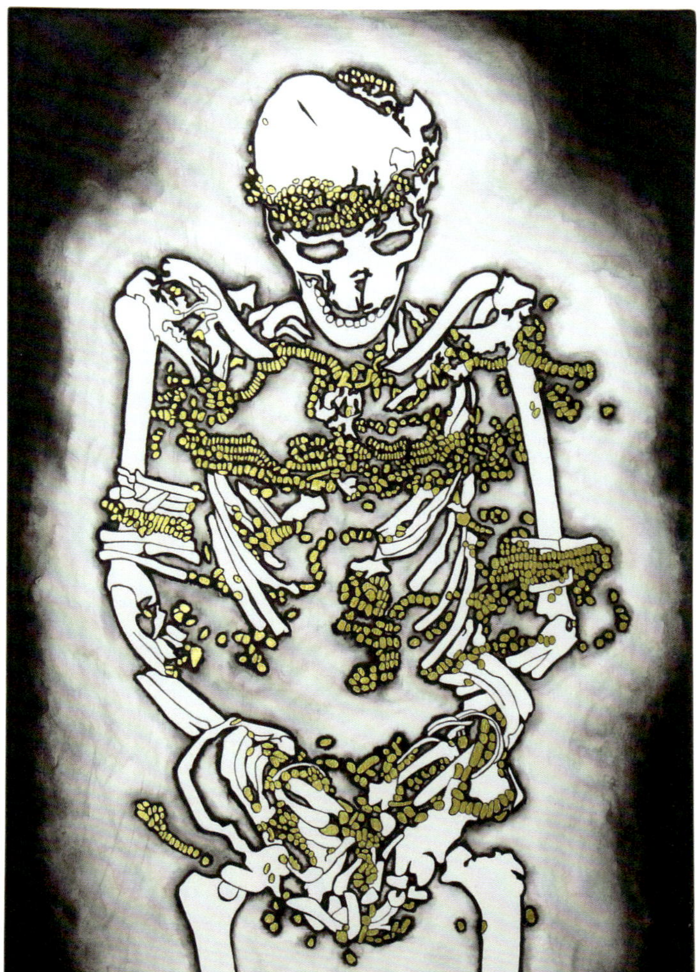

Jason Clay Lewis
Warrior, 2006
Acrylic and gold leaf on canvas, 45.7 x 30.5 cm
Jason Clay Lewis

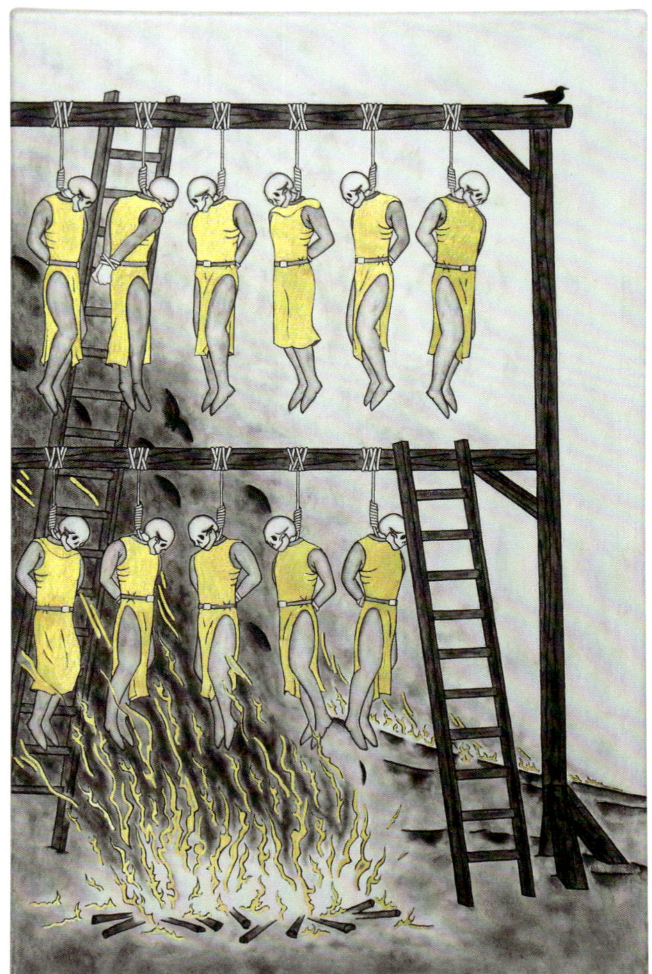

Jason Clay Lewis
Torture panel VI, 2006/07
Acrylic and gold leaf on canvas, 43.2 x 26.7 cm
Jason Clay Lewis

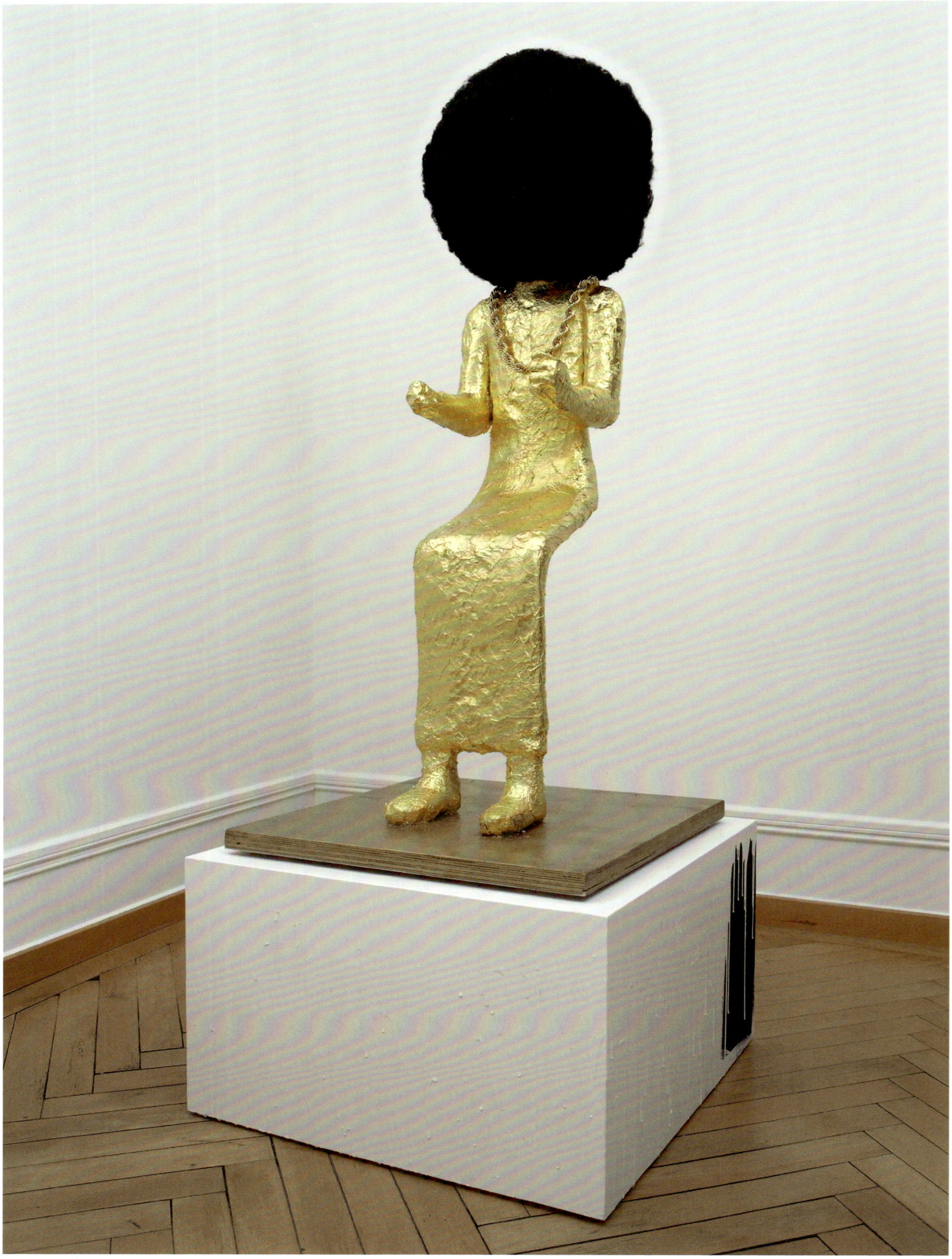

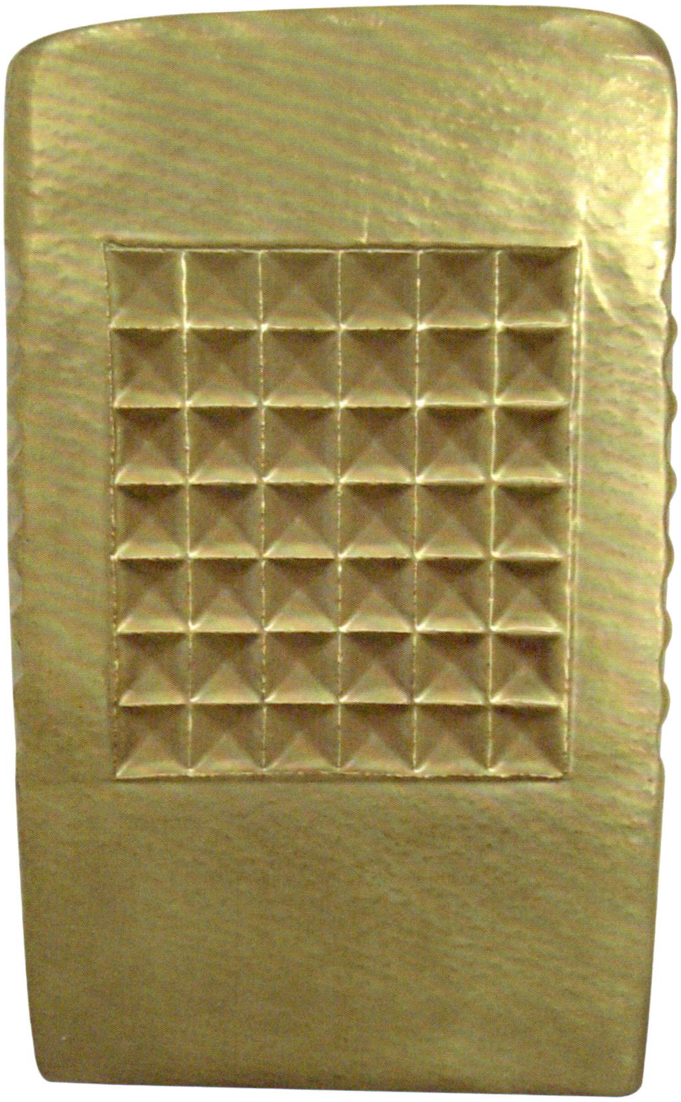

Heinz Mack
Mack K40, 1997
Polished gold, 35 x 20 x 6 cm
Galerie Walker Schloss Ebenau Rosental

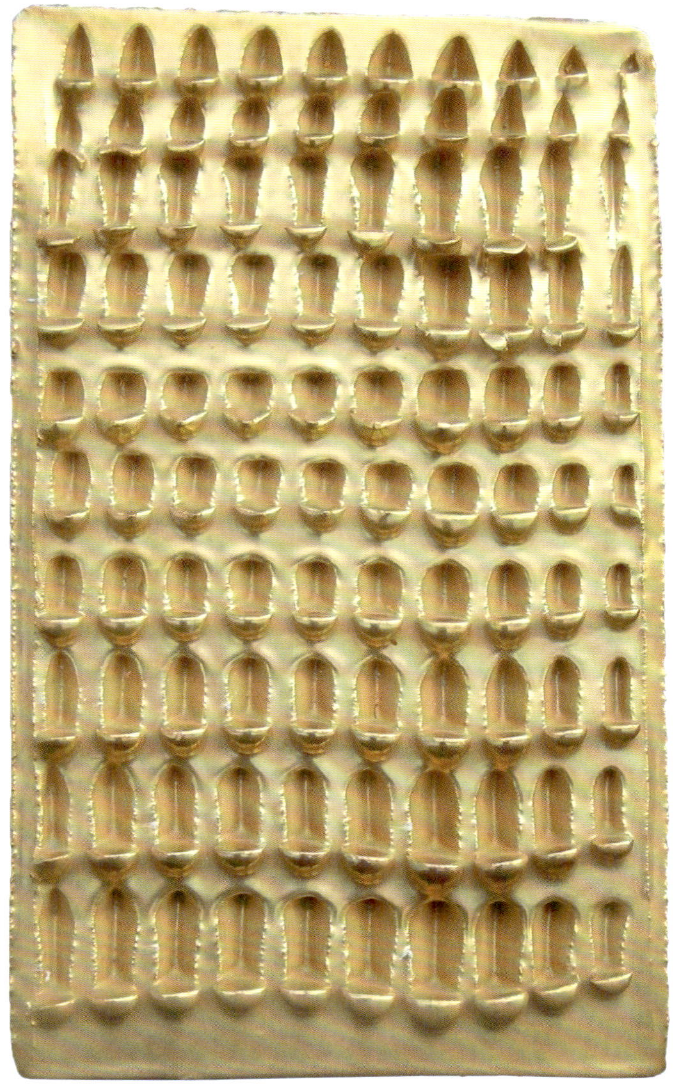

Heinz Mack
Mack K41, 1997
Polished gold, glazed, 32 x 19 x 4.5 cm
Galerie Walker Schloss Ebenau Rosental

Edward Lipski
King, 2007, Gold leaf, synthetic hair, gold chain, wood,
114 x 60 x 60 cm
Private collection, Switzerland

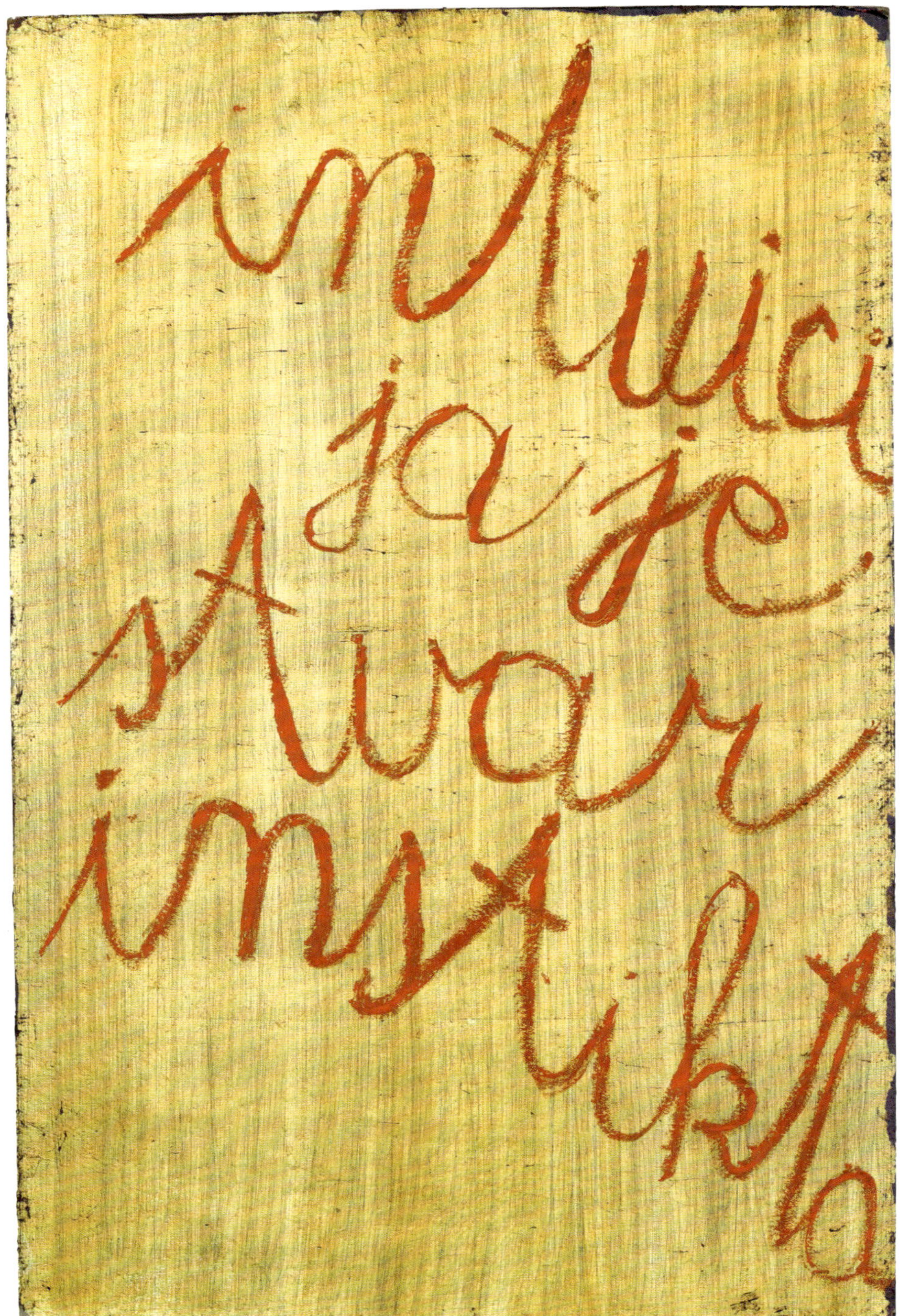

Dimitrije Bašičević Mangelos
Intuicija je stvar instinkta (Intuition is a matter of instinct),
1977/78
Tempera and gold paint on board
66.4 x 45 cm
Collection Flick, Salzburg

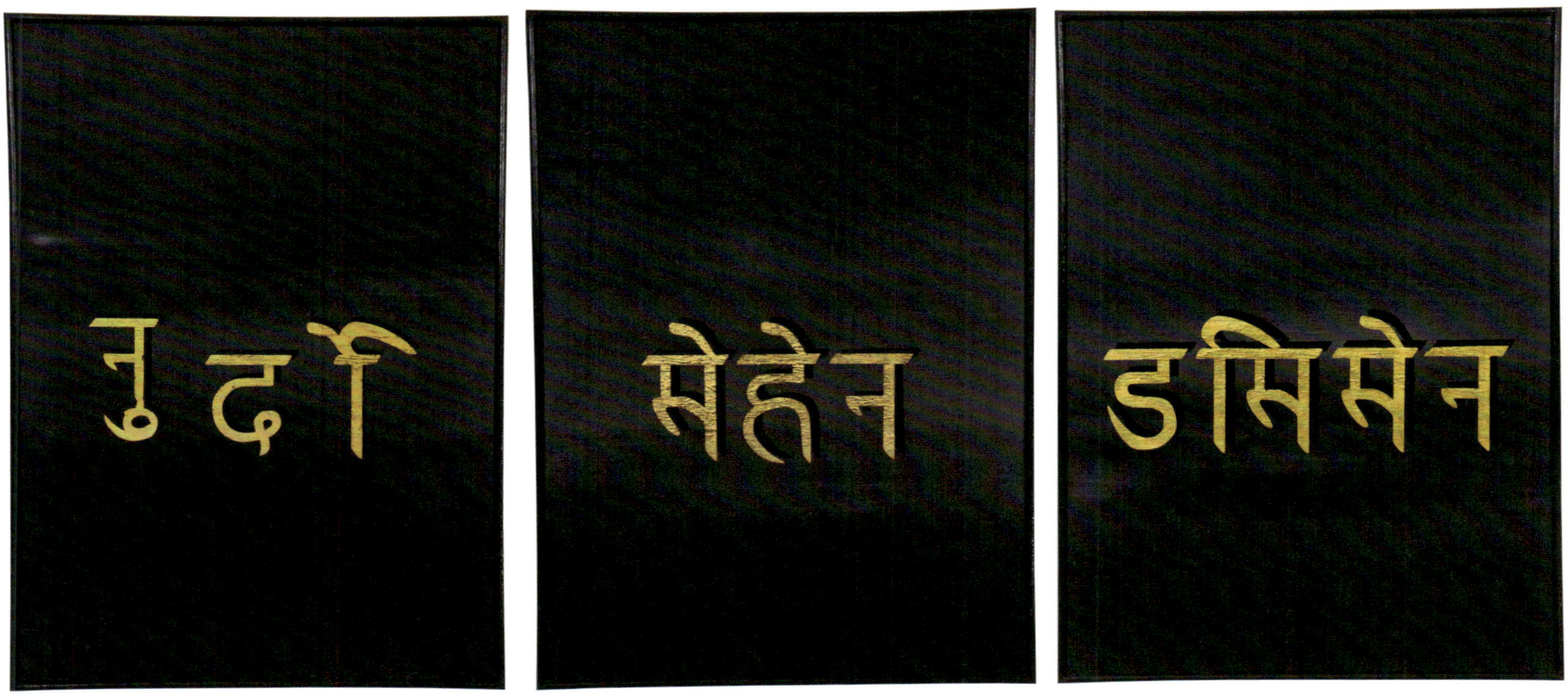

Franz Mayrhofer
Devanagari from the series "Nur Wissende sehen", 2001
Gold leaf and acrylic on wood, each: 70 x 50 cm
Dr. Jeanette Mayrhofer-Berger, Dr. Franz Mayrhofer

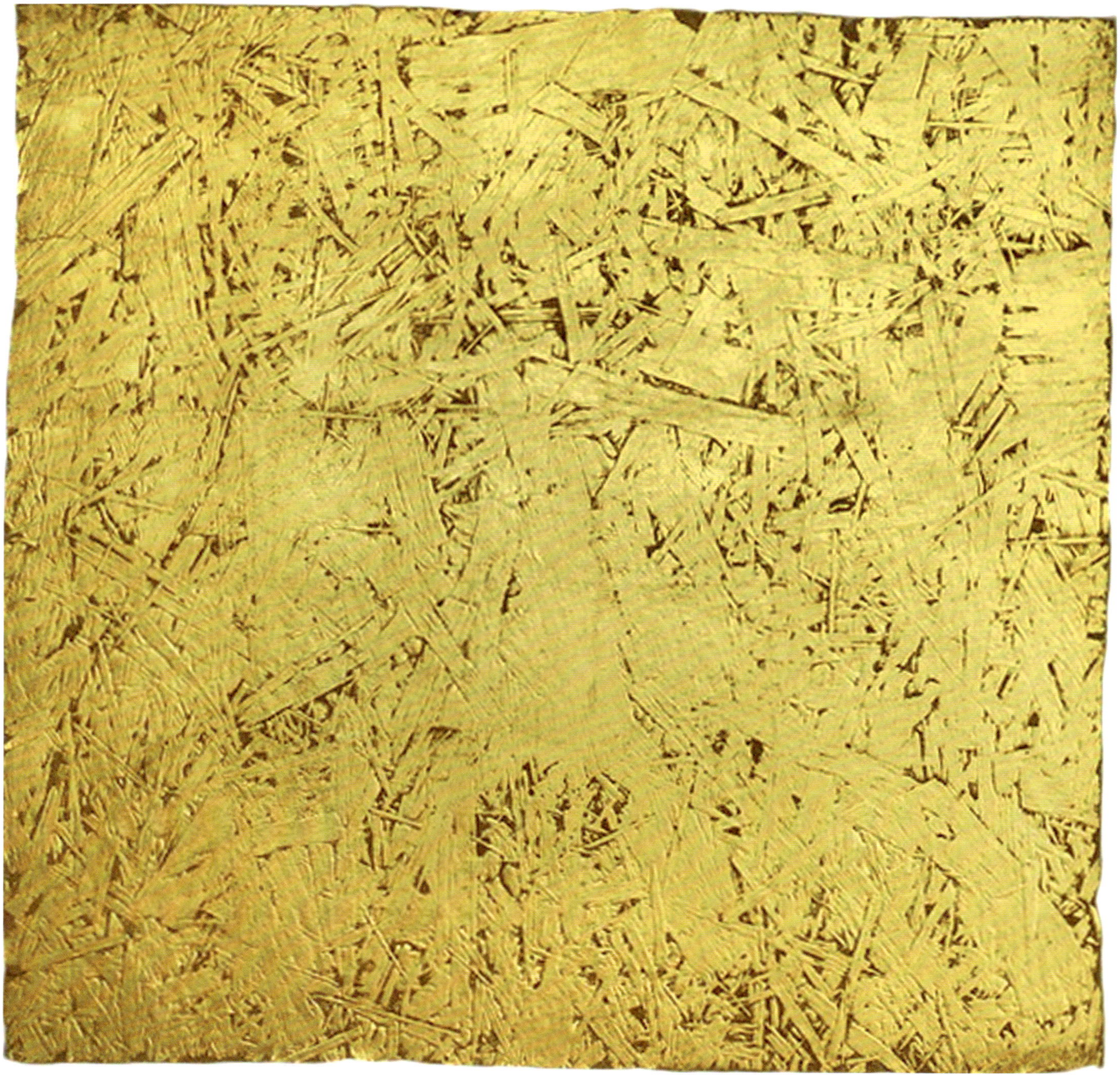

Winfried Muthesius
Hommage à Vincent, 2011
Gold leaf on wood, 50 x 50 cm
Private collection

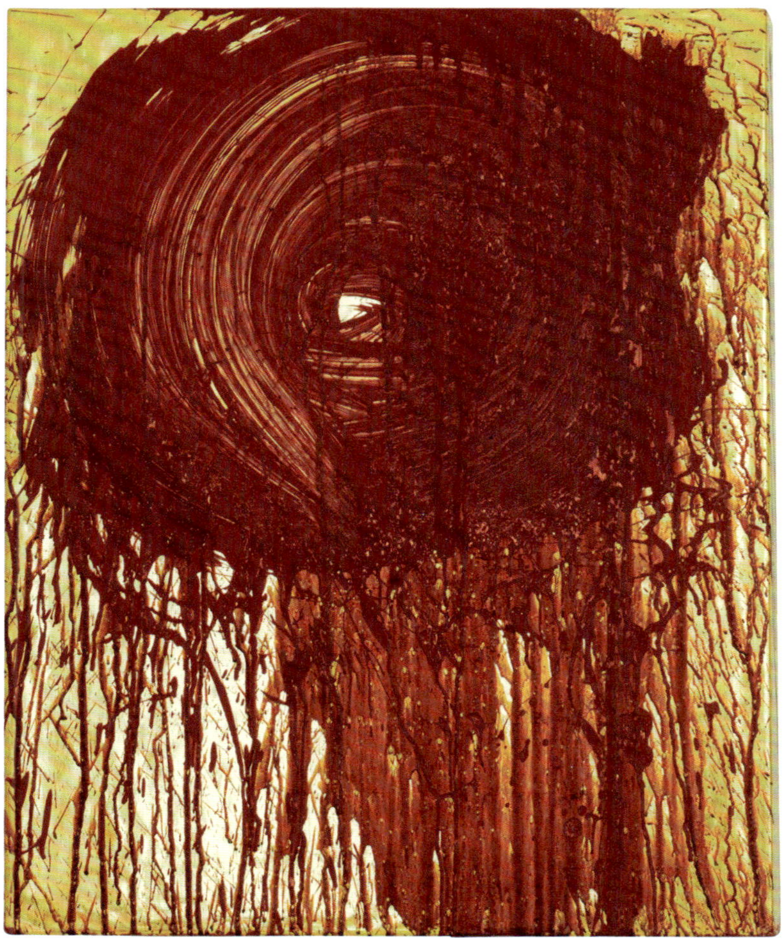

Hermann Nitsch, Hermann Staudinger
n. t., 2011
Blood and painting on oil gilding, 100 x 80 cm
Hermann Nitsch

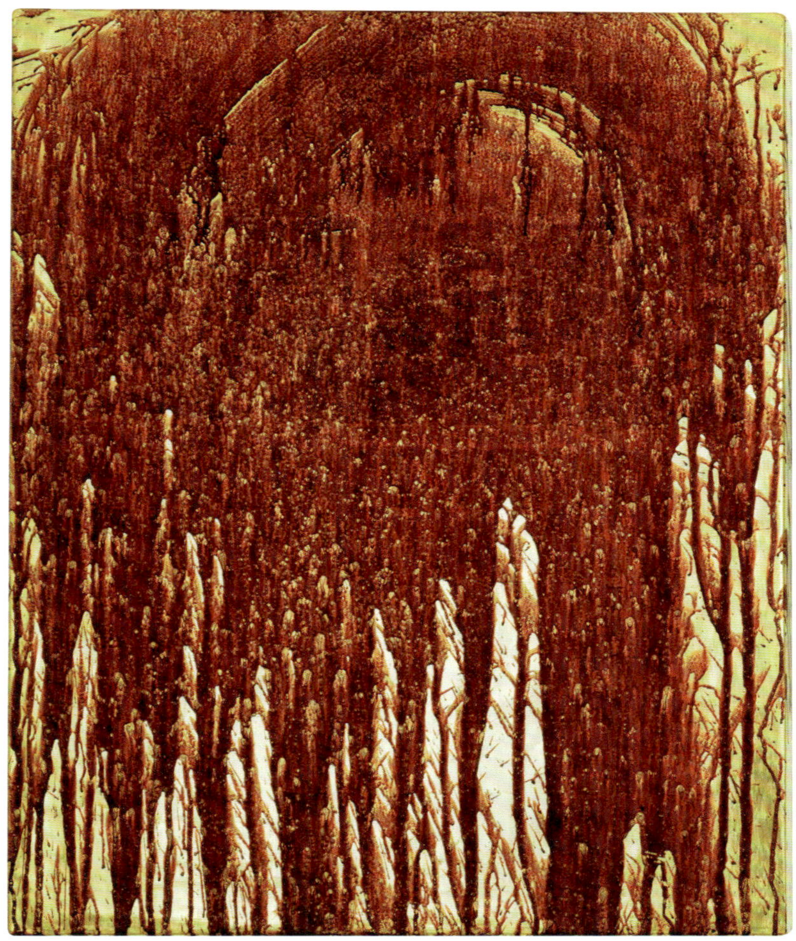

Hermann Nitsch, Hermann Staudinger
n. t., 2011
Blood and painting on oil gilding, 100 x 80 cm
Hermann Nitsch

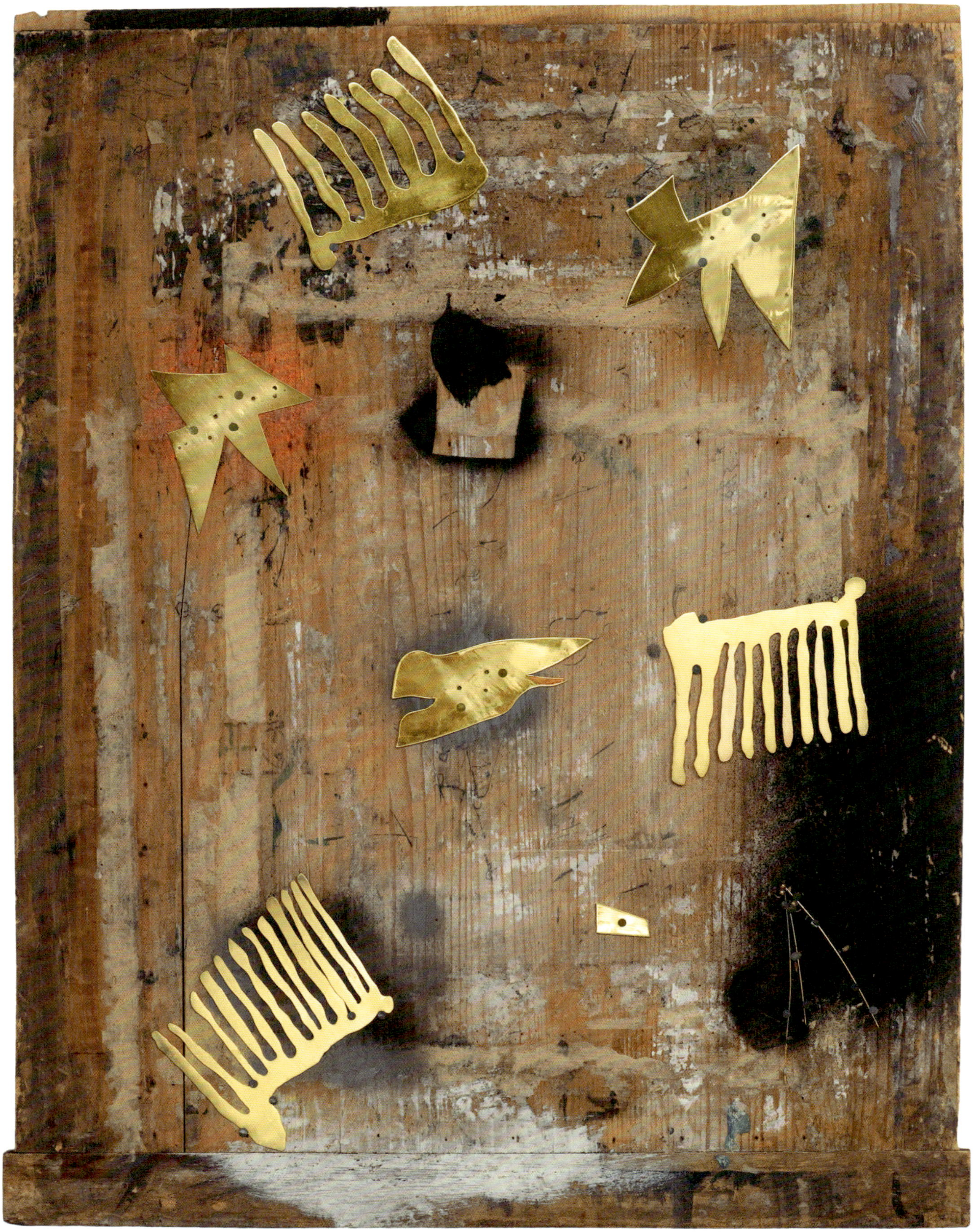

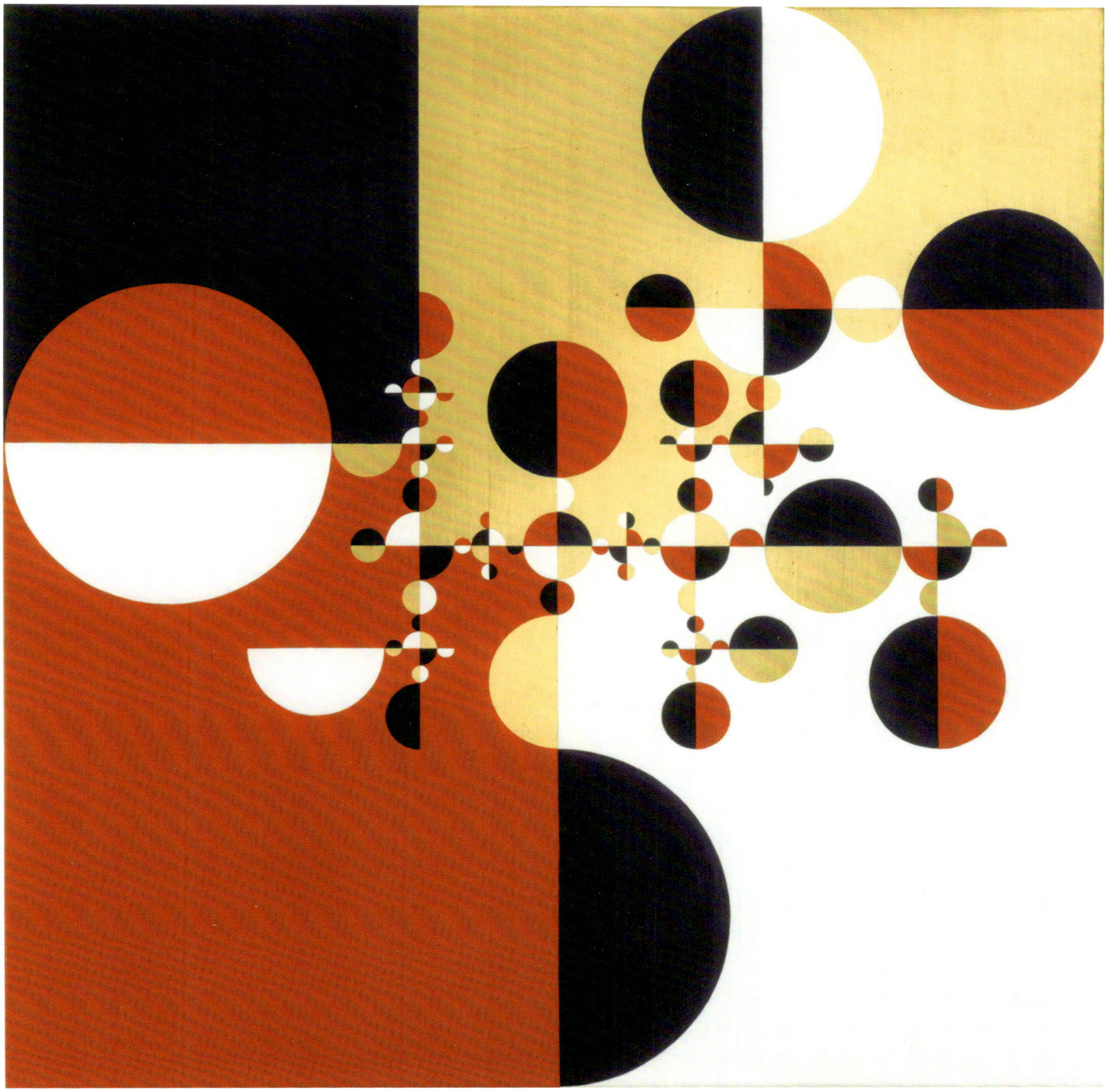

Gabriel Orozco
Samurai Tree 6C", 2006
Egg tempera and gold leaf on red cedar wood, 55 x 55 cm
Collection Titze

Oswald Oberhuber
Untitled, 1950s
Collage, metals, mixed media, wood, 68 x 52.5 cm
Galerie Maier, Innsbruck

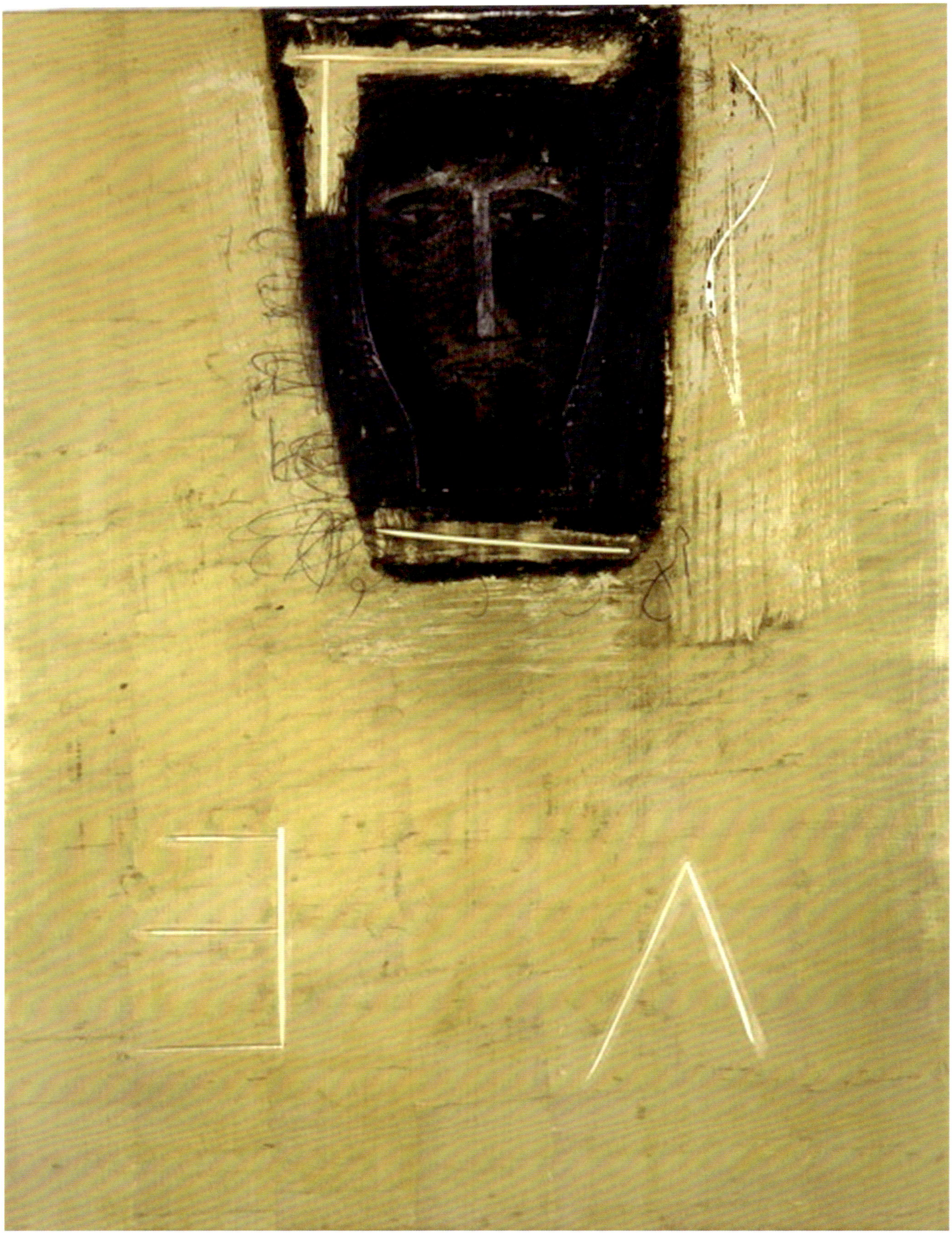

Mimmo Paladino
Thema n.°4, 2000
Mixed media on panel, 80 x 60 cm
Courtesy Galleria Christian STEIN

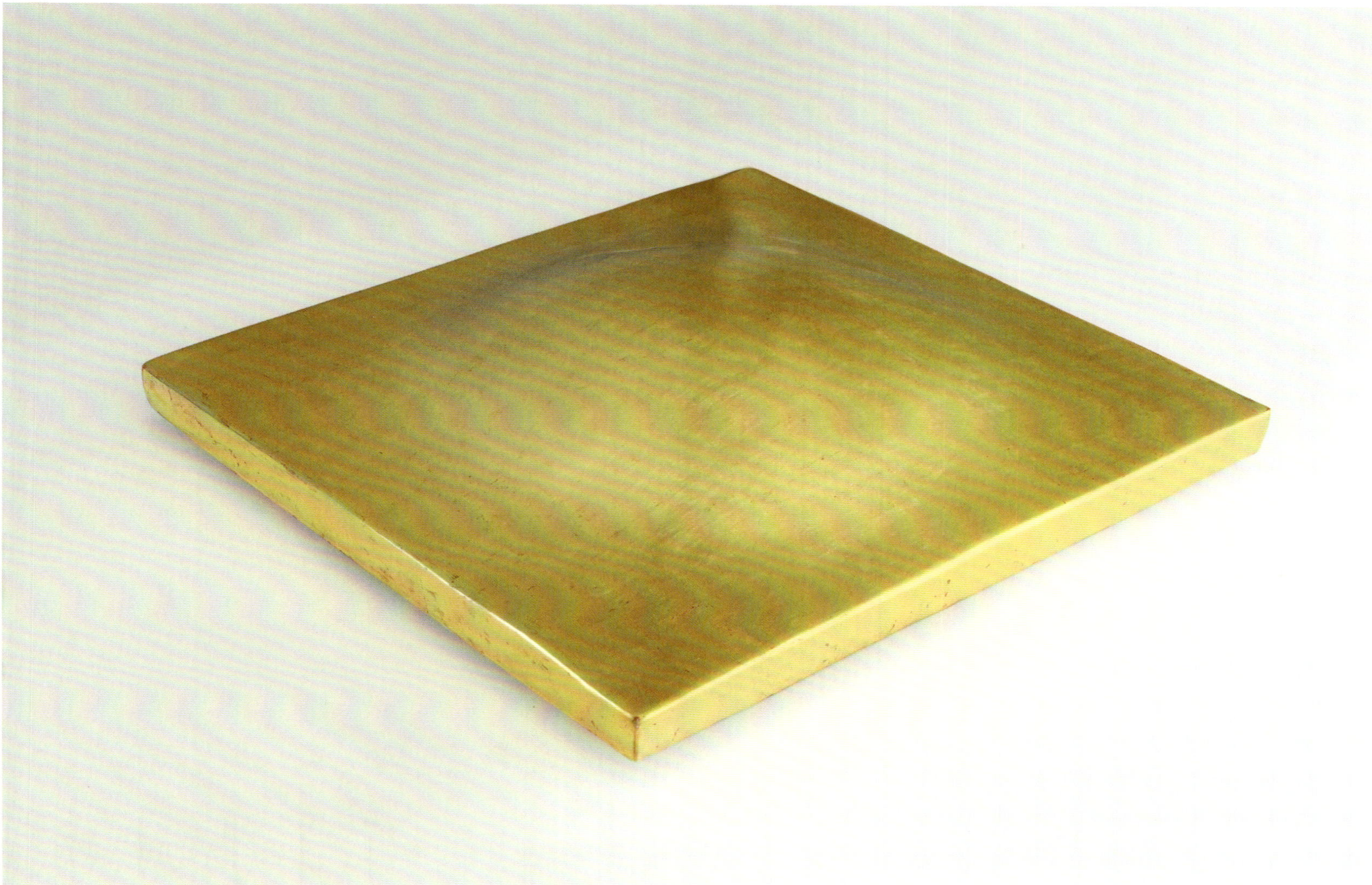

Eva Pliem
Atem 5, from the series of *Bäuche*, 2009
Bole gilding on plaster, c. 43.5 x 41 cm
Courtesy the artist

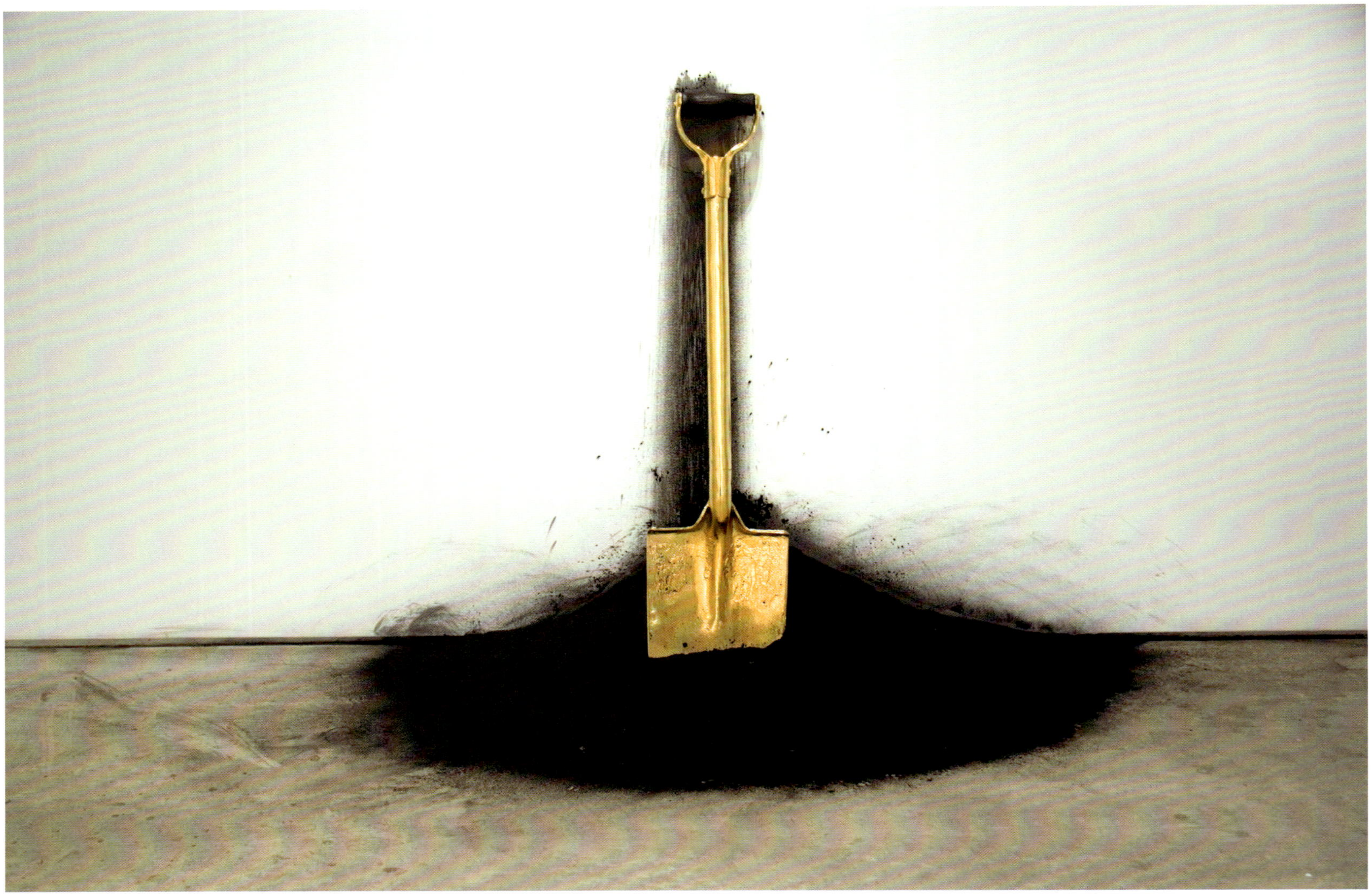

Robin Rhode
Spade, 2007
Bronze-gilt and charcoal, 88 x 21 x 4 cm
Collection Goetz

Edouard Sacaillan
Ascetic and the Armenian Alphabet, 2007
Vinyl, oil and gold paint on canvas, 204 x 133 cm
Private Collection, Athens, Courtesy Kalfayan Galleries, Athens—Thessaloniki

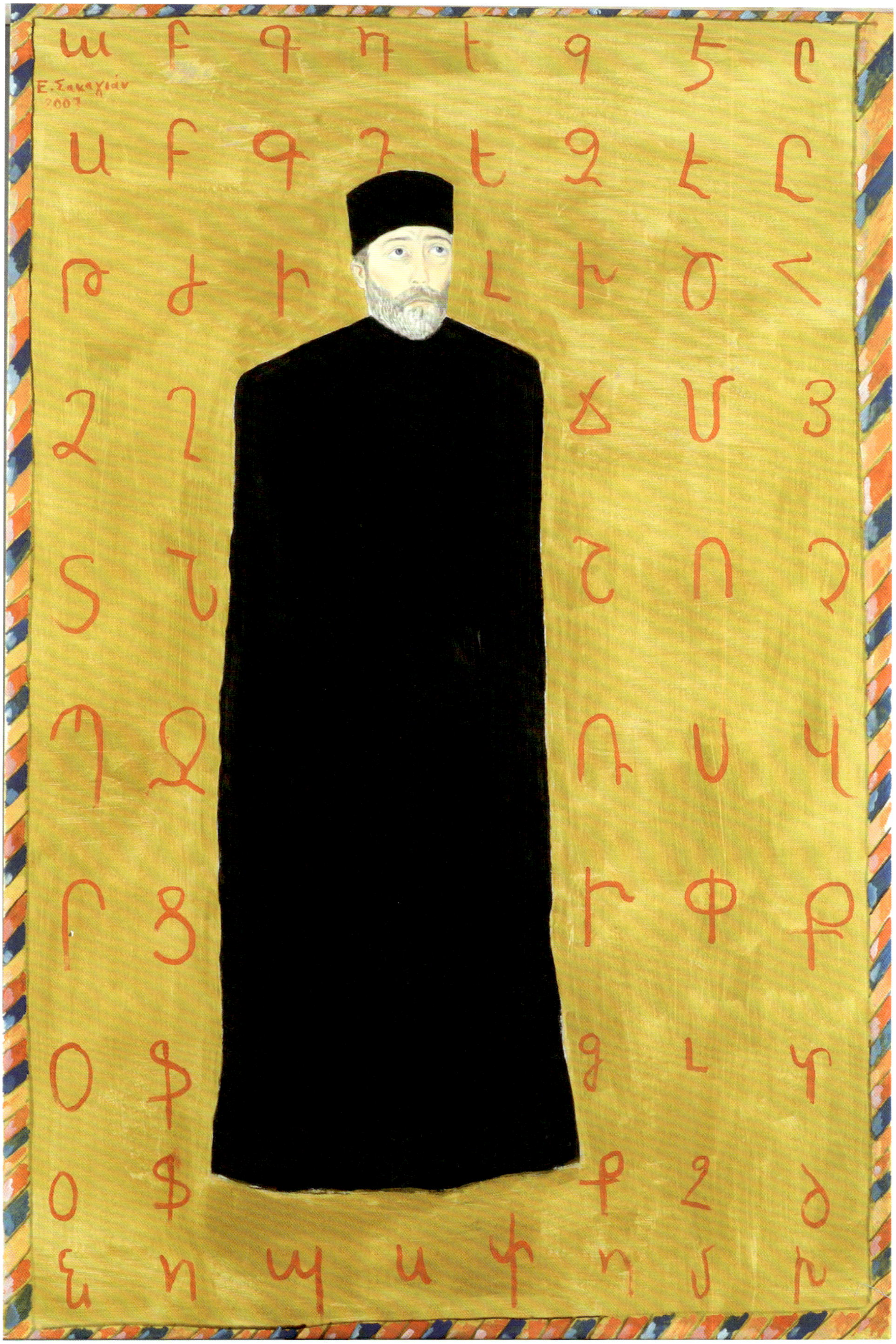
Ε. Σακαγιάν
2007

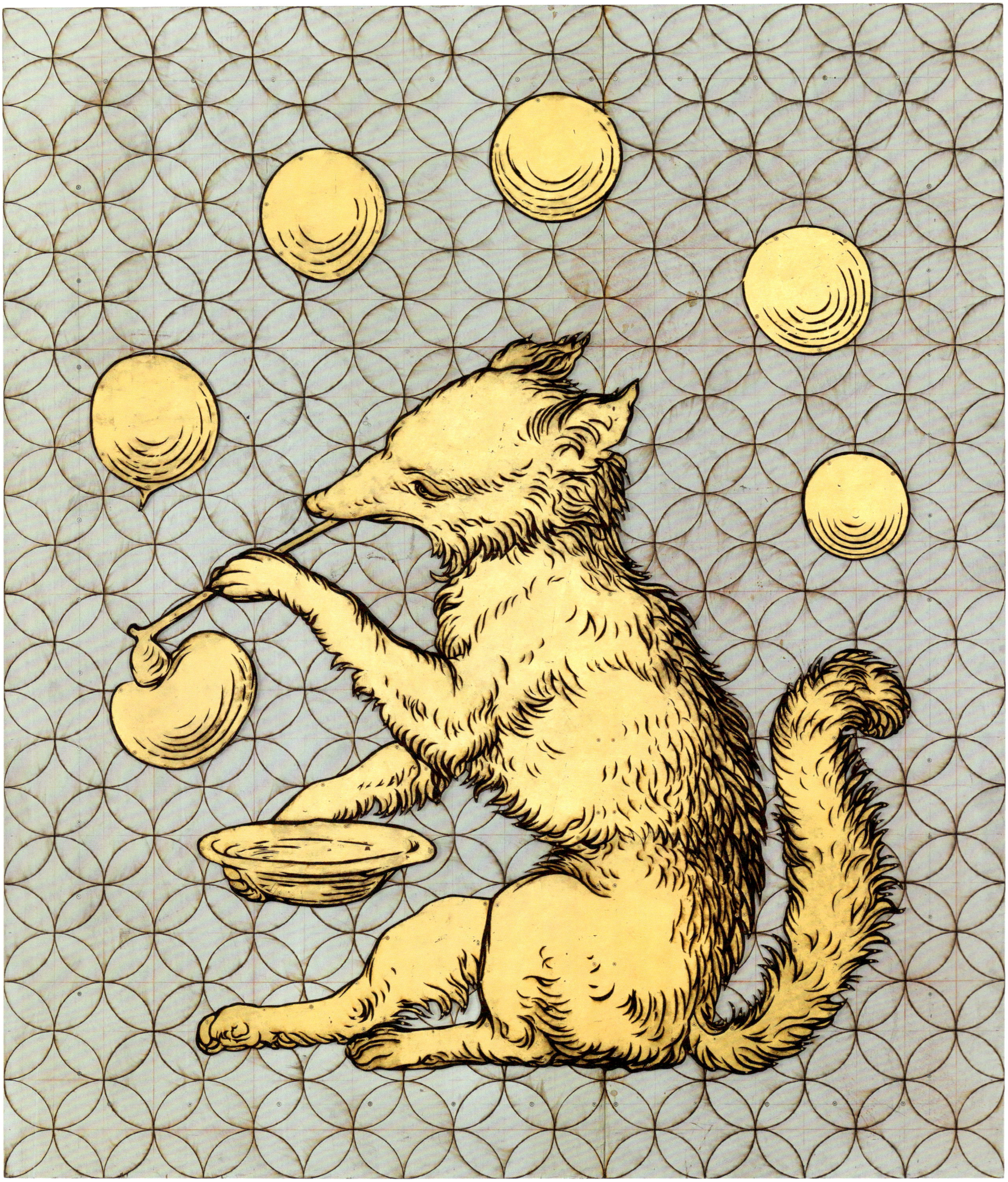

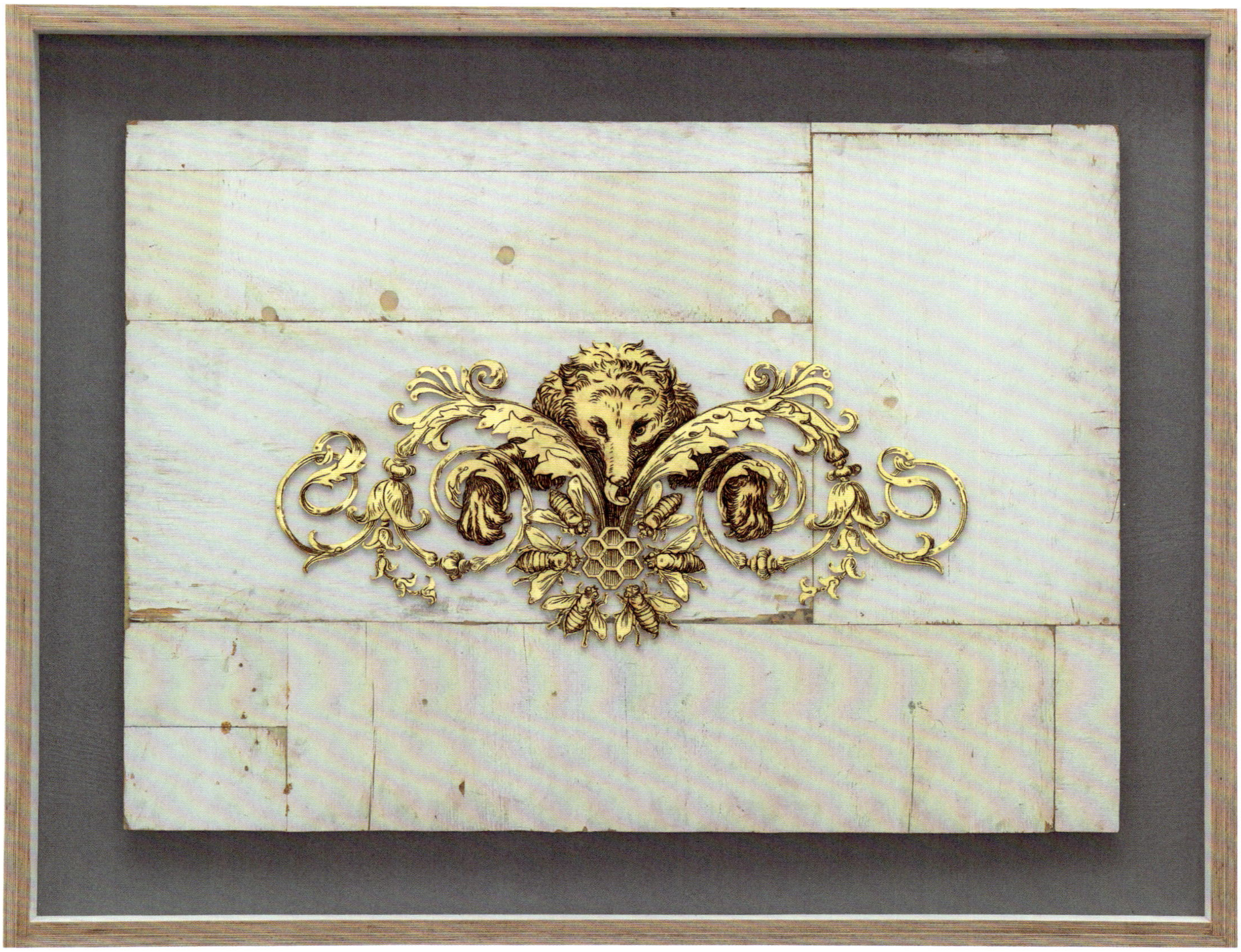

Tom Sachs
Reineke Fuchs: Honeybear, 2008
Plywood, wood, resin, gold leaf, synthetic polymer paint and iron components, 123.5 x 158.4 x 11.4 cm
Galerie Thaddaeus Ropac, Salzburg · Paris

Tom Sachs
Reineke Fuchs: Bubble Fox, 2008
Plywood, synthetic polymer paint and gold leaf, 243.8 x 203.2 x 12.7 cm
Galerie Thaddaeus Ropac, Salzburg · Paris

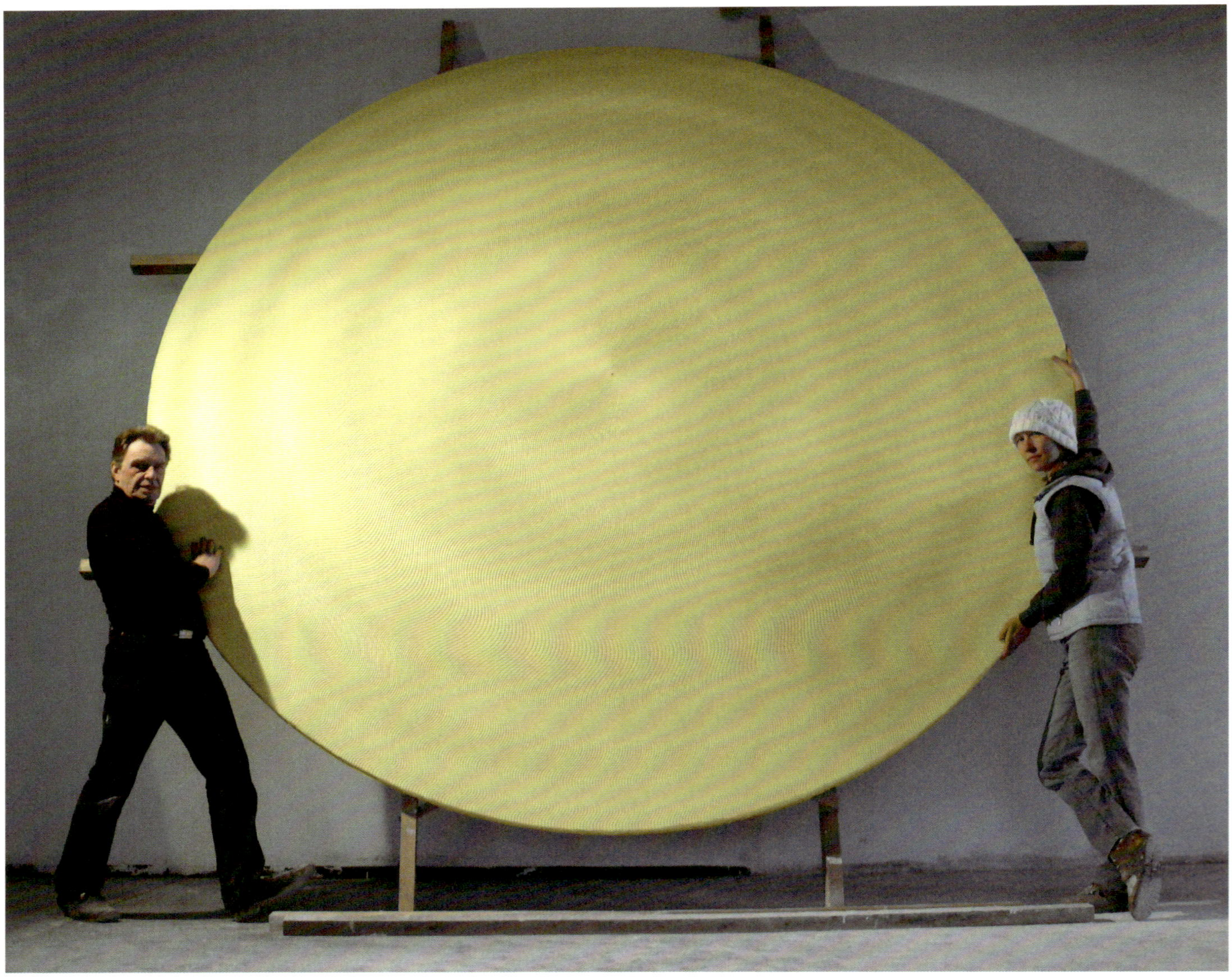

Walter Schnabl
Schild des Apollon, 2011
Plaster-chalk ground on canvas, 23-carat gilding, diameter 3.33 m
Walter Schnabl

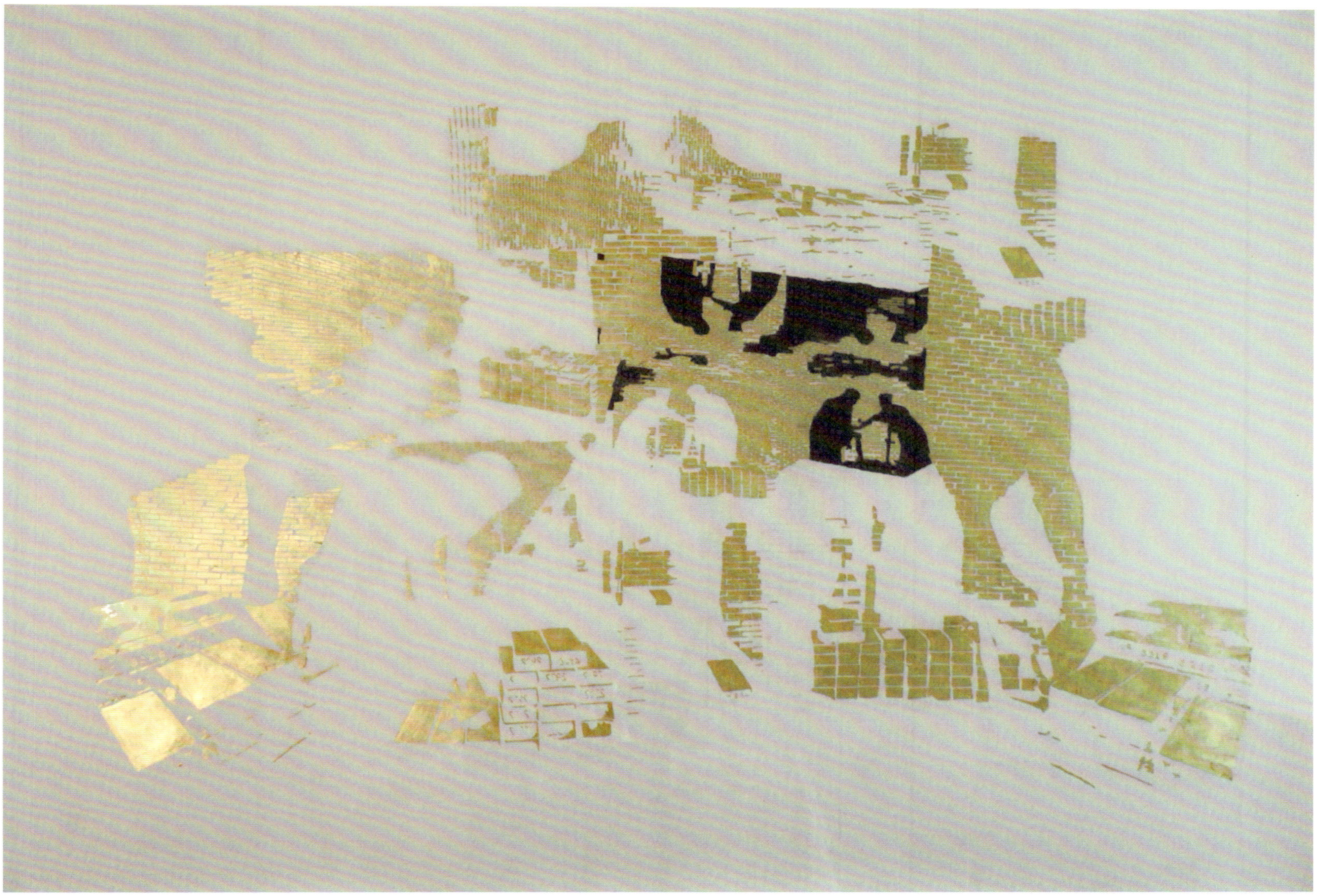

Lucy Skaer
Solid Ground-Tarsal, 2006
Pencil and gold leaf on Fabriano paper, 140 x 200 cm
Collection of the Baloise Group, Basel

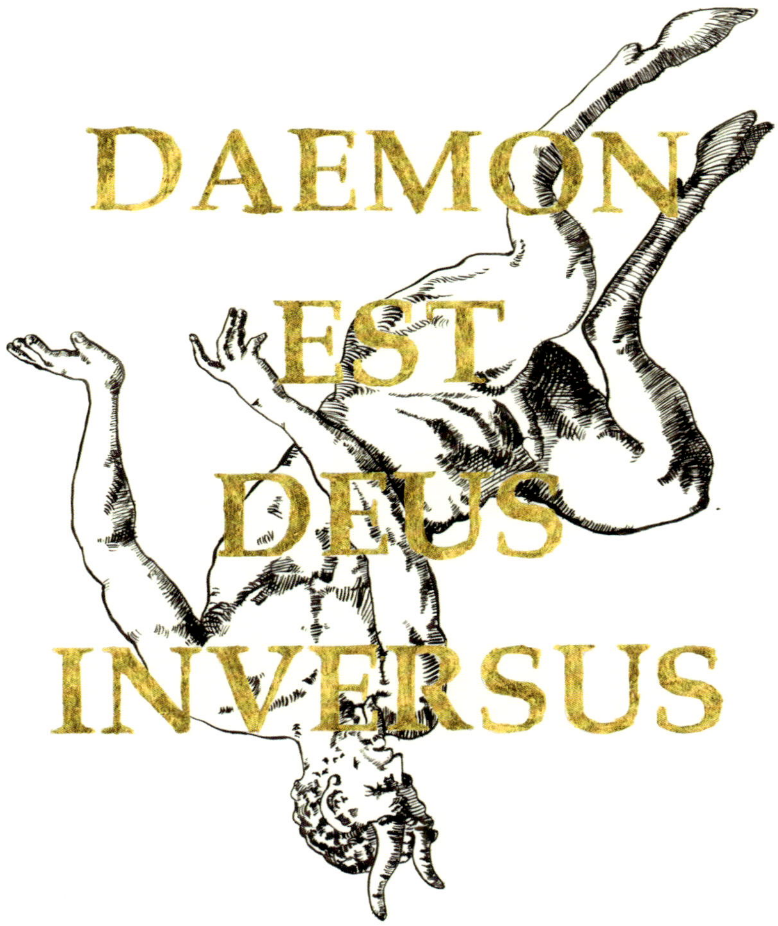

Panos Tsagaris
Untitled, 2010
Gold leaf (23-carat) and ink on paper, 76 x 56 cm
Courtesy Kalfayan Galleries, Athens—Thessaloniki

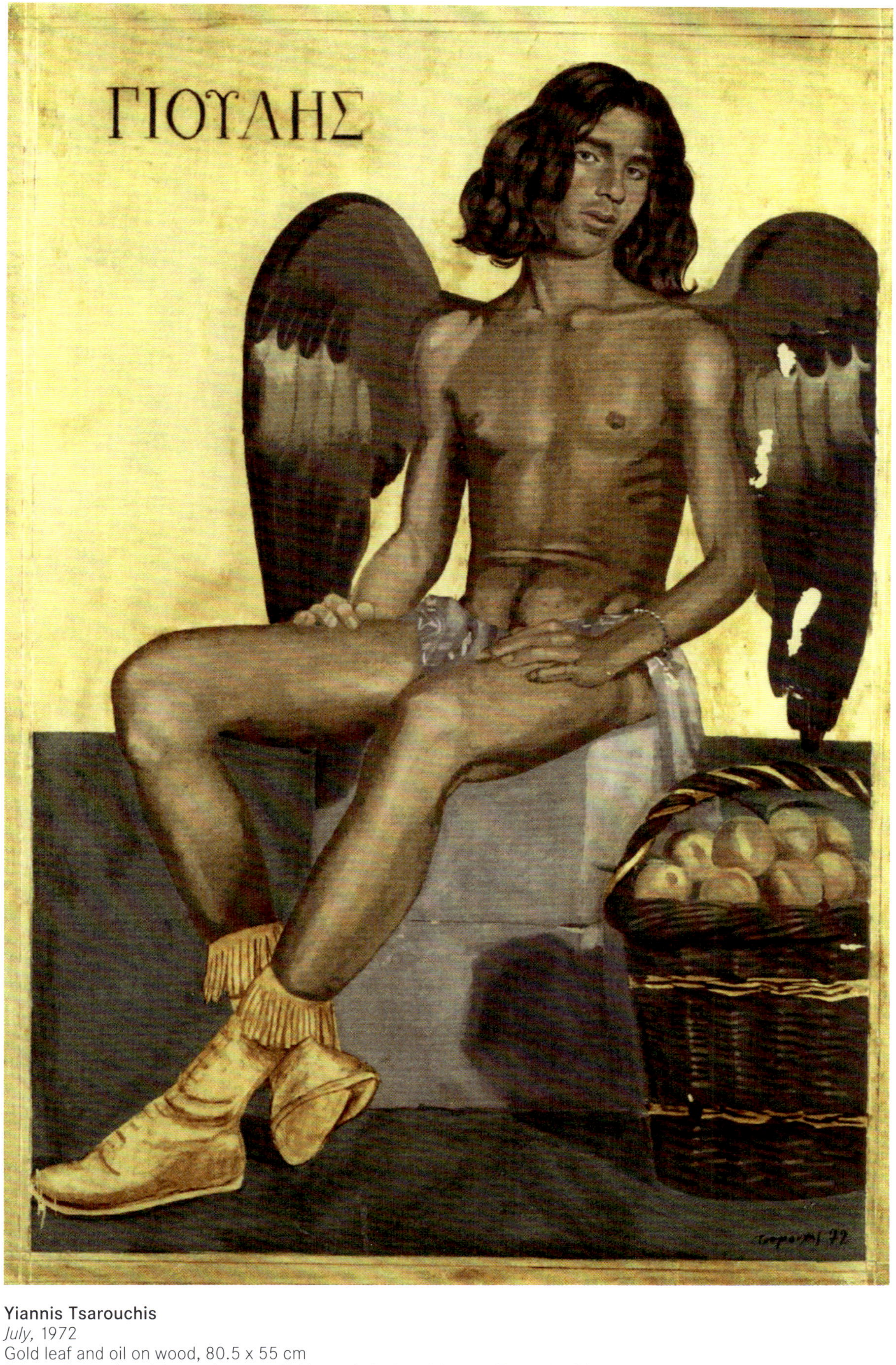

Yiannis Tsarouchis
July, 1972
Gold leaf and oil on wood, 80.5 x 55 cm
Private Collection, Athens, Courtesy Kalfayan Galleries, Athens—Thessaloniki

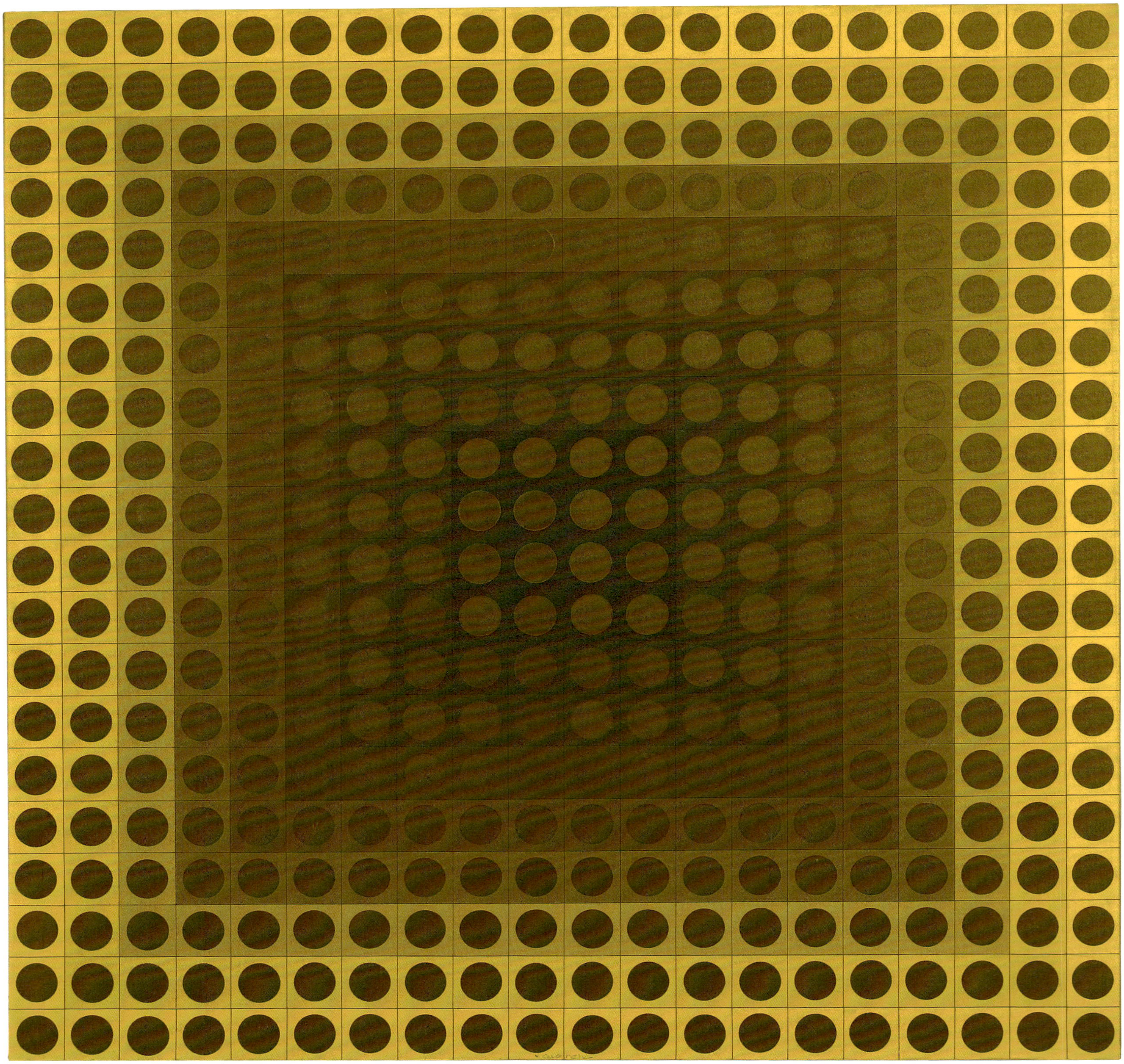

Victor Vasarely
Taller, 1965
Gold leaf and oil tempera on wood, 80 x 80 cm
Museum Ritter, Waldenbuch (G)

Robert Weber
Für mich deine Liebe, für meine Arbeit deinen Respekt, 2005
Box (Listro—Ethiopian shoe-shine box, 24-carat gilded, on base), wood, gold leaf, acrylic glass, base: 28.5 x 17 x 19 cm, height of base: 140 cm
Robert Weber

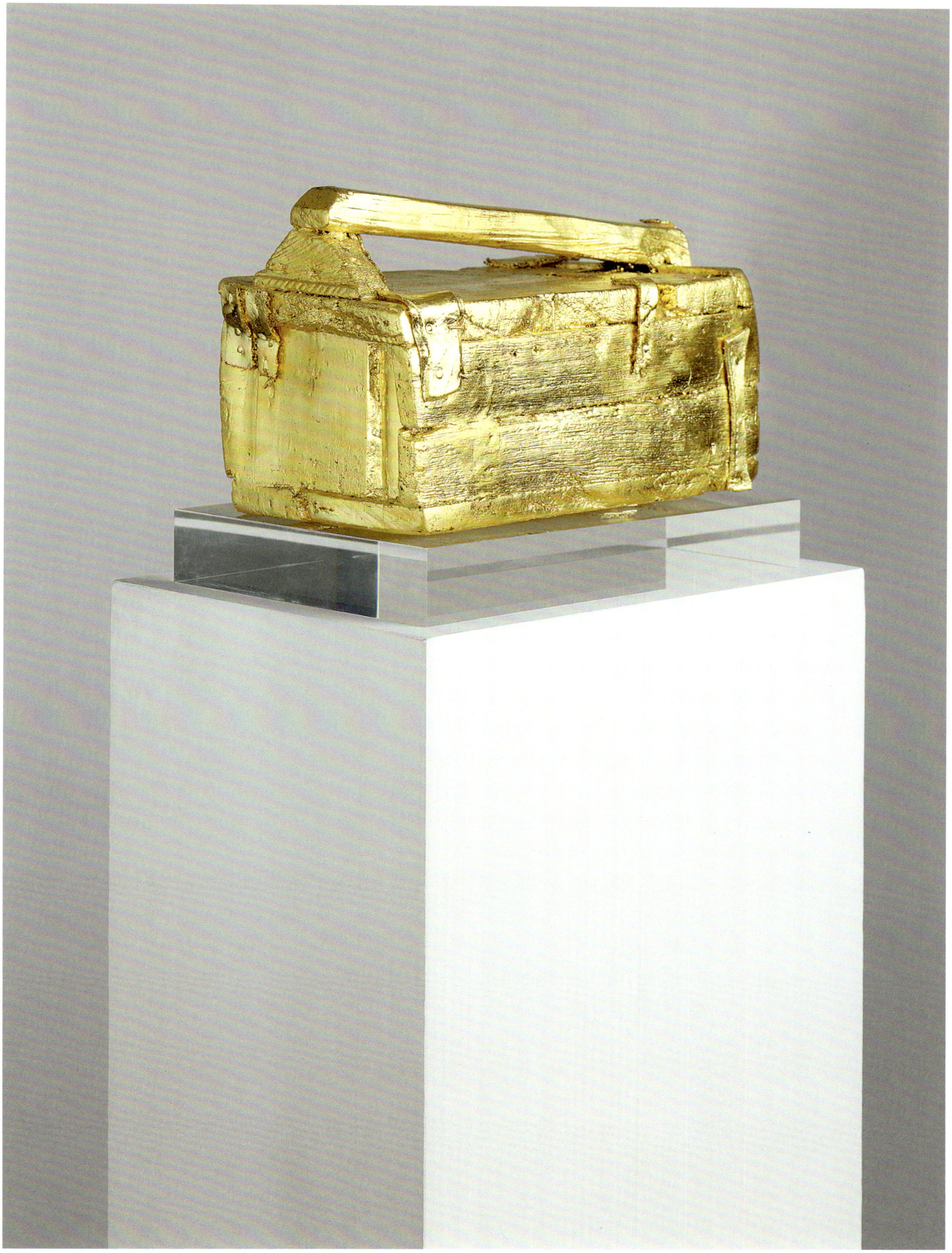

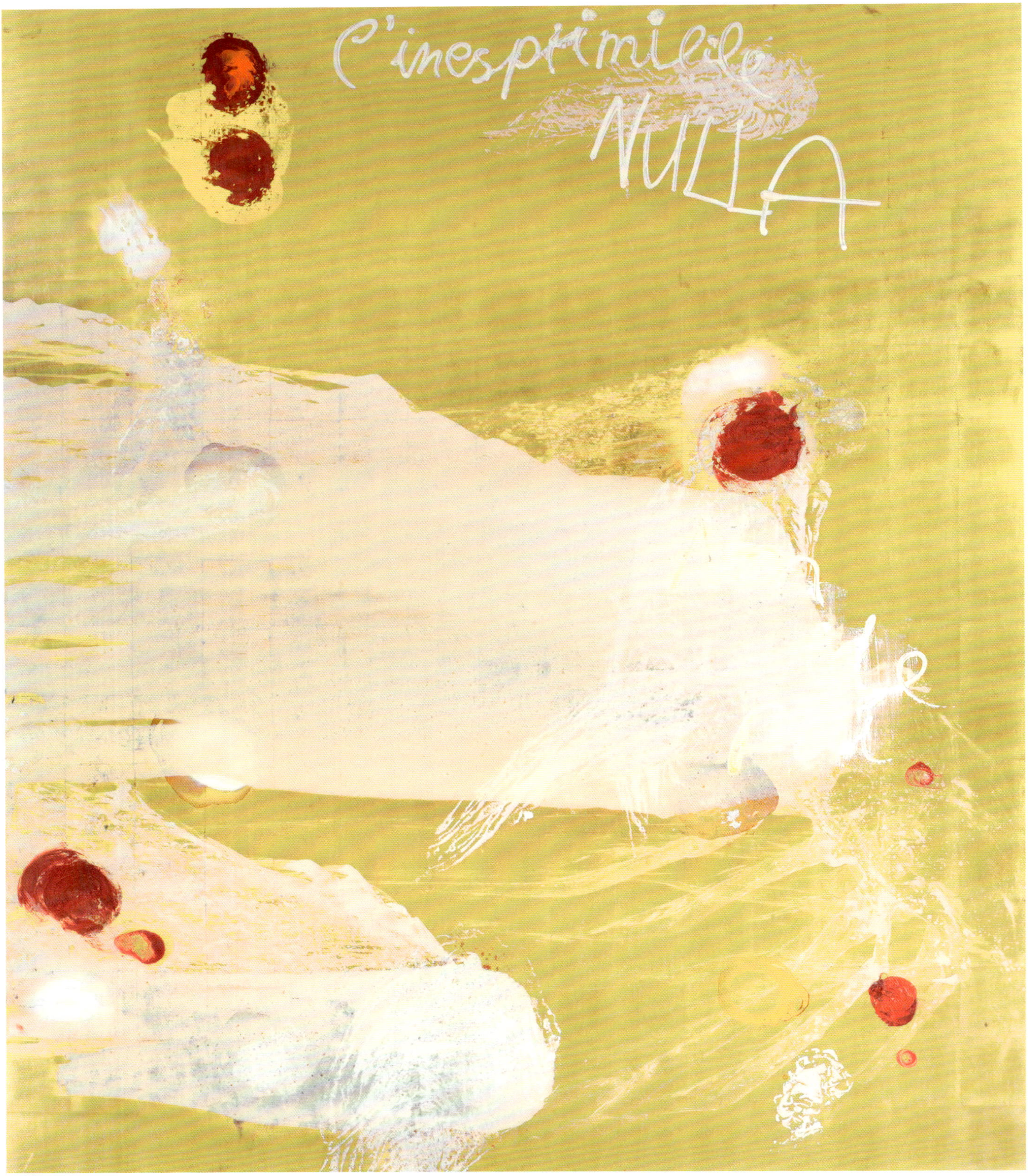

Robert Weber
L'inesprimibile nulla, Non si vede piú (The inexpressible nothing, nothing more is to be seen), 2006
Oil and emulsions on gold ground and wood, 230 x 200 cm
Robert Weber

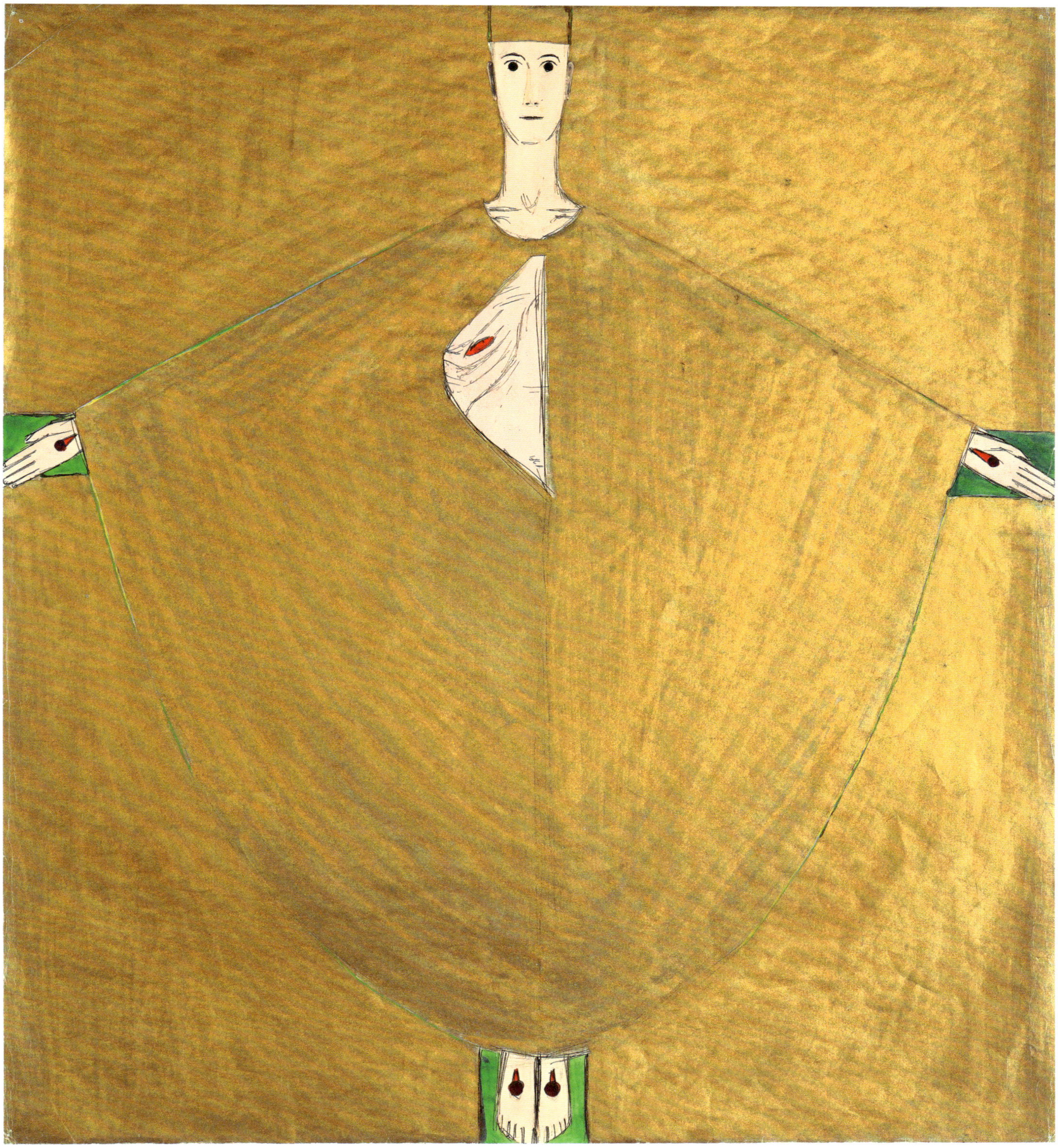

Max Weiler
Christ the King (design for a mosaic in the Ignaz Seipel church in Vienna), 1935
Gold, watercolours, India ink on paper, 78.5 x 70 cm
Private collection

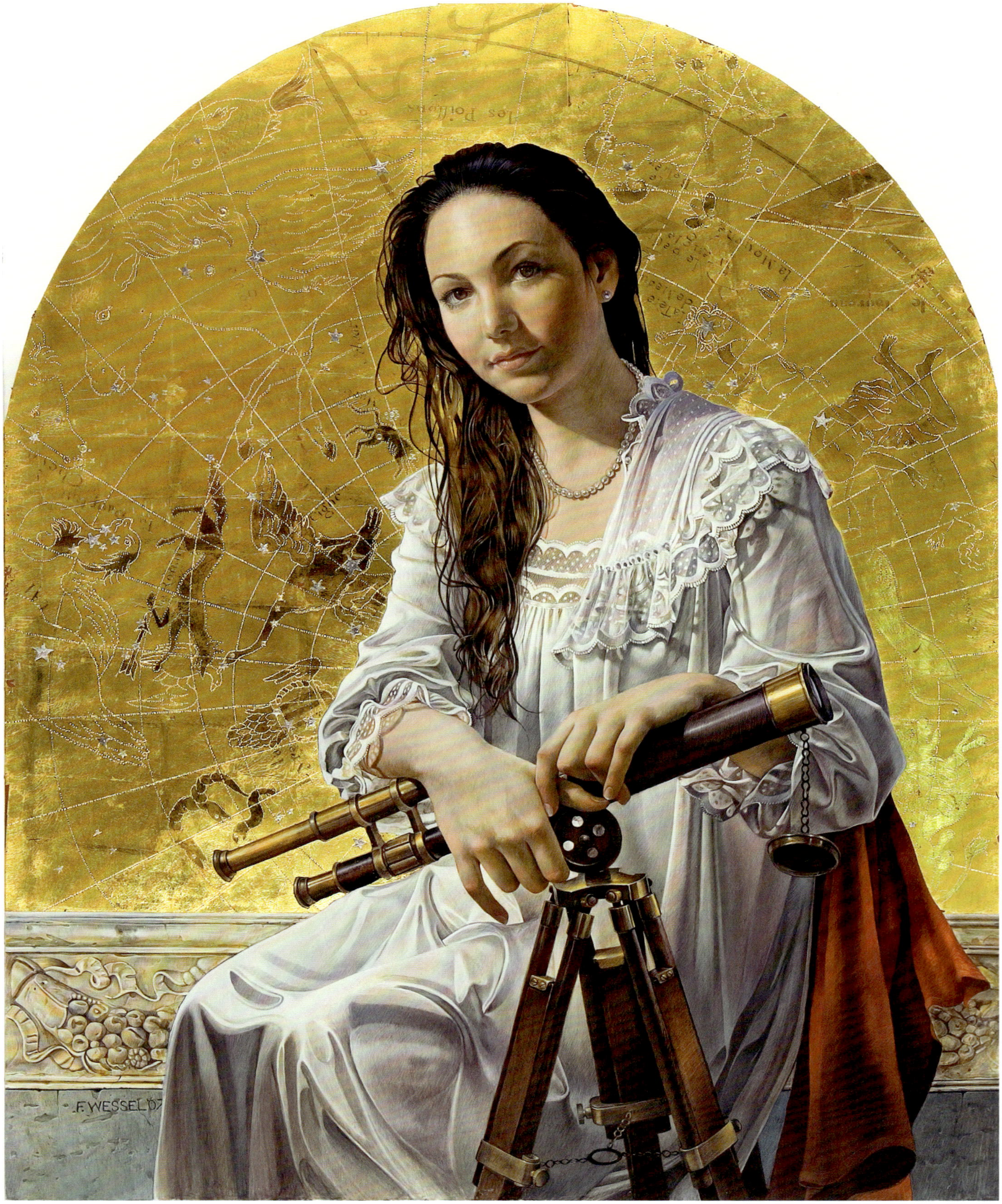

Fred Wessel
Becca (Telescope), 2007
Egg tempera, gold leaf and platinum foil, 75 x 60 cm
Dr. Michael Wood

Franz West
Passstück (Hommage à ...), 1983–1985
Plaster, papier mâché and gold leaf, 118 x 24 x 26 cm
Courtesy Galerie Elisabeth und Klaus Thoman, Innsbruck

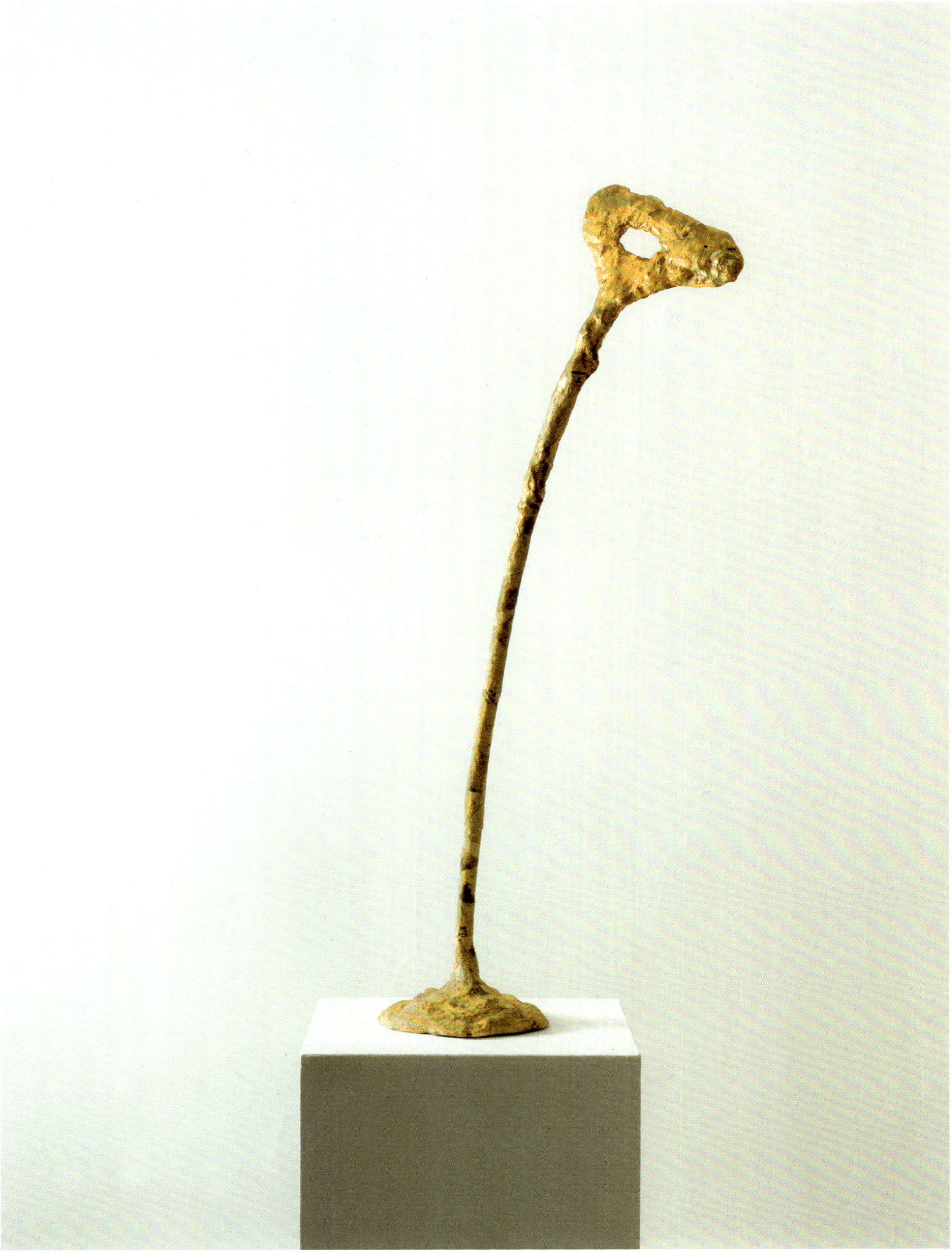

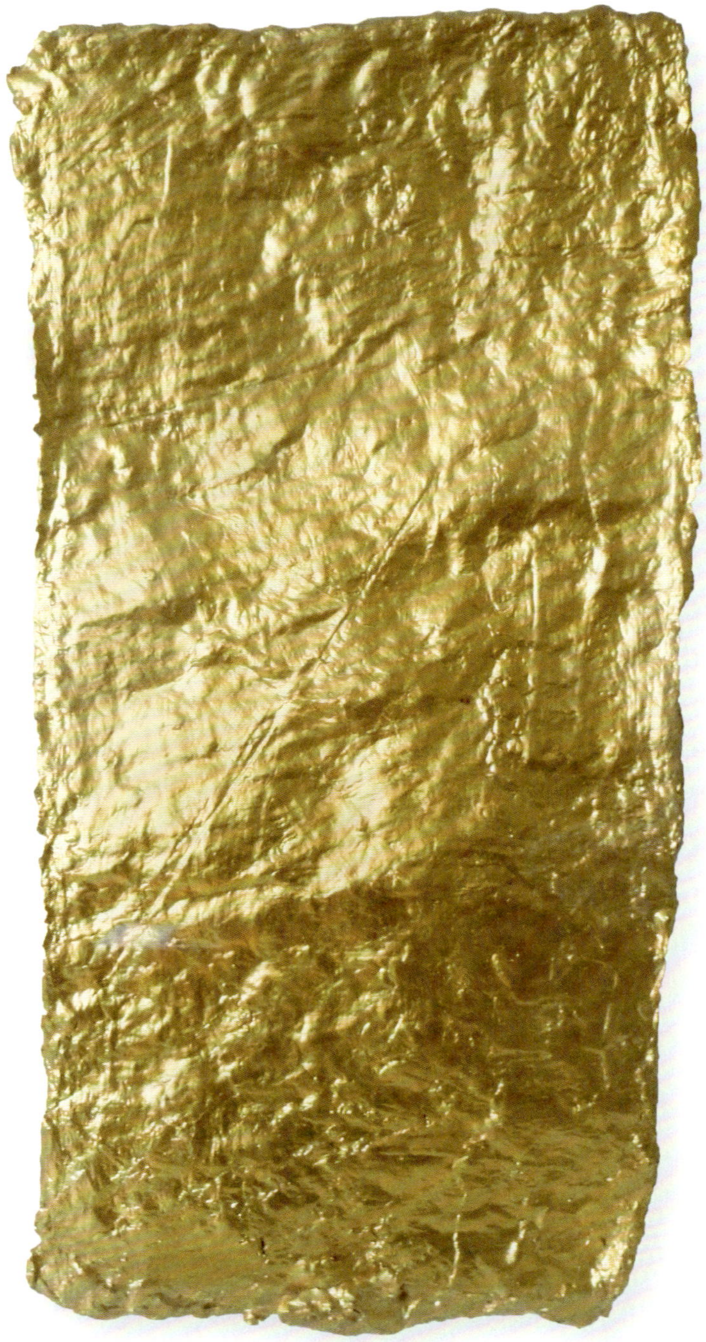

Franz West
Untitled, 1987
Formica board, plaster and gold leaf, 82 x 40 cm
Private Collection

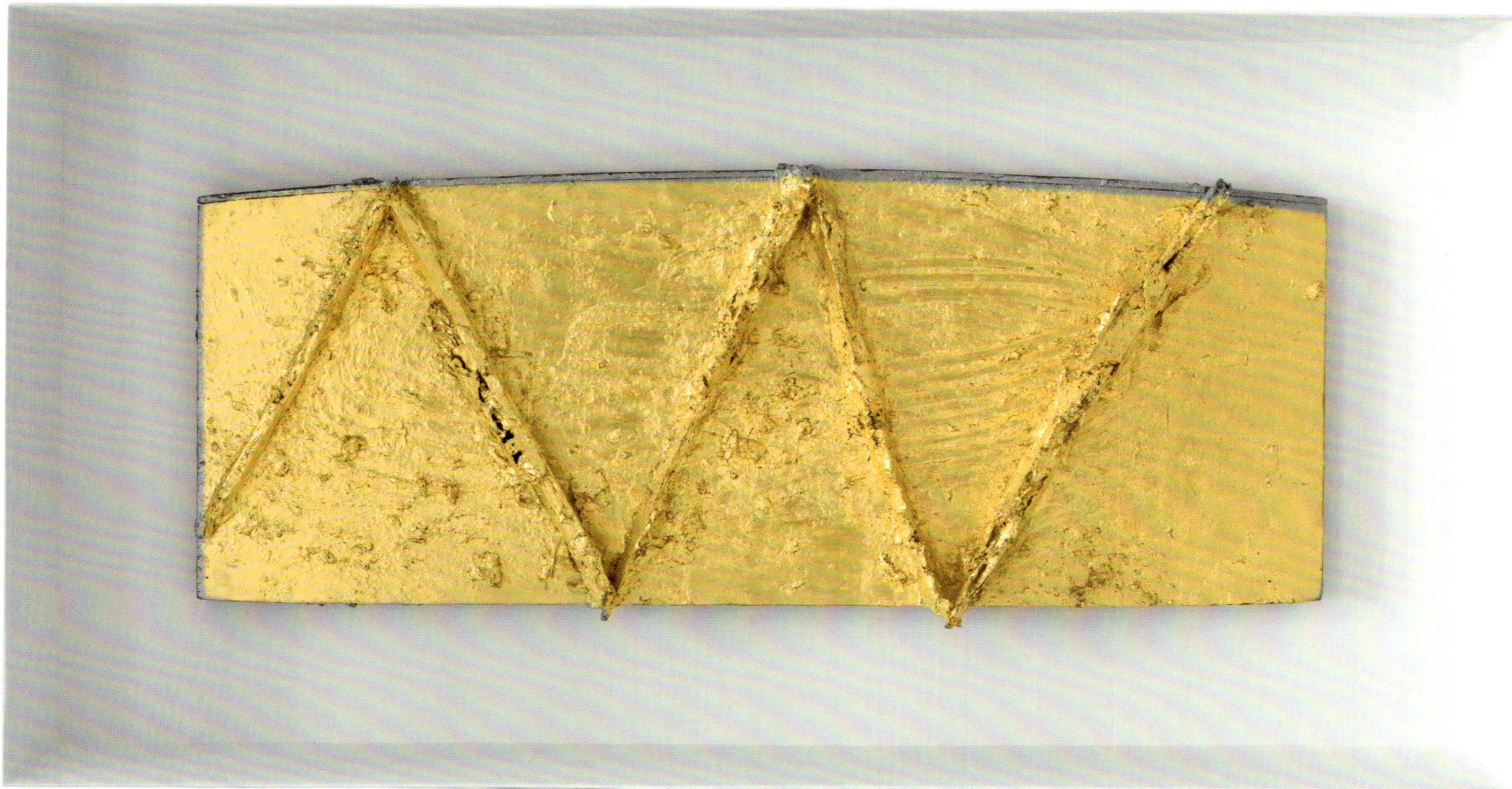

Franz West
Abgabe, 1987
Wood, papier mâché, gold foil, 36 x 90 x 9 cm
Galerie Meyer Kainer

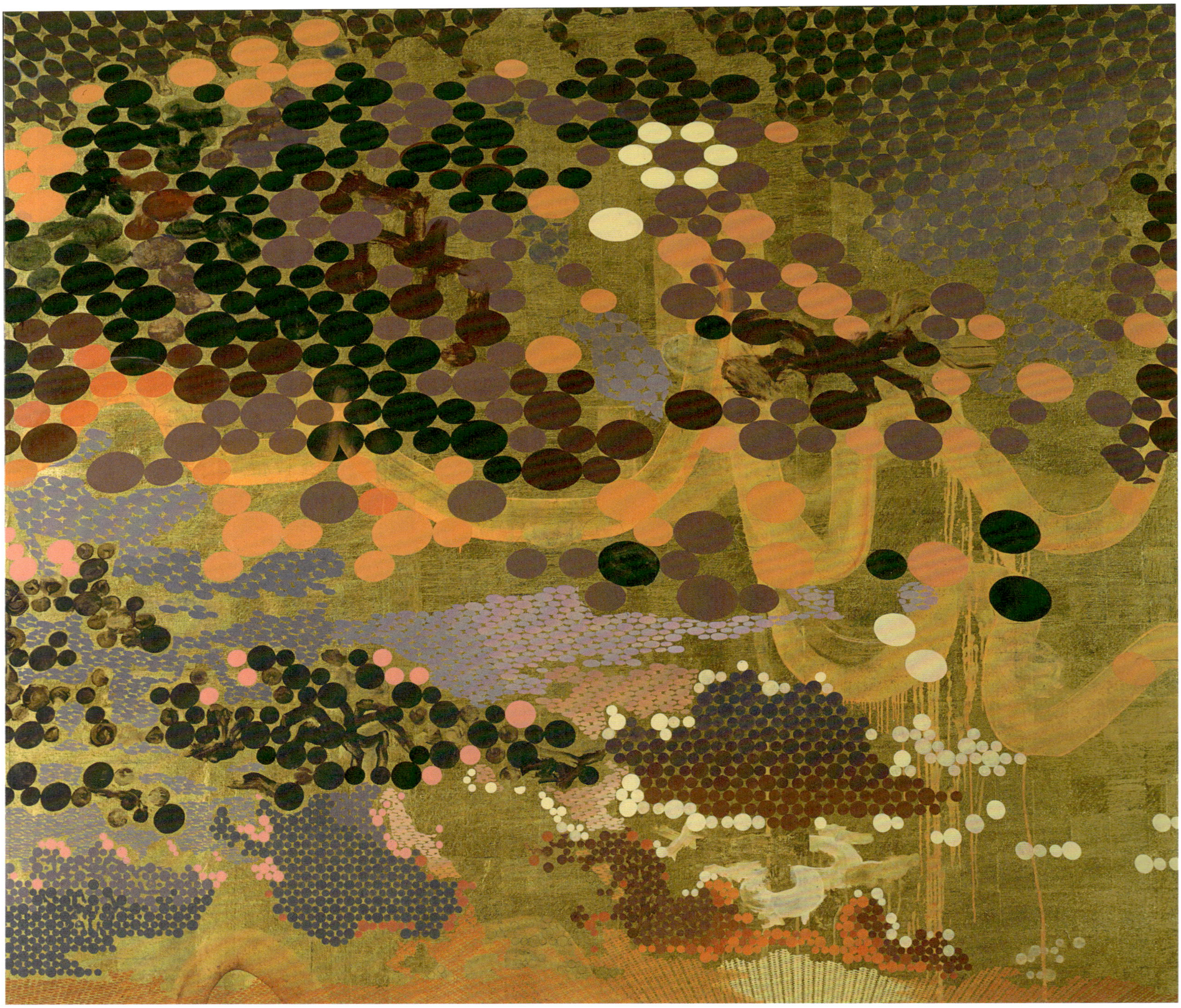

Toby Ziegler
The Hedonistic Imperative (2nd version), 2006
Oil, gold leaf and pencil on canvas, 210 x 242.5 cm
Tate: Presented by David Gorton 2006

Appendix

Artists Index

Artworks

List of exhibited Works

Carlos Aires
Black Requiem, 2010
Polychrome and gold leaf plaster and fibreglass sculpture, golden disco globe, wooden base
110 x 40 x 40 cm
Courtesy Mario Mauroner Contemporary Art Vienna

Abbas Akhavan
Islands: Units E 11 [zone 2], E 12 [zone 2], D 11 [zone 3], D 12 [zone 3]
Gold leaf on drybuild board
each 60 x 60 cm
The Third Line, Dubai

Abbas Akhavan
Herd #1, 2010
22 carat gold on Xerox copy
29.7 x 21 cm
The Third Line, Dubai

Abbas Akhavan
Dates, 2010
22-carat gold on Xerox copy
29.7 x 21 cm
The Third Line, Dubai

Abbas Akhavan
Palm, 2010
22-carat gold on Xerox copy
29.7 x 21 cm
The Third Line, Dubai

Abbas Akhavan
Palms, 2010
22-carat gold on Xerox copy
29.7 x 21 cm
The Third Line, Dubai

Franz Josef Altenburg
Frame, 2005
Ceramic, polished gold
26.5 x 22.5 x 6 cm
Courtesy the artist

Franz Josef Altenburg
Frame, 2006
Ceramic, polished gold
26.5 x 22.5 x 6 cm
Courtesy the artist

Johannes Angerbauer
Kristalltag, 10. November 1998
Silkscreen print, security glass, steelframe, gilt with 23–carat gold
'Kristalltag' audiofile: hammerstroke, stroke and crackle
50 x 50 cm
Courtesy the artist

Johannes Angerbauer
Social gold Kiss, 2011
Corridor installation
3 panes of toughened glass with a gold layer
Courtesy the artist

Stephen Antonakos
Arimathea, 2001
Gold leaf on wood and neon light
122 x 122 x 15 cm
Courtesy Kalfayan Galleries, Athens—Thessaloniki

Ian Anüll
Reden ist Silber, Schweigen ist gold, 2004
Screenprint on MDF, upper surface gilded
25 x 25 x 7 cm
Courtesy Edition 5, Erstfeld, Switzerland

John M. Armleder
Illico, 2011
Silver-gilt
7 x 23 x 27 cm
Galerie Andrea Caratsch, Zurich

John M. Armleder
Siro, 2011
Acrylic on canvas with gilded anti-pigeon barbed wire
180 x 160 cm
Galerie Andrea Caratsch, Zurich

Pupils of Rudolf Bacher: Karl Borschke, Karl Maria May, Leopold Zlabinger, Andreas Harsch, Ferdinand Kitt, Anton Schellerer (gilder)
Altar-Triptychon, pre-1912
Oil on tempera on wood, partly gilded
Central section: 211 x 154 cm; wings: each 211 x 68 cm
Belvedere, Vienna

Stephan Balkenhol
Mann auf goldener Kugel, 2004
Poplar wood, painted, gilded with gold leaf
190 x 65 x 45 cm
Private collection, Germany

Georg Baselitz
De Chiricos Selbstbildnis, 1995
Gold leaf and oil on canvas
200 x 162 cm
Private collection, Hamburg

Willi Baumeister
Mauerbild mit Metallen, 1923
Oil, gold leaf and board on canvas
89 x 61 cm
Stiftung Kunstsammlung Nordrhein-Westfalen, Düsseldorf

Joachim Baur
Lungot, harvest, 1992/95
Glass frame, bean tendrils, bean pods, gilded seeds, digital LCD display
100 x 30 x 14.5 cm
Kunstsammlung/Kulturvermittlung Steiermark

Michael Bergt
Tree of Life, 2009
Gold leaf on wood
40.6 x 33 cm
Yolanda and Abram Eisenstein, USA

Michael Bergt
By The Horns, 2009
Gold leaf on wood
31.1 x 25.4 cm
Peter Selig, San Antonio, Texas

Michael Bergt
Venus and Mars, 2009
Egg tempera and gold leaf on wood
22.9 x 29.2 cm
Courtesy the artist, via Jane Sauer Gallery, Santa Fe, New Mexico

Michael Bergt
Balancing Act, 2009
Egg tempera and gold leaf on wood
50.8 x 38.1 cm
Courtesy the artist, via Jane Sauer Gallery, Santa Fe, New Mexico

Michael Bergt
Icarus, 2009
Gold leaf on wood
45.1 x 35.6 cm
Courtesy the artist, via Jane Sauer Gallery, Santa Fe, New Mexico

Mosè Bianchi
Paolo & Francesca, um 1877
Watercolour and gold on board
64.5 × 85.5 cm
Comune di Milano, Galleria d'Arte Moderna, Milano

Hans Bischoffshausen
Invasion, 1973
Cell cement, gold, red lacquer
60.5 x 60.5 cm
Galerie Walker Schloss Ebenau Rosental

Jill Biskin
Saint Catherine, 1992
Egg tempera and gold leaf on wood
16 x 12 cm
Courtesy of King & Spalding

Jill Biskin
The Elements, 2003
Egg tempera and gold leaf on wood
16 x 20 cm
Jill Biskin

William Blake
King Sebert, the North Front of his Monument, 1775
Watercolours, ink, mosaic gold, pencil on paper
36.8 x 28 cm
By permission of the Society of Antiquaries of London

Arik Brauer
Pan, 2009
Oil, gold leaf on board
105 x 160 cm
Private collection, Arik Brauer

Thomas Brix
Ariola, 2002
Acrylic, gold leaf on chipboard
60 x 60 cm
Thomas Brix

Guillaume Bruère
Bananagold, 2010
Gold leaf on banana peel
Variable Maße
Courtesy the artist and Galerie Heike Curtze, Vienna/Berlin

Chris Burden
Gold Bullets - Pointies and Roundies, 2003
10 22-carat gold cartridges, 2 cabinets (wood, Perspex)
each 26 x 14.6 x 15.9 cm
Galerie Krinzinger, Vienna

Pam Burnley-Schol
Transfiguration: Cabbage and Steel, 2010
Oil and gold leaf on wood
61 x 61 x 5.1 cm
Pamela Blake Burnley-Schol

Michael Burges
Reverse Glass Painting No. 10, 2011
Acrylic, Perspex, 23-carat gold leaf, aluminium
180 x 150 cm
LS LandskronSchneidzik Galerie und Kunstagentur

James Lee Byars
IS, 1991
Wooden sphere, gilded
Diameter 33 cm
Collection Titze

Sarah Cain
Breaking Light, 2010
Acrylic, gold leaf and silver leaf on canvas and Masonite plate
259.1 x 205.1 cm
Courtesy the artist and Anthony Meier Fine Arts, San Francisco

Theres Cassini
Pflasterstein, 2011
Gilt cobblestone
8 x 8 x 8 cm
Private Collection

Gor Chahal (Гор Чахал)
Fat Cross, 2008
Gold on canvas
17 x 18 cm
Gor Chahal

Małgorzata Chodakowska
Seherin, 2009
Bronze-gilt
Height 187 cm
LS LandskronSchneidzik Galerie und Kunstagentur

George Condo
The Linden Boy, 2002
Bronze-gilt
21 x 18 x 19 cm
Private collection, Courtesy Galerie Bruno Bischofberger, Zurich

George Condo
The old Water Nymph, 2002
Bronze-gilt
31 x 23 x 22.5 cm
Private collection, Courtesy Galerie Bruno Bischofberger, Zurich

George Condo
Young Warrior, 2002
Bronze-gilt
22.5 x 17.5 x 20.5 cm
Private collection, Courtesy Galerie Bruno Bischofberger, Zurich

George Condo
The Old Sea Hag, 2002
Bronze-gilt
27 x 16 x 20.5 cm
Private collection, Courtesy Galerie Bruno Bischofberger, Zurich

George Condo
The Alexandrian Women, 2002
Bronze-gilt
32 x 22 x 26 cm
Private collection, Courtesy Galerie Bruno Bischofberger, Zurich

George Condo
The Crying Girl, 2002
Bronze-gilt
26 x 19.5 x 23 cm
Private collection, Courtesy Galerie Bruno Bischofberger, Zurich

George Condo
Pan, 2002
Bronze-gilt
22 x 17 x 19.5 cm
Private collection, Courtesy Galerie Bruno Bischofberger, Zurich

George Condo
The Virgin, 2002
Bronze-gilt
26.5 x 20 x 21 cm
Private collection, Courtesy Galerie Bruno Bischofberger, Zurich

George Condo
Atlanta, 2002
Bronze-gilt
15 x 11 x 13 cm
Private collection, Courtesy Galerie Bruno Bischofberger, Zurich

George Condo
Lamentation, 2002
Bronze-gilt
29 x 20 x 17.5 cm
Private collection, Courtesy Galerie Bruno Bischofberger, Zurich

George Condo
Dionysus, 2002
Bronze-gilt
23.5 x 20.5 x 20 cm
Private collection, Courtesy Galerie Bruno Bischofberger, Zurich

George Condo
Perseus, 2002
Bronze-gilt
23 x 20.5 x 19 cm
Private collection, Courtesy Galerie Bruno Bischofberger, Zurich

George Condo
The Alcoholic, 2002
Bronze-gilt
36 x 27 x 21.5 cm
Private collection, Courtesy Galerie Bruno Bischofberger, Zurich

George Condo
The Frenchman, 2002
Bronze-gilt
23 x 17 x 20 cm
Private collection, Courtesy Galerie Bruno Bischofberger, Zurich

George Condo
The Senate Council, 2002
Bronze-gilt
30 x 20 x 22.5 cm
Private collection, Courtesy Galerie Bruno Bischofberger, Zurich

George Condo
Tristessa, 2002
Bronze-gilt
21 x 15.5 x 18 cm
Private collection, Courtesy Galerie Bruno Bischofberger, Zurich

William Cordova
Wholesellers, retailers & bullshitters, 2004
Gold leaf, graphite pencil and ink on paper
137.2 x 274.3 cm
Collection Peters-Messer, Viersen

William Cordova
Untitled, 2010
Gold leaf and collage in vintage frame
42 x 56 cm
Private Collection, Germany, ARNDT, Berlin

Johann Georg Danninger
Prunktisch, 1800–1810
Metal, bronze, partly gilded, inlays, Pietra dura
82 x 91 cm
MAK—Österreichisches Museum für angewandte Kunst/Gegenwartskunst, Vienna

Jiři Georg Dokoupil
Madonna, 1986
Acrylic and gold foil on linen
180.5 x 140.5 cm
Galerie Andrea Caratsch, Zurich

Johannes Domenig
Untitled, 2008
Bole gilding on wood, rose thorns
150 x 120 x 5 cm
In the artist's possession

Mary Frances Dondelinger
Canary in the Coal Mine, 2010
Egg tempera and 23-carat gold on panel
20.3 x 25.4 cm
In the private collection of G. Todd and R. Craig

Mary Frances Dondelinger
Nativity, 2008
Egg tempera, 23 and 24-carat gold on wood panels
43.2 x 72.4 cm
Courtesy the artist

Mary Frances Dondelinger
Kuenzler Hedgehog Cactus, 2010
Egg tempera, 23-carat gold leaf, gluten glue and acrylic on cardboard
Diameter 25.4 cm
Courtesy the artist

Mary Frances Dondelinger
Ords Kangaroo Rat, 2009
Egg tempera, 23-carat gold leaf, gluten glue and acrylic on cardboard
Diameter 25.4 cm
Courtesy the artist

Godfried Donkor
Three Madonnas, 2009
Collage/giclée-print on heavy paper, hand-finished with real gold leaf, GOLD EDITION, 4/30
100 x 70 cm
Collection Kempen, Aachen

Godfried Donkor
Slave To Champ, 2009
Collage/giclée-print on heavy paper, hand-finished with real gold leaf, GOLD EDITION, 4/30
100 x 70 cm
Collection Kempen, Aachen

Godfried Donkor
Birth of Venus, 2009
Collage/giclée-print on heavy paper, hand-finished with real gold leaf, GOLD EDITION, 4/30
100 x 70 cm
Collection Kempen, Aachen

Godfried Donkor
St Georg Stevenson vs. St Jack Broughton, 2010
Oil and gold leaf on canvas
200 x 200 cm
ARTCO Galerie Herzogenrath Germany

Christian Eckart
Eidolon # 1101, 1989/2000
23-carat gold leaf on poplar wood
213.4 x 243.8 x 35.6 cm
Galerie Thaddaeus Ropac, Salzburg·Paris

Christian Eckart
Detail Painting, No. 9, 1986
Acrylic and gold leaf on plywood
39.5 x 35.5 cm
Collection Clo and Marcel Fleiss, Paris

Rudolf H. Eisenmenger
Design for the safety curtain of the Staatsoper, 1954
Mixed media, gold leaf
49 x 57 cm
Private collection

Manfred Erjautz
D OR, 1962/2012
Solari Udine flip clock, gilded and dis-figured
34.5 x 28 x 12 cm
Courtesy the artist

Jan Fabre
Golden holy dung beetle, 2011
Silicone bronze, 24-carat gold
90 x 45 x 45 cm
Courtesy Mario Mauroner Contemporary Art Gallery, Vienna

Jan Fabre
Golden human brain with angel wings, 2011
Silicone bronze, 24-carat gold
22 x 30.5 x 26 cm
Courtesy Mario Mauroner Contemporary Art Gallery, Vienna

Luciano Fabro
Tamerlano, 1969
Bronze-gilt
26 x 19 x 5.3 cm
Courtesy Collection Goetz

Helmut Federle
Untitled, 2002
Screenprint, gold leaf on aluminium
193 x 116 cm
Collection Sigrid and Franz Wojda, Vienna

Norbert Fleischmann
expand II, 2008
Screenprint on glass, gold leaf, framed
Edition 1/3 bis 3/3, + 1e.a + 1p.p.
Werkstatt Stalzer, Vienna
37.5 x 166 cm
Norbert Fleischmann

Norbert Fleischmann
call, 2011
Acrylic, gold paint on lined canvas, framed
53 x 49.5 cm
Norbert Fleischmann

Norbert Fleischmann
summon, 2011
Resin oil on gold ground, wooden construction
30 x 50 cm
Norbert Fleischmann

Sylvie Fleury
Gucci Handcuffs, 2001/02
Gilded metal
Edition 11/25
23 x 9 cm
Galerie Thaddaeus Ropac, Salzburg·Paris

Sylvie Fleury
Ladder, 2007
23.5-carat gold
Edition 2/8
Height 196 cm
Galerie Thaddaeus Ropac, Salzburg·Paris

Sylvie Fleury
Yes to All, 2004
Steel with painted 24-carat gold plate
Edition 4/8
71.7 x 56.5 cm
Private collection, Salzburg

Günther Förg
Untitled, 1986
Acrylic, gold leaf on wood
41 x 60 cm
Grässlin Collection, St. Georgen

Ernst Fuchs
Die Geheimnisse des hochheiligen Rosenkranzes, linkes Bild des Triptychon, 1958–1961
Mixed media, gold leaf on goat-skins
300 x 300 cm
Erzdiözese Wien, Hetzendorf parish church

Makoto Fujimara
Interior Castles, 2008
Japanese gold leaf and mineral pigment on koudama paper mounted on canvas
121.9 x 121.9 cm
Courtesy of Dillon Gallery New York

Gajin Fujita
East vs. West, 2010
Platinum foil, gold leaf, acrylic, paint spray and paint marker on wood
12 panels, each 213.4 x 55.9 cm (in total 213.4 x 670.6 cm)
Private collection, Courtesy of L. A. Louver, Venice, California

Gajin Fujita
The Saints, 2008
Gold leaf, acrylic, spray paint and paint marker on wood
182.9 x 274.3 cm
U.S. Biennial Inc., New York, Prospect New Orleans collection

Jakob Gasteiger
Klondike, 2011
Gilded aluminium
Height 62 cm
In the artist's possession

General Idea
AIDS, 1987
Acrylic, gold leaf on canvas
76 x 76 cm
Collection Katharina Faerber, Geneva

Robert Ginder
Haven, 1991
Transfer gold leaf on wood
102.9 x 134.6 cm
Lora Schlesinger Gallery, Santa Monica, L.A./New York

Robert Ginder
June, 1993
Oil, gold leaf on wood
83.9 x 30.5 cm
Lora Schlesinger Gallery, Santa Monica, L.A./New York

Robert Ginder
Love, 1997
Oil, gold leaf on wood
101.6 x 91.4 cm
Lora Schlesinger Gallery, Santa Monica, L.A./New York

Bruno Gironcoli
Untitled (Blattähre), 2003
Gilded cast aluminium
55 x 49 x 42 cm
Courtesy Galerie Elisabeth and Klaus Thoman, Innsbruck

Yoko Grandsagne
Zen Garten, 2010
Mixed media, gold leaf on canvas
195 x 390 cm
Yoko Grandsagne

Greek (Cretan painter)
Hinged altarpiece (triptych) with Philoxenia in the central section, 16th cent.
Wood, painted
25 x 34 cm (open)
Kunsthistorisches Museum, Vienna, Kunstkammer

Karl Haas
Buckelpokal, 1895
Galvanoplastic, copper-gilt
Height 34 cm, Diameter 11 cm
MAK—Österreichisches Museum für angewandte Kunst/ Gegenwartskunst, Vienna

Richard Hamilton
The "Gold" Guggenheim, 1965–1966
High relief, gold leaf, fibreglass and cellulose
121.8 x 122.4 x 20.5 cm
Louisiana Museum of Modern Art
Donation: The Museum Foundation

Barkley Hendricks
Fela: Amen, Amen, Amen, Amen, 2002
Oil and variegated metal leaf on canvas, wooden frame, armature, 27 pairs of high heels
169.6 x 118.7 cm
Private Collection

Lori Hersberger
Intergalactic Totem, 2011
Steel, gold
120 x 50 x 25 cm
Galerie Thaddaeus Ropac, Salzburg·Paris

David Hewson
Moon Goddess, 2000
23-carat gold water gilt on wood
68.6 x 68.6 cm
Susan R. Alexander

Friedensreich Hundertwasser
460 Hommage au tachisme, 1961
Mixed media: egg tempera, oil and gold leaf on packing paper primed with chalk and polyvinyl, mounted on jute with PV
195 x 130 cm
Die Hundertwasser Gemeinnützige Privatstiftung, Vienna

Marcello Jori
ti adoro, 2005
Gold leaf on wood
104 x 120 cm
Private collection, Prato

Jürgen Kadow
Untitled, 2005
Oil, gold leaf on canvas
80 x 100 cm
Private collection, Germany

Josef Kern
Ein Sommer in Bäumen, 2006
Wood, gold leaf
202 x 142 cm
Private collection

Josef Kern
Julia schläft, 2006
Wood, gold leaf
185 x 172 cm
In the artist's possession

Yves Klein
Valeur or (Monochrom gold), 1960
Gold leaf on plywood
59 x 44.5 cm
Kunstmuseen Krefeld

Imi Knoebel
Menningebild 37/3, 1976/2010
Transfer gold leaf on wood
190.7 x 351 x 9 cm
Private collection, Sindelfingen

Cornelius Kolig
Goldrahmen, 1986–2011
Aluminium, polyurethane, real gold paint
120 x 150 x 33 cm
Cornelius Kolig

Cornelius Kolig
Ehrenzeichen, 1977
Cast aluminium, real gold paint
Diameter 22 cm
Cornelius Kolig

Kolkoz
Marine, 2009
Gilded frame elements
87.5 x 147.5 x 11.5 cm
Courtesy Galerie Perrotin, Paris

Milan Kunc
Recycled World, 2001
Oil and gold leaf on canvas
200 x 240 cm
Milan Kunc

Milan Kunc
Vitamin Attack, 2005
Oil and gold leaf on canvas
130 x 150 cm
Milan Kunc

Milan Kunc
Cross the Badland, 1994
Oil and gold leaf on canvas
190 x 220 cm
Milan Kunc

Robert Kushner
White Lilac, Lilac Lilac, 2004
Oil, acrylic, gold leaf, glitter and paper on Japanese screen
170.8 x 187.3 cm (open)
Private collection, Courtesy DC Moore Gallery, New York

Alicja Kwade
Kohle, 2006
Cast bronze of coal briquette, gilded, 24-carat real gold, 77 parts
33 x 34 x 35.8 cm
Private collection, Stuttgart

Eduart Lebiedzki
Design for the mosaic frieze in the portico of the Reichsrat building in Vienna, 1900
Oil on canvas
79 x 77 cm
Wien Museum

Eduart Lebiedzki
Design for the mosaic frieze in the portico of the Reichsrat building in Vienna, 1900
Oil on canvas
73 x 136 cm
Wien Museum

Eduart Lebiedzki
Design for the mosaic frieze in the portico of the Reichsrat building in Vienna, 1907
Oil on canvas
79 x 77 cm
Wien Museum

Frédéric Léglise
Charline, 2007
Oil, gold leaf and gloss paint on canvas
115 x 88.5 cm
Collection David Marcel Fleiss, Galerie 1900–2000, Paris

Rudolf Leitner-Gründberg
Signaturschriftbild, 2010
Gold leaf, oil, packing paper, painted gilt solid-wood frame (part of the artwork), behind glass
114 x 149 cm

Rudolf Leitner-Gründberg
Die Lösung des Welträtsels ist die Hinwendung zu unserer höchsten Möglichkeit, 2002
Triptych—gold leaf, oil, solid wood case
131 x 90 x 23.5 cm

Sherrie Levine
Fountain (after Marcel Duchamp: A.P.), 1991
Cast bronze
36.8 x 36.2 x 63.5 cm
Collection Walker Art Center, Minneapolis
T. B. Walker Acquisition Fund, 1992

Jason Clay Lewis
Girl with dragon, 2007/08
Acrylic and gold leaf on canvas
121.9 x 91.4 cm
Jason Clay Lewis

Jason Clay Lewis
Warrior, 2006
Acrylic and gold leaf on canvas
45.7 x 30.5 cm
Jason Clay Lewis

Jason Clay Lewis
Torture panel VI, 2006/07
Acrylic and gold leaf on canvas
43.2 x 26.7 cm
Jason Clay Lewis

Edward Lipski
King, 2007
Gold leaf, synthetic hair, gold chain, wood
114 x 60 x 60 cm
Private collection, Switzerland

Hung Liu
Polly with Mandarin Ducks, 2008
Mixed media on wood
104.1 x 104.1 cm
Turner Carroll Gallery, Santa Fe/Dallas

Heinz Mack
Mack K40, 1997
Polished gold
35 x 20 x 6 cm
Galerie Walker Schloss Ebenau Rosental

Heinz Mack
Mack K41, 1997
Polished gold, glazed
32 x 19 x 4.5 cm
Galerie Walker Schloss Ebenau Rosental

Dimitrije Bašičević Mangelos
Intuicija je stvar instinkta (Intuition is a matter of instinct), 1977/78
Tempera and gold paint on board
Frame: 66.4 x 45 cm
Collection Flick, Salzburg

Franz Mayrhofer
Devanagari, 2001
From the series "Nur Wissende sehen"
Gold leaf and acrylic on wood
70 x 50 cm
Dr. Jeanette Mayrhofer-Berger, Dr. Franz Mayrhofer

Herbert Mehler
Stella della Zucca, 2008
Corten-steel, gilded
150 x 43 x 43 cm
LS LandskronSchneidzik Galerie und Kunstagentur

Herbert Mehler
Ondella, 2007
Corten-steel, gilded
33 x 50 x 50 cm
LS LandskronSchneidzik Galerie und Kunstagentur

Herbert Mehler
Fiamma, 2006/2011
Corten-steel, gilded
Height 225 cm
LS LandskronSchneidzik Galerie und Kunstagentur

Frans van Mieris the Elder
The Doctor's Visit, 1657
Oil on copper-gilt
34 x 27 cm
Kunsthistorisches Museum, Vienna, Gemäldegalerie

Gary Milek
November Study #2, undated
Egg tempera and gold leaf on wood
33 x 53.3 cm
Painting on loan from Janet and Stephen Flanders of Norwich, Vermont

Patricia Miranda
Siena Landscape Series #1, 2005
Egg tempera and 22-carat gold leaf on panel
22.9 x 15.2 cm
Patricia Miranda

Patricia Miranda
Siena Landscape Series #3, 2005
Egg tempera and 22-carat gold leaf on panel
30.5 x 24.4 cm
Patricia Miranda

Patricia Miranda
Siena Landscape Series #4, 2005
Egg tempera and 22-carat gold leaf on panel
15.2 x 22.9 cm
Patricia Miranda

Paul Morrison
Pyxide, 2010
24-carat gold leaf and acrylic on linen
72 x 54 x 4 cm
© the Artist/Galerie Bob van Orsouw, Zurich

Paul Morrison
Rachilla, 2010
24-carat gold leaf and acrylic on linen
72 x 54 x 4 cm
© the Artist/Galerie Bob van Orsouw, Zurich

Wojciech Mucha
Triptych, 2006
Oil and gold leaf on wood
78 x 200 cm (open, with frame)
84.5 x 105 x 12 cm (closed)
Private collection/estate of W. Mucha, Geneva

Peter Murphy
Kurt Cobain Nirvanamind, 2009
Egg tempera and gold leaf on wood
92 x 48 cm
Peter Murphy, Courtesy the artist

Peter Murphy
Jimi Hendrix, Are you experienced?, 2008
Egg tempera and gold leaf on wood
92 x 48 cm
Peter Murphy, Courtesy the artist

Winfried Muthesius
Hommage à Vincent, 2011
Gold leaf on wood
50 x 50 cm
Private collection

Pete Nawara
St. Peter, 2004
Mixed media on canvas
180 x 203 cm
Pete Nawara

Hermann Nitsch
n. t., 2011
Blood and paint on bole gilding
100 x 80 cm
Hermann Nitsch

Hermann Nitsch
n. t., 2011
Blood and paint on oil gilding
100 x 80 cm
Hermann Nitsch

Not Vital
The Golden Calf, 2001
18-carat gold
71 x 85 x 29 cm
Private collection

Oswald Oberhuber
Untitled, 1950s
Collage, metals, mixed media, wood
68 x 52.5 cm
Galerie Maier, Innsbruck

Emil Orlik
Japanese Garden, 1901/02
Tempera and bronze-gilt on paper (?), silk
37.5 x 54 cm
Belvedere, Vienna

Gabriel Orozco
Samurai Tree 6C", 2006
Egg tempera and gold leaf on red cedar wood
55 x 55 cm
Collection Titze

Mimmo Paladino
Thema n.°4, 2000
Mixed media on panel
80 x 60 cm
Courtesy Galleria Christian STEIN

Pepo Pichler
Boy with Halo, 2002
Wood, gold leaf, steel
98 x 40 x 34 cm
Private collection

Pepo Pichler
Vache sabot jongleur, 2010
Wood, plaster, gold leaf, concrete
35 x 18.5 x 18.5 cm
Private collection

Eva Pliem
from the series of *Bäuche: Atem 5, 6* and *7,* 2009
Bole gilding on plaster
c. 43.5 x 41 cm
Courtesy the artist

Marc Quinn
A Beautiful Continuation, 2009
Chrome-plated bronze and gold leaf
110 x 63 x 52 cm
Galerie Thaddaeus Ropac, Salzburg·Paris

Kurt Regschek
Triptych, 1960
Mixed media on gold leaf and wood
25 x 32 cm
Private collection

Karl Manfred Rennertz
Goldenes Boot, 2002
Tulip tree, gold leaf, blackened in fire
Length 2.70 m
Property of the artist

Robin Rhode
Spade, 2007
Bronze-gilt and charcoal
88 x 21 x 4 cm
Collection Goetz

Gerhard Richter
Kreuz, 1996
Gold
19.5 x 19.5 x 1.5 cm
Winterthur, Kunstmuseum Winterthur

Gerwald Rockenschaub
Painting, 2009
MDF, gold leaf
10 x 10 x 3 cm
Courtesy Georg Kargl Fine Arts, Vienna and
Gallery Mehdi Chouakri, Berlin

Gerwald Rockenschaub
Untitled, 2008
Gold leaf frame, acrylic glass
40 x 40 x 5 cm
Courtesy Galerie Vera Munro

Gerwald Rockenschaub
Untitled, 2008
Gold leaf frame, acrylic glass
40 x 40 x 5 cm
Courtesy Galerie Vera Munro

Edouard Sacaillan
Ascetic and the Armenian Alphabet, 2007
Vinyl, oil and gold paint on canvas
204 x 133 cm
Private Collection, Athens
Courtesy Kalfayan Galleries, Athens—Thessaloniki

Tom Sachs
Reineke Fuchs: Bubble Fox, 2008
Plywood, synthetic polymer paint and gold leaf
243.8 x 203.2 x 12.7 cm
Galerie Thaddaeus Ropac, Salzburg·Paris

Tom Sachs
Reineke Fuchs: Honeybear, 2008
Plywood, wood, resin, gold leaf, synthetic polymer paint and iron components
123.5 x 158.4 x 11.4 cm
Galerie Thaddaeus Ropac, Salzburg·Paris

Salzburg Painter
Christ on the Cross between the Virgin and SS John, Catharine and Sebastian, c. 1470
Painting on spruce wood
94.5 x 126.5 cm
Belvedere, Vienna

Elisabeth von Samsonow
Elektra, 2009
Lime wood, gold leaf, oil glaze
Height 300 cm
In the artist's possession, Courtesy the artist

Carme Sanglas
Solstice, 2008
Mixed media and gold leaf on paper
56 x 56 cm
Carme Sanglas

Carme Sanglas
Factory (Fàbrica), 2005
Mixed media and gold leaf on paper
177 x 150 cm
Carme Sanglas

Carme Sanglas
Paths and Roads (Camins i carreres), 2003
Mixed media, silver leaf and gold leaf on paper
60 x 71 cm
Carme Sanglas

Walter Schnabl
Schild des Apollon, 2011
Plaster-chalk ground on canvas, 23-carat gilding
Diameter 3.33 m
Walter Schnabl

Gundula Schulze Eldowy
Das lebendige Bild, Konstantinopel, 1997
Photograph with gold leaf
c. 189 x 130 cm
Private collection

Gustav Seyfferth
Roman Autumn Festival, c. 1875/76
Oil, gold leaf on canvas
124 x 159 cm, frame richly decorated, c. 20 cm broad
Neue Galerie Graz at the Universalmuseum Joanneum

Lucy Skaer
Solid Ground-Tarsal, 2006
Pencil and gold leaf on Fabriano paper
140 x 200 cm
Collection of the Baloise Group, Basel

Nedko Solakov
A Middle-Aged Story, 2006
Acrylic and India ink on gilded lime wood
61 x 60 x 8 cm
Collection of Aaron and Barbara Levine, Washington D. C, ARNDT, Berlin

Nedko Solakov
The Leader, 2010
Acrylic and India ink on gilded, carved lime wood
64 x 61 x 4.5 cm
Courtesy GALLERIA CONTINUA, San Gimignano/Beijing/Le Moulin

Nedko Solakov
The Tunnel, 2009
Acrylic and India ink on gilded, carved lime wood
60 x 60 x 4 cm
Courtesy GALLERIA CONTINUA, San Gimignano/Beijing/Le Moulin

Nedko Solakov
The Denial, 2009
Acrylic and India ink on gilded, carved lime wood
54.5 x 70 x 8.5 cm
Courtesy GALLERIA CONTINUA, San Gimignano/Beijing/Le Moulin

Nedko Solakov
Middle Fingers, 2009
Acrylic and India ink on gilded, carved lime wood
43 x 88 x 7 cm
Private Collection, Milan
Courtesy GALLERIA CONTINUA, San Gimignano/Beijing/Le Moulin

Nedko Solakov
The Water Pyramid, 2008
Acrylic and India ink on carved lime wood
45 x 69 x 8 cm
Courtesy the artist, ARNDT, Berlin

Nedko Solakov
The Investment (the money pouch), 2008
Acrylic and India ink on carved lime wood
77 x 83 x 9.5 cm
Courtesy private collection, Australia

Nedko Solakov
An Anus Story, 2005
Acrylic and India ink on gilded, carved lime wood
61 x 61 x 8 cm
Collezione Nunzia e Vittorio Gaddi, Italia

Hermann Staudinger
Gargano, 2010
Ed. 10
Digital print on 23-carat gold on wood
56 x 100 cm
Courtesy the artist

Hermann Staudinger
Fossanova, 2011
Ed. 10
Digital print on acrylic glass, mounted on 23-carat gilded wood
72 x 41 cm
Courtesy the artist

Giandomenico Tiepolo
Jacopo Porto is appointed Governor of Vicenza in 1022, c. 1757
Fresco in grisaille and gold leaf, mounted on canvas
262 x 183 cm
Arp Museum Bahnhof Rolandseck/Collection Rau for UNICEF

Giandomenico Tiepolo
Donato Porto Becomes a Venetian Patrician in 1379, c. 1757
Fresco in grisaille and gold leaf, mounted on canvas
262 x 183 cm
Arp Museum Bahnhof Rolandseck/Collection Rau for UNICEF

Panos Tsagaris
Untitled, 2010
Gold leaf (23-carat) and ink on paper
76 x 56 cm
Courtesy Kalfayan Galleries, Athens—Thessaloniki

Yiannis Tsarouchis
July, 1972
Gold leaf and oil on wood
80.5 x 55 cm
Private Collection, Athens
Courtesy Kalfayan Galleries, Athens—Thessaloniki

Unknown artist
Mummy portrait of y young man, c. AD 120–130
Encaustic on wood with gold leaf
41 x 20.2 cm
Ägyptisches Museum und Papyrussammlung,
Staatliche Museen zu Berlin

Unknown artist
House sign 'Zum Auge Gottes'
(from house 9, Nussdorferstrasse 75), 18th cent.
Fire-gilded copper
Wien Museum, Vienna

Unknown artist
So-called Marian Vestments
One of the total of 17 items
2 chasubles, 4 dalmatics, 1 cope, 3 stoles, 4 maniples,
2 bursas, 1 pall
Objects of two different materials: one white with polychrome floral decoration, the other blue with opulent gold decoration
Parish of St Augustine, Vienna

Victor Vasarely
Taller, 1965
Gold leaf and oil tempera on wood
80 x 80 cm
Museum Ritter, Waldenbuch (G)

Tabitha Vevers
Eden (Expulsion), 2008
Oil and gold leaf on Ivorine
35.6 x 38.1 cm
Private collection

Tabitha Vevers
Flying Dream (In the Dunes II), 2011
Oil and gold leaf on galvanized steel
25.4 x 30.5 cm
Private collection

Tabitha Vevers
Shiva (While We Dance), 2010
Oil and gold on Mylar foil
24.1 x 29.2 cm
Kim Manocherian

Tabitha Vevers
Bananaman, 2010
Oil and gold on Mylar foil
27.9 x 36.2 cm
Tabitha Vevers

Jan Maarten Voskuil
Gilded Frame, 2002
Wood, gold leaf
145 x 115 x 60 cm
Courtesy the artist

Otto Wagner
One of the eight original marble slabs of the Karlsplatz Pavilion,
1898/99
Carrara marble, partly gilded
Wien Museum

Andy Warhol,
Still life, 1957/58
India ink and gold leaf on turquoise paper
22 x 24.8 cm
Private collection

Robert Weber
Für mich deine Liebe, für meine Arbeit deinen Respekt, 2005
Box (Listro—Ethiopian shoe-shine box, 24-carat gilded, on base)
Wood, gold leaf, acrylic glass, base
28.5 x 17 x 19 cm
Height of base 140 cm
Robert Weber

Robert Weber
L'inesprimibile nulla, Non si vede piú (The inexpressible nothing, nothing more is to be seen), 2006
Oil and emulsions on gold ground and wood
230 x 200 cm
Robert Weber

Max Weiler
Christ the King (design for a mosaic in the Ignaz Seipel church in Vienna), 1935
Gold, watercolours, India ink on paper
78.5 x 70 cm
Private collection

Fred Wessel
Sienese Icon (Dahlia), 1997
Tempera with gold leaf
71.1 x 71.1 cm
Private collection

Fred Wessel
Angel's Trumpet, 1998
Egg tempera and gold leaf
25.4 x 17.8 cm
Sheldon Bigler, Henrico, VA

Fred Wessel
Venetian Icon (Ginger), 1998
Tempera with gold leaf
25.4 x 17.8 cm
Linda B. Neely

Fred Wessel
Maize, 1993
Tempera with gold leaf
48.3 x 55.9 cm
Collection of the artist

Fred Wessel
Becca (Telescope), 2007
Egg tempera, gold leaf and platinum foil
75 x 60 cm
Dr. Michael Wood

Fred Wessel
Shells, 2006
Egg tempera, gold leaf
17.8 x 22.9 cm
Arden Gallery, Boston, Massachusetts

Franz West
Untitled, 1987
Gilt papier mâché
82 x 40 cm
Collection Geyer, Vienna

Franz West
Abgabe, 1987
Wood, papier mâché, gold foil
36 x 90 x 9 cm
Galerie Meyer Kainer

Franz West
Passstück (Hommage à …), 1983–1985
Plaster, papier mâché and gold leaf
118 x 24 x 26 cm
Courtesy Galerie Elisabeth und Klaus Thoman, Innsbruck

Clemens Wolf
Opened Space (Golden Fence), 2010
24-carat gold leaf, metal fence, concrete blocks, bolt cutter
200 x 380 x 150 cm
Galerie Steinek

Erwin Wurm
"*Goldknödel*", 1989
Silver-gilt
Diameter 7 cm
Private collection

Toby Ziegler
The Hedonistic Imperative (2nd version), 2006
Öl, gold leaf and pencil on canvas
210 x 242.5 cm
Tate: Presented by David Gorton 2006

A Word of Thanks

We should like to thank all the artists, lenders, collectors, museums and authors who have contributed to this endeavour. In particular, we should like to mention the following for their generosity and assistance:

Eveline Baumgartner (Galerie Bob van Orsouw, Zürich)
Barbara Bergmann (Schauwerk Sindelfingen)
Dieter Blume (University of Jena)
Sarah Granatir Bryan (Anthony Meier Fine Arts, San Francisco)
Julie Burchardi (Galerie Arndt, Berlin)
Andrea Caratsch (Galerie Andrea Caratsch, Zürich)
Galerie Heike Curtze Vienna-Berlin
Gudrun Danzer (Neue Galerie, Graz)
Paolo Di Buono (Vatican Mosaic Studio)
Julia Dudkiewicz (Society of Antiquaries of London)
Elizabeth East (L. A. Louver, Venice, CA, USA)
Katharina Faerber (Vandoeuvres, Switzerland)
Elein Fleiss (Lisbon)
Marcel Fleiss (Galerie 1900–2000, Paris)
Corinne Michaela Flick (London, Salzburg)
Matthias Frehner (Kunstmuseum Bern)
Andrea Fürst (Hundertwasser Gemeinnützige Privatstiftung, Vienna)
Vittorio Gaddi (Lucca)
Jenny Gibbs (The Arts Foundation, Wellington, New Zealand)
Ingvild Goetz (Munich)
Dorothee Golz (Vienna)
Katinka Gratzer (Vienna)
Detlev Gretenkort (Büro Georg Baselitz, Munich)
Martina Grießer-Stermscheg (Vienna)
Isabelle Guggenheim (Baloise Group, Basel)
Joram Harel (Hundertwasser Gemeinnützige Privatstiftung, Vienna)
Anne Harmssen (Museumslandschaft Hessen, Kassel)
Martin Hentschel (Kunstmuseen Krefeld)
Isabella Herzig (Galerie Ropac, Salzburg)
Walter Holzer (Vienna)
Marli Hoppe-Ritter (Museum Ritter, Waldenbuch, Germany)
Ayaka Ishizaki (Mizuma Art Gallery, Tokyo)
Galerie Martin Janda (Vienna)
Roswitha Juffinger (Residenzgalerie Salzburg)
Arsen & Roupen Kalfayan (Kalfayan Galleries, Athens-Thessaloniki)
Galerie Kargl (Vienna)
Edelbert Köb (Vienna)
Johann König (Berlin)
Oliver Kornhoff (Landes-Stiftung Arp Museum Bahnhof Rolandseck, Remagen)
Karola Kraus (Vienna)
Walter Krause (Vienna)
Sarah Kronschläger (Galerie Heike Curtze Vienna)
Anette Kruszynski (Stiftung Kunstsammlung Nordrhein-Westfalen, Düsseldorf)
Inge Landskron (LS LandskronSchneidzik, Galerie und Kunstagentur, Nuremberg)
Heidi Lange (DC Moore Gallery, New York)
Aaron & Barbara Levine (Washington, DC, USA)
Suzanne Leydag (Landes-Stiftung Arp Museum Bahnhof Rolandseck, Remagen)
Elisabeth Madlener (Galerie nächst St. Stephan, Vienna)
Mario Mauroner (Mario Mauroner Contemporary Art, Vienna)
Michaela und Christoph Marx (Munich)
Galerie Maier (Innsbruck)
Joachim Melchers (ARTCO Galerie, Herzogenrath)
Stephanie Mercado (Lora Schlesinger Gallery, Santa Monica, CA, USA)
Maria Missbach (Perchtoldsdorf)
Franziska Muliar (Grossenzersdorf)
Galerie Vera Munro (Hamburg)
Ulrike Neutatz (Galerie Ropac, Salzburg)
Ruth und Jürg Nyffeler (Edition 5, Erstfeld, Switzerland)
Florian Peters-Messer (fpmcollection Viersen, Berlin)
Verusca Piazzesi (Galleria Continua San Gimignano Siena, Italy)
Vicki Politis (Kalfayan Galleries, Athens-Thessaloniki)
Lisa Polten (Galerie Arndt Berlin)
Judith Radlegger (Mario Mauroner Contemporary Art, Vienna)
Katie Rashid, (Jack Shainman Gallery, New York)
Lisl Regschek (Vienna)
Florian Rehn (Galerie Max Hetzler, Berlin)
Rik Reinking (Hamburg)
Galerie Thaddaeus Ropac (Salzburg, Paris)
Ulrik Runeberg (Landesmuseum für Kunst und Kulturgeschichte Münster)
Peter Schaufler (Schauwerk Sindelfingen)
Edward Schexnayder (Jack Shainman Gallery, New York)
Anne Schloen (Cologne)
Regina Schneidzik (LS LandskronSchneidzik, Galerie und Kunstagentur, Nuremberg)
Gabriele Schober (Plattform für junge Kunst GmbH, Vienna)
Karl Schütz (Kunsthistorisches Museum, Vienna)
Dieter Schwarz (Kunstmuseum Winterthur)
Martin Stangl (Kunsthaus Vienna)
Galerie Steinek (Vienna)
Christa Steinle (Neue Galerie, Graz)
Ursula Storch (Wien Museum)
Galerie Elisabeth & Klaus Thoman (Innsbruck)
Anne Lambert Tracht (ConsultArt, Inc Atlanta, USA)
Stephan Urbaschek (Sammlung Goetz, Munich)
Judith Walker (Galerie Walker, Ebenau, Carinthia)
Katrina Weber (The Third Line, Dubai)
Yvonne Weiler (Vienna)
Barbara Willert (Museum Ritter, Waldenbuch)
Galerie Hubert Winter (Vienna)
Sigrid und Franz Wojda (Vienna)

Authors

Tatjana Bayerová
heads the chemical laboratory at the Institute for Conservation and Restoration at the University of Applied Art in Vienna, where she also teaches.

Jürgen Borchhardt
studied at the universities of Heidelberg and Berlin. He was in charge of classical archaeology at the German Archaeological Institute in Istanbul from 1966 to 1970. From 1965 to 1968 he was the head of the Field Survey Project in Myra (southern Turkey) and from 1969 to bis 2001 in the same position at the excavations in Limyra (Vilayet Antalya). He has taught at the universities of Frankfurt, Vienna (1982–2001) and Giessen. Currently he lives as Emeritus Professor of Classical Archaeology in Vienna.

Markus Castor
studied art history, philosophy and classical archaeology in Trier, Freiburg and Konstanz. He took his doctorate in 1991 with a dissertation on colour and space in the work of Diego Velázquez. After a lectureship at the Art-historical Institute at the University of Freiburg, he worked at the Technical University of Dresden on the courtly art of the Renaissance and Baroque. From 2002 he was Chargé de Recherche at the German Forum for Art History, and since 2006 has been Directeur de Recherche in Paris, working on the history of scholarship and art in post-medieval Europe.

Joachim Driller
is professor of the history of art, design and architecture at the university of Coburg. He has published a great deal on the architecture, art and design of the nineteenth and twentieth centuries. Most recently he contributed to the exhibition and catalogue *Richard Neutra in Europa. Bauten und Projekte 1960–1970*, Museum MARTa Herford 2010.

Ulrike Gehring
studied history, art history, and German language and literature in Frankfurt and Paris. She held a scholarship at the Institut de l'Histoire de l'Art et d'Archéologie, Sorbonne IV (Paris). From 2001 to 2003 she was an academic assistant at the Centre for Art and Media Technology. In 2003 she curated and organized the exhibition *Obsessive Malerei* (MNK | ZKM). At the same time she taught part-time at the universities of Frankfurt and Heidelberg. In 2003 she was appointed junior professor of art history at the University of Trier, focusing on modern and contemporary art and the new media. In 2008 she held a scholarship at the German Historical Institute in London. Since 2009 she has been professor of art history at the University of Trier. From 2006 to 2011 she co-edited the journal *Kritische Berichte. Zeitschrift für Kunst- und Kulturwissenschaften*.

Lisa Gräber
is an academic assistant in the department of stone restoration at the Institute of Conservation and Restoration at the University of Applied Art in Vienna.

Bodo Hell
lives in Vienna and on the Dachstein. He works in prose, radio, theatre, writing in the public space, texts on the visual arts, photography, film, music, exhibitions, alpine-meadow farming, and books, most recently: *Admont Abscondita* (with Norbert Trummer), 2008; *Nothelfer*, 2008/2010; *Herbe Garbe, Weiberkittel* (with Elsbeth Wallnöfer, P. & W. Kubelka), 2008; *Immergrün* (with Linda Wolfsgruber), 2011.

Agnes Husslein-Arco,
has been since 2007 the Director of the Belvedere in Vienna. She is an art historian, the curator of numerous exhibitions on classical modern and contemporary art, and the author and editor of scholarly publications. In 1981 she opened the Vienna branch of Sotheby's, of which she was the managing director until 2000. In addition she also managed, from 1988, the Sotheby's branches in Budapest and Prague. In the 1990s she was Director of European Development in the Solomon R. Guggenheim Museum, from 2001 to 2003 the Director of the Rupertinum in Salzburg, and from 2003 to 2005 the founding Director of the Museum der Moderne in Salzburg. From 2002 to 2004 she was in charge if building up the Museum Moderner Kunst Kärnten (MMKK).

Judith Kern
works as an academic assistant in the department of painting restoration at the Institute of Conservation and Restoration at the University of Applied Art in Vienna.

Alexander Klee
has worked at the Belvedere since 2010 as the curator for the second half of the nineteenth and the early twentieth centuries. An art historian he was a founding member of the Adolf Hölzel Foundation in Stuttgart, and has prepared the catalogue raisonné of the works of Adolf Hölzel. His focus of research lies in the nineteenth and twentieth centuries, with numerous works on Hölzel, Georg Karl Pfahler and classical modern artists.

Regina Knaller
works as an academic assistant in the department of textile restoration at the Institute of Conservation and Restoration at the University of Applied Art in Vienna.

Gabriela Krist
heads the Institute of Conservation and Restoration at the University of Applied Art in Vienna.

Veronika Loiskandl
works as an academic assistant in the department of painting restoration at the Institute of Conservation and Restoration at the University of Applied Art in Vienna.

Marija Milcin
works as an academic assistant in the department of stone restoration at the Institute of Conservation and Restoration at the University of Applied Art in Vienna.

Linda Pelzmann
is head of the Department of Economic Psychology at the University of Klagenfurt, Austria. From 1992 to 1999 she was visiting professor of economic psychology at the University of Economics, Vienna. She has had working visits and visiting professorships in Bangkok (Thailand), New Zealand, Sofia (Bulgaria), Ljubljana (Slovenia), Chicago (USA) and Cambridge (USA). From 1999 to 2001 she was Visiting Scholar and Research Professor at Harvard University, Cambridge (USA), where her research project was 'Mass Manufactured Will'.

Veronika Pirker-Aurenhammer
has been curator of the medieval collection at the Belvedere since 2006. She was responsible for the rearrangement of the collection with its open depository, and has curated a number of exhibitions. Before that, she spent many years researching at the Öster-reichische Akademie der Wissenschaften, the Österreichische Nationalbibliothek, the Federal Heritage Office, and at the Institut für Kunstgeschichte in Vienna. She has published widely, mainly in the field art in the Middle Ages and Early Modern period. Since 2009 she has been on the Committee of the Association of Austrian Art-historians.

Eva Putzgruber
works as an academic assistant in the department of object restoration at the Institute of Conservation and Restoration at the University of Applied Art in Vienna.

Michael Viktor Schwarz
studied art history in Freiburg. From 1986 to 1990 he was engaged in research at the Bibliotheca Hertziana in Rome, and in 1991 obtained his professorial qualification in Freiburg. From 1991 to 1998 he taught at Freiburg, Tübingen and Trier, In 1998 was appointed professor first in Dresden and then in Vienna. Since 2006 he has been Dean of the Faculty of History and Cultural Studies in Vienna. He has published widely on medieval Italy, Modern Art in Germany, Max Beckmann and the visual media in Christian worship.

Agnes Szökrön-Michl
works as an academic assistant in the department of painting restoration at the Institute of Conservation and Restoration at the University of Applied Art in Vienna.

Thomas Zaunschirm
was Professor of Art History at the universities of Freiburg (1989–1994) and Duisburg-Essen (1995–2007). He has written books on Marcel Duchamp, Gustav Klimt, Jeff Koons, Arnold Schönberg, Diego Velázquez and others.
He has curated exhibitions including *Wiener Diwan—Sigmund Freud heute* (20er Haus 1989), *Im Zentrum der Welt* (Kunstverein Kärnten 1992) and *Die Farben Schwarz* (Joanneum Graz 1999). He lives in Vienna.

Glossary

Amalgamating: painting the cleaned and de-greased metal surface with mercury in order to improve the adhesion of the → gold amalgam during → fire gilding.

Aurum musivum: powder consisting of crystalline tin (IV) sulphide with binding agents, which was painted on as a gold substitute.

Aurum: Latin word for the precious metal gold.

Bole gilding: gold-leaf gilding technique for which a specially prepared coloured earth, the bole, is used as a substrate or foundation for the gold leaf.

Bole: slurry of cleaned coloured earth with a high clay-content.

Brocade: woven fabric with precious metal threads.

Bronzing: bound brass or bronze powder painted on a base to imitate gold.

Chasing: depressions created in the freshly applied gold leaf to produce a decorative design. If the substrate is already too hard, the gold leaf tears.

EDS (Energy-dispersive X-ray spectroscopy): an analytical technique that uses the x-rays generated by a → scanning electron microscope.

Embroidery: ornamentation created by adding vegetable, animal or mineral threads to a support material, such as a woven item or leather. There are a large number of embroidery techniques.

Engraving: here, technique whereby contours and patterns are carved into the gesso substrate.

Fire gilding: gilding produced by applying a → gold amalgam on a metal base.

Galvanic gilding: gilding produced by the precipitation of gold dissolved in an electrolyte on to a metal object using an electric current.

Gesso: a mixture of animal-glue binder and chalk; used as a substrate for bole gilding.

Gilder's tip: soft brush made from squirrel hair used to apply the gold leaf.

Glaze: here, layer that produces a smooth, non-porous base on a stone surface prior to gilding.

Gold amalgam: gold-and-mercury alloy.

Mosaic gold: gold pigment made of pulverized gold leaf. The gold powder, which was bound (for example) by egg white or gum arabic was filled into shells and sold in a dried state.

Oil size: specially prepared drying oil, usually linseed oil, which serves as an adhesive agent between the substrate material and the gold leaf.

Pastiglia technique: this technique produces raised sculpted patterns by adding to the chalk base.

Pressbrokat: imitation of brocade, produced in a negative-positive process with the help of tin foil, subsequently applied to the base, and, depending on the pattern, gilded with gold leaf and painted.

Repoussage: as → chasing, but applied from the reverse side.

Scanning electron microscope: a microscope that uses electron beams to produce an image. The electron beam scans the area under investigation, producing very high-resolution images that are magnified by up to a million-fold. During the scanning process the x-rays generated have an element-specific energy or wavelength. This makes a qualitative and quantitative element analysis possible.

Spinning: rotating a fibre to produce a thread.

Sublimation: here, evaporating mercury by exposing it to heat after applying gold amalgam during the process of fire-gilding. This also creates the bond between the gold and the metal base during fire-gilding.

Tablet weaving: special weaving style used to manufacture narrow textile items that are produced using cards or tablets with holes in them. The threads usually run through holes in the corners of several tablets and are generally fixed at both ends. Apart from additional decorative weft threads, the weft is invisible.

Picture credits

Essays

p. 6 © Belvedere, Vienna (© Belvedere, Vienna); p. 12, Fig. 1: © Belvedere, Vienna (Photo: Markus Fellinger), Fig. 2: © Gallery Victoria Miro, Fig.. 3: © 2012 Richard Wright / Photography by Mike Bruce. Courtesy of Gagosian Gallery; p. 13, Fig.. 4: © Kiefer Auktionen, Pforzheim, Fig.5: © bpk | Hamburger Kunsthalle | Elke Walford; p. 15, Fig. 6: © Staaatliche Kunsthalle Karlsruhe 2011 / Photo by A. Fischer/H. Kohler, Fig. 7: © Kunsthistorisches Museum, Vienna; p. 17, Fig. 8: Scan from: Hans Peter Riese, *Kasimir Sewerinowitsch Malewitsch* , Reinbek bei Hamburg 2002, Fig. 9: © Courtesy Noel Barrett Auctions, Carversville, PA, Fig. 10: © Philadelphia Museum of Art, The Louise and Walter Arensberg Collection, 1950 / © Succession Marcel Duchamp/ VBK, Vienna 2012; p. 18, Fig. 11: Scan from: Diana Küppers [Hrsg.], *Künstlerschmuck der Avantgarde*, Ausst. Kat. Museum für Angewandte Kunst Köln 2009, Stiftung Wilhelm Lehmbruck-Museum, Duisburg 2009/2010 / © Succession Picasso/ VBK, Vienna 2012; p. 20, Fig. 12: © Winfried Muthesius Photo: Jörg Kuß, Berlin, Fig. 13: Scan from: Die Furche, 9.4.2009, p. 1.; p. 29, Fig. 1: Scan from: Albrecht Dürer, Underweysung der Messung 1538; p. 30, Fig. 2: © Belvedere, Vienna, Fig. 3: © Belvedere, Vienna, Fig. 4: © Private collection, Vienna; p. 32, Fig. 5: © Belvedere, Vienna; p. 33, Fig. 6: © Belvedere, Vienna, Fig. 7: © Österreichische Nationalbibliothek, Bildarchiv; p. 34, Fig. 8: © Erzabtei St. Peter, Salzburg; p. 35, Fig. 9: © Photo Scala, Florence - courtesy of the Ministero Beni e Att. Culturali, Fig. 10: © Belvedere, Vienna; p. 39, Fig. 1: © Institut für Realienkunde—ÖAW; p. 40, Fig. 2 © Belvedere, Vienna, Fig. 3: © Belvedere, Vienna; p. 41, Fig. 4: © Belvedere, Vienna; p. 43, Fig. 5: © Belvedere, Vienna; Fig. 6: © Belvedere, Vienna; p. 44, Fig. 7: © Belvedere, Vienna (Photo: Edgar Knaack) p. 44, Fig. 8: © Belvedere, Vienna (Photo: Edgar Knaack) p. 46, 47, Fig. 10, 11: © Belvedere, Vienna; p. 53, Fig. 1: © Samuel H. Kress Collection, Image courtesy of the National Gallery of Art, Washington; p. 57-59, Fig. 1-4: © Belvedere, Vienna; p. 60, Fig. 5: © Wolfgang Staudt; p. 61, Fig. 6: © Heeresgeschichtliches Museum, Vienna; p. 62, Fig. 7: © Franz Svoboda - www.fotocommunity.de/pc/pc/mypics/11067/display/25667207; p. 65, Fig. 1: © The Frank Lloyd Wright Fdn, AZ / Art Resource, NY/Scala, Florence / VBK, Vienna 2012 ; p. 66, Fig. 2: © Chris Little; p. 67, Fig. 3: © Chicago History Museum; p. 68, Fig. 4: © Alexander Klee, Fig. 5: © MAK; p. 71, Fig. 1: © Photo: Wolfgang Träger; p. 71, Fig. 2: Private collection; p. 72, Fig. 3: © Private collection / VBK, Vienna 2012; p. 73, Fig. 4: © Louisiana Museum of Modern Art, Denmark. Deponering: Museumsfonden af 7. december 1966 / VBK, Vienna 2012, Fig. 5: © Rodtschenko-Archiv Moscow / VBK, Vienna 2012; p. 74, Fig. 6: © Solomon R. Guggenheim Museum New York / Untitled Press inc./VBK, Vienna 2012, Fig. 7: ©Photo: Stefan Altenburger Photography, Zurich © Roni Horn; p. 75, Fig. 8: © The Andy Warhol Foundation for the Visual Artp.Inc./ VBK, Vienna 2012; p. 79-81, Fig. 1-3: © Wien Museum; p. 82, Fig. 4: Scan from: David Fontana, *Die verborgene Sprache der Symbole. Ein illustrierter Führer durch die Welt der Symbole*, Munich 1994; p. 83, Fig. 5: © Private collection; p. 84, Fig. 6: © bpk / Lutz Braun; Fig. 7: Scan from: Percy Schramm, *Sphaira, Globus, Reichsapfel*, Stuttgart 1958; p. 85, Fig. 8: © Kunsthistorisches Museum, Vienna, Fig. 9: © Scan from: *Land der Bibel. Schätze aus dem Bible Lands Museum Jerusalem*, Ausst. Kat. Kunsthistorisches Museum Vienna 1997/1998; p. 86, Fig. 10: © bpk/SBB; p. 89, Fig. 1: © Bibliothèque nationale de France; p. 90, Fig. 2: © Markus Castor; p. 91, Fig. 3: © RMN (Château de Versailles) / Gérard Blot; p. 92, Fig. 4: © RMN (Château de Versailles) / J. Derenne; p. 93, Fig. 5: © Markus Castor; p. 94, Fig. 6: © Markus Castor, Fig. 7: Gemeinfreies Bild; p. 95, Fig. 8: © RMN (Château de Versailles) / Gérard Blot; p. 96, Fig. 9: © Markus Castor; p. 97, Fig. 10: © Markus Castor; p. 101-104, Fig. 1-7: © Bodo Hell: p. 111, Fig. 1: © Bildarchiv Foto Marburg; p. 113, Fig. 2: Private collection; p. 116 -117, Fig. 1: © Universität für angewandte Kunst Vienna / Institut für Konservierung und Restaurierung / Judith Kern, Fig. 2: © Universität für angewandte Kunst Vienna / Institut für Konservierung und Restaurierung / Lisa Gräber, Fig. 3: © Universität für angewandte Kunst Vienna / Institut für Konservierung und Restaurierung, Fig. 4: © Universität für angewandte Kunst Vienna / Institut für Konservierung und Restaurierung; p. 123, Fig. 1: © Bildarchiv Stefano Fabi, Rome; p. 124, Fig. 2: Gemeinfreies Bild; p. 125, Fig. 3: © Whitworth Art Gallery, The University of Manchester; p. 127, Fig. 4: © Peggy Guggenheim Collection, Venice (Solomon R. Guggenheim Foundation N.Y)/ © VBK, Vienna 2012; p. 128, Fig. 5: © bpk | Museum der bildenden Künste, Leipzig | Ursula Gerstenberger / © VBK, Vienna 2012; p. 129, Fig. 6: Photo © mumok/Lisa Rastl, 2002 / © Chris Burden; p. 130, Fig. 7: Photo: Courtesy of Jack Shainman Gallery, NY / © Subodh Gupta; p. 174, Fig. 1: © Veneranda Biblioteca Ambrosiana/DeAgostini Picture Library/Scala, Florence; p. 200, Fig. 1: © The Third Line, Dubai; p. 233, Fig. 1: © The Lazarides Galleries / Paul Insect

Plates

p. 133 © bpk/Ägyptisches Museum und Papyrussammlung, SMB/ Georg Niedermeiser; p. 134 © Gundula Schulze Eldowy; p. 135: © Kunsthistorisches Museum, Vienna; p. 136-137 © Arp Museum Bahnhof Rolandseck/Sammlung Rau für UNICEF; p. 138 © Society of Antiquaries of London, p. 139 © Galleria d'Arte Moderna di Milano; p. 140 © Neue Galerie Graz am Universalmuseum Joanneum, Photo: N. Lackner, UMJ; p. 141 Belvedere, Vienna; p. 142 © Stiftung Kunstsammlung Nordrhein-Westfalen, Düsseldorf, Photo: Walter Klein, Düsseldorf / © VBK, Vienna 2012; p. 143 © Louisiana Museum of Modern Art / © VBK, Vienna 2012; p. 144 © Courtesy Galerie Krinzinger, Vienna; p. 145 © Private collection; p. 147 © Private collection; p. 153 © Franz Josef Altenburg; p. 154, 155 © Courtesy of the artist; p. 156 © Kunstmuseum Bern, Estate James Lee Byars, Courtesy Gallery Michael Werner, Märkisch Wilmersdorf, Berlin & New York; p. 157 © Photo: Francois Fernandes © Collection Titze, Estate James Lee Byars, Courtesy Gallery Michael Werner, Märkisch Wilmersdorf, Berlin & New York; p. 158 © Courtesy ARNDT, Berlin; p. 159 © Galerie Thaddaeus Ropac; p. 160/161

© Norbert Fleischmann; p. 162 © Josef Kern; p. 163 © Cornelius Kolig / © VBK, Vienna 2012; p. 164 © Courtesy Galerie Perrotin, Paris, Photography: Quentin Beun; p. 165 © Courtesy Galerie Vera Munro; p. 166 © Photo: Bernd Borchardt, courtesy ARNDT, Berlin; p. 167 © Courtesy of the artist and ARNDT, Berlin; p. 168-170 © Courtesy GALLERIA CONTINUA, San Gimignano/Beijing/Le Moulin (The Denial: Photo: Ela Bialkowska; The Leader: Photo: Oak Taylor-Smith); p. 170 unten © Photo: ARNDT, Berlin; p. 171 © Jan Maarten Voskuil; p. 176 © Collection of Dame Jenny Gibbs; p. 177 © Joachim Baur; p. 178 © Thomas Brix /© VBK, Vienna 2012; p. 179 © Guillaume Bruère; p. 180 © Pamela Blake Burnley-Schol; p. 181 © Private collection; p. 182 © Courtesy of the artist; p. 183 © Manfred Erjautz, Photo: Caroline Heider / © VBK, Vienna 2012; p. 184: © Galerie Thaddaeus Ropac; p. 185 © OK Harris Works of Art, New York, NY; p. 186: © Courtesy of the artist; p. 187 © Milan Kunc, Photo by Martin Polák, Galerie Caesar archive, Olomouc, Czech Republic; p. 188 © Photo: Courtesy DC Moore Gallery, New York; p. 189 © LS LandskronSchneidzik Galerie und Kunstagentur / © VBK, Vienna 2012; p. 190/191 © the Artist / Courtesy Galerie Bob van Orsouw, Zurich; p. 192 © Galerie Thaddaeus Ropac; p. 193 © Private collection; p. 194 © Courtesy of the artist; p. 195 © Arden Gallery, Boston, Massachusets; p. 196 © Courtesy of the artist; p. 197 © Private collection; p. 202/203 © The Third Line, Dubai; p. 204 © Courtesy of Dillon Gallery New York; p. 205: © Jakob Gasteiger/ © VBK, Vienna 2012; p. 206/207 © OK Harris Works of Art New York, NY; p. 208 © Jürgen Kadow; p. 209 oben © Courtesy Galeria Gian Enzo Sperone, New York, Photo by Martin Polák, Galerie Caesar archive, Olomouc, Czech Republic; p. 209 unten © Milan Kunc, Photo by Martin Polák, Galerie Caesar archive, Olomouc, Czech Republic; p. 210/211 © LS LandskronSchneidzik Galerie und Kunstagentur / © VBK, Vienna 2012; p. 212 © Gary Milek; p. 213 © Patricia Miranda; p. 214 © Karl Manfred Rennertz /© VBK, Vienna 2012; p. 215-217 © Carme Sanglas; p. 218/219 © Hermann Staudinger; p. 220/221 © Tabitha Vevers; p. 224-226 © Vienna Museum ; p. 227 © Kunstmuseum Winterthur; p. 228 © Cornelius Kolig/VBK, Vienna 2012; p. 229 © Courtesy Georg Kargl Fine Arts, Vienna and Galerie Mehdi Chouakri, Berlin; p. 235 © Courtesy Kalfayan Galleries, Athens—Thessaloniki; p. 236 © Galerie Andrea Caratsch, Zurich; p. 237-239 © Belvedere, Vienna; p. 240 © Courtesy Gor Chahal; p. 241 © Galerie Andrea Caratsch, Zurich / © VBK, Vienna 2012; p. 242-243 © Collection Kempen, Aachen, Photo: courtesy ARTCO Galerie Herzogenrath Deutschland; p. 244 © ARTCO Galerie Herzogenrath Deutschland; p. 245 © Ernst Fuchs; p. 246-247 © Yoko Grandsagne; p. 248 © Photo: Courtesy of the artist and Jack Shainman Gallery, NY; p. 249 © Galerie Heike Curtze Vienna-Berlin, Photo: © Barbara Leitner-Szapáry, p. 250-251 © Peter Murphy; p. 252 © Private collection/Estate W. Mucha, Geneva, Photo: Loan Oei; p. 253 © Pete Nawara; p. 254 © Pepo Pichler; p. 255 ©Private collection; p. 260-263 © Private Collection, Courtesy Galerie Bruno Bischofberger, Zurich / © VBK, Vienna 2012; p. 264 © Courtesy the artist; p. 265 © Galerie Steinek; p. 266 © Kunsthistorisches Museum, Vienna; p. 267 © Mary Frances Dondelinger; p. 268/269 © Courtesy Mario Mauroner Contemporary Art Gallery, Vienna, Photo: Pat Verbruggen, © Angelos bvba/Jan Fabre / © VBK, Vienna 2012; p. 270 © Photo: Courtesy L.A. Louver, Venice, CA; p. 271 © Courtesy Mizuma Art Gallery, Tokyo; p. 277 © Courtesy Mario Mauroner Contemporary Art Vienna; p. 278 © Johannes Angerbauer/ © VBK, Vienna 2012; p. 279 © Edition 5, Erstfeld, Geneva; p. 280 © Galerie Andrea Caratsch, Zurich; p. 281 © Photo: Galerie Thaddaeus Ropac / © VBK, Vienna 2012; p. 282 © Villa Grisebach Auktionen, Berlin / VBK, Vienna 2012; p. 283 Galerie Walker Schloss Ebenau Rosental; p. 284 © Courtesy of King & Spalding; p. 285 © Jill Biskin; p. 286 © Arik Brauer; p. 287 © LS LandskronSchneidzik Galerie und Kunstagentur; p. 288 © Courtesy of the artist and Anthony Meier Fine Arts, San Francisco; p. 289 © LS LandskronSchneidzik Galerie und Kunstagentur; p. 290 © Collection Peters-Messer, Viersen, Photo: Studio Hadler, Viersen; p. 291 © Johannes Domenig, Photo: Imre Cserjan, Vienna / © VBK, Vienna 2012; p. 292 © Mary Frances Dondelinger; p. 293 © Collection Clo and Marcel Fleiss, Paris; p. 294 © Private collection; p. 295 ©Courtesy Collection Goetz, Photo: © Philipp Schönborn, München; p. 296 © Collection Sigrid and Franz Wojda, Vienna / © VBK, Vienna 2012; p. 297-298 © Galerie Thaddaeus Ropac; p. 299 © Collection Grässlin, St. Georgen; p. 300/301: © Courtesy L. A. Louver, Venice, Kalifornien; p. 302 © Collection Katharina Faerber, Geneva, Photo: General Idea; p. 303 © Courtesy Galerie Elisabeth and Klaus Thoman, Innsbruck; p. 304 © Galerie Thaddaeus Ropac; p. 305 © Die Hundertwasser Gemeinnutzige Privatstiftung, Vienna; p. 306 © Courtesy of the artist; p. 307 © Josef Kern; p. 308 © Kunstmuseum Krefeld / © VBK, Vienna 2012; p. 309 © Courtesy Private collection Stuttgart; p. 310 Collection David Marcel Fleiss, Galerie 1900-2000, Paris; p. 311 © Courtesy Turner Carroll Gallery, Santa Fe, New Mexico; p. 312 © Rudolf Leitner-Gründberg, Photo: © Barbara Leitner-Szapáry; p. 313 © Collection Walker Art Center, Minneapolis, T. B. Walker Acquisition Fund, 1992; p. 314/315 © Jason Clay Lewis; p. 316 © Photo: Courtesy Galerie Bob van Orsouw, Zurich; p. 317 Galerie Walker Schloss Ebenau Rosental / © VBK, Vienna 2012; p. 318 © Collection Flick, Salzburg, Photo: Ulrich Ghezzi; p. 319 © Dr. Jeanette Mayrhofer-Berger, Dr. Franz Mayrhofer, p. 320 © Private collection, Photo: Jörg Kuß, Berlin / © VBK, Vienna 2012; p. 321 © Hermann Nitsch and Hermann Staudinger; p. 322 © Galerie Maier, Innsbruck, Photo: Watzek Photografie, Innsbruck; p. 323 © Collection Titze; p. 324 © Courtesy Galleria Christian STEIN / © VBK, Vienna 2012; p. 325 © Mag. Eva Pliem; p. 326 © Collection Goetz, Photo: Courtesy carlier | gebauer; p. 327 © Courtesy Kalfayan Galleries, Athens—Thessaloniki; p. 328/329 © Galerie Thaddaeus Ropac; p. 330 © Walter Schnabl; p. 331 © Collection of the Baloise Group, Basel; p. 332 © Courtesy Kalfayan Galleries, Athens—Thessaloniki; p. 333 © Courtesy Kalfayan Galleries, Athens—Thessaloniki © Yannis Tsarouchis Foundation; p. 334 © Museum Ritter, Waldenbuch, Photo: Gerhard Sauer / © VBK, Vienna 2012; p. 335/336 © Robert Weber; p. 337 © Yvonne Weiler; p. 338 © Fred Wessel; p. 339 © Courtesy Galerie Elisabeth and Klaus Thoman, Innsbruck; p. 340 © Private collection; p. 341 © Courtesy Galerie Meyer Kainer, Photo: Tina Herzl; p. 342 © Tate, London, presented by David Gorton 2006

Colophon

This book is published on the occasion of the exhibition
Gold from March 15 to June 17 2012 at the Belvedere, Vienna.

Director: Agnes Husslein-Arco
Curator: Thomas Zaunschirm
Curatorial Assistant: Markus Fellinger
Exhibition Management: Tatjana Gawron

Publication
Editors: Agnes Husslein Arco, Thomas Zaunschirm
Publication Management: Ute Stadlbauer
Picture editing: Markus Fellinger
Project Management: Claudia Wagner, Starnberg
Translation: Josephine Cordero Sapien, Newton Abbot,
Naomi Barker, Bovey Tracey
Copy-Editing: Michael Scuffil, Cologne
Graphic design and typesetting: Peter Baldinger
Lithography: Pixelstorm, Vienna
Paper: Luxo Samt New 150g/qm
Set in: Corporate S BQ
Production: Rainer Arnold
Printed and bound by Firmengruppe Appl, aprinta druck GmbH,
Wemding
Printed in Germany

Bibliographic information published by the Deutsche Nationalbibliothek
The Deutsche Nationalbibliothek lists this publication in the Deutsche Nationalbibliografie; detailed bibliographic data are available in the Internet at http://dnb.d-nb.de.

ISBN: 978-3-7774-5361-3 (Museum)
ISBN: 978-3-7774-4981-4 (Retail)

www.hirmerpublishers.com

The editor would like to express her sincere gratitude to Antal P. de Bekessy, whose generous assistance in the context of the *Friends of the Belvedere* has made this English version of the catalogue possible.